MW00586245

E
L

EVERYMAN'S LIBRARY

EVERYMAN,

I WILL GO WITH THEE,

AND BE THY GUIDE,

IN THY MOST NEED

TO GO BY THY SIDE

JOHN MUIR

SELECTED WRITINGS

WITH AN INTRODUCTION
BY TERRY TEMPEST WILLIAMS

EVERYMAN'S LIBRARY
Alfred A. Knopf New York London Toronto

377

THIS IS A BORZOI BOOK
PUBLISHED BY ALFRED A. KNOPF

First included in Everyman's Library, 2017

www.randomhouse.com/everymans
www.everymanslibrary.co.uk

ISBN 978-1-101-90762-2 (US)
978-1-84159-377-7 (UK)

A CIP catalogue reference for this book is available from the British Library

Library of Congress Cataloging-in-Publication Data

Names: Muir, John, 1838–1914.
Title: Selected writings / John Muir; with an introduction by Terry Tempest Williams.
Description: New York: Alfred A. Knopf, [2017] | Series: Everyman's library | "A Borzoi book."
Identifiers: LCCN 2016028378 | ISBN 9781101907627 (hardcover : alk. paper)
Subjects: LCSH: Muir, John, 1838–1914—Childhood and youth. | Conservationists—United States—Biography. | Conservationists—Scotland—Biography. | Naturalists—Scotland—Biography. | Naturalists—United States—Biography. | Natural history—West (U.S.) | Natural history—United States.
Classification: LCC QH31.M9 A348 2017 | DDC 333.72092/2 [B]--dc23
LC record available at https://lccn.loc.gov/2016028378

Book design by Barbara de Wilde and Carol Devine Carson

Typeset in the UK by Input Data Services, 17 King Square, Bridgwater, Somerset TA6 3DJ
Printed and bound in Germany by GGP Media GmbH, Pössneck

JOHN MUIR

CONTENTS

CONTENTS

JOHN MUIR

INTRODUCTION

Earth—Planet—Universe

John Muir is a man I would have loved to have met on the trail. I would have enjoyed walking with him through Tuolumne Meadows in his beloved Yosemite listening to him discuss each wildflower by name; tell stories of each peak he climbed and the weather on that day; what he saw and how he felt. I wonder if he would have ranted and raved or kindly addressed and advocated for these wildlands in his lifelong pursuit to protect them.

He might have said as he did in his essay, "The Wild Parks and Forest Reservations of the West," published in his book, *Our National Parks*: "Thousands of tired, nerve-shaken, over-civilized people are beginning to find out that going to the mountains is going home; that wildness is a necessity; and that mountain parks and reservations are useful not only as fountains of timber and irrigating rivers, but as fountains of life."

And I would have agreed with him.

I like to imagine that he could walk with me now in the red rock desert of the Colorado Plateau where Utah, Colorado, New Mexico, and Arizona share a common boundary point in what is known as the Four Corners. We would stand on the South Rim of the Grand Canyon in shared awe, where he once stood and proclaimed that it was "as if you had found it after death, on some other star; so incomparably lovely and grand and supreme is it above all the other cañons in our fire-moulded, earthquake-shaken, rain-washed, wave-washed, river and glacier sculptured world."

We might have discussed a wild life versus a domesticated one, and he would have exclaimed, "I have been too long wild" without any thought of changing his passionate stance toward the virtues of a life lived outside.

And then, I would have asked him to visit Big Flats outside of Moab, Utah, now a series of oil and gas drilling sites that look like monstrous, mechanical ravens with their heads rising up and down as they peck on carrion in the red sand on the edge of Canyonlands National Park, which now feels like an annex for the fossil fuel industry.

He might have expressed a longing for a reprieve in the cool alpine air of the Flathead Reserve, now Glacier National Park in Montana, where he said, "Give a month at least to this precious reserve. The time will not be taken from the sum of your life. Instead of shortening, it will indefinitely lengthen it and make you truly immortal." And I would have had to describe to him that out of the some 150 glaciers visible during the seventy-six years of his life, a century after he'd left this Earth only 25 remain, and that glaciologists now predict the glaciers will be gone in fifteen years.

I would sit down with Mr. Muir in the shade of a juniper tree and speak of our warming planet, warming from an increased use of carbon through our excesses of driving cars, traveling by plane, and our societal and global dependence on coal, oil, and all manner of fossil fuels. We would speak of a population of billions and rising.

Perhaps he would inquire about water, being the citizen scientist he was, forever curious, always two steps ahead mentally, and say something to the effect of, "Oil is optional, water is not," as the photographer Edward Burtynsky recently said after having spent a lifetime taking pictures of mined and spent landscapes, where toxins fan out at a river's delta like a blood-red hand on the planet.

And then, I can envision we would time-travel back to Yosemite Valley during the days that Congress shut down the government, from October 1 to 16, 2013, because they could not agree on how to fund our national debt. During this dispute, our national parks were closed to the public. The lands rested. The deer and bears and hawks returned momentarily to the uncommon quiet of wild nature and to savor the silences. But the resourceful Muir would have found a way in, and we would have walked quietly, joyously up the trail to Vernal Falls and baptize ourselves once again into the Church of Awe and cleanse ourselves momentarily of grief. We would have sat very near the falls, wet from the ecstatic spray, and he would have recounted the time when he watched the moon through the veil of water as he braved his own annihilation just for the experience of standing behind the vertical torrent. Neither one of us would be able to tell if our tears were of joy or sadness, both of us knowing they spring from the same source: a love of beauty and all things wild.

What I would ask John Muir at that tender moment is, "When you wrote 'John Muir–Earth–Planet–Universe' inside the cover of your notebook, how did you know we were made of stardust? How did you understand a planetary consciousness when a hundred years later we are just beginning to comprehend your prescient words, 'When we try to pick out anything by itself, we find it hitched to everything else in the Universe?'"

Donald Worster, the historian and author of *A Passion for Nature: The Life of John Muir*, sees the Scottish-born naturalist's legacy as a spiritual one. At a forum about Muir at Stanford University he said:

> We should think about John Muir as the inventor of a new American religion. I don't mean to put America in there exclusively, because people all over the world have responded to him ...
>
> I think he is a religious prophet, and a lot of people in this country have followed him or followed those ideas ... He's important for the great work he did for getting our national parks and forests and wildlife refuges and wild places and beauty in our consciousness and concerned about saving them. I think he's also important ... as a kind of measuring stick for understanding where we were as a people and where we ought to be today ...

Consider me one of Muir's followers, even disciples, of his wild joy, walking behind him on the path of wilderness protection. I am grateful that Everyman's Library has chosen to publish this John Muir Reader in the nick of time.

*

When I was in college I worked at Sam Weller's used bookstore in Salt Lake City, Utah. Sam knew I loved nature and nature writing. He took me aside one day after work.

"I want to show you something," he said. We walked up the stairs into his office. "We just picked up a library from an estate sale. I thought you would appreciate this." On his desk were eight volumes of a set of John Muir's writings. "Two of the volumes are missing," Sam said. "But you don't come across things like this very often." Always the businessman, Sam said, "I'll give it to you for a good price."

Sam's "good price" meant I worked free for a month, but it was worth it to a budding naturalist who cherished Muir. Missing was a two-volume biography, *The Life and Letters of John Muir*, by William Frederic Bade (1923, 1924), which Sam helped me to locate.

This special ten-volume set was published by Houghton Mifflin in Boston between 1916 and 1924. They printed 750 sets. Ours was number #551. These books, bound in green buckram with brown leather patches bordered in gold glued to their spines that bear the name "JOHN MUIR," were respectfully placed alongside Carson, Abbey, and Stegner. The first volume had a handwritten manuscript fragment glued to the first page before the title appeared: hence "The Manuscript Edition." Muir's own script, penned in sepia ink, read "learn that as these [landscapes] we now behold have succeeded those of the pre-glacial age, so they in turn are withering and vanishing to be succeeded by those yet unborn. The sun was wheeling far to the west when I began to seek a way home. I first scanned the western spurs."

I found another version of these words again in Volume IV, *The Mountains of California*, in chapter four describing Muir's heroic (or ill-conceived) first ascent of Mount Ritter, 13,300 feet, in Yosemite.

As a young writer, I was fascinated to see how he had edited this snippet of writing. The word "landscapes" crossed out in pencil remained omitted in the final text. He substituted "those" for "others," as in "others yet unborn." And the last sentence in the final version of *The Mountains of California* was greatly expanded, editing out "when I began to seek a way home." The phrase, "the sun was wheeling far to the west" which began the original sentence found its new place in the middle of the polished construction: "But in the midst of these fine lessons and landscapes, I had to remember that the sun was wheeling far to the west, while a new way down the mountain had to be discovered to some point on the timber line where I could have a fire; for I had not even burdened myself with a coat." The word "home" used in the fragment never appeared again. The last sentence of the manuscript fragment became the first phrase in the sentence that followed the expanded one: I first scanned the western spurs, hoping some way might appear through which I might reach the northern glacier, and cross its snout; or pass around the lake into which it flows, and thus strike my morning track.

The manuscript fragment glued inside John Muir's collected works highlights both the Yosemite naturalist's prowess as a gentleman mountaineer and his passion and verve for all things wild on the page and in the world. But it also showcases his sensitivity to language through his word choices, and these revisions showed his genuine skill as a writer. He taught me as a young writer that you can change your mind on the page—even change it often—until the right word appears and sticks. The process of revision is not only the cornerstone of good writing but smart living. For Muir, climbing Mount Ritter was also an exercise in revision. The path he planned and imagined taking was not the path he took. Nevertheless he did summit the peak.

*

John Muir authored sixteen books and countless articles. His most enduring legacy may be found in literature. For this gift, I would turn to him on the trail and thank him. His voice, often rhapsodic and, at times, ecstatic, was urgent in his resounding call for wilderness protection. He believed in geology and the language of glaciers. He fought against the damming of the Hetch Hetchy Valley in Yosemite, and lost, having to endure the flooding of "his temple." Even so, his grief was borne through the endless wells of solace, in his own ground truthing of the Tuolumne Meadows.

His words survive him. "Everybody needs beauty as well as bread, places to play in and pray in, where Nature may heal and cheer and give strength to body and soul alike."

John Muir knew the world from the soles of his feet upward. "All the world seems a church and the mountains altars," he wrote. Nothing exists in isolation. Nature is holy. The naturalist gleaned his spirituality in part from the great sequoias:

> It took more than three thousand years to make some of the oldest of the Sequoias,—trees that are still standing in perfect strength and beauty, waving and singing in the mighty forests of the Sierra. Through all the eventful centuries since Christ's time, and long before that, God has cared for these trees, saved them from drought, disease, avalanches, and a thousand storms; but he cannot save them from sawmills and fools.

Muir speaks in direct language with an eye toward politics. Here we can both see the remnants of religiosity of John Muir's upbringing where his father, Daniel, was an ardent preacher for the Disciples of Christ (which Muir strongly rejected) and his own brand of mysticism, where he recognized a world animated by the spirit of creation.

*

Muir thus deeply understood that the world is interconnected and interrelated, that our evolution as a species, physically and spiritually, does not go unnoticed by the universe around us. And yet, despite his intimate knowledge of these things, Muir could not have predicted the future. His vantage point on the flanks of Mount Ritter in 1872, the same year Yellowstone became America's first national park, could not have prepared him for the scale of changes piled on to the Earth by modernity from his era to ours—neither Muir nor God could save the Earth from "fools." And how could he ever have imagined our present appetite of an expanding global population and the carbon load now weighing heavy on all of us?

We don't need to denounce John Muir's legacy, as some critics have suggested, believing his vision belongs to a privileged, sexist, racist white man of the early twentieth century. We need to broaden it. We need to deepen Muir's ethos of the wild and expand our thinking about how environmental issues and social issues must be seen as issues of justice in a world increasingly weighted toward the advantage of the privileged.

Muir's call for wilderness in the past may become a clarion call for wilderness in the future, one on which our survival as a species rests. We now know that if we are to ensure a livable future for our children and their children, we must keep close to 80 percent of all remaining fossil fuels in the ground. This means keeping the public in our public lands where America's protected wilderness is found.

Detractors of wilderness in the past few decades have said, "Wilderness has become irrelevant before it has become resolved," suggesting the protection of wilderness no longer matters because wilderness itself is a human construct. But with the era of climate change upon us, we are now recognizing that many of our most valuable carbon banks beneath the ground, insurance policies for the future, are found within our protected wilderness areas, national parks, and refuges. Protecting our public lands has taken on an added layer of urgency and import. These open spaces are also reservoirs for our spirit, where awe and majesty are fueled.

I like to imagine John Muir in conversation with the imminent biologist E. O. Wilson.

Oh, how I would love to eavesdrop on the two of them discussing Wilson's Half-Earth Theory, the overwhelming evidence that our capacity to flourish as a species is in direct relationship to leaving half of the Earth in a state of wildness.

When I first read John Muir's description of climbing a conifer during a snowstorm and riding its top while all around him windblown trees crashed to the forest floor, I wanted to love and live that passionately in the world.

> After cautiously casting about, I made choice of the tallest of a group of Douglas Spruces that were growing close together like a tuft of grass, no one of which seemed likely to fall unless all the rest fell with it. Though comparatively young, they were about 100 feet high, and their lithe, brushy tops were rocking and swirling in wild ecstasy. Being accustomed to climb trees in making botanical studies, I experienced no difficulty in reaching the top of this one, and never before did I enjoy so noble an exhilaration of motion. The slender tops fairly flapped and swished in the passionate torrent, bending and swirling

backward and forward, round and round, tracing indescribable combinations of vertical and horizontal curves, while I clung with muscles firm braced, like a bobolink on a reed.

Awe defines John Muir. Perhaps it began for him in 1863, sitting beneath a black locust tree on the Wisconsin University campus where he was a student. His friend Griswold showed him a flower from that tree and told him it belonged to the pea family. "But how can that be," Muir recounted in *The Story of My Boyhood Home and Youth*, "when the pea is a weak, clinging, straggling herb, and the locust is a big, thorny hardwood tree?" This insight stabbed deeply into Muir's consciousness and changed the course of his life. "This fine lesson charmed me and sent me flying to the woods and meadows in wild enthusiasm." Griswold's "fine lesson" also injected him with a case of wanderlust rarely seen since. He walked a thousand miles from Indianapolis to the Gulf of Mexico. He sailed to Cuba and from there to Panama, where he crossed the isthmus and sailed up the west coast of the continent to San Francisco. He walked across the San Joaquin Valley and first glimpsed the "Mighty Sierra":

> Then it seemed to me the Sierra should be called not the Nevada or Snowy Range, but the Range of Light. And after ten years spent in the heart of it, rejoicing and wondering, bathing in its glorious floods of light, seeing the sunbursts of morning among the icy peaks, the noonday radiance on the trees and rocks and snow, the flush of the alpenglow, and a thousand dashing waterfalls with their marvelous abundance of irised spray, it still seems above all others the Range of Light.

Along with the locust and the pea, this journey opened wide the lens through which he experienced the world. Muir both believed in God's creation and supported Darwin's theory of evolution. "The awe with which Muir viewed every part of the natural world—floods and storms as much as anything else—was," according to historian Terry Gifford in his book, *Reconnecting with John Muir*, "intended in his writings to induce respect, humility, and, ultimately, conservation." It became his bridge between seemingly opposing banks of thought.

INTRODUCTION

In the nineteenth century, the American West was on the verge of selling itself to the highest bidder, and I believe we stand at this threshold again today. Only, now, there is so much more at stake—the very health and wealth of the planet. John Muir and the community that surrounded him, including President Theodore Roosevelt, stopped this wholesale destruction of wild America through brave acts of conservation, be it the creation of the National Park Service or the importance of wild words on the page, which showed us that wilderness everywhere matters not only to the human spirit, but to life itself in all its evolutionary processes. We are not the only species that lives and loves and breathes on this planet. Earth is terrain of transformation: dynamic, indifferent, and acutely personal. By that I mean a mountain stands its ground, regardless of whether we choose to approach it; but as John Muir has shown us repeatedly, if we do approach the mountain, it is we who are moved.

Awe, I believe, set Muir apart during his lifetime. Awe is his currency today.

May the words between these covers reach us now and inspire another generation to fall in love with wild nature, to care for it, to know that wilderness is not optional but central to our survival in the centuries to come. May Muir's writing remind us how to embrace this beautiful, broken world once again with an open heart, especially in the midst of what we stand to lose as temperatures rise. And may the planetary consciousness John Muir found in all life—from the wonder of a pea plant to the grandeur of great rivers of ice, now receding glaciers—inspire us to engage in the great work of ecstatic observation, relentless wandering, and joyous-fierce actions on behalf of the Earth. May we use our own curiosity as a compass.

With this reader in hand, we have a map—John Muir walks with us.

Terry Tempest Williams

SELECT BIBLIOGRAPHY

BADÈ, WILLIAM FREDERIC, *The Life and Letters of John Muir*, 2 vols, Houghton Mifflin, 1923–4.

EHRLICH, GRETEL, *John Muir: Nature's Visionary*, National Geographic Society, 2000.

GIFFORD, TERRY, ed., *John Muir: His Life and Letters and Other Writings*, Bâton Wicks, London; The Mountaineers, Seattle, 1996.

JONES, HOLWAY R., *Join Muir and the Sierra Club: The Battle for Yosemite*, Sierra Club, 1965.

MATTHIESSEN, F. O., *American Renaissance: Art and Expression in the Age of Emerson and Whitman*, Oxford University Press, 1941.

SARGENT, SHIRELY, *John Muir in the Yosemite*, Flying Spur Press, 1971.

WOLFE, LINNIE MARSH, ed., *John of the Mountains: The Unpublished Journals of John Muir*, Houghton Mifflin, 1938.

WOLFE, LINNIE MARSH, *Son of the Wilderness: The Life of John Muir*, Knopf, 1945.

WORSTER, DONALD, *A Passion for Nature: The Life of John Muir*, Oxford University Press, 2008.

CHRONOLOGY

DATE	AUTHOR'S LIFE	LITERARY CONTEXT
1838	Born April 21 in Dunbar, Scotland.	Tocqueville: *De la démocratie en Amérique* (1835) translated, published in US. Lyell: *Elements of Geology*. Audubon: *The Birds of America*.
1839		Darwin: *The Voyage of the* Beagle.
1841		Cooper: *The Deerslayer*. Emerson: *Essays* (I).
1846		Melville: *Typee*.
1848		Marx and Engels: *Communist Manifesto*.
1849	Moves to America with his family, settling in southern Wisconsin near the Fox River.	Death of Poe. Parkman: *The Oregon Trail*.
1850		Hawthorne: *The Scarlet Letter*.
1851		Melville: *Moby-Dick*.
1852		Stowe: *Uncle Tom's Cabin*.
1854		Thoreau: *Walden*.
1855		Whitman: *Leaves of Grass* (1st ed.)
1856		Burton: *First Footsteps in East Africa*. Ruskin: *Modern Painters* (IV).
1858		
1859		Darwin: *The Origin of Species*.
1860	First exhibits inventions (clock and thermometer) at State Agricultural Fair; deemed a "genius" by judges.	Hawthorne: *The Marble Faun*. Eliot: *The Mill on the Floss*. Whittier: *Snow-Bound*.
1861	Enrolls in University of Wisconsin for one semester, taking geology and chemistry as a first formal introduction to scientific study, as well as Latin and Greek. Invents a study desk that holds a book in place for reading, then replaces it with a new book. Unable to return to school in the fall due to straitened finances, works as a district schoolteacher near Madison.	Dickens: *Great Expectations*.

HISTORICAL EVENTS

Underground Railroad for runaway slaves set up in Ohio. Deportation of Cherokee Nation from Georgia. Chartist movement launched in England and Scotland.

Famine in Scottish Highlands (to 1848) prompts a further wave of Highland Clearances.
US makes huge territorial gains at the end of the Mexican War, including California, New Mexico and Arizona. Year of Revolutions in Europe.
California Gold Rush underway (to 1855); widespread violence against American Indian tribes leading to catastrophic decline in population.
US Department of Interior established.
Compromise of 1850: California admitted to the Union as a free state.
Louis Napoleon's *coup d'état* in France. Great Exhibition in London.

Speke discovers Lake Victoria, which he correctly identifies as the source of the River Nile.
First commercial oil well in US (Titus, Pennsylvania).
Abraham Lincoln elected US President. Pony Express founded.

Kansas admitted to the Union as a free state. Outbreak of Civil War in US. Western Union completes first transcontinental telegraph line across North America. Apache Wars (to 1900). Italy united under Victor Emmanuel II. Tsar Alexander II emancipates the serfs in Russia.

JOHN MUIR

DATE	AUTHOR'S LIFE	LITERARY CONTEXT
1862	Returns to university.	
1863	Leaves the university for the "University of the Wilderness."	Longfellow: *Tales of a Wayside Inn.* Bates: *The Naturalist on the River Amazons.*
1864	Travels to Canada to escape Civil War draft, signed into law by President Lincoln (to 1866).	Death of Hawthorne. Marsh: *Man and Nature.* Thoreau: *The Maine Woods.* Verne: *Journey to the Centre of the Earth.*
1865		Whitman: *Drum Taps.* Carroll: *Alice's Adventures in Wonderland.*
1866		
1867	An accident in a factory in Indianapolis where Muir works leads to an eye injury and temporary blindness, which takes weeks to recover. Plans a trip to South America, and takes a one-thousand-mile walk to the Gulf of Mexico, keeping a journal of the trip.	Marx: *Das Kapital* (I).
1868	Spends the summer exploring California's Central Valley region while waiting to return to South America, writing his first California journals. Takes a job as a shepherd of 1,800 sheep.	Harte: "The Luck of Roaring Camp." Collins: *The Moonstone.*
1869	First summer in the Sierra.	Alcott: *Little Women.* Harte: "The Outcasts of Poker Flat." Wallace: *The Malay Archipelago.*
1870	First controversy over his claims that glaciers shaped the landscape of the Yosemite.	Bowles: *Our New West.*
1871	Ralph Waldo Emerson visits Muir in Yosemite. Muir's visit to Hetch Hetchy Valley. Publishes first essay from California "Yosemite Glaciers" in December in the *New York Tribune.*	Darwin: *The Descent of Man.* Burroughs: *Wake-Robin.*
1872	Writes "A Geologist's Winter Walk." Publishes "Yosemite	Stanley: *How I Found Livingstone.*

HISTORICAL EVENTS

US Department of Agriculture established. Lincoln calls it "the people's department" given 90 percent of Americans are farmers. Passage of the Homestead Act to encourage western settlement.
Lincoln issues Emancipation Proclamation (amendment to constitution passed 1865). Battle of Gettysburg.

US Civil War rages on. Lincoln re-elected President. Lincoln signs a bill giving Yosemite Valley and Mariposa Big Tree Grove to California as state park lands, as the US's first act of wilderness preservation. Massacre of Cheyenne and Arapaho Indians at Sand Creek, Colorado. Louis Pasteur establishes connection between bacteria and disease.
End of Civil War. Assassination of President Lincoln. Southern Democrat Andrew Johnson becomes President; favours a policy of Reconstruction lenient towards the South.
Civil Rights Bill passed by Congress in spite of presidential veto.
Building of the first elevated railroad (New York). Alaska purchased from Russia. British North America Act establishing self-governing dominion of Canada.

Failed impeachment of President Johnson. Gladstone's first Liberal ministry in Britain (to 1874).

First Pacific Railroad completed, linking East and West coasts. Suez Canal opens. Ulysses S. Grant becomes US President (to 1877). Elizabeth Cady Stanton founds National Women's Suffrage Organization.

US and Britain sign Treaty of Washington settling US–Canada boundary.

Yellowstone National Park established. Grant is re-elected.

DATE	AUTHOR'S LIFE	LITERARY CONTEXT
1872 *cont.*	Valley in Flood," "Twenty Hill Hollow," and "Living Glaciers of California" in *The Overland Monthly.*	King: *Mountaineering in the Sierra Nevada.* Twain: *Roughing It.*
1873	Biggest output of journal writing exploring eastern Yosemite, Tuolumne, Kings Canyon, and Mt. Whitney, where he notes abuse to mountain sheep.	Verne: *Round the World in Eighty Days.*
1874	Meets Louisa Wanda (Louie) Strentzel, via friends and mentors Ezra and Jeanne Carr. Finishes a series of articles on glaciers for *The Overland Monthly* as a regular contributor.	Hardy: *Far From the Madding Crowd.*
1875		Powell: *Exploration of the Colorado River of the West and Its Tributaries.*
1876	Publishes "God's First Temples" in *Sacramento Record-Union.*	Twain: *Tom Sawyer.* Henry James: *Roderick Hudson.*
1877		
1879	Becomes engaged to Louie prior to leaving on first six-month Alaska trip in June. Discovers Glacier Bay and Muir Glacier.	Stevenson: *Travels with a Donkey in the Cévennes.*
1880	Marries Louie April 14. Returns to Alaska in July, exploring glaciers with a missionary and his dog, Stickeen.	Twain: *A Tramp Abroad.*
1881	First daughter, Anna Wanda, born March 25. Third trip to Alaska. Starts ranching and fruit-farming in California, which he continues for eight years.	James: *The Portrait of a Lady*; *Washington Square.*
1882		Death of Emerson. Howells: A Modern Instance.
1883		Twain: *Life on the Mississippi.* Stevenson: *Treasure Island.*
1884		Twain: *Adventures of Huckleberry Finn.* Ruskin: *The Storm-Cloud of the Nineteenth Century.*

HISTORICAL EVENTS

Remington manufacture first typewriter. Hansen discovers leprosy bacillus.

Disraeli British Prime Minister (to 1880). Impressionists' first exhibition in Paris. Arbor Day first observed.

A short-lived Civil Rights Act guarantees equal access, regardless of color, to public transport and recreation, and protects the right to serve on juries.

Great Sioux War (to 1877); US troops defeated at the Battle of Little Bighorn. Colorado becomes US state. Alexander Graham Bell patents the telephone.
Henry Morton Stanley reaches the mouth of the Congo River, having crossed Africa.
United States Geological Survey established.

Gladstone again British Prime Minister (to 1885). First Boer War.

President Garfield inaugurated (March) and assassinated (September). Succeeded by Chester Arthur (to 1885).

Jesse James shot.

Brooklyn Bridge opened. Supreme Court declares Civil Rights Act (1875) unconstitutional. Krakatoa volcano erupts in Indonesia, killing over 36,000 and scattering ash into the atmosphere.

DATE	AUTHOR'S LIFE	LITERARY CONTEXT
1885	Father dies in October.	Howells: *The Rise of Silas Lapham*.
1886	Second daughter, Helen, born January 23.	James: *The Bostonians*.
1887		Jewett: *A White Heron*.
1888	Contributes to and edits anthology *Picturesque California*.	Kipling: *Plain Tales from the Hills*. Doughty: *Travels in Arabia Deserta*.
1889	Urged by *The Century Magazine*'s associate editor Robert Underwood Johnson to write about protecting Yosemite.	Stevenson: *The Master of Ballantrae*.
1890	Solo expedition over Muir Glacier for ten days (fourth trip to Alaska). Congress creates Yosemite National Park thanks to influence of Muir and Johnson. Campaigns for Kings Canyon National Park (est. 1939) and sees establishment of Sequoia National Park.	Dickinson: *Poems*.
1892	Cofounds the Sierra Club, of which he is President until his death.	Gilman: *The Yellow Wallpaper*. Kipling: *Barrack-Room Ballads*.
1893	Tours Europe, visiting Scotland, England, Ireland, Norway, Switzerland and France.	
1894	Publishes first book, *The Mountains of California*.	Twain: *The Tragedy of Pudd'nhead Wilson*. Chopin: *Bayou Folk*. Kipling: *The Jungle Book* (I).
1895		Crane: *The Red Badge of Courage*. Wells: *The Time Machine*.
1896	Fifth Alaska trip.	
1897	Granted honorary degree from University of Wisconsin. Sixth Alaska trip.	Chopin: *A Night in Acadie*. Robinson: *The Children of the Night*.
1898		James: *The Turn of the Screw*.
1899	Seventh Alaska trip.	Chopin: *The Awakening*. Norris: *McTeague*.
1900		Dreiser: *Sister Carrie*. Conrad: *Lord Jim*. Freud: *The Interpretation of Dreams*.

HISTORICAL EVENTS

Cleveland administration (to 1889). US Biological Survey established.

Last major war with Indians ends with Geronimo's capture.

Dawes Act enforces assimilation of American Indians by depriving them of their communal tribal lands in exchange for individual allotments.
Dunlop patents pneumatic tyre.

Large central area of the Indian Territory opened for settlement; major land rush. First Kodak camera using roll film manufactured. Harrison administration (to 1893).

US population reaches 63 million. Idaho and Wyoming become US states.

Cleveland re-elected for another term, defeating President Harrison.
Ellis Island opens as an immigration station.

Income tax declared unconstitutional in US. Uprisings against Spanish rule in Cuba.

McKinley elected (serves to 1901). Klondike Gold Rush (to 1899).

Spanish–American War. Discovery of radium.
Gold discovered in Alaska. Philippine–American War (to 1902). Second Boer War (to 1902). Aspirin invented.
US population reaches 76 milllion. Planck's quantum theory.

DATE	AUTHOR'S LIFE	LITERARY CONTEXT
1901	Publishes *Our National Parks.*	Conan Doyle: *The Hound of the Baskervilles.* Kipling: *Kim.*
1902		James: *The Wings of the Dove.* Conrad: *Heart of Darkness.*
1903	Meets with President Theodore Roosevelt on camping trip in Yosemite, where they lay the foundation for Roosevelt's conservation programs. Embarks on a world tour (to 1904) to study trees, visiting Europe, Russia, the Middle East, the Far East, Australia and New Zealand.	James: *The Ambassadors.* Du Bois: *The Souls of Black Folk.* London: *The Call of the Wild.*
1905	Louie dies August 6.	Wharton: *The House of Mirth.* Scott: *The Voyage of the* Discovery.
1906		Burroughs: *Camping and Tramping with Roosevelt.*
1907	Begins fight to save Hetch Hetchy Valley in Yosemite National Park from damming.	Adams: *The Education of Henry Adams.* Conrad: *The Secret Agent.*
1909	Publishes *Stickeen.*	Stein: *Three Lives.*
1910		Burroughs: *In the Catskills.*
1911	Publishes *My First Summer in the Sierra.* Granted honorary degree from Yale. Takes year-long trip to South America and Africa, including traveling up the Amazon to study the rainforest.	Wharton: *Ethan Frome.* Pound: *Canzoni.*
1912	Returns home from trip in March. Publishes *The Yosemite.*	Dreiser: *The Financier.* E. R. Burroughs: *Tarzan of the Apes.*
1913	Loses battle for the Hetch Hetchy Valley. Publishes *The Story of My Boyhood and Youth.* Granted honorary degree by University of California.	Lawrence: *Sons and Lovers.* Wharton: *The Custom of the Country.* Frost: *A Boy's Will.*
1914	Dies in Los Angeles December 24 from pneumonia.	Frost: *North of Boston.*
1916	*A Thousand-Mile Walk to the Gulf*, the first of 60 volumes of journals, published posthumously.	
1919		

HISTORICAL EVENTS

Theodore Roosevelt becomes US President. Marconi transmits messages across the Atlantic. Death of Queen Victoria.

Coal strike in the US (May–October).

Wright brothers' first successful powered flight. First National Bird Reserve opens on Pelican Island, Florida.

First Russian Revolution. Einstein's first theory of relativity.

Antiquities Act protects historic monuments and landmarks in US. The great San Francisco earthquake and fire fuel the city's desire to dam Hetch Hetchy Valley for a new water supply.

US immigration reaches annual peak of 1,285,000. Peace Conference at the Hague attempts but fails to halt the arms race. Oklahoma admitted as a US state.

Taft administration (to 1913). Henry Ford's Model T car. Robert Edwin Peary reaches North Pole.

US House of Representatives votes in favour of direct election of senators. Mexican Civil War. Roald Amundsen reaches South Pole.

New Mexico and Arizona become US states. Sinking of the *Titanic*.

Wilson administration (to 1921). Amory Show, New York, introduces Post-Impressionist art to US.

President Wilson creates National Park Service.

World War I (to 1918). US maintains neutrality. Panama Canal opens.

The National Parks Association (re-named the National Parks Conservation Association in 1970) founded with support of Sierra Club.

JOHN MUIR
SELECTED WRITINGS

THE STORY OF MY BOYHOOD AND YOUTH

I

A BOYHOOD IN SCOTLAND

Earliest Recollections—The "Dandy Doctor" Terror—Deeds of Daring—The Savagery of Boys—School and Fighting—Birds'-nesting.

WHEN I WAS a boy in Scotland I was fond of everything that was wild, and all my life I've been growing fonder and fonder of wild places and wild creatures. Fortunately around my native town of Dunbar, by the stormy North Sea, there was no lack of wildness, though most of the land lay in smooth cultivation. With red-blooded playmates, wild as myself, I loved to wander in the fields to hear the birds sing, and along the seashore to gaze and wonder at the shells and seaweeds, eels and crabs in the pools among the rocks when the tide was low; and best of all to watch the waves in awful storms thundering on the black headlands and craggy ruins of the old Dunbar Castle when the sea and the sky, the waves and the clouds, were mingled together as one. We never thought of playing truant, but after I was five or six years old I ran away to the seashore or the fields almost every Saturday, and every day in the school vacations except Sundays, though solemnly warned that I must play at home in the garden and back yard, lest I should learn to think bad thoughts and say bad words. All in vain. In spite of the sure sore punishments that followed like shadows, the natural inherited wildness in our blood ran true on its glorious course as invincible and unstoppable as stars.

My earliest recollections of the country were gained on short walks with my grandfather when I was perhaps not over three years old. On one of these walks grandfather took me to Lord Lauderdale's gardens, where I saw figs growing against a sunny wall and tasted some of them, and got as many apples to eat as I wished. On another memorable walk in a hayfield, when we sat down to rest on one of the haycocks I heard a sharp, prickly, stinging cry, and, jumping up eagerly, called grandfather's attention to it. He said he heard only the wind, but I insisted on digging into the hay and turning it over until we discovered the source of the strange exciting sound,—a mother field mouse with half a dozen naked young hanging to her teats. This to me was a wonderful discovery. No hunter could have been more excited on discovering a bear and her cubs in a wilderness den.

I was sent to school before I had completed my third year. The first schoolday was doubtless full of wonders, but I am not able to recall any of them. I remember the servant washing my face and getting soap in my eyes, and mother hanging a little green bag with my first book in it around my neck so I would not lose it, and its blowing back in the sea-wind like a flag. But before I was sent to school my grandfather, as I was told, had taught me my letters from shop signs across the street. I can remember distinctly how proud I was when I had spelled my way through the little first book into the second, which seemed large and important, and so on to the third. Going from one book to another formed a grand triumphal advancement, the memories of which still stand out in clear relief.

The third book contained interesting stories as well as plain reading- and spelling-lessons. To me the best story of all was "Llewellyn's Dog," the first animal that comes to mind after the needle-voiced field mouse. It so deeply interested and touched me and some of my classmates that we read it over and over with aching hearts, both in and out of school and shed bitter tears over the brave faithful dog, Gelert, slain by his own master, who imagined that he had devoured his son because he came to him all bloody when the boy was lost, though he had saved the child's life by killing a big wolf. We have to look far back to learn how great may be the capacity of a child's heart for sorrow and sympathy with animals as well as with human friends and neighbors. This auld-lang-syne story stands out in the throng of old schoolday memories as clearly as if I had myself been one of that Welsh hunting-party—heard the bugles blowing, seen Gelert slain, joined in the search for the lost child, discovered it at last happy and smiling among the grass and bushes beside the dead, mangled wolf, and wept with Llewellyn over the sad fate of his noble, faithful dog friend.

Another favorite in this book was Southey's poem "The Inchcape Bell," a story of a priest and a pirate. A good priest in order to warn seamen in dark stormy weather hung a big bell on the dangerous Inchcape Rock. The greater the storm and higher the waves, the louder rang the warning bell, until it was cut off and sunk by wicked

Ralph the Rover. One fine day, as the story goes, when the bell was ringing gently, the pirate put out to the rock, saying, "I'll sink that bell and plague the Abbot of Aberbrothok." So he cut the rope, and down went the bell "with a gurgling sound; the bubbles rose and burst around," etc. Then "Ralph the Rover sailed away; he scoured the seas for many a day; and now, grown rich with plundered store, he steers his course for Scotland's shore." Then came a terrible storm with cloud darkness and night darkness and high roaring waves. "Now where we are," cried the pirate, "I cannot tell, but I wish I could hear the Inchcape bell." And the story goes on to tell how the wretched rover "tore his hair," and "curst himself in his despair," when "with a shivering shock" the stout ship struck on the Inchcape Rock, and went down with Ralph and his plunder beside the good priest's bell. The story appealed to our love of kind deeds and of wildness and fair play.

A lot of terrifying experiences connected with these first school-days grew out of crimes committed by the keeper of a low lodging-house in Edinburgh, who allowed poor homeless wretches to sleep on benches or the floor for a penny or so a night, and, when kind Death came to their relief, sold the bodies for dissection to Dr. Hare of the medical school. None of us children ever heard anything like the original story. The servant girls told us that "Dandy Doctors," clad in long black cloaks and supplied with a store of sticking-plaster of wondrous adhesiveness, prowled at night about the country lanes and even the town streets, watching for children to choke and sell. The Dandy Doctor's business method, as the servants explained it, was with lightning quickness to clap a sticking-plaster on the face of a scholar, covering mouth and nose, preventing breathing or crying for help, then pop us under his long black cloak and carry us to Edinburgh to be sold and sliced into small pieces for folk to learn how we were made. We always mentioned the name "Dandy Doctor" in a fearful whisper, and never dared venture out of doors after dark. In the short winter days it got dark before school closed, and in cloudy weather we sometimes had difficulty in finding our way home unless a servant with a lantern was sent for us; but during the Dandy Doctor period the school was closed earlier, for if detained until the usual hour the teacher could not get us to leave the school-room. We would rather stay all night supperless than dare the mysterious doctors supposed to be lying in wait for us. We had to go up a hill called the Davel Brae that lay between the schoolhouse and the main street. One evening just before dark, as we were running up the hill, one of the boys shouted, "A Dandy Doctor! A Dandy Doctor!" and we all fled pellmell back into the schoolhouse to the astonishment of Mungo Siddons, the teacher. I can remember to this day the amused look on the good dominie's face as he stared and tried to guess what had got into us, until one of the older boys breathlessly explained that there was an awful big Dandy Doctor on the Brae and we couldna gang hame. Others corroborated the dreadful news. "Yes! We saw him, plain as onything, with his lang black cloak to hide us in, and some of us thought we saw a sticken-plaister ready in his hand." We were in such a state of fear and trembling that the teacher saw he wasn't going to get rid of us without going himself as leader. He went only a short distance, however, and turned us over to the care of the two biggest scholars, who led us to the top of the Brae and then left us to scurry home and dash into the door like pursued squirrels diving into their holes.

Just before school skaled (closed), we all arose and sang the fine hymn "Lord, dismiss us with Thy blessing." In the spring when the swallows were coming back from their winter homes we sang—

> "Welcome, welcome, little stranger,
> Welcome from a foreign shore;
> Safe escaped from many a danger . . ."

and while singing we all swayed in rhythm with the music. "The Cuckoo," that always told his name in the spring of the year, was another favorite song, and when there was nothing in particular to call to mind any special bird or animal, the songs we sang were widely varied, such as

> "The whale, the whale is the beast for me,
> Plunging along through the deep, deep sea."

But the best of all was "Lord, dismiss us with Thy blessing," though at that time the most significant part I fear was the first three words.

With my school lessons father made me learn hymns and Bible verses. For learning "Rock of Ages" he gave me a penny, and I thus became suddenly rich. Scotch boys are seldom spoiled with money.

We thought more of a penny those economical days than the poorest American schoolboy thinks of a dollar. To decide what to do with that first penny was an extravagantly serious affair. I ran in great excitement up and down the street, examining the tempting goodies in the shop windows before venturing on so important an investment. My playmates also became excited when the wonderful news got abroad that Johnnie Muir had a penny, hoping to obtain a taste of the orange, apple, or candy it was likely to bring forth.

At this time infants were baptized and vaccinated a few days after birth. I remember very well a fight with the doctor when my brother David was vaccinated. This happened, I think, before I was sent to school. I couldn't imagine what the doctor, a tall, severe-looking man in black, was doing to my brother, but as mother, who was holding him in her arms, offered no objection, I looked on quietly while he scratched the arm until I saw blood. Then, unable to trust even my mother, I managed to spring up high enough to grab and bite the doctor's arm, yelling that I wasna gan to let him hurt my bonnie brither, while to my utter astonishment mother and the doctor only laughed at me. So far from complete at times is sympathy between parents and children, and so much like wild beasts are baby boys, little fighting, biting, climbing pagans.

Father was proud of his garden and seemed always to be trying to make it as much like Eden as possible, and in a corner of it he gave each of us a little bit of ground for our very own in which we planted what we best liked, wondering how the hard dry seeds could change into soft leaves and flowers and find their way out to the light; and, to see how they were coming on, we used to dig up the larger ones, such as peas and beans, every day. My aunt had a corner assigned to her in our garden which she filled with lilies, and we all looked with the utmost respect and admiration at that precious lily-bed and wondered whether when we grew up we should ever be rich enough to own one anything like so grand. We imagined that each lily was worth an enormous sum of money and never dared to touch a single leaf or petal of them. We really stood in awe of them. Far, far was I then from the wild lily gardens of California that I was destined to see in their glory.

When I was a little boy at Mungo Siddons's school a flower-show was held in Dunbar, and I saw a number of the exhibitors carrying large handfuls of dahlias, the first I had ever seen. I thought them marvelous in size and beauty and, as in the case of my aunt's lilies, wondered if I should ever be rich enough to own some of them.

Although I never dared to touch my aunt's sacred lilies, I have good cause to remember stealing some common flowers from an apothecary, Peter Lawson, who also answered the purpose of a regular physician to most of the poor people of the town and adjacent country. He had a pony which was considered very wild and dangerous, and when he was called out of town he mounted this wonderful beast, which, after standing long in the stable, was frisky and boisterous, and often to our delight reared and jumped and danced about from side to side of the street before he could be persuaded to go ahead. We boys gazed in awful admiration and wondered how the druggist could be so brave and able as to get on and stay on that wild beast's back. This famous Peter loved flowers and had a fine garden surrounded by an iron fence, through the bars of which, when I thought no one saw me, I oftentimes snatched a flower and took to my heels. One day Peter discovered me in this mischief, dashed out into the street and caught me. I screamed that I wouldna steal any more if he would let me go. He didn't say anything but just dragged me along to the stable where he kept the wild pony, pushed me in right back of its heels, and shut the door. I was screaming, of course, but as soon as I was imprisoned the fear of being kicked quenched all noise. I hardly dared breathe. My only hope was in motionless silence. Imagine the agony I endured! I did not steal any more of his flowers. He was a good hard judge of boy nature.

I was in Peter's hands some time before this, when I was about two and a half years old. The servant girl bathed us small folk before putting us to bed. The smarting soapy scrubbings of the Saturday nights in preparation for the Sabbath were particularly severe, and we all dreaded them. My sister Sarah, the next older than me, wanted the long-legged stool I was sitting on awaiting my turn, so she just tipped me off. My chin struck on the edge of the bath-tub, and, as I was talking at the time, my tongue happened to be in the way of my teeth when they were closed by the blow, and a deep gash was cut on the side of it, which bled profusely. Mother came running at the noise I made, wrapped me up, put me in the servant girl's arms and told her to run with me through the garden and out by a back way to Peter Lawson to have something done to stop the bleeding. He simply pushed a wad of cotton into my mouth after soaking it in some brown astringent stuff, and told me to be sure to keep my mouth shut and all would soon be well. Mother put me to bed, calmed my fears, and told me to lie still and sleep like a gude bairn. But just as I was dropping off to sleep I swallowed the bulky wad of medicated cotton and with it, as I imagined, my tongue also. My screams over so great a loss brought mother, and when she anxiously took me in her arms and inquired what was the matter, I told her that I had swallowed my tongue. She only laughed at me, much to my astonishment, when I expected that she would bewail the awful loss her boy had sustained. My sisters, who were older than I, oftentimes said when I happened to be talking too much, "It's a pity you hadn't swallowed at least half of that long tongue of yours when you were little."

It appears natural for children to be fond of water, although the Scotch method of making every duty dismal contrived to make necessary bathing for health terrible to us. I well remember among the awful experiences of childhood being taken by the servant to the seashore when I was between two and three years old, stripped at the side of a deep pool in the rocks, plunged into it among crawling crawfish and slippery wriggling snake-like eels, and drawn up gasping and shrieking only to be plunged down again and again. As the time approached for this terrible bathing, I used to hide in the darkest corners of the house, and oftentimes a long search was required to find me. But after we were a few years older, we enjoyed bathing with other boys as we wandered along the shore, careful, however, not to get into a pool that had an invisible boy-devouring monster at the bottom of it. Such pools, miniature maelstroms, were called "sookin-in-goats" and were well known to most of us. Nevertheless we never ventured into any pool on strange parts of the coast before we had thrust a stick into it. If the stick were not pulled out of our hands, we boldly entered and enjoyed plashing and ducking long ere we had learned to swim.

One of our best playgrounds was the famous old Dunbar Castle, to which King Edward fled after his defeat at Bannockburn. It was built more than a thousand years ago, and though we knew little of its history, we had heard many mysterious stories of the battles fought about its walls, and firmly believed that every bone we found in the ruins belonged to an ancient warrior. We tried to see who could climb highest on the crumbling peaks and crags, and took chances that no cautious mountaineer would try. That I did not fall and finish my rock-scrambling in those adventurous boyhood days seems now a reasonable wonder.

Among our best games were running, jumping, wrestling, and scrambling. I was so proud of my skill as a climber that when I first heard of hell from a servant girl who loved to tell its horrors and warn us that if we did anything wrong we would be cast into it, I always insisted that I could climb out of it. I imagined it was only a sooty pit with stone walls like those of the castle, and I felt sure there must be chinks and cracks in the masonry for fingers and toes. Anyhow the terrors of the horrible place seldom lasted long beyond the telling; for natural faith casts out fear.

Most of the Scotch children believe in ghosts, and some under peculiar conditions continue to believe in them all through life. Grave ghosts are deemed particularly dangerous, and many of the most credulous will go far out of their way to avoid passing through or near a graveyard in the dark. After being instructed by the servants in the nature, looks, and habits of the various black and white ghosts, boowuzzies, and witches we often speculated as to whether they could run fast, and tried to believe that we had a good chance to get away from most of them. To improve our speed and wind, we often took long runs into the country. Tam o' Shanter's mare outran a lot of witches,—at least until she reached a place of safety beyond the keystone of the bridge,—and we thought perhaps we also might be able to outrun them.

Our house formerly belonged to a physician, and a servant girl told us that the ghost of the dead doctor haunted one of the unoccupied rooms in the second story that was kept dark on account of a heavy window-tax. Our bedroom was adjacent to the ghost room, which had in it a lot of chemical apparatus,—glass tubing, glass and brass retorts, test-tubes, flasks, etc.,—and we thought that those strange articles were still used by the old dead doctor in compounding physic. In the long summer days David and I were put to bed several hours before sunset. Mother tucked us in carefully, drew the curtains of the big old-fashioned bed, and told us to lie still and sleep like gude bairns; but we were usually out of bed, playing games of daring called "scootchers," about as soon as our loving mother reached the foot of the stairs, for we couldn't lie still, however hard we might try. Going into the ghost room was regarded as a very great scootcher. After venturing in a few steps and rushing back in terror, I used to dare David to go as far without getting caught.

The roof of our house, as well as the crags and walls of the old castle, offered fine mountaineering exercise. Our bedroom was lighted by a dormer window. One night I opened it in search of good scootchers and hung myself out over the slates, holding on to the sill, while the wind was making a balloon of my nightgown. I then dared David to try the adventure, and he did. Then I went out again and hung by one hand, and David did the same. Then I hung by one finger, being careful not to slip, and he did that too. Then I stood on the sill and examined the edge of the left wall of the window, crept up the slates along its side by slight finger-holds, got astride of the roof, sat there a few minutes looking at the scenery over the garden wall while the wind was howling and threatening to blow me off, then managed to slip down, catch hold of the sill, and get safely back into the room. But before attempting this scootcher, recognizing its dangerous character, with commendable caution I warned David that in case I should happen to slip I would grip the rain-trough when I was going over the eaves and hang on, and that he must then run fast downstairs and tell father to get a ladder for me, and tell him to be quick because I would soon be tired hanging dangling in the wind by my hands. After my return from this capital scootcher, David, not to be outdone, crawled up to the top of the window-roof, and got bravely astride of it; but in trying to return he lost courage and began to greet (to cry), "I canna get doon. Oh, I canna get doon." I leaned out of the window and shouted encouragingly, "Dinna greet, Davie, dinna greet, I'll help ye doon. If you greet, fayther will hear, and gee us baith an awfu' skelping." Then, standing on the sill and holding on by one hand to the window-casing, I directed him to slip his feet down within reach, and, after securing a good hold, I jumped inside and dragged him in by his heels. This finished scootcher-scrambling for the night and frightened us into bed.

In the short winter days, when it was dark even at our early bedtime, we usually spent the hours before going to sleep playing voyages around the world under the bed-clothing. After mother had carefully covered us, bade us good-night and gone downstairs, we set out on our travels. Burrowing like moles, we visited France, India, America, Australia, New Zealand, and all the places we had ever heard of; our travels never ending until we fell asleep. When mother came to take a last look at us, before she went to bed, to see that we were covered, we were oftentimes covered so well that she had difficulty in finding us, for we were hidden in all sorts of positions where sleep happened to overtake us, but in the morning we always found ourselves in good order, lying straight like gude bairns, as she said.

Some fifty years later, when I visited Scotland, I got one of my Dunbar schoolmates to introduce me to the owners of our old home, from whom I obtained permission to go upstairs to examine our bedroom window and judge what sort of adventure getting on its roof must have been, and with all my after experience in mountaineering, I found that what I had done in daring boyhood was now beyond my skill.

Boys are often at once cruel and merciful, thoughtlessly hard-hearted and tender-hearted, sympathetic, pitiful, and kind in ever changing contrasts. Love of neighbors, human or animal, grows up amid savage traits, coarse and fine. When father made out to get us securely locked up in the back yard to prevent our shore and field wanderings, we had to play away the comparatively dull time as best we could. One of our amusements was hunting cats without seriously hurting them. These sagacious animals knew, however, that, though not very dangerous, boys were not to be trusted. One time in particular I remember, when we began throwing stones at an experienced old Tom, not wishing to hurt him much, though he was a tempting mark. He soon saw what we were up to, fled to the stable, and climbed to the top of the hay manger. He was still within range, however, and we kept the stones flying faster and faster, but he just blinked and played possum without wincing either at our best shots or at the noise we made. I happened to strike him pretty hard with a good-sized pebble, but he still blinked and sat still as if without feeling. "He must be mortally wounded," I said, "and now we must kill him to put him out of pain," the savage in us rapidly growing with indulgence. All took heartily to this sort of cat mercy and began throwing the heaviest stones we could manage, but that old fellow knew what characters we were, and just as we imagined him mercifully dead he evidently thought the play was becoming too serious and that it was time to retreat; for suddenly with a wild whirr and gurr of energy he launched himself over our heads, rushed across the yard in a blur of speed, climbed to the roof of another building and over the garden wall, out of pain and bad company, with all his lives wideawake and in good working order.

After we had thus learned that Tom had at least nine lives, we tried to verify the common saying that no matter how far cats fell they always landed on their feet unhurt. We caught one in our back yard, not Tom but a smaller one of manageable size, and somehow got him smuggled up to the top story of the house. I don't know how in the world we managed to let go of him, for as soon as we opened the window and held him over the sill he knew his danger and made violent efforts to scratch and bite his way back into the room; but we determined to carry the thing through, and at last managed to drop him. I can remember to this day how the poor creature in danger of his life strained and balanced as he was falling and managed to alight on his feet. This was a cruel thing for even wild boys to do, and we never tried the experiment again, for we sincerely pitied the poor fellow when we saw him creeping slowly away, stunned and frightened, with a swollen black and blue chin.

Again—showing the natural savagery of boys—we delighted in dog-fights, and even in the horrid red work of slaughter-houses, often running long distances and climbing over walls and roofs to see a pig killed, as soon as we heard the desperately earnest squealing. And if the butcher was good-natured, we begged him to let us get a near view of the mysterious insides and to give us a bladder to blow up for a foot-ball.

But here is an illustration of the better side of boy nature. In our back yard there were three elm trees and in the one nearest the house a pair of robin-redbreasts had their nest. When the young were almost able to fly, a troop of the celebrated "Scottish Grays" visited Dunbar, and three or four of the fine horses were lodged in our stable. When the soldiers were polishing their swords and helmets, they happened to notice the nest, and just as they were leaving, one of them climbed the tree and robbed it. With sore sympathy we watched the young birds as the hard-hearted robber pushed them one by one beneath his jacket,—all but two that jumped out of the nest and tried to fly, but they were easily caught as they fluttered on the ground, and were hidden away with the rest. The distress of the bereaved parents, as they hovered and screamed over the frightened crying children they so long had loved and sheltered and fed, was pitiful to see; but the shining soldier rode grandly away on his big gray horse, caring only for the few pennies the young songbirds would bring and the beer they would buy, while we all, sisters and brothers, were crying and sobbing. I remember, as if it happened this day, how my heart fairly ached and choked me. Mother put us to bed and tried to comfort us, telling us that the little birds would be well fed and grow big, and soon learn to sing in pretty cages; but again and again we rehearsed the sad story of the poor bereaved birds and their frightened children, and could not be comforted. Father came into the room when we were half asleep and still sobbing, and I heard mother telling him that, "a' the bairns' hearts were broken over the robbing of the nest in the elm."

After attaining the manly, belligerent age of five or six years, very few of my schooldays passed without a fist fight, and half a dozen was no uncommon number. When any classmate of our own age questioned our rank and standing as fighters, we always made haste to settle the matter at a quiet place on the Davel Brae. To be a "gude fechter" was our highest ambition, our dearest aim in life in or out of school. To be a good scholar was a secondary consideration, though we tried hard to hold high places in our classes and gloried in being Dux. We fairly reveled in the battle stories of glorious William Wallace and Robert the Bruce, with which every breath of Scotch air is saturated, and of course we were all going to be soldiers. On the Davel Brae battleground we often managed to bring on something like real war, greatly more exciting than personal combat. Choosing leaders, we divided into two armies. In winter damp snow furnished plenty of ammunition to make the thing serious, and in summer sand and grass sods. Cheering and shouting some battle-cry such as "Bannockburn! Bannockburn! Scotland forever! The Last War in India!" we were led bravely on. For heavy battery work we stuffed our Scotch blue bonnets with snow and sand, sometimes mixed with gravel, and fired them at each other as cannon-balls.

Of course we always looked eagerly forward to vacation days and thought them slow in coming. Old Mungo Siddons gave us a lot of gooseberries or currants and wished us a happy time. Some sort of special closing-exercises—singing, recitations, etc.—celebrated the great day, but I remember only the berries, freedom from school work, and opportunities for runaway rambles in the fields and along the wave-beaten seashore.

An exciting time came when at the age of seven or eight years I left the auld Davel Brae school for the grammar school. Of course I had a terrible lot of fighting to do, because a new scholar had to meet every one of his age who dared to challenge him, this being the common introduction to a new school. It was very strenuous for the first month or so, establishing my fighting rank, taking up new studies, especially Latin and French, getting acquainted with new classmates and the master and his rules. In the first few Latin and French lessons the new teacher, Mr. Lyon, blandly smiled at our comical blunders, but pedagogical weather of the severest kind quickly set in, when for every mistake, everything short of perfection, the taws was promptly applied. We had to get three lessons every day in Latin, three in French, and as many in English, besides spelling, history, arithmetic, and geography. Word lessons in particular, the wouldst-couldst-shouldst-have-loved kind, were kept up, with much warlike thrashing, until I had committed the whole of the French, Latin, and English grammars to memory, and in connection with reading-lessons we were called on to recite parts of them with the rules over and over again, as if all the regular and irregular incomprehensible verb stuff was poetry. In addition to all this, father made me learn so many Bible verses every day that by the time I was eleven years of age I had about three fourths of the Old Testament and all of the New by heart and by sore flesh. I could recite the New Testament from the beginning of Matthew to the end of Revelation without a single stop. The dangers of cramming and of making scholars study at home instead of letting their little brains rest were never heard of in those days. We carried our schoolbooks home in a strap every night and committed to memory our next day's lessons before we went to bed, and to do that we had to bend our attention as closely on our tasks as lawyers on great million-dollar cases. I can't conceive of anything that would now enable me to concentrate my attention more fully than when I was a mere stripling boy, and it was all done by whipping,—thrashing in general. Old-fashioned Scotch teachers spent no time in seeking short roads to knowledge, or in trying any of the new-fangled psychological methods so much in vogue nowadays. There was nothing said about making the seats easy or the lessons easy. We were simply driven pointblank against our books like soldiers against the enemy, and sternly ordered, "Up and at 'em. Commit your lessons to memory!" If we failed in any part, however slight, we were whipped; for the grand, simple, all-sufficing Scotch discovery had been made that there was a close connection between the skin and the memory, and that irritating the skin excited the memory to any required degree.

Fighting was carried on still more vigorously in the high school than in the common school. Whenever any one was challenged, either the challenge was allowed or it was decided by a battle on the seashore, where with stubborn enthusiasm we battered each other as if we had not been sufficiently battered by the teacher. When we were so fortunate as to finish a fight without getting a black eye, we usually escaped a thrashing at home and another next morning at school, for other traces of the fray could be easily washed off at a well on the church brae, or concealed, or passed as results of playground accidents; but a black eye could never be explained away from downright fighting. A good double thrashing was the inevitable penalty, but all without avail; fighting went on without the slightest abatement, like natural storms; for no punishment less than death could quench the ancient inherited belligerence burning in our pagan blood. Nor could we be made to believe it was fair that father and teacher should thrash us so industriously for our good, while begrudging us the pleasure of thrashing each other for our good. All these various thrashings, however, were admirably influential in developing not only memory but fortitude as well. For if we did not endure our school punishments and fighting pains without flinching and making faces, we were mocked on the playground, and public opinion on a Scotch playground was a powerful agent in controlling behavior; therefore we at length managed to keep our features in smooth repose while enduring pain that would try anybody but an American Indian. Far from feeling that we were called on to endure too much pain, one of our playground games was thrashing each other with whips about two feet long made from the tough, wiry stems of a species of polygonum fastened together in a stiff, firm braid. One of us handing two of these whips to a companion to take his choice, we stood up close together and thrashed each other on the legs until one succumbed to the intolerable pain and thus lost the game. Nearly all of our playground games were strenuous,—shin-battering shinny, wrestling, prisoners' base, and dogs and hares,—all augmenting in no slight degree our lessons in fortitude.

Moreover, we regarded our punishments and pains of every sort as training for war, since we were all going to be soldiers. Besides single combats we sometimes assembled on Saturdays to meet the scholars of another school, and very little was required for the growth of strained relations, and war. The immediate cause might be nothing more than a saucy stare. Perhaps the scholar stared at would insolently inquire, "What are ye glowerin' at, Bob?" Bob would reply, "I'll look where I hae a mind and hinder me if ye daur." "Weel, Bob," the outraged stared-at scholar would reply, "I'll soon let ye see whether I daur or no!" and give Bob a blow on the face. This opened the battle, and every good scholar belonging to either school was drawn into it. After both sides were sore and weary, a strong-lunged warrior would be heard above the din of battle shouting, "I'll tell ye what we'll dae wi' ye. If ye'll let us alane we'll let ye alane!" and the school war ended as most wars between nations do; and some of them begin in much the same way.

Notwithstanding the great number of harshly enforced rules, not very good order was kept in school in my time. There were two schools within a few rods of each other, one for mathematics, navigation, etc., the other, called the grammar school, that I attended. The masters lived in a big freestone house within eight or ten yards of the schools, so that they could easily step out for anything they wanted or send one of the scholars. The moment our master disappeared, perhaps for a book or a drink, every scholar left his seat and his lessons, jumped on top of the benches and desks or crawled beneath them, tugging, rolling, wrestling, accomplishing in a minute a depth of disorder and din unbelievable save by a Scottish scholar. We even carried on war, class against class, in those wild, precious minutes. A watcher gave the alarm when the master opened his house-door to return, and it was a great feat to get into our places before he entered, adorned in awful majestic authority, shouting "Silence!" and striking resounding blows with his cane on a desk or on some unfortunate scholar's back.

Forty-seven years after leaving this fighting school, I returned on a visit to Scotland, and a cousin in Dunbar introduced me to a minister who was acquainted with the history of the school, and obtained for me an invitation to dine with the new master. Of course I gladly accepted, for I wanted to see the old place of fun and pain, and the battleground on the sands. Mr. Lyon, our able teacher and thrasher,

I learned, had held his place as master of the school for twenty or thirty years after I left it, and had recently died in London, after preparing many young men for the English Universities. At the dinner-table, while I was recalling the amusements and fights of my old schooldays, the minister remarked to the new master, "Now, don't you wish that you had been teacher in those days, and gained the honor of walloping John Muir?" This pleasure so merrily suggested showed that the minister also had been a fighter in his youth. The old freestone school building was still perfectly sound, but the carved, ink-stained desks were almost whittled away.

The highest part of our playground back of the school commanded a view of the sea, and we loved to watch the passing ships and, judging by their rigging, make guesses as to the ports they had sailed from, those to which they were bound, what they were loaded with, their tonnage, etc. In stormy weather they were all smothered in clouds and spray, and showers of salt scud torn from the tops of the waves came flying over the playground wall. In those tremendous storms many a brave ship foundered or was tossed and smashed on the rocky shore. When a wreck occurred within a mile or two of the town, we often managed by running fast to reach it and pick up some of the spoils. In particular I remember visiting the battered fragments of an unfortunate brig or schooner that had been loaded with apples, and finding fine unpitiful sport in rushing into the spent waves and picking up the red-cheeked fruit from the frothy, seething foam.

All our school-books were extravagantly illustrated with drawings of every kind of sailing-vessel, and every boy owned some sort of craft whittled from a block of wood and trimmed with infinite pains,—sloops, schooners, brigs, and full-rigged ships, with their sails and string ropes properly adjusted and named for us by some old sailor. These precious toy craft with lead keels we learned to sail on a pond near the town. With the sails set at the proper angle to the wind, they made fast straight voyages across the pond to boys on the other side, who readjusted the sails and started them back on the return voyages. Oftentimes fleets of half a dozen or more were started together in exciting races.

Our most exciting sport, however, was playing with gunpowder. We made guns out of gas-pipe, mounted them on sticks of any shape, clubbed our pennies together for powder, gleaned pieces of lead here and there and cut them into slugs, and, while one aimed, another applied a match to the touchhole. With these awful weapons we wandered along the beach and fired at the gulls and solan-geese as they passed us. Fortunately we never hurt any of them that we knew of. We also dug holes in the ground, put in a handful or two of powder, tamped it well around a fuse made of a wheat-stalk, and, reaching cautiously forward, touched a match to the straw. This we called making earthquakes. Oftentimes we went home with singed hair and faces well peppered with powder-grains that could not be washed out. Then, of course, came a correspondingly severe punishment from both father and teacher.

Another favorite sport was climbing trees and scaling garden-walls. Boys eight or ten years of age could get over almost any wall by standing on each other's shoulders, thus making living ladders. To make walls secure against marauders, many of them were finished on top with broken bottles imbedded in lime, leaving the cutting edges sticking up; but with bunches of grass and weeds we could sit or stand in comfort on top of the jaggedest of them.

Like squirrels that begin to eat nuts before they are ripe, we began to eat apples about as soon as they were formed, causing, of course, desperate gastric disturbances to be cured by castor oil. Serious were the risks we ran in climbing and squeezing through hedges, and, of course, among the country folk we were far from welcome. Farmers passing us on the roads often shouted by way of greeting: "Oh, you vagabonds! Back to the toon wi' ye. Gang back where ye belang. You're up to mischief, Ise warrant. I can see it. The gamekeeper'll catch ye, and maist like ye'll a' be hanged some day."

Breakfast in those auld-lang-syne days was simple oatmeal porridge, usually with a little milk or treacle, served in wooden dishes called "luggies," formed of staves hooped together like miniature tubs about four or five inches in diameter. One of the staves, the lug or ear, a few inches longer than the others, served as a handle, while the number of luggies ranged in a row on a dresser indicated the size of the family. We never dreamed of anything to come after the porridge, or of asking for more. Our portions were consumed in about a couple of minutes; then off to school. At noon we came racing home ravenously hungry. The midday meal, called dinner, was usually vegetable broth, a small piece of boiled mutton, and barley-meal scone. None of us liked the barley scone bread, therefore we got all we wanted of it, and in desperation had to eat it, for we were always hungry, about as hungry after as before meals. The evening meal was called "tea" and was served on our return from school. It consisted, as far as we children were concerned, of half a slice of white bread without butter, barley scone, and warm water with a little milk and sugar in it, a beverage called "content," which warmed but neither cheered nor inebriated. Immediately after tea we ran across the street with our books to Grandfather Gilrye, who took pleasure in seeing us and hearing us recite our next day's lessons. Then back home to supper, usually a boiled potato and piece of barley scone. Then family worship, and to bed.

Our amusements on Saturday afternoons and vacations depended mostly on getting away from home into the country, especially in the spring when the birds were calling loudest. Father sternly forbade David and me from playing truant in the fields with plundering wanderers like ourselves, fearing we might go on from bad to worse, get hurt in climbing over walls, caught by gamekeepers, or lost by falling over a cliff into the sea. "Play as much as you like in the back yard and garden," he said, "and mind what you'll get when you forget and disobey." Thus he warned us with an awfully stern countenance, looking very hard-hearted, while naturally his heart was far from hard, though he devoutly believed in eternal punishment for bad boys both here and hereafter. Nevertheless, like devout martyrs of wildness, we stole away to the seashore or the green, sunny fields with almost religious regularity, taking advantage of opportunities when father was very busy, to join our companions, oftenest to hear the birds sing and hunt their nests, glorying in the number we had discovered and called our own. A sample of our nest chatter was something like this: Willie Chisholm would proudly exclaim—"I ken (know) seventeen nests, and you, Johnnie, ken only fifteen."

"But I wouldna gie my fifteen for your seventeen, for five of mine are larks and mavises. You ken only three o' the best singers."

"Yes, Johnnie, but I ken six goldies and you ken only one. Maist of yours are only sparrows and linties and robin-redbreasts."

Then perhaps Bob Richardson would loudly declare that he "kenned mair nests than onybody, for he kenned twenty-three, with about fifty eggs in them and mair than fifty young birds—maybe a hundred. Some of them naething but raw gorblings but lots of them as big as their mithers and ready to flee. And aboot fifty craw's nests and three fox dens."

"Oh, yes, Bob, but that's no fair, for naebody counts craw's nests and fox holes, and then you live in the country at Bellehaven where ye have the best chance."

"Yes, but I ken a lot of bumbee's nests, baith the red-legged and the yellow-legged kind."

"Oh, wha cares for bumbee's nests!"

"Weel, but here's something! Ma father let me gang to a fox hunt, and man, it was grand to see the hounds and the lang-legged horses lowpin the dykes and burns and hedges!"

The nests, I fear, with the beautiful eggs and young birds, were prized quite as highly as the songs of the glad parents, but no Scotch boy that I know of ever failed to listen with enthusiasm to the songs of the skylarks. Oftentimes on a broad meadow near Dunbar we stood for hours enjoying their marvelous singing and soaring. From the grass where the nest was hidden the male would suddenly rise, as straight as if shot up, to a height of perhaps thirty or forty feet, and, sustaining himself with rapid wing-beats, pour down the most delicious melody, sweet and clear and strong, overflowing all bounds, then suddenly he would soar higher again and again, ever higher and higher, soaring and singing until lost to sight even on perfectly clear days, and oftentimes in cloudy weather "far in the downy cloud," as the poet says.

To test our eyes we often watched a lark until he seemed a faint speck in the sky and finally passed beyond the keenest-sighted of us all. "I see him yet!" we would cry, "I see him yet!" "I see him yet!" "I see him yet!" as he soared. And finally only one of us would be left to claim that he still saw him. At last he, too, would have to admit that the singer had soared beyond his sight, and still the music came pouring down to us in glorious profusion, from a height far above our vision, requiring marvelous power of wing and marvelous power of voice, for that rich, delicious, soft, and yet clear music was distinctly heard long after the bird was out of sight. Then, suddenly ceasing, the glorious singer would appear, falling like a bolt straight down to his nest, where his mate was sitting on the eggs.

It was far too common a practice among us to carry off a young lark just before it could fly, place it in a cage, and fondly, laboriously feed it. Sometimes we succeeded in keeping one alive for a year or two, and when awakened by the spring weather it was pitiful to see the quivering imprisoned soarer of the heavens rapidly beating its wings and singing as though it were flying and hovering in the air like its parents. To keep it in health we were taught that we must supply it with a sod of grass the size of the bottom of the cage, to make the poor bird feel as though it were at home on its native meadow,—a meadow perhaps a foot or at most two feet square. Again and again it would try to hover over that miniature meadow from its miniature sky just underneath the top of the cage. At last, conscience-stricken, we carried the beloved prisoner to the meadow west of Dunbar where it was born, and, blessing its sweet heart, bravely set it free, and our exceeding great reward was to see it fly and sing in the sky.

In the winter, when there was but little doing in the fields, we organized running-matches. A dozen or so of us would start out on races that were simply tests of endurance, running on and on along a public road over the breezy hills like hounds, without stopping or getting tired. The only serious trouble we ever felt in these long races was an occasional stitch in our sides. One of the boys started the story that sucking raw eggs was a sure cure for the stitches. We had hens in our back yard, and on the next Saturday we managed to swallow a couple of eggs apiece, a disgusting job, but we would do almost anything to mend our speed, and as soon as we could get away after taking the cure we set out on a ten or twenty mile run to prove its worth. We thought nothing of running right ahead ten or a dozen miles before turning back; for we knew nothing about taking time by the sun, and none of us had a watch in those days. Indeed, we never cared about time until it began to get dark. Then we thought of home and the thrashing that awaited us. Late or early, the thrashing was sure, unless father happened to be away. If he was expected to return soon, mother made haste to get us to bed before his arrival. We escaped the thrashing next morning, for father never felt like thrashing us in cold blood on the calm holy Sabbath. But no punishment, however sure and severe, was of any avail against the attraction of the fields and woods. It had other uses, developing memory, etc., but in keeping us at home it was of no use at all. Wildness was ever sounding in our ears, and Nature saw to it that besides school lessons and church lessons some of her own lessons should be learned, perhaps with a view to the time when we should be called to wander in wildness to our heart's content. Oh, the blessed enchantment of those Saturday runaways in the prime of the spring! How our young wondering eyes reveled in the sunny, breezy glory of the hills and the sky, every particle of us thrilling and tingling with the bees and glad birds and glad streams! Kings may be blessed; we were glorious, we were free,—school cares and scoldings, heart thrashings and flesh thrashings alike, were forgotten in the fullness of Nature's glad wildness. These were my first excursions,—the beginnings of lifelong wanderings.

II

A NEW WORLD

Stories of America—Glorious News—Crossing the Atlantic—The New Home—A Baptism in Nature—New Birds—The Adventures of Watch—Scotch Correction—Marauding Indians.

OUR GRAMMAR-SCHOOL reader, called, I think, "Maccoulough's Course of Reading," contained a few natural-history sketches that excited me very much and left a deep impression, especially a fine description of the fish hawk and the bald eagle by the Scotch ornithologist Wilson, who had the good fortune to wander for years in the American woods while the country was yet mostly wild. I read his description over and over again, till I got the vivid picture he drew by heart,—the long-winged hawk circling over the heaving waves, every motion watched by the eagle perched on the top of a crag or dead tree; the fish hawk poising for a moment to take aim at a fish and plunging under the water; the eagle with kindling eye spreading his wings ready for instant flight in case the attack should prove successful; the hawk emerging with a struggling fish in his talons, and proud flight; the eagle launching himself in pursuit; the wonderful wing-work in the sky, the fish hawk, though encumbered with his prey, circling higher, higher, striving hard to keep above the robber eagle; the eagle at length soaring above him, compelling him with a cry of despair to drop his hard-won prey; then the eagle steadying himself for a moment to take aim, descending swift as a lightning-bolt, and seizing the falling fish before it reached the sea.

Not less exciting and memorable was Audubon's wonderful story of the passenger pigeon, a beautiful bird flying in vast flocks that darkened the sky like clouds, countless millions assembling to rest and sleep and rear their young in certain forests, miles in length and breadth, fifty or a hundred nests on a single tree; the overloaded branches bending low and often breaking; the farmers gathering from far and near, beating down countless thousands of the young and old birds from their nests and roosts with long poles at night, and in the morning driving their bands of hogs, some of them brought from farms a hundred miles distant, to fatten on the dead and wounded covering the ground.

In another of our reading-lessons some of the American forests were described. The most interesting of the trees to us boys was the sugar maple, and soon after we had learned this sweet story we heard everybody talking about the discovery of gold in the same wonder-filled country.

One night, when David and I were at grandfather's fireside solemnly learning our lessons as usual, my father came in with news, the most wonderful, most glorious, that wild boys ever heard. "Bairns," he said, "you needna learn your lessons the nicht, for we're gan to America the morn!" No more grammar, but boundless woods full of mysterious good things; trees full of sugar, growing in ground full of gold; hawks, eagles, pigeons, filling the sky; millions of birds' nests, and no gamekeepers to stop us in all the wild, happy land. We were utterly, blindly glorious. After father left the room, grandfather gave David and me a gold coin apiece for a keepsake, and looked very serious, for he was about to be deserted in his lonely old age. And when we in fullness of young joy spoke of what we were going to do, of the wonderful birds and their nests that we should find, the sugar and gold, etc., and promised to send him a big box full of that tree sugar packed in gold from the glorious paradise over the sea, poor lonely grandfather, about to be forsaken, looked with downcast eyes on the floor and said in a low, trembling, troubled voice, "Ah, poor laddies, poor laddies, you'll find something else ower the sea forbye gold and sugar, birds' nests and freedom fra lessons and schools. You'll find plenty hard, hard work." And so we did. But nothing he could say could cloud our joy or abate the fire of youthful, hopeful, fearless adventure. Nor could we in the midst of such measureless excitement see or feel the shadows and sorrows of his darkening old age. To my schoolmates, met that night on the street, I shouted the glorious news, "I'm gan to Amaraka the morn!" None could believe it. I said, "Weel, just you see if I am at the skule the morn!"

Next morning we went by rail to Glasgow and thence joyfully sailed away from beloved Scotland, flying to our fortunes on the wings of the winds, carefree as thistle seeds. We could not then know what we were leaving, what we were to encounter in the New

World, nor what our gains were likely to be. We were too young and full of hope for fear or regret, but not too young to look forward with eager enthusiasm to the wonderful schoolless bookless American wilderness. Even the natural heart-pain of parting from grandfather and grandmother Gilrye, who loved us so well, and from mother and sisters and brother was quickly quenched in young joy. Father took with him only my sister Sarah (thirteen years of age), myself (eleven), and brother David (nine), leaving my eldest sister, Margaret, and the three youngest of the family, Daniel, Mary, and Anna, with mother, to join us after a farm had been found in the wilderness and a comfortable house made to receive them.

In crossing the Atlantic before the days of steamships, or even the American clippers, the voyages made in old-fashioned sailing-vessels were very long. Ours was six weeks and three days. But because we had no lessons to get, that long voyage had not a dull moment for us boys. Father and sister Sarah, with most of the old folk, stayed below in rough weather, groaning in the miseries of seasickness, many of the passengers wishing they had never ventured in "the auld rockin'creel," as they called our bluff-bowed, wave-beating ship, and, when the weather was moderately calm, singing songs in the evenings,—"The Youthful Sailor Frank and Bold," "Oh, why left I my hame, why did I cross the deep," etc. But no matter how much the old tub tossed about and battered the waves, we were on deck every day, not in the least seasick, watching the sailors at their rope-hauling and climbing work; joining in their songs, learning the names of the ropes and sails, and helping them as far as they would let us; playing games with other boys in calm weather when the deck was dry, and in stormy weather rejoicing in sympathy with the big curly-topped waves.

The captain occasionally called David and me into his cabin and asked us about our schools, handed us books to read, and seemed surprised to find that Scotch boys could read and pronounce English with perfect accent and knew so much Latin and French. In Scotch schools only pure English was taught, although not a word of English was spoken out of school. All through life, however well educated, the Scotch spoke Scotch among their own folk, except at times when unduly excited on the only two subjects on which Scotchmen get much excited, namely religion and politics. So long as the controversy went on with fairly level temper, only gude braid

Scots was used, but if one became angry, as was likely to happen, then he immediately began speaking severely correct English, while his antagonist, drawing himself up, would say: "Weel, there's na use pursuing this subject ony further, for I see ye hae gotten to your English."

As we neared the shore of the great new land, with what eager wonder we watched the whales and dolphins and porpoises and sea-birds, and made the good-natured sailors teach us their names and tell us stories about them!

There were quite a large number of emigrants aboard, many of them newly married couples, and the advantages of the different parts of the New World they expected to settle in were often discussed. My father started with the intention of going to the backwoods of Upper Canada. Before the end of the voyage, however, he was persuaded that the States offered superior advantages, especially Wisconsin and Michigan, where the land was said to be as good as in Canada and far more easily brought under cultivation; for in Canada the woods were so close and heavy that a man might wear out his life in getting a few acres cleared of trees and stumps. So he changed his mind and concluded to go to one of the Western States.

On our wavering westward way a grain-dealer in Buffalo told father that most of the wheat he handled came from Wisconsin; and this influential information finally determined my father's choice. At Milwaukee a farmer who had come in from the country near Fort Winnebago with a load of wheat agreed to haul us and our formidable load of stuff to a little town called Kingston for thirty dollars. On that hundred-mile journey, just after the spring thaw, the roads over the prairies were heavy and miry, causing no end of lamentation, for we often got stuck in the mud, and the poor farmer sadly declared that never, never again would he be tempted to try to haul such a cruel, heart-breaking, wagon-breaking, horse-killing load, no, not for a hundred dollars. In leaving Scotland, father, like many other home-seekers, burdened himself with far too much luggage, as if all America were still a wilderness in which little or nothing could be bought. One of his big iron-bound boxes must have weighed about four hundred pounds, for it contained an old-fashioned beam-scales with a complete set of cast-iron counter-weights, two of them fifty-six pounds each, a twenty-eight, and so on down to a single pound. Also a lot of iron wedges, carpenter's tools, and so forth, and at Buffalo, as if on the very edge of the wilderness, he gladly added to his burden a big cast-iron stove with pots and pans, provisions enough for a long siege, and a scythe and cumbersome cradle for cutting wheat, all of which he succeeded in landing in the primeval Wisconsin woods.

A land-agent at Kingston gave father a note to a farmer by the name of Alexander Gray, who lived on the border of the settled part of the country, knew the section-lines, and would probably help him to find a good place for a farm. So father went away to spy out the land, and in the mean time left us children in Kingston in a rented room. It took us less than an hour to get acquainted with some of the boys in the village; we challenged them to wrestle, run races, climb trees, etc., and in a day or two we felt at home, carefree and happy, notwithstanding our family was so widely divided. When father returned he told us that he had found fine land for a farm in sunny open woods on the side of a lake, and that a team of three yoke of oxen with a big wagon was coming to haul us to Mr. Gray's place.

We enjoyed the strange ten-mile ride through the woods very much, wondering how the great oxen could be so strong and wise and tame as to pull so heavy a load with no other harness than a chain and a crooked piece of wood on their necks, and how they could sway so obediently to right and left past roadside trees and stumps when the driver said *haw* and *gee*. At Mr. Gray's house, father again left us for a few days to build a shanty on the quarter-section he had selected four or five miles to the westward. In the mean while we enjoyed our freedom as usual, wandering in the fields and meadows, looking at the trees and flowers, snakes and birds and squirrels. With the help of the nearest neighbors the little shanty was built in less than a day after the rough bur-oak logs for the walls and the white-oak boards for the floor and roof were got together.

To this charming hut, in the sunny woods, overlooking a flowery glacier meadow and a lake rimmed with white water-lilies, we were hauled by an ox-team across trackless carex swamps and low rolling hills sparsely dotted with round-headed oaks. Just as we arrived at the shanty, before we had time to look at it or the scenery about it, David and I jumped down in a hurry off the load of household goods, for we had discovered a blue jay's nest, and in a minute or so we were up the tree beside it, feasting our eyes on the beautiful green eggs and beautiful birds,—our first memorable discovery. The handsome birds had not seen Scotch boys before and made a desperate screaming as if we were robbers like themselves; though we left the eggs untouched, feeling that we were already beginning to get rich, and wondering how many more nests we should find in the grand sunny woods. Then we ran along the brow of the hill that the shanty stood on, and down to the meadow, searching the trees and grass tufts and bushes, and soon discovered a bluebird's and a woodpecker's nest, and began an acquaintance with the frogs and snakes and turtles in the creeks and springs.

This sudden plash into pure wildness—baptism in Nature's warm heart—how utterly happy it made us! Nature streaming into us, wooingly teaching her wonderful glowing lessons, so unlike the dismal grammar ashes and cinders so long thrashed into us. Here without knowing it we still were at school; every wild lesson a love lesson, not whipped but charmed into us. Oh, that glorious Wisconsin wilderness! Everything new and pure in the very prime of the spring when Nature's pulses were beating highest and mysteriously keeping time with our own! Young hearts, young leaves, flowers, animals, the winds and the streams and the sparkling lake, all wildly, gladly rejoicing together!

Next morning, when we climbed to the precious jay nest to take another admiring look at the eggs, we found it empty. Not a shell-fragment was left, and we wondered how in the world the birds were able to carry off their thin-shelled eggs either in their bills or in their feet without breaking them, and how they could be kept warm while a new nest was being built. Well, I am still asking these questions. When I was on the Harriman Expedition I asked Robert Ridgway, the eminent ornithologist, how these sudden flittings were accomplished, and he frankly confessed that he didn't know, but guessed that jays and many other birds carried their eggs in their mouths; and when I objected that a jay's mouth seemed too small to hold its eggs, he replied that birds' mouths were larger than the narrowness of their bills indicated. Then I asked him what he thought they did with the eggs while a new nest was being prepared. He didn't know; neither do I to this day. A specimen of the many puzzling problems presented to the naturalist.

We soon found many more nests belonging to birds that were not half so suspicious. The handsome and notorious blue jay plunders the nests of other birds and of course he could not trust us. Almost all the others—brown thrushes, bluebirds, song sparrows, kingbirds, hen-hawks, nighthawks, whip-poor-wills, woodpeckers, etc.—simply tried to avoid being seen, to draw or drive us away, or paid no attention to us.

We used to wonder how the woodpeckers could bore holes so perfectly round, true mathematical circles. We ourselves could not have done it even with gouges and chisels. We loved to watch them feeding their young, and wondered how they could glean food enough for so many clamorous, hungry, unsatisfiable babies, and how they managed to give each one its share; for after the young grew strong, one would get his head out of the door-hole and try to hold possession of it to meet the food-laden parents. How hard they worked to support their families, especially the red-headed and speckledy woodpeckers and flickers; digging, hammering on scaly bark and decaying trunks and branches from dawn to dark, coming and going at intervals of a few minutes all the livelong day!

We discovered a hen-hawk's nest on the top of a tall oak thirty or forty rods from the shanty and approached it cautiously. One of the pair always kept watch, soaring in wide circles high above the tree, and when we attempted to climb it, the big dangerous-looking bird came swooping down at us and drove us away.

We greatly admired the plucky kingbird. In Scotland our great ambition was to be good fighters, and we admired this quality in the handsome little chattering flycatcher that whips all the other birds. He was particularly angry when plundering jays and hawks came near his home, and took pains to thrash them not only away from the nest-tree but out of the neighborhood. The nest was usually built on a bur oak near a meadow where insects were abundant, and where no undesirable visitor could approach without being discovered. When a hen-hawk hove in sight, the male immediately set off after him, and it was ridiculous to see that great, strong bird hurrying away as fast as his clumsy wings would carry him, as soon as he saw the little, waspish kingbird coming. But the kingbird easily overtook him, flew just a few feet above him, and with a lot of chattering, scolding notes kept diving and striking him on the back of the head until tired; then he alighted to rest on the hawk's broad shoulders, still scolding and chattering as he rode along, like an angry boy pouring out vials of wrath. Then, up and at him again with his sharp bill; and after he had thus driven and ridden his big enemy a mile or so from the nest, he went home to his mate, chuckling and bragging as if trying to tell her what a wonderful fellow he was.

This first spring, while some of the birds were still building their nests and very few young ones had yet tried to fly, father hired a Yankee to assist in clearing eight or ten acres of the best ground for a field. We found new wonders every day and often had to call on this Yankee to solve puzzling questions. We asked him one day if there was any bird in America that the kingbird couldn't whip. What about the sandhill crane? Could he whip that long-legged, long-billed fellow?

"A crane never goes near kingbirds' nests or notices so small a bird," he said, "and therefore there could be no fighting between them." So we hastily concluded that our hero could whip every bird in the country except perhaps the sandhill crane.

We never tired listening to the wonderful whip-poor-will. One came every night about dusk and sat on a log about twenty or thirty feet from our cabin door and began shouting "Whip poor Will! Whip poor Will!" with loud emphatic earnestness. "What's that? What's that?" we cried, when this startling visitor first announced himself. "What do you call it?"

"Why, it's telling you its name," said the Yankee. "Don't you hear it and what he wants you to do? He says his name is 'Poor Will' and he wants you to whip him, and you may if you are able to catch him." Poor Will seemed the most wonderful of all the strange creatures we had seen. What a wild, strong, bold voice he had, unlike any other we had ever heard on sea or land!

A near relative, the bull-bat, or nighthawk, seemed hardly less wonderful. Towards evening scattered flocks kept the sky lively as they circled around on their long wings a hundred feet or more above the ground, hunting moths and beetles, interrupting their rather slow but strong, regular wing-beats at short intervals with quick quivering strokes while uttering keen, squeaky cries something like *pfee, pfee*, and every now and then diving nearly to the ground with a loud ripping, bellowing sound, like bull-roaring, suggesting its name; then turning and gliding swiftly up again. These fine wild gray birds, about the size of a pigeon, lay their two eggs on bare ground without anything like a nest or even a concealing bush or grass-tuft. Nevertheless they are not easily seen, for they are colored like the ground. While sitting on their eggs, they depend so much upon not being noticed that if you are walking rapidly ahead they allow you to step within an inch or two of them without flinching. But if they see by your looks that you have discovered them, they leave their eggs or young, and, like a good many other birds, pretend that they are sorely wounded, fluttering and rolling over on the ground and gasping as if dying, to draw you away. When pursued we were surprised to find that just when we were on the point of overtaking them they were always able to flutter a few yards farther, until they had led us about a quarter of a mile from the nest; then, suddenly getting well, they quietly flew home by a roundabout way to their precious babies or eggs, o'er a' the ills of life victorious, bad boys among the worst. The Yankee took particular pleasure in encouraging us to pursue them.

Everything about us was so novel and wonderful that we could hardly believe our senses except when hungry or while father was thrashing us. When we first saw Fountain Lake Meadow, on a sultry evening, sprinkled with millions of lightning-bugs throbbing with light, the effect was so strange and beautiful that it seemed far too marvelous to be real. Looking from our shanty on the hill, I thought that the whole wonderful fairy show must be in my eyes; for only in fighting, when my eyes were struck, had I ever seen anything in the least like it. But when I asked my brother if he saw anything strange in the meadow he said, "Yes, it's all covered with shaky fire-sparks." Then I guessed that it might be something outside of us, and applied to our all-knowing Yankee to explain it. "Oh, it's nothing but lightnin'-bugs," he said, and kindly led us down the hill to the edge of the fiery meadow, caught a few of the wonderful bugs, dropped them into a cup, and carried them to the shanty, where we watched them throbbing and flashing out their mysterious light at regular intervals, as if each little passionate glow were caused by the beating of a heart. Once I saw a splendid display of glow-worm light in the foothills of the Himalayas, north of Calcutta, but glorious as it appeared in pure starry radiance, it was far less impressive than the extravagant abounding, quivering, dancing fire on our Wisconsin meadow.

Partridge drumming was another great marvel. When I first heard the low, soft, solemn sound I thought it must be made by some strange disturbance in my head or stomach, but as all seemed serene within, I asked David whether he heard anything queer. "Yes," he said, "I hear something saying *boomp*, *boomp*, *boomp*, and I'm wondering at it." Then I was half satisfied that the source of the mysterious sound must be in something outside of us, coming perhaps from the ground or from some ghost or bogie or woodland fairy. Only after long watching and listening did we at last discover it in the wings of the plump brown bird.

The love-song of the common jack snipe seemed not a whit less mysterious than partridge drumming. It was usually heard on cloudy evenings, a strange, unearthly, winnowing, spiritlike sound, yet easily heard at a distance of a third of a mile. Our sharp eyes soon detected the bird while making it, as it circled high in the air over the meadow with wonderfully strong and rapid wing-beats, suddenly descending and rising, again and again, in deep, wide loops; the tones being very low and smooth at the beginning of the descent, rapidly increasing to a curious little whirling storm-roar at the bottom, and gradually fading lower and lower until the top was reached. It was long, however, before we identified this mysterious wing-singer as the little brown jack snipe that we knew so well and had so often watched as he silently probed the mud around the edges of our meadow stream and spring-holes, and made short zigzag flights over the grass uttering only little short, crisp quacks and chucks.

The love-songs of the frogs seemed hardly less wonderful than those of the birds, their musical notes varying from the sweet, tranquil, soothing peeping and purring of the hylas to the awfully deep low-bass blunt bellowing of the bullfrogs. Some of the smaller species have wonderfully clear, sharp voices and told us their good Bible names in musical tones about as plainly as the whip-poor-will. *Isaac, Isaac*; *Yacob, Yacob*; *Israel, Israel*; shouted in sharp, ringing, far-reaching tones, as if they had all been to school and severely drilled in elocution. In the still, warm evenings, big bunchy bullfrogs bellowed, *Drunk! Drunk! Drunk! Jug o' rum! Jug o' rum!* and early in the spring, countless thousands of the commonest species, up to the throat in cold water, sang in concert, making a mass of music, such as it was, loud enough to be heard at a distance of more than half a mile.

Far, far apart from this loud marsh music is that of the many species of hyla, a sort of soothing immortal melody filling the air like light.

We reveled in the glory of the sky scenery as well as that of the woods and meadows and rushy, lily-bordered lakes. The great thunder-storms in particular interested us, so unlike any seen in Scotland, exciting awful, wondering admiration. Gazing awe-stricken, we watched the upbuilding of the sublime cloud-mountains,—glowing, sun-beaten pearl and alabaster cumuli, glorious in beauty and majesty and looking so firm and lasting that birds, we thought, might build their nests amid their downy bosses; the black-browed storm-clouds marching in awful grandeur across the landscape, trailing broad gray sheets of hail and rain like vast cataracts, and ever and anon flashing down vivid zigzag lightning followed by terrible crashing thunder. We saw several trees shattered, and one of them, a punky old oak, was set on fire, while we wondered why all the trees and everybody and everything did not share the same fate, for oftentimes the whole sky blazed. After sultry storm days, many of the nights were darkened by smooth black apparently structureless cloud-mantles which at short intervals were illumined with startling suddenness to a fiery glow by quick, quivering lightning-flashes, revealing the landscape in almost noonday brightness, to be instantly quenched in solid blackness.

But those first days and weeks of unmixed enjoyment and freedom, reveling in the wonderful wildness about us, were soon to be mingled with the hard work of making a farm. I was first put to burning brush in clearing land for the plough. Those magnificent brush fires with great white hearts and red flames, the first big, wild outdoor fires I had ever seen, were wonderful sights for young eyes. Again and again, when they were burning fiercest so that we could hardly approach near enough to throw on another branch, father put them to awfully practical use as warning lessons, comparing their heat with that of hell, and the branches with bad boys. "Now, John," he would say,—"now, John, just think what an awful thing it would be to be thrown into that fire:—and then think of hellfire, that is so many times hotter. Into that fire all bad boys, with sinners of every sort who disobey God, will be cast as we are casting branches into this brush fire, and although suffering so much, their sufferings will never never end, because neither the fire nor the sinners can die." But those terrible fire lessons quickly faded away in the blithe wilderness air; for no fire can be hotter than the heavenly fire of faith and hope that burns in every healthy boy's heart.

Soon after our arrival in the woods some one added a cat and puppy to the animals father had bought. The cat soon had kittens, and it was interesting to watch her feeding, protecting, and training them. After they were able to leave their nest and play, she went out hunting and brought in many kinds of birds and squirrels for them, mostly ground squirrels (spermophiles), called "gophers" in Wisconsin. When she got within a dozen yards or so of the shanty, she announced her approach by a peculiar call, and the sleeping kittens immediately bounced up and ran to meet her, all racing for the first bite of they knew not what, and we too ran to see what she brought. She then lay down a few minutes to rest and enjoy the enjoyment of her feasting family, and again vanished in the grass and flowers, coming and going every half-hour or so. Sometimes she brought in birds that we had never seen before, and occasionally a flying squirrel, chipmunk, or big fox squirrel. We were just old enough, David and I, to regard all these creatures as wonders, the strange inhabitants of our new world.

The pup was a common cur, though very uncommon to us, a black and white short-haired mongrel that we named "Watch." We always gave him a pan of milk in the evening just before we knelt in family worship, while daylight still lingered in the shanty. And, instead of attending to the prayers, I too often studied the small wild creatures playing around us. Field mice scampered about the cabin as though it had been built for them alone, and their performances were very amusing. About dusk, on one of the calm, sultry nights so grateful to moths and beetles, when the puppy was lapping his milk, and we were on our knees, in through the door came a heavy broad-shouldered beetle about as big as a mouse, and after it had droned and boomed round the cabin two or three times, the pan of milk, showing white in the gloaming, caught its eyes, and, taking good aim, it alighted with a slanting, glinting plash in the middle of the pan like a duck alighting in a lake. Baby Watch, having never before seen anything like that beetle, started back, gazing in dumb astonishment and fear at the black sprawling monster trying to swim. Recovering somewhat from his fright, he began to bark at the creature, and ran round and round his milk-pan, wouf-woufing, gurring, growling, like an old dog barking at a wild-cat or a bear. The natural astonishment and curiosity of that boy dog getting his first entomological lesson in this wonderful world was so immoderately funny that I had great difficulty in keeping from laughing out loud.

Snapping turtles were common throughout the woods, and we were delighted to find that they would snap at a stick and hang on like bull-dogs; and we amused ourselves by introducing Watch to them, enjoying his curious behavior and theirs in getting acquainted with each other. One day we assisted one of the smallest of the turtles to get a good grip of poor Watch's ear. Then away he rushed, holding his head sidewise, yelping and terror-stricken, with the strange bug-like reptile biting hard and clinging fast,—a shameful amusement even for wild boys.

As a playmate Watch was too serious, though he learned more than any stranger would judge him capable of, was a bold, faithful watch-dog, and in his prime a grand fighter, able to whip all the other dogs in the neighborhood. Comparing him with ourselves, we soon learned that although he could not read books he could read faces, was a good judge of character, always knew what was going on and what we were about to do, and liked to help us. We could run nearly as fast as he could, see about as far, and perhaps hear as well, but in sense of smell his nose was incomparably better than ours. One sharp winter morning when the ground was covered with snow, I noticed that when he was yawning and stretching himself after leaving his bed he suddenly caught the scent of something that excited him, went round the corner of the house, and looked intently to the westward across a tongue of land that we called West Bank, eagerly questioning the air with quivering nostrils, and bristling up as though he felt sure that there was something dangerous in that direction and had actually caught sight of it. Then he ran toward the Bank, and I followed him, curious to see what his nose had discovered. The top of the Bank commanded a view of the north end of our lake and meadow, and when we got there we saw an Indian hunter with a long spear, going from one muskrat cabin to another, approaching cautiously, careful to make no noise, and then suddenly thrusting his spear down through the house. If well aimed, the spear went through the poor beaver rat as it lay cuddled up in the snug nest it had made for itself in the fall with so much far-seeing care, and when the hunter felt the spear quivering, he dug down the mossy hut with his tomahawk and secured his prey,—the flesh for food, and the skin to sell for a dime or so. This was a clear object lesson on dogs' keenness of scent. That

Indian was more than half a mile away across a wooded ridge. Had the hunter been a white man, I suppose Watch would not have noticed him.

When he was about six or seven years old, he not only became cross, so that he would do only what he liked, but he fell on evil ways, and was accused by the neighbors who had settled around us of catching and devouring whole broods of chickens, some of them only a day or two out of the shell. We never imagined he would do anything so grossly undoglike. He never did at home. But several of the neighbors declared over and over again that they had caught him in the act, and insisted that he must be shot. At last, in spite of tearful protests, he was condemned, and executed. Father examined the poor fellow's stomach in search of sure evidence, and discovered the heads of eight chickens that he had devoured at his last meal. So poor Watch was killed simply because his taste for chickens was too much like our own. Think of the millions of squabs that preaching, praying men and women kill and eat, with all sorts of other animals great and small, young and old, while eloquently discoursing on the coming of the blessed peaceful, bloodless millennium! Think of the passenger pigeons that fifty or sixty years ago filled the woods and sky over half the continent, now exterminated by beating down the young from the nests together with the brooding parents, before they could try their wonderful wings; by trapping them in nets, feeding them to hogs, etc. None of our fellow mortals is safe who eats what we eat, who in any way interferes with our pleasures, or who may be used for work or food, clothing or ornament, or mere cruel, sportish amusement. Fortunately many are too small to be seen, and therefore enjoy life beyond our reach. And in looking through God's great stone books made up of records reaching back millions and millions of years, it is a great comfort to learn that vast multitudes of creatures, great and small and infinite in number, lived and had a good time in God's love before man was created.

The old Scotch fashion of whipping for every act of disobedience or of simple, playful forgetfulness was still kept up in the wilderness, and of course many of those whippings fell upon me. Most of them were outrageously severe, and utterly barren of fun. But here is one that was nearly all fun.

Father was busy hauling lumber for the frame house that was to be got ready for the arrival of my mother, sisters, and brother, left behind in Scotland. One morning, when he was ready to start for another load, his ox-whip was not to be found. He asked me if I knew anything about it. I told him I didn't know where it was, but Scotch conscience compelled me to confess that when I was playing with it I had tied it to Watch's tail, and that he ran away, dragging it through the grass, and came back without it. "It must have slipped off his tail," I said, and so I didn't know where it was. This honest, straightforward little story made father so angry that he exclaimed with heavy, foreboding emphasis: "The very deevil's in that boy!" David, who had been playing with me and was perhaps about as responsible for the loss of the whip as I was, said never a word, for he was always prudent enough to hold his tongue when the parental weather was stormy, and so escaped nearly all punishment. And, strange to say, this time I also escaped, all except a terrible scolding, though the thrashing weather seemed darker than ever. As if unwilling to let the sun see the shameful job, father took me into the cabin where the storm was to fall, and sent David to the woods for a switch. While he was out selecting the switch, father put in the spare time sketching my play-wickedness in awful colors, and of course referred again and again to the place prepared for bad boys. In the midst of this terrible word-storm, dreading most the impending thrashing, I whimpered that I was only playing because I couldn't help it; didn't know I was doing wrong; wouldn't do it again, and so forth. After this miserable dialogue was about exhausted, father became impatient at my brother for taking so long to find the switch; and so was I, for I wanted to have the thing over and done with. At last, in came David, a picture of open-hearted innocence, solemnly dragging a young bur-oak sapling, and handed the end of it to father, saying it was the best switch he could find. It was an awfully heavy one, about two and a half inches thick at the butt and ten feet long, almost big enough for a fence-pole. There wasn't room enough in the cabin to swing it, and the moment I saw it I burst out laughing in the midst of my fears. But father failed to see the fun and was very angry at David, heaved the bur-oak outside and passionately demanded his reason for fetching "sic a muckle rail like that instead o' a switch? Do ye ca' that a switch? I have a gude mind to thrash you insteed o' John." David, with demure, downcast eyes, looked preternaturally righteous, but as usual prudently answered never a word.

It was a hard job in those days to bring up Scotch boys in the way they should go; and poor overworked father was determined to do it if enough of the right kind of switches could be found. But this time, as the sun was getting high, he hitched up old Tom and Jerry and made haste to the Kingston lumber-yard, leaving me unscathed and as innocently wicked as ever; for hardly had father got fairly out of sight among the oaks and hickories, ere all our troubles, hell-threatenings, and exhortations were forgotten in the fun we had lassoing a stubborn old sow and laboriously trying to teach her to go reasonably steady in rope harness. She was the first hog that father bought to stock the farm, and we boys regarded her as a very wonderful beast. In a few weeks she had a lot of pigs, and of all the queer, funny, animal children we had yet seen, none amused us more. They were so comic in size and shape, in their gait and gestures, their merry sham fights, and the false alarms they got up for the fun of scampering back to their mother and begging her in most persuasive little squeals to lie down and give them a drink.

After her darling short-snouted babies were about a month old, she took them out to the woods and gradually roamed farther and farther from the shanty in search of acorns and roots. One afternoon we heard a rifle-shot, a very noticeable thing, as we had no near neighbors, as yet. We thought it must have been fired by an Indian on the trail that followed the right bank of the Fox River between Portage and Packwaukee Lake and passed our shanty at a distance of about three quarters of a mile. Just a few minutes after that shot was heard, along came the poor mother rushing up to the shanty for protection, with her pigs, all out of breath and terror-stricken. One of them was missing, and we supposed of course that an Indian had shot it for food. Next day, I discovered a blood-puddle where the Indian trail crossed the outlet of our lake. One of father's hired men told us that the Indians thought nothing of levying this sort of black-mail whenever they were hungry. The solemn awe and fear in the eyes of that old mother and those little pigs I never can forget; it was as unmistakable and deadly a fear as I ever saw expressed by any human eye, and corroborates in no uncertain way the oneness of all of us.

III

LIFE ON A WISCONSIN FARM

Humanity in Oxen—Jack, the Pony—Learning to Ride—Nob and Nell—Snakes—Mosquitoes and their Kin—Fish and Fishing—Considering the Lilies—Learning to Swim—A Narrow Escape from Drowning and a Victory—Accidents to Animals.

COMING DIRECT FROM school in Scotland while we were still hopefully ignorant and far from tame,—notwithstanding the unnatural profusion of teaching and thrashing lavished upon us,—getting acquainted with the animals about us was a never-failing source of wonder and delight. At first my father, like nearly all the backwoods settlers, bought a yoke of oxen to do the farm work, and as field after field was cleared, the number was gradually increased until we had five yoke. These wise, patient, plodding animals did all the ploughing, logging, hauling, and hard work of every sort for the first four or five years, and, never having seen oxen before, we looked at them with the same eager freshness of conception as we did at the wild animals. We worked with them, sympathized with them in their rest and toil and play, and thus learned to know them far better than we should had we been only trained scientific naturalists. We soon learned that each ox and cow and calf had individual character. Old white-faced Buck, one of the second yoke of oxen we owned, was a notably sagacious fellow. He seemed to reason sometimes almost like ourselves. In the fall we fed the cattle lots of pumpkins and had to split them open so that mouthfuls could be readily broken off. But Buck never waited for us to come to his help. The others, when they were hungry and impatient, tried to break through the hard rind with their teeth, but seldom with success if the pumpkin was full grown. Buck never wasted time in this mumbling, slavering way, but crushed them with his head. He went to the pile, picked out a good one, like a boy choosing an orange or apple, rolled it down on to the open ground, deliberately kneeled in front of it, placed his broad, flat brow on top of it, brought his weight hard down and crushed it, then quietly arose and went on with his meal in comfort. Some would call this "instinct," as if so-called "blind instinct" must necessarily make an ox stand on its head to break pumpkins when its teeth got sore, or when nobody came with an ax to split them. Another fine ox showed his skill when hungry by opening all the fences that stood in his way to the corn-fields.

The humanity we found in them came partly through the expression of their eyes when tired, their tones of voice when hungry and calling for food, their patient plodding and pulling in hot weather, their long-drawn-out sighing breath when exhausted and suffering like ourselves, and their enjoyment of rest with the same grateful looks as ours. We recognized their kinship also by their yawning like ourselves when sleepy and evidently enjoying the same peculiar pleasure at the roots of their jaws; by the way they stretched themselves in the morning after a good rest; by learning languages,—Scotch, English, Irish, French, Dutch,—a smattering of each as required in the faithful service they so willingly, wisely rendered; by their intelligent, alert curiosity, manifested in listening to strange sounds; their love of play; the attachments they made; and their mourning, long continued, when a companion was killed.

When we went to Portage, our nearest town, about ten or twelve miles from the farm, it would oftentimes be late before we got back, and in the summer-time, in sultry, rainy weather, the clouds were full of sheet lightning which every minute or two would suddenly illumine the landscape, revealing all its features, the hills and valleys, meadows and trees, about as fully and clearly as the noonday sunshine; then as suddenly the glorious light would be quenched, making the darkness seem denser than before. On such nights the cattle had to find the way home without any help from us, but they never got off the track, for they followed it by scent like dogs. Once, father, returning late from Portage or Kingston, compelled Tom and Jerry, our first oxen, to leave the dim track, imagining they must be going wrong. At last they stopped and refused to go farther. Then father unhitched them from the wagon, took hold of Tom's tail, and was thus led straight to the shanty. Next morning he set out to seek his wagon and found it on the brow of a steep hill above an impassable swamp. We learned less from the cows, because we did not enter so far into their lives, working with them, suffering heat and cold, hunger and thirst, and almost deadly weariness with them; but none with natural charity could fail to sympathize with them in their love for their calves, and to feel that it in no way differed from the divine mother-love of a woman in thoughtful, self-sacrificing care; for they would brave every danger, giving their lives for their offspring. Nor could we fail to sympathize with their awkward, blunt-nosed baby calves, with such beautiful, wondering eyes looking out on the world and slowly getting acquainted with things, all so strange to them, and awkwardly learning to use their legs, and play and fight.

Before leaving Scotland, father promised us a pony to ride when we got to America, and we saw to it that this promise was not forgotten. Only a week or two after our arrival in the woods he bought us a little Indian pony for thirteen dollars from a store-keeper in Kingston who had obtained him from a Winnebago or Menominee Indian in trade for goods. He was a stout handsome bay with long black mane and tail, and, though he was only two years old, the Indians had already taught him to carry all sorts of burdens, to stand without being tied, to go anywhere over all sorts of ground fast or slow, and to jump and swim and fear nothing,—a truly wonderful creature, strangely different from shy, skittish, nervous, superstitious civilized beasts. We turned him loose, and, strange to say, he never ran away from us or refused to be caught, but behaved as if he had known Scotch boys all his life; probably because we were about as wild as young Indians.

One day when father happened to have a little leisure, he said, "Noo, bairns, rin doon the meadow and get your powny and learn to ride him." So we led him out to a smooth place near an Indian mound back of the shanty, where father directed us to begin. I mounted for the first memorable lesson, crossed the mound, and set out at a slow walk along the wagon-track made in hauling lumber; then father shouted: "Whup him up, John, whup him up! Make him gallop; gallopin' is easier and better than walkin' or trottin'." Jack was willing, and away he sped at a good fast gallop. I managed to keep my balance fairly well by holding fast to the mane, but could not keep from bumping up and down, for I was plump and elastic and so was Jack; therefore about half of the time I was in the air.

After a quarter of a mile or so of this curious transportation, I cried, "Whoa, Jack!" The wonderful creature seemed to understand Scotch, for he stopped so suddenly I flew over his head, but he stood perfectly still as if that flying method of dismounting were the regular way. Jumping on again, I bumped and bobbed back along the grassy, flowery track, over the Indian mound, cried, "Whoa, Jack!" flew over his head, and alighted in father's arms as gracefully as if it were all intended for circus work.

After going over the course five or six times in the same free, picturesque style, I gave place to brother David, whose performances were much like my own. In a few weeks, however, or a month, we were taking adventurous rides more than a mile long out to a big meadow frequented by sandhill cranes, and returning safely with wonderful stories of the great long-legged birds we had seen, and how on the whole journey away and back we had fallen off only five or six times. Gradually we learned to gallop through the woods without roads of any sort, bareback and without rope or bridle, guiding only by leaning from side to side or by slight knee pressure. In this free way we used to amuse ourselves, riding at full speed across a big "kettle" that was on our farm, without holding on by either mane or tail.

These so-called "kettles" were formed by the melting of large detached blocks of ice that had been buried in moraine material thousands of years ago when the ice-sheet that covered all this region was receding. As the buried ice melted, of course the moraine material above and about it fell in, forming hopper-shaped hollows, while the grass growing on their sides and around them prevented the rain and wind from filling them up. The one we performed in was perhaps seventy or eighty feet wide and twenty or thirty feet deep; and without a saddle or hold of any kind it was not easy to keep from slipping over Jack's head in diving into it, or over his tail climbing out. This was fine sport on the long summer Sundays when we were able to steal away before meeting-time without being seen. We got very warm and red at it, and oftentimes poor Jack, dripping with sweat like his riders, seemed to have been boiled in that kettle.

In Scotland we had often been admonished to be bold, and this advice we passed on to Jack, who had already got many a wild lesson from Indian boys. Once, when teaching him to jump muddy streams, I made him try the creek in our meadow at a place where it is about twelve feet wide. He jumped bravely enough, but came down with a grand splash hardly more than halfway over. The water was only about a foot in depth, but the black vegetable mud half afloat was unfathomable. I managed to wallow ashore, but poor Jack sank deeper and deeper until only his head was visible in the black abyss, and his Indian fortitude was desperately tried. His foundering so suddenly in the treacherous gulf recalled the story of the Abbot of Aberbrothok's bell, which went down with a gurgling sound while bubbles rose and burst around. I had to go to father for help. He tied a long hemp rope brought from Scotland around Jack's neck, and Tom and Jerry seemed to have all they could do to pull him out. After which I got a solemn scolding for asking the "puir beast to jump intil sic a salt bottomless place."

We moved into our frame house in the fall, when mother with the rest of the family arrived from Scotland, and, when the winter snow began to fly, the bur-oak shanty was made into a stable for Jack. Father told us that good meadow hay was all he required, but we fed him corn, lots of it, and he grew very frisky and fat. About the middle of winter his long hair was full of dust and, as we thought, required washing. So, without taking the frosty weather into account, we gave him a thorough soap and water scouring, and as we failed to get him rubbed dry, a row of icicles formed under his belly. Father happened to see him in this condition and angrily asked what we had been about. We said Jack was dirty and we had washed him to make him healthy. He told us we ought to be ashamed of ourselves, "soaking the puir beast in cauld water at this time o' year"; that when we wanted to clean him we should have sense enough to use the brush and curry-comb.

In summer Dave or I had to ride after the cows every evening about sundown, and Jack got so accustomed to bringing in the drove that when we happened to be a few minutes late he used to go off alone at the regular time and bring them home at a gallop. It used to make father very angry to see Jack chasing the cows like a shepherd dog, running from one to the other and giving each a bite on the rump to keep them on the run, flying before him as if pursued by wolves. Father would declare at times that the wicked beast had the deevil in him and would be the death of the cattle. The corral and barn were just at the foot of a hill, and he made a great display of the drove on the home stretch as they walloped down that hill with their tails on end.

One evening when the pell-mell Wild West show was at its wildest, it made father so extravagantly mad that he ordered me to "Shoot Jack!" I went to the house and brought the gun, suffering most horrible mental anguish, such as I suppose unhappy Abraham felt when commanded to slay Isaac. Jack's life was spared, however, though I can't tell what finally became of him. I wish I could. After father bought a span of work horses he was sold to a man who said he was going to ride him across the plains to California. We had him, I think, some five or six years. He was the stoutest, gentlest, bravest little horse I ever saw. He never seemed tired, could canter all day with a man about as heavy as himself on his back, and feared nothing. Once fifty or sixty pounds of beef that was tied on his back slid over his shoulders along his neck and weighed down his head to the ground, fairly anchoring him; but he stood patient and still for half an hour or so without making the slightest struggle to free himself, while I was away getting help to untie the pack-rope and set the load back in its place.

As I was the eldest boy I had the care of our first span of work horses. Their names were Nob and Nell. Nob was very intelligent, and even affectionate, and could learn almost anything. Nell was entirely different; balky and stubborn, though we managed to teach her a good many circus tricks; but she never seemed to like to play with us in anything like an affectionate way as Nob did. We turned them out one day into the pasture, and an Indian, hiding in the brush that had sprung up after the grass fires had been kept out, managed to catch Nob, tied a rope to her jaw for a bridle, rode her to Green Lake, about thirty or forty miles away, and tried to sell her for fifteen dollars. All our hearts were sore, as if one of the family had been lost. We hunted everywhere and could not at first imagine what had become of her. We discovered her track where the fence was broken down, and, following it for a few miles, made sure the track was Nob's; and a neighbor told us he had seen an Indian riding fast through the woods on a horse that looked like Nob. But we could find no farther trace of her until a month or two after she was lost, and we had given up hope of ever seeing her again. Then we learned that she had been taken from an Indian by a farmer at Green Lake because he saw that she had been shod and had worked in harness. So when the Indian tried to sell her the farmer said: "You are a thief. That is a white man's horse. You stole her."

"No," said the Indian, "I brought her from Prairie du Chien and she has always been mine."

The man, pointing to her feet and the marks of the harness, said: "You are lying. I will take that horse away from you and put her in my pasture, and if you come near it I will set the dogs on you." Then he advertised her. One of our neighbors happened to see the advertisement and brought us the glad news, and great was our rejoicing when father brought her home. That Indian must have treated her with terrible cruelty, for when I was riding her through the pasture several years afterward, looking for another horse that we wanted to catch, as we approached the place where she had been captured she stood stock still gazing through the bushes, fearing the Indian might still be hiding there ready to spring; and she was so excited that she trembled, and her heartbeats were so loud that I could hear them distinctly as I sat on her back, *boomp*, *boomp*, *boomp*, like the drumming of a partridge. So vividly had she remembered her terrible experiences.

She was a great pet and favorite with the whole family, quickly learned playful tricks, came running when we called, seemed to know everything we said to her, and had the utmost confidence in our friendly kindness.

We used to cut and shock and husk the Indian corn in the fall, until a keen Yankee stopped overnight at our house and among other labor-saving notions convinced father that it was better to let it stand, and husk it at his leisure during the winter, then turn in the cattle to eat the leaves and trample down the stalks, so that they could be ploughed under in the spring. In this winter method each of us took two rows and husked into baskets, and emptied the corn on the ground in piles of fifteen to twenty basketfuls, then loaded it into the wagon to be hauled to the crib. This was cold, painful work, the temperature being oftentimes far below zero and the ground covered with dry, frosty snow, giving rise to miserable crops of chilblains and frosted fingers,—a sad change from the merry Indian-summer husking, when the big yellow pumpkins covered the cleared fields;—golden corn, golden pumpkins, gathered in the hazy golden weather. Sad change, indeed, but we occasionally got some fun out of the nipping, shivery work from hungry prairie chickens, and squirrels and mice that came about us.

The piles of corn were often left in the field several days, and while loading them into the wagon we usually found field mice in them,—big, blunt-nosed, strong-scented fellows that we were taught to kill just because they nibbled a few grains of corn. I used to hold one while it was still warm, up to Nob's nose for the fun of seeing her make faces and snort at the smell of it; and I would say: "Here, Nob," as if offering her a lump of sugar. One day I offered her an extra fine, fat, plump specimen, something like a little woodchuck, or muskrat, and to my astonishment, after smelling it curiously and doubtfully, as if wondering what the gift might be, and rubbing it back and forth in the palm of my hand with her upper lip, she deliberately took it into her mouth, crunched and munched and chewed it fine and swallowed it, bones, teeth, head, tail, everything. Not a single hair of that mouse was wasted. When she was chewing it she nodded and grunted, as though critically tasting and relishing it.

My father was a steadfast enthusiast on religious matters, and, of course, attended almost every sort of church-meeting, especially revival meetings. They were occasionally held in summer, but mostly in winter when the sleighing was good and plenty of time available. One hot summer day father drove Nob to Portage and back, twenty-four miles over a sandy road. It was a hot, hard, sultry day's work, and she had evidently been over-driven in order to get home in time for one of these meetings. I shall never forget how tired and wilted she looked that evening when I unhitched her; how she drooped in her stall, too tired to eat or even to lie down. Next morning it was plain that her lungs were inflamed; all the dreadful symptoms were just the same as my own when I had pneumonia. Father sent for a Methodist minister, a very energetic, resourceful man, who was a blacksmith, farmer, butcher, and horse-doctor as well as minister; but all his gifts and skill were of no avail. Nob was doomed. We bathed her head and tried to get her to eat something, but she couldn't eat, and in about a couple of weeks we turned her loose to let her come around the house and see us in the weary suffering and loneliness of the shadow of death. She tried to follow us children, so long her friends and workmates and playmates. It was awfully touching. She had several hemorrhages, and in the forenoon of her last day, after she had had one of her dreadful spells of bleeding and gasping for breath, she came to me trembling, with beseeching, heartbreaking looks, and after I had bathed her head and tried to soothe and pet her, she lay down and gasped and died. All the family gathered about her, weeping, with aching hearts. Then dust to dust.

She was the most faithful, intelligent, playful, affectionate, human-like horse I ever knew, and she won all our hearts. Of the many advantages of farm life for boys one of the greatest is the gaining a real knowledge of animals as fellow-mortals, learning to respect them and love them, and even to win some of their love. Thus god-like sympathy grows and thrives and spreads far beyond the teachings of churches and schools, where too often the mean, blinding, loveless doctrine is taught that animals have neither mind nor soul, have no rights that we are bound to respect, and were made only for man, to be petted, spoiled, slaughtered, or enslaved.

At first we were afraid of snakes, but soon learned that most of them were harmless. The only venomous species seen on our farm were the rattlesnake and the copperhead, one of each. David saw the rattler, and we both saw the copperhead. One day, when my brother came in from his work, he reported that he had seen a snake that made a queer buzzy noise with its tail. This was the only rattlesnake seen on our farm, though we heard of them being common on limestone hills eight or ten miles distant. We discovered the copperhead when we were ploughing, and we saw and felt at the first long, fixed, half-charmed, admiring stare at him that he was an awfully dangerous fellow. Every fibre of his strong, lithe, quivering body, his burnished copper-colored head, and above all his fierce, able eyes, seemed to be overflowing full of deadly power, and bade us beware. And yet it is only fair to say that this terrible, beautiful reptile showed no disposition to hurt us until we threw clods at him and tried to head him off from a log fence into which he was trying to escape. We were barefooted and of course afraid to let him get very near, while we vainly battered him with the loose sandy clods of the freshly ploughed field to hold him back until we could get a stick. Looking us in the eyes after a moment's pause, he probably saw we were afraid, and he came right straight at us, snapping and looking terrible, drove us out of his way, and won his fight.

Out on the open sandy hills there were a good many thick burly blow snakes, the kind that puff themselves up and hiss. Our Yankee declared that their breath was very poisonous and that we must not go near them. A handsome ringed species common in damp, shady places was, he told us, the most wonderful of all the snakes, for if chopped into pieces, however small, the fragments would wriggle themselves together again, and the restored snake would go on about its business as if nothing had happened. The commonest kinds were the striped slender species of the meadows and streams, good swimmers, that lived mostly on frogs.

Once I observed one of the larger ones, about two feet long, pursuing a frog in our meadow, and it was wonderful to see how fast the legless, footless, wingless, finless hunter could run. The frog, of course, knew its enemy and was making desperate efforts to escape to the water and hide in the marsh mud. He was a fine, sleek yellow muscular fellow and was springing over the tall grass in wide-arching jumps. The green-striped snake, gliding swiftly and steadily, was keeping the frog in sight and, had I not interfered, would probably have tired out the poor jumper. Then, perhaps, while digesting and enjoying his meal, the happy snake would himself be swallowed frog and all by a hawk. Again, to our astonishment, the small specimens were attacked by our hens. They pursued and pecked away at them until they killed and devoured them, oftentimes quarreling over the division of the spoil, though it was not easily divided.

We watched the habits of the swift-darting dragonflies, wild bees, butterflies, wasps, beetles, etc., and soon learned to discriminate between those that might be safely handled and the pinching or stinging species. But of all our wild neighbors the mosquitoes were the first with which we became very intimately acquainted.

The beautiful meadow lying warm in the spring sunshine, outspread between our lily-rimmed lake and the hill-slope that our shanty stood on, sent forth thirsty swarms of the little gray, speckledy, singing, stinging pests; and how tellingly they introduced themselves! Of little avail were the smudges that we made on muggy evenings to drive them away; and amid the many lessons which they insisted upon teaching us we wondered more and more at the extent of their knowledge, especially that in their tiny, flimsy bodies room could be found for such cunning palates. They would drink their fill from brown, smoky Indians, or from old white folk flavored with tobacco and whiskey, when no better could be had. But the surpassing fineness of their taste was best manifested by their enthusiastic appreciation of boys full of lively red blood, and of girls in full bloom fresh from cool Scotland or England. On these it was pleasant to witness their enjoyment as they feasted. Indians, we were told, believed that if they were brave fighters they would go after death to a happy country abounding in game, where there were no mosquitoes and no cowards. For cowards were driven away by themselves to a miserable country where there was no game fit to eat, and where the sky was always dark with huge gnats and mosquitoes as big as pigeons.

We were great admirers of the little black water-bugs. Their whole lives seemed to be play, skimming, swimming, swirling, and waltzing together in little groups on the edge of the lake and in the meadow springs, dancing to music we never could hear. The long-legged skaters, too, seemed wonderful fellows, shuffling about on top of the water, with air-bubbles like little bladders tangled under their hairy feet; and we often wished that we also might be shod in the same way to enable us to skate on the lake in summer as well as in icy winter. Not less wonderful were the boatmen, swimming on their backs, pulling themselves along with a pair of oar-like legs.

Great was the delight of brothers David and Daniel and myself when father gave us a few pine boards for a boat, and it was a memorable day when we got that boat built and launched into the lake. Never shall I forget our first sail over the gradually deepening water, the sunbeams pouring through it revealing the strange plants covering the bottom, and the fishes coming about us, staring and wondering as if the boat were a monstrous strange fish.

The water was so clear that it was almost invisible, and when we floated slowly out over the plants and fishes, we seemed to be miraculously sustained in the air while silently exploring a veritable fairyland.

We always had to work hard, but if we worked still harder we were occasionally allowed a little spell in the long summer evenings about sundown to fish, and on Sundays an hour or two to sail quietly without fishing-rod or gun when the lake was calm. Therefore we gradually learned something about its inhabitants,—pickerel, sunfish, black bass, perch, shiners, pumpkin-seeds, ducks, loons, turtles, muskrats, etc. We saw the sunfishes making their nests in little openings in the rushes where the water was only a few feet deep, ploughing up and shoving away the soft gray mud with their noses, like pigs, forming round bowls five or six inches in depth and about two feet in diameter, in which their eggs were deposited. And with what beautiful, unweariable devotion they watched and hovered over them and chased away prowling spawn-eating enemies that ventured within a rod or two of the precious nest!

The pickerel is a savage fish endowed with marvelous strength and speed. It lies in wait for its prey on the bottom, perfectly motionless like a waterlogged stick, watching everything that moves, with fierce, hungry eyes. Oftentimes when we were fishing for some other kinds over the edge of the boat, a pickerel that we had not noticed would come like a bolt of lightning and seize the fish we had caught before we could get it into the boat. The very first pickerel that I ever caught jumped into the air to seize a small fish dangling on my line, and, missing its aim, fell plump into the boat as if it had dropped from the sky.

Some of our neighbors fished for pickerel through the ice in midwinter. They usually drove a wagon out on the lake, set a large number of lines baited with live minnows, hung a loop of the lines over a small bush planted at the side of each hole, and watched to see the loops pulled off when a fish had taken the bait. Large quantities of pickerel were often caught in this cruel way.

Our beautiful lake, named Fountain Lake by father, but Muir's Lake by the neighbors, is one of the many small glacier lakes that adorn the Wisconsin landscapes. It is fed by twenty or thirty meadow springs, is about half a mile long, half as wide, and surrounded by low finely-modeled hills dotted with oak and hickory, and meadows full of grasses and sedges and many beautiful orchids and ferns. First there is a zone of green, shining rushes, and just beyond the rushes a zone of white and orange water-lilies fifty or sixty feet wide forming a magnificent border. On bright days, when the lake was rippled by a breeze, the lilies and sun-spangles danced together in radiant beauty, and it became difficult to discriminate between them.

On Sundays, after or before chores and sermons and Bible-lessons, we drifted about on the lake for hours, especially in lily time, getting finest lessons and sermons from the water and flowers, ducks, fishes, and muskrats. In particular we took Christ's advice and devoutly "considered the lilies"—how they grow up in beauty out of gray lime mud, and ride gloriously among the breezy sun-spangles. On our way home we gathered grand bouquets of them to be kept fresh all the week. No flower was hailed with greater wonder and admiration by the European settlers in general—Scotch, English, and Irish—than this white water-lily (*Nymphæa odorata*). It is a magnificent plant, queen of the inland waters, pure white, three or four inches in diameter, the most beautiful, sumptuous, and deliciously fragrant of all our Wisconsin flowers. No lily garden in civilization we had ever seen could compare with our lake garden.

The next most admirable flower in the estimation of settlers in this part of the new world was the pasque-flower or wind-flower (*Anemone patens* var. *nuttalliana*). It is the very first to appear in the spring, covering the cold gray-black ground with cheery blossoms. Before the ax or plough had touched the "oak openings" of Wisconsin, they were swept by running fires almost every autumn after the grass became dry. If from any cause, such as early snow-storms or late rains, they happened to escape the autumn fire besom, they were likely to be burned in the spring after the snow melted. But whether burned in the spring or fall, ashes and bits of charred twigs and grass stems made the whole country look dismal. Then, before a single grass-blade had sprouted, a hopeful multitude of large hairy, silky buds about as thick as one's thumb came to light, pushing up through the black and gray ashes and cinders, and before these buds were fairly free from the ground they opened wide and displayed purple blossoms about two inches in diameter, giving beauty for ashes in glorious abundance. Instead of remaining in the ground waiting for warm weather and companions, this admirable plant seemed to be in haste to rise and cheer the desolate landscape. Then at its leisure, after other plants had come to its help, it spread its leaves and grew up to a height of about two or three feet. The spreading leaves formed a whorl on the ground, and another about the middle of the stem as an involucre, and on the top of the stem the silky, hairy long-tailed seeds formed a head like a second flower. A little church was established among the earlier settlers and the meetings at first were held in our house. After working hard all the week it was difficult for boys to sit still through long sermons without falling asleep, especially in warm weather. In this drowsy trouble the charming anemone came to our help. A pocketful of the pungent seeds industriously nibbled while the discourses were at their dullest kept us awake and filled our minds with flowers.

The next great flower wonders on which we lavished admiration, not only for beauty of color and size, but for their curious shapes, were the cypripediums, called "lady's-slippers" or "Indian moccasins." They were so different from the familiar flowers of old Scotland. Several species grew in our meadow and on shady hillsides,—yellow, rose-colored, and some nearly white, an inch or more in diameter, and shaped exactly like Indian moccasins. They caught the eye of all the European settlers and made them gaze and wonder like children. And so did calopogon, pogonia, spiranthes, and many other fine plant people that lived in our meadow. The beautiful Turk's-turban (*Lilium superbum*) growing on stream-banks was rare in our neighborhood, but the orange lily grew in abundance on dry ground beneath the bur-oaks and often brought Aunt Ray's lily-bed in Scotland to mind. The butterfly-weed, with its brilliant scarlet flowers, attracted flocks of butterflies and made fine masses of color. With autumn came a glorious abundance and variety of asters, those beautiful plant stars, together with goldenrods, sunflowers, daisies, and liatris of different species, while around the shady margin of the meadow many ferns in beds and vaselike groups spread their beautiful fronds, especially the osmundas (*O. claytoniana*, *regalis*, and *cinnamomea*) and the sensitive and ostrich ferns.

Early in summer we feasted on strawberries, that grew in rich beds beneath the meadow grasses and sedges as well as in the dry sunny woods. And in different bogs and marshes, and around their borders on our own farm and along the Fox River, we found dewberries and cranberries, and a glorious profusion of huckleberries, the fountain-heads of pies of wondrous taste and size, colored in the heart like sunsets. Nor were we slow to discover the value of the hickory trees yielding both sugar and nuts. We carefully counted the different kinds on our farm, and every morning when we could steal a few minutes before breakfast after doing the chores, we visited the trees that had been wounded by the ax, to scrape off and enjoy the thick white delicious syrup that exuded from them, and gathered the nuts as they fell in the mellow Indian summer, making haste to get a fair share with the sapsuckers and squirrels. The hickory makes fine masses of color in the fall, every leaf a flower, but it was the sweet sap and sweet nuts that first interested us. No harvest in the Wisconsin woods was ever gathered with more pleasure and care. Also, to our delight, we found plenty of hazelnuts, and in a few places abundance of wild apples. They were desperately sour, and we used to fill our pockets with them and dare each other to eat one without making a face,—no easy feat.

One hot summer day father told us that we ought to learn to swim. This was one of the most interesting suggestions he had ever offered, but precious little time was allowed for trips to the lake, and he seldom tried to show us how. "Go to the frogs," he said, "and they will give you all the lessons you need. Watch their arms and legs and see how smoothly they kick themselves along and dive and come up. When you want to dive, keep your arms by your side or over your head, and kick, and when you want to come up, let your legs drag and paddle with your hands."

We found a little basin among the rushes at the south end of the lake, about waist-deep and a rod or two wide, shaped like a sunfish's nest. Here we kicked and plashed for many a lesson, faithfully trying to imitate frogs; but the smooth, comfortable sliding gait of our amphibious teachers seemed hopelessly hard to learn. When we tried to kick frog-fashion, down went our heads as if weighted with lead the moment our feet left the ground. One day it occurred to me to hold my breath as long as I could and let my head sink as far as it liked without paying any attention to it, and try to swim under the water instead of on the surface. This method was a great success, for at the very first trial I managed to cross the basin without touching bottom, and soon learned the use of my limbs. Then, of course, swimming with my head above water soon became so easy that it seemed perfectly natural. David tried the plan with the same success. Then we began to count the number of times that we could swim around the basin without stopping to rest, and after twenty or thirty rounds failed to tire us, we proudly thought that a little more practice would make us about as amphibious as frogs.

On the fourth of July of this swimming year one of the Lawson boys came to visit us, and we went down to the lake to spend the great warm day with the fishes and ducks and turtles. After gliding about on the smooth mirror water, telling stories and enjoying the company of the happy creatures about us, we rowed to our bathing-pool, and David and I went in for a swim, while our companion fished from the boat a little way out beyond the rushes. After a few turns in the pool, it occurred to me that it was now about time to try deep water. Swimming through the thick growth of rushes and lilies was somewhat dangerous, especially for a beginner, because one's arms and legs might be entangled among the long, limber stems; nevertheless I ventured and struck out boldly enough for the boat, where the water was twenty or thirty feet deep. When I reached the end of the little skiff I raised my right hand to take hold of it to surprise Lawson, whose back was toward me and who was not aware of my approach; but I failed to reach high enough, and, of course, the weight of my arm and the stroke against the over-leaning stern of the boat shoved me down and I sank, struggling, frightened and confused. As soon as my feet touched the bottom, I slowly rose to the surface, but before I could get breath enough to call for help, sank back again and lost all control of myself. After sinking and rising I don't know how many times, some water got into my lungs and I began to drown. Then suddenly my mind seemed to clear. I remembered that I could swim under water, and, making a desperate struggle toward the shore, I reached a point where with my toes on the bottom I got my mouth above the surface, gasped for help, and was pulled into the boat.

This humiliating accident spoiled the day, and we all agreed to keep it a profound secret. My sister Sarah had heard my cry for help, and on our arrival at the house inquired what had happened. "Were you drowning, John? I heard you cry you couldna get oot." Lawson made haste to reply, "Oh, no! He was juist haverin (making fun)."

I was very much ashamed of myself, and at night, after calmly reviewing the affair, concluded that there had been no reasonable cause for the accident, and that I ought to punish myself for so nearly losing my life from unmanly fear. Accordingly at the very first opportunity, I stole away to the lake by myself, got into my boat, and instead of going back to the old swimming-bowl for further practice, or to try to do sanely and well what I had so ignominiously failed to do in my first adventure, that is, to swim out through the rushes and lilies, I rowed directly out to the middle of the lake, stripped, stood up on the seat in the stern, and with grim deliberation took a header and dove straight down thirty or forty feet, turned easily, and, letting my feet drag, paddled straight to the surface with my hands as father had at first directed me to do. I then swam round the boat, glorying in my suddenly acquired confidence and victory over myself, climbed into it, and dived again, with the same triumphant success. I think I went down four or five times, and each time as I made the dive-spring shouted aloud, "Take that!" feeling that I was getting most gloriously even with myself.

Never again from that day to this have I lost control of myself in water. If suddenly thrown overboard at sea in the dark, or even while asleep, I think I would immediately right myself in a way some would call "instinct," rise among the waves, catch my breath, and try to plan what would better be done. Never was victory over self more complete. I have been a good swimmer ever since. At a slow gait I think I could swim all day in smooth water moderate in temperature. When I was a student at Madison, I used to go on long swimming-journeys, called exploring expeditions, along the south shore of Lake Mendota, on Saturdays, sometimes alone, sometimes with another amphibious explorer by the name of Fuller.

My adventures in Fountain Lake call to mind the story of a boy who in climbing a tree to rob a crow's nest fell and broke his leg, but as soon as it healed compelled himself to climb to the top of the tree he had fallen from.

Like Scotch children in general we were taught grim self-denial, in season and out of season, to mortify the flesh, keep our bodies in subjection to Bible laws, and mercilessly punish ourselves for every fault imagined or committed. A little boy, while helping his sister to drive home the cows, happened to use a forbidden word. "I'll have to tell fayther on ye," said the horrified sister. "I'll tell him that ye said a bad word." "Weel," said the boy, by way of excuse, "I couldna help the word comin' into me, and it's na waur to speak it oot than to let it rin through ye."

A Scotch fiddler playing at a wedding drank so much whiskey that on the way home he fell by the roadside. In the morning he was ashamed and angry and determined to punish himself. Making haste to the house of a friend, a gamekeeper, he called him out, and requested the loan of a gun. The alarmed gamekeeper, not liking the fiddler's looks and voice, anxiously inquired what he was going to do with it. "Surely," said he, "you're no gan to shoot yoursel." "No-o," with characteristic candor replied the penitent fiddler, "I dinna think that I'll juist exactly kill mysel, but I'm gaun to tak a dander doon the burn (brook) wi' the gun and gie mysel a deevil o' a fleg (fright)."

One calm summer evening a red-headed woodpecker was drowned in our lake. The accident happened at the south end, opposite our memorable swimming-hole, a few rods from the place where I came so near being drowned years before. I had returned to the old home during a summer vacation of the State University, and, having made a beginning in botany, I was, of course, full of enthusiasm and ran eagerly to my beloved pogonia, calopogon, and cypripedium gardens, osmunda ferneries, and the lake lilies and pitcher-plants.

A little before sundown the day-breeze died away, and the lake, reflecting the wooded hills like a mirror, was dimpled and dotted and streaked here and there where fishes and turtles were poking out their heads and muskrats were scuffing themselves along with their flat tails making glittering tracks. After lingering a while, dreamily recalling the old, hard, half-happy days, and watching my favorite red-headed woodpeckers pursuing moths like regular flycatchers, I swam out through the rushes and up the middle of the lake to the north end and back, gliding slowly, looking about me, enjoying the scenery as I would in a saunter along the shore, and studying the habits of the animals as they were explained and recorded on the smooth glassy water.

On the way back, when I was within a hundred rods or so of the end of my voyage, I noticed a peculiar plashing disturbance that could not, I thought, be made by a jumping fish or any other inhabitant of the lake; for instead of low regular out-circling ripples such as are made by the popping up of a head, or like those raised by the quick splash of a leaping fish, or diving loon or muskrat, a continuous struggle was kept up for several minutes ere the outspreading, interfering ring-waves began to die away. Swimming hastily to the spot to try to discover what had happened, I found one of my woodpeckers floating motionless with outspread wings. All was over. Had I been a minute or two earlier, I might have saved him. He had glanced on the water I suppose in pursuit of a moth, was unable to rise from it, and died struggling, as I nearly did at this same spot. Like me he seemed to have lost his mind in blind confusion and fear. The water was warm, and had he kept still with his head a little above the surface, he would sooner or later have been wafted ashore. The best aimed flights of birds and man "gang aft agley," but this was the first case I had witnessed of a bird losing its life by drowning.

Doubtless accidents to animals are far more common than is generally known. I have seen quails killed by flying against our house when suddenly startled. Some birds get entangled in hairs of their own nests and die. Once I found a poor snipe in our meadow that was unable to fly on account of difficult egg-birth. Pitying the poor mother, I picked her up out of the grass and helped her as gently as I could, and as soon as the egg was born she flew gladly away. Oftentimes I have thought it strange that one could walk through the woods and mountains and plains for years without seeing a single blood-spot. Most wild animals get into the world and out of it without being noticed. Nevertheless we at last sadly learn that they are all subject to the vicissitudes of fortune like ourselves. Many birds lose their lives in storms. I remember a particularly severe Wisconsin winter, when the temperature was many degrees below zero and the snow was deep, preventing the quail, which feed on the ground, from getting anything like enough of food, as was pitifully shown by a flock I found on our farm frozen solid in a thicket of oak sprouts. They were in a circle about a foot wide, with their heads outward, packed close together for warmth. Yet all had died without a struggle, perhaps more from starvation than frost. Many small birds lose their lives in the storms of early spring, or even summer. One mild spring morning I picked up more than a score out of the grass and flowers, most of them darling singers that had perished in a sudden storm of sleety rain and hail.

In a hollow at the foot of an oak tree that I had chopped down one cold winter day, I found a poor ground squirrel frozen solid in its snug grassy nest, in the middle of a store of nearly a peck of wheat it had carefully gathered. I carried it home and gradually thawed and warmed it in the kitchen, hoping it would come to life like a pickerel I caught in our lake through a hole in the ice, which, after being frozen as hard as a bone and thawed at the fireside, squirmed itself out of the grasp of the cook when she began to scrape it, bounced off the table, and danced about on the floor, making wonderful springy jumps as if trying to find its way back home to the lake. But for the poor spermophile nothing I could do in the way of revival was of any avail. Its life had passed away without the slightest struggle, as it lay asleep curled up like a ball, with its tail wrapped about it.

IV

A PARADISE OF BIRDS

Bird Favorites—The Prairie Chickens—Waterfowl—
A Loon on the Defensive—Passenger Pigeons.

THE WISCONSIN OAK openings were a summer paradise for song birds, and a fine place to get acquainted with them; for the trees stood wide apart, allowing one to see the happy homeseekers as they arrived in the spring, their mating, nest-building, the brooding and feeding of the young, and, after they were full-fledged and strong, to see all the families of the neighborhood gathering and getting ready to leave in the fall. Excepting the geese and ducks and pigeons nearly all our summer birds arrived singly or in small draggled flocks, but when frost and falling leaves brought their winter homes to mind they assembled in large flocks on dead or leafless trees by the side of a meadow or field, perhaps to get acquainted and talk the thing over. Some species held regular daily meetings for several weeks before finally setting forth on their long southern journeys. Strange to say, we never saw them start. Some morning we would find them gone. Doubtless they migrated in the night time. Comparatively few species remained all winter, the nut-hatch, chickadee, owl, prairie chicken, quail, and a few stragglers from the main flocks of ducks, jays, hawks, and bluebirds. Only after the country was settled did either jays or bluebirds winter with us.

The brave, frost-defying chickadees and nut-hatches stayed all the year wholly independent of farms and man's food and affairs.

With the first hints of spring came the brave little bluebirds, darling singers as blue as the best sky, and of course we all loved them. Their rich, crispy warbling is perfectly delightful, soothing and cheering, sweet and whisperingly low, Nature's fine love touches, every note going straight home into one's heart. And withal they are hardy and brave, fearless fighters in defense of home. When we boys approached their knot-hole nests, the bold little fellows kept scolding and diving at us and tried to strike us in the face, and oftentimes we were afraid they would prick our eyes. But the boldness of the little housekeepers only made us love them the more.

None of the bird people of Wisconsin welcomed us more heartily than the common robin. Far from showing alarm at the coming of settlers into their native woods, they reared their young around our gardens as if they liked us, and how heartily we admired the beauty and fine manners of these graceful birds and their loud cheery song of *Fear not, fear not, cheer up, cheer up*. It was easy to love them for they reminded us of the robin redbreast of Scotland. Like the bluebirds they dared every danger in defense of home, and we often wondered that birds so gentle could be so bold and that sweet-voiced singers could so fiercely fight and scold.

Of all the great singers that sweeten Wisconsin one of the best known and best loved is the brown thrush or thrasher, strong and able without being familiar, and easily seen and heard. Rosy purple evenings after thunder-showers are the favorite song-times, when the winds have died away and the steaming ground and the leaves and flowers fill the air with fragrance. Then the male makes haste to the top-most spray of an oak tree and sings loud and clear with delightful enthusiasm until sundown, mostly I suppose for his mate sitting on the precious eggs in a brush heap. And how faithful and watchful and daring he is! Woe to the snake or squirrel that ventured to go nigh the nest! We often saw him diving on them, pecking them about the head and driving them away as bravely as the kingbird drives away hawks. Their rich and varied strains make the air fairly quiver. We boys often tried to interpret the wild ringing melody and put it into words.

After the arrival of the thrushes came the bobolinks, gushing, gurgling, inexhaustible fountains of song, pouring forth floods of sweet notes over the broad Fox River meadows in wonderful variety and volume, crowded and mixed beyond description, as they hovered on quivering wings above their hidden nests in the grass. It seemed marvelous to us that birds so moderate in size could hold so much of this wonderful song stuff. Each one of them poured forth music enough for a whole flock, singing as if its whole body, feathers and all, were made up of music, flowing, glowing, bubbling melody interpenetrated here and there with small scintillating prickles and spicules. We never became so intimately acquainted with the bobolinks as with the thrushes, for they lived far out on the broad Fox River meadows, while the thrushes sang on the tree-tops around every home. The bobolinks were among the first of our great singers to leave us in the fall, going apparently direct to the rice-fields of the Southern States, where they grew fat and were slaughtered in countless numbers for food. Sad fate for singers so purely divine.

One of the gayest of the singers is the redwing blackbird. In the spring, when his scarlet epaulets shine brightest, and his little modest gray wife is sitting on the nest, built on rushes in a swamp, he sits on a nearby oak and devotedly sings almost all day. His rich simple strain is *baumpalee*, *baumpalee*, or *bobalee* as interpreted by some. In summer, after nesting cares are over, they assemble in flocks of hundreds and thousands to feast on Indian corn when it is in the milk. Scattering over a field, each selects an ear, strips the husk down far enough to lay bare an inch or two of the end of it, enjoys an exhilarating feast, and after all are full they rise simultaneously with a quick birr of wings like an old-fashioned church congregation fluttering to their feet when the minister after giving out the hymn says, "Let the congregation arise and sing." Alighting on nearby trees, they sing with a hearty vengeance, bursting out without any puttering prelude in gloriously glad concert, hundreds or thousands of exulting voices with sweet gurgling *baumpalees* mingled with chippy vibrant and exploding globules of musical notes, making a most enthusiastic, indescribable joy-song, a combination unlike anything to be heard elsewhere in the bird kingdom; something like bagpipes, flutes, violins, pianos, and human-like voices all bursting and bubbling at once. Then suddenly some one of the joyful congregation shouts Chirr! Chirr! and all stop as if shot.

The sweet-voiced meadowlark with its placid, simple song of *peery-eery-ódical* was another favorite, and we soon learned to admire the Baltimore oriole and its wonderful hanging nests, and the scarlet tanager glowing like fire amid the green leaves.

But no singer of them all got farther into our hearts than the little speckle-breasted song sparrow, one of the first to arrive and begin nest-building and singing. The richness, sweetness, and pathos of this small darling's song as he sat on a low bush often brought tears to our eyes.

The little cheery, modest chickadee midget, loved by every innocent boy and girl, man and woman, and by many not altogether innocent, was one of the first of the birds to attract our attention, drawing nearer and nearer to us as the winter advanced, bravely singing his faint silvery, lisping, tinkling notes ending with a bright *dee, dee, dee!* however frosty the weather.

The nut-hatches, who also stayed all winter with us, were favorites with us boys. We loved to watch them as they traced the bark-furrows of the oaks and hickories head downward, deftly flicking off loose scales and splinters in search of insects, and braving the coldest weather as if their little sparks of life were as safely warm in winter as in summer, unquenchable by the severest frost. With the help of the chickadees they made a delightful stir in the solemn winter days, and when we were out chopping we never ceased to wonder how their slender naked toes could be kept warm when our own were so painfully frosted though clad in thick socks and boots. And we wondered and admired the more when we thought of the little midgets sleeping in knot-holes when the temperature was far below zero, sometimes thirty-five degrees below, and in the morning, after a minute breakfast of a few frozen insects and hoarfrost crystals, playing and chatting in cheery tones as if food, weather, and everything was according to their own warm hearts. Our Yankee told us that the name of this darling was Devil-downhead.

Their big neighbors the owls also made good winter music, singing out loud in wild, gallant strains bespeaking brave comfort, let the frost bite as it might. The solemn hooting of the species with the widest throat seemed to us the very wildest of all the winter sounds.

Prairie chickens came strolling in family flocks about the shanty, picking seeds and grasshoppers like domestic fowls, and they became still more abundant as wheat- and corn-fields were multiplied, but also wilder, of course, when every shotgun in the country was aimed at them. The booming of the males during the mating-season was one of the loudest and strangest of the early spring sounds, being easily heard on calm mornings at a distance of a half or three fourths of a mile. As soon as the snow was off the ground, they assembled in flocks of a dozen or two on an open spot, usually on the side of a ploughed field, ruffled up their feathers, inflated the curious colored sacks on the sides of their necks, and strutted about with queer gestures something like turkey gobblers, uttering strange loud, rounded, drumming calls,—*boom! boom! boom!* interrupted by choking sounds. My brother Daniel caught one while she was sitting on her nest in our corn-field. The young are just like domestic chicks, run with the mother as soon as hatched, and stay with her until autumn, feeding on the ground, never taking wing unless disturbed. In winter, when full-grown, they assemble in large flocks, fly about sundown to selected roosting-places on tall trees, and to feeding-places in the morning,—unhusked corn-fields, if any are to be found in the neighborhood, or thickets of dwarf birch and willows, the buds of which furnish a considerable part of their food when snow covers the ground.

The wild rice-marshes along the Fox River and around Pucaway Lake were the summer homes of millions of ducks, and in the Indian summer, when the rice was ripe, they grew very fat. The magnificent mallards in particular afforded our Yankee neighbors royal feasts almost without price, for often as many as a half-dozen were killed at a shot, but we seldom were allowed a single hour for hunting and so got very few. The autumn duck season was a glad time for the Indians also, for they feasted and grew fat not only on the ducks but on the wild rice, large quantities of which they gathered as they glided through the midst of the generous crop in canoes, bending down handfuls over the sides, and beating out the grain with small paddles.

The warmth of the deep spring fountains of the creek in our meadow kept it open all the year, and a few pairs of wood ducks, the most beautiful, we thought, of all the ducks, wintered in it. I well remember the first specimen I ever saw. Father shot it in the creek during a snow-storm, brought it into the house, and called us around him, saying: "Come, bairns, and admire the work of God displayed in this bonnie bird. Naebody but God could paint feathers like these. Juist look at the colors, hoo they shine, and hoo fine they overlap and blend thegether like the colors o' the rainbow." And we all agreed that never, never before had we seen so awfu' bonnie a bird. A pair nested every year in the hollow top of an oak stump about fifteen feet high that stood on the side of the meadow, and we used to wonder how they got the fluffy young ones down from the nest and across the meadow to the lake when they were only helpless, featherless midgets; whether the mother carried them to the water on her back or in her mouth. I never saw the thing done or found anybody who had until this summer, when Mr. Holabird, a keen observer, told me that he once saw the mother carry them from the nest tree in her mouth, quickly coming and going to a nearby stream, and in a few minutes get them all together and proudly sail away.

Sometimes a flock of swans were seen passing over at a great height on their long journeys, and we admired their clear bugle notes, but they seldom visited any of the lakes in our neighborhood, so seldom that when they did it was talked of for years. One was shot by a blacksmith on a millpond with a long-range Sharp's rifle, and many of the neighbors went far to see it.

The common gray goose, Canada honker, flying in regular harrow-shaped flocks, was one of the wildest and wariest of all the large birds that enlivened the spring and autumn. They seldom ventured to alight in our small lake, fearing, I suppose, that hunters might be concealed in the rushes, but on account of their fondness for the young leaves of winter wheat when they were a few inches high, they often alighted on our fields when passing on their way south, and occasionally even in our corn-fields when a snow-storm was blowing, and they were hungry and wing-weary, with nearly an inch of snow on their backs. In such times of distress we used to pity them, even while trying to get a shot at them. They were exceedingly cautious and circumspect; usually flew several times round the adjacent thickets and fences to make sure that no enemy was near before settling down, and one always stood on guard, relieved from time to time, while the flock was feeding. Therefore there was no chance to creep up on them unobserved; you had to be well hidden before the flock arrived. It was the ambition of boys to be able to shoot these wary birds. I never got but two, both of them at one so-called lucky shot. When I ran to pick them up, one of them flew away, but as the poor fellow was sorely wounded he didn't fly far. When I caught him after a short chase, he uttered a piercing cry of terror and despair, which the leader of the flock heard at a distance of about a hundred rods. They had flown off in frightened disorder, of course, but had got into the regular harrow-shape order when the leader heard the cry, and I shall never forget how bravely he left his place at the head of the flock and hurried back screaming and struck at me in trying to save his companion. I dodged down and held my hands over my head, and thus escaped a blow of his elbows. Fortunately I had left my gun at the fence, and the life of this noble bird was spared after he had risked it in trying to save his wounded friend or neighbor or family relation. For so shy a bird boldly to attack a hunter showed wonderful sympathy and courage. This is one of my strangest hunting experiences.

Never before had I regarded wild geese as dangerous, or capable of such noble self-sacrificing devotion.

The loud clear call of the handsome bob-whites was one of the pleasantest and most characteristic of our spring sounds, and we soon learned to imitate it so well that a bold cock often accepted our challenge and came flying to fight. The young run as soon as they are hatched and follow their parents until spring, roosting on the ground in a close bunch, heads out ready to scatter and fly. These fine birds were seldom seen when we first arrived in the wilderness, but when wheat-fields supplied abundance of food they multiplied very fast, although oftentimes sore pressed during hard winters when the snow reached a depth of two or three feet, covering their food, while the mercury fell to twenty or thirty degrees below zero. Occasionally, although shy on account of being persistently hunted, under pressure of extreme hunger in the very coldest weather when the snow was deepest they ventured into barnyards and even approached the doorsteps of houses, searching for any sort of scraps and crumbs, as if piteously begging for food. One of our neighbors saw a flock come creeping up through the snow, unable to fly, hardly able to walk, and while approaching the door several of them actually fell down and died; showing that birds, usually so vigorous and apparently independent of fortune, suffer and lose their lives in extreme weather like the rest of us, frozen to death like settlers caught in blizzards. None of our neighbors perished in storms, though many had feet, ears, and fingers frost-nipped or solidly frozen.

As soon as the lake ice melted, we heard the lonely cry of the loon, one of the wildest and most striking of all the wilderness sounds, a strange, sad, mournful, unearthly cry, half laughing, half wailing. Nevertheless the great northern diver, as our species is called, is a brave, hardy, beautiful bird, able to fly under water about as well as above it, and to spear and capture the swiftest fishes for food. Those that haunted our lake were so wary none was shot for years, though every boy hunter in the neighborhood was ambitious to get one to prove his skill. On one of our bitter cold New Year holidays I was surprised to see a loon in the small open part of the lake at the mouth of the inlet that was kept from freezing by the warm spring water. I knew that it could not fly out of so small a place, for these heavy birds have to beat the water for half a mile or so before they can get fairly on the wing. Their narrow, finlike wings are very small as compared with the weight of the body and are evidently made for flying through water as well as through the air, and it is by means of their swift flight through the water and the swiftness of the blow they strike with their long, spear-like bills that they are able to capture the fishes on which they feed. I ran down the meadow with the gun, got into my boat, and pursued that poor winter-bound straggler. Of course he dived again and again, but had to come up to breathe, and I at length got a quick shot at his head and slightly wounded or stunned him, caught him, and ran proudly back to the house with my prize. I carried him in my arms; he didn't struggle to get away or offer to strike me, and when I put him on the floor in front of the kitchen stove, he just rested quietly on his belly as noiseless and motionless as if he were a stuffed specimen on a shelf, held his neck erect, gave no sign of suffering from any wound, and though he was motionless, his small black eyes seemed to be ever keenly watchful. His formidable bill, very sharp, three or three and a half inches long, and shaped like a pickax, was held perfectly level. But the wonder was that he did not struggle or make the slightest movement. We had a tortoise-shell cat, an old Tom of great experience, who was so fond of lying under the stove in frosty weather that it was difficult even to poke him out with a broom; but when he saw and smelled that strange big fishy, black and white, speckledy bird, the like of which he had never before seen, he rushed wildly to the farther corner of the kitchen, looked back cautiously and suspiciously, and began to make a careful study of the handsome but dangerous-looking stranger. Becoming more and more curious and interested, he at length advanced a step or two for a nearer view and nearer smell; and as the wonderful bird kept absolutely motionless, he was encouraged to venture gradually nearer and nearer until within perhaps five or six feet of its breast. Then the wary loon, not liking Tom's looks in so near a view, which perhaps recalled to his mind the plundering minks and muskrats he had to fight when they approached his nest, prepared to defend himself by slowly, almost imperceptibly drawing back his long pickax bill, and without the slightest fuss or stir held it level and ready just over his tail. With that dangerous bill drawn so far back out of the way, Tom's confidence in the stranger's peaceful intentions seemed almost complete, and, thus encouraged, he at last ventured forward with wondering, questioning eyes and quivering nostrils until he was only eighteen or twenty inches from the loon's smooth white breast. When the beautiful bird, apparently as peaceful and inoffensive as a flower, saw that his hairy yellow enemy had arrived at the right distance, the loon, who evidently was a fine judge of the reach of his spear, shot it forward quick as a lightning-flash, in marvelous contrast to the wonderful slowness of the preparatory poising, backward motion. The aim was true to a hair-breadth. Tom was struck right in the centre of his forehead, between the eyes. I thought his skull was cracked. Perhaps it was. The sudden astonishment of that outraged cat, the virtuous indignation and wrath, terror, and pain, are far beyond description. His eyes and screams and desperate retreat told all that. When the blow was received, he made a noise that I never heard a cat make before or since; an awfully deep, condensed, screechy, explosive *Wuck!* as he bounced straight up in the air like a bucking bronco; and when he alighted after his spring, he rushed madly across the room and made frantic efforts to climb up the hard-finished plaster wall. Not satisfied to get the width of the kitchen away from his mysterious enemy, for the first time that cold winter he tried to get out of the house, anyhow, anywhere out of that loon-infested room. When he finally ventured to look back and saw that the barbarous bird was still there, tranquil and motionless in front of the stove, he regained command of some of his shattered senses and carefully commenced to examine his wound. Backed against the wall in the farthest corner, and keeping his eye on the outrageous bird, he tenderly touched and washed the sore spot, wetting his paw with his tongue, pausing now and then as his courage increased to glare and stare and growl at his enemy with looks and tones wonderfully human, as if saying: "You confounded fishy, unfair rascal! What did you do that for? What had I done to you? Faithless, legless, long-nosed wretch!" Intense experiences like the above bring out the humanity that is in all animals. One touch of nature, even a cat-and-loon touch, makes all the world kin.

It was a great memorable day when the first flock of passenger pigeons came to our farm, calling to mind the story we had read about them when we were at school in Scotland. Of all God's feathered people that sailed the Wisconsin sky, no other bird seemed to us so wonderful. The beautiful wanderers flew like the winds in flocks of millions from climate to climate in accord with the weather, finding their food—acorns, beechnuts, pine-nuts, cranberries, strawberries, huckleberries, juniper berries, hackberries, buckwheat, rice, wheat, oats, corn—in fields and forests thousands of miles apart. I have seen flocks streaming south in the fall so large that they were flowing over from horizon to horizon in an almost continuous stream all day long, at the rate of forty or fifty miles an hour, like a mighty river in the sky, widening, contracting, descending like falls and cataracts, and rising suddenly here and there in huge ragged masses like high-plashing spray. How wonderful the distances they flew in a day—in a year—in a lifetime! They arrived in Wisconsin in the spring just after the sun had cleared away the snow, and alighted in the woods to feed on the fallen acorns that they had missed the previous autumn. A comparatively small flock swept thousands of acres perfectly clean of acorns in a few minutes, by moving straight ahead with a broad front. All got their share, for the rear constantly became the van by flying over the flock and alighting in front, the entire flock constantly changing from rear to front, revolving something like a wheel with a low buzzing wing roar that could be heard a long way off. In summer they feasted on wheat and oats and were easily approached as they rested on the trees along the sides of the field after a good full meal, displaying beautiful iridescent colors as they moved their necks backward and forward when we went very near them. Every shotgun was aimed at them and everybody feasted on pigeon pies, and not a few of the settlers feasted also on the beauty of the wonderful birds. The breast of the male is a fine rosy red, the lower part of the neck behind and along the sides changing from the red of the breast to gold, emerald green and rich crimson. The general color of the upper parts is grayish blue, the under parts white. The extreme length of the bird is about seventeen inches; the finely modeled slender tail about eight inches, and extent of wings twenty-four inches. The females are scarcely less beautiful. "Oh, what bonnie, bonnie birds!" we exclaimed over the first that fell into our hands. "Oh, what colors! Look at their breasts, bonnie as roses, and at their necks aglow wi' every color juist like the wonderfu' wood ducks. Oh, the bonnie, bonnie creatures, they beat a'! Where did they a' come fra, and where are they a' gan? It's awfu' like a sin to kill them!" To this some smug, practical old sinner would remark: "Aye, it's a peety, as ye say, to kill the bonnie things, but they were made to be killed, and sent for us to eat as the quails were sent to God's chosen people, the Israelites, when they were starving in the desert ayont the Red Sea. And I must confess that meat was never put up in neater, handsomer-painted packages."

In the New England and Canada woods beechnuts were their best and most abundant food, farther north, cranberries and huckleberries. After everything was cleaned up in the north and winter was coming on, they went south for rice, corn, acorns, haws, wild grapes, crab-apples, sparkle-berries, etc. They seemed to require more than half of the continent for feeding-grounds, moving from one table to another, field to field, forest to forest, finding something ripe and wholesome all the year round. In going south in the fine Indian-summer weather they flew high and followed one another, though the head of the flock might be hundreds of miles in advance. But against head winds they took advantage of the inequalities of the ground, flying comparatively low. All followed the leader's ups and downs over hill and dale though far out of sight, never hesitating at any turn of the way, vertical or horizontal that the leaders had taken, though the largest flocks stretched across several States, and belts of different kinds of weather.

There were no roosting- or breeding-places near our farm, and I never saw any of them until long after the great flocks were exterminated. I therefore quote, from Audubon's and Pokagon's vivid descriptions.

"Toward evening," Audubon says, "they depart for the roosting-place, which may be hundreds of miles distant. One on the banks of Green River, Kentucky, was over three miles wide and forty long."

"My first view of it," says the great naturalist, "was about a fortnight after it had been chosen by the birds, and I arrived there nearly two hours before sunset. Few pigeons were then to be seen, but a great many persons with horses and wagons and armed with guns, long poles, sulphur pots, pine pitch torches, etc., had already established encampments on the borders. Two farmers had driven upwards of three hundred hogs a distance of more than a hundred miles to be fattened on slaughtered pigeons. Here and there the people employed in plucking and salting what had already been secured were sitting in the midst of piles of birds. Dung several inches thick covered the ground. Many trees two feet in diameter were broken off at no great distance from the ground, and the branches of many of the tallest and largest had given way, as if the forest had been swept by a tornado.

"Not a pigeon had arrived at sundown. Suddenly a general cry arose—'Here they come!' The noise they made, though still distant, reminded me of a hard gale at sea passing through the rigging of a close-reefed ship. Thousands were soon knocked down by the pole-men. The birds continued to pour in. The fires were lighted and a magnificent as well as terrifying sight presented itself. The pigeons pouring in alighted everywhere, one above another, until solid masses were formed on the branches all around. Here and there the perches gave way with a crash, and falling destroyed hundreds beneath, forcing down the dense groups with which every stick was loaded; a scene of uproar and conflict. I found it useless to speak or even to shout to those persons nearest me. Even the reports of the guns were seldom heard, and I was made aware of the firing only by seeing the shooters reloading. None dared venture within the line of devastation. The hogs had been penned up in due time, the picking up of the dead and wounded being left for the next morning's employment. The pigeons were constantly coming in and it was after midnight before I perceived a decrease in the number of those that arrived. The uproar continued all night, and anxious to know how far the sound reached I sent off a man who, returning two hours after, informed me that he had heard it distinctly three miles distant.

"Toward daylight the noise in some measure subsided; long before objects were distinguishable the pigeons began to move off in a direction quite different from that in which they had arrived the evening before, and at sunrise all that were able to fly had disappeared. The howling of the wolves now reached our ears, and the foxes, lynxes, cougars, bears, coons, opossums, and polecats were seen sneaking off, while eagles and hawks of different species, accompanied by a crowd of vultures, came to supplant them and enjoy a share of the spoil.

"Then the authors of all this devastation began their entry amongst the dead, the dying and mangled. The pigeons were picked up and piled in heaps until each had as many as they could possibly dispose of, when the hogs were let loose to feed on the remainder.

"The breeding-places are selected with reference to abundance of food, and countless myriads resort to them. At this period the note of the pigeon is coo coo coo, like that of the domestic species but much shorter. They caress by billing, and during incubation the male supplies the female with food. As the young grow, the tyrant of creation appears to disturb the peaceful scene, armed with axes to chop down the squab-laden trees, and the abomination of desolation and destruction produced far surpasses even that of the roosting places."

Pokagon, an educated Indian writer, says: "I saw one nesting-place in Wisconsin one hundred miles long and from three to ten miles wide. Every tree, some of them quite low and scrubby, had from one to fifty nests on each. Some of the nests overflow from the oaks to the hemlock and pine woods. When the pigeon hunters attack the breeding-places they sometimes cut the timber from thousands of acres. Millions are caught in nets with salt or grain for bait, and schooners, sometimes loaded down with the birds, are taken to New York where they are sold for a cent apiece."

V

YOUNG HUNTERS

American Head-Hunters—Deer—A Resurrected Woodpecker—Muskrats—Foxes and Badgers—A Pet Coon—Bathing—Squirrels—Gophers—A Burglarious Shrike.

IN THE OLDER eastern States it used to be considered great sport for an army of boys to assemble to hunt birds, squirrels, and every other unclaimed, unprotected live thing of shootable size. They divided into two squads, and, choosing leaders, scattered through the woods in different directions, and the party that killed the greatest number enjoyed a supper at the expense of the other. The whole neighborhood seemed to enjoy the shameful sport especially the farmers afraid of their crops. With a great air of importance, laws were enacted to govern the gory business. For example, a gray squirrel must count four heads, a woodchuck six heads, common red squirrel two heads, black squirrel ten heads, a partridge five heads, the larger birds, such as whip-poor-wills and nighthawks two heads each, the wary crows three, and bob-whites three. But all the blessed company of mere songbirds, warblers, robins, thrushes, orioles, with nut-hatches, chickadees, blue jays, woodpeckers, etc., counted only one head each. The heads of the birds were hastily wrung off and thrust into the game-bags to be counted, saving the bodies only of what were called game, the larger squirrels, bob-whites, partridges, etc. The blood-stained bags of the best slayers were soon bulging full. Then at a given hour all had to stop and repair to the town, empty their dripping sacks, count the heads, and go rejoicing to their dinner. Although, like other wild boys, I was fond of shooting, I never had anything to do with these abominable head-hunts. And now the farmers having learned that birds are their friends wholesale slaughter has been abolished.

We seldom saw deer, though their tracks were common. The Yankee explained that they traveled and fed mostly at night, and hid in tamarack swamps and brushy places in the daytime, and how the Indians knew all about them and could find them whenever they were hungry.

Indians belonging to the Menominee and Winnebago tribes occasionally visited us at our cabin to get a piece of bread or some matches, or to sharpen their knives on our grindstone, and we boys watched them closely to see that they didn't steal Jack. We wondered at their knowledge of animals when we saw them go direct to trees on our farm, chop holes in them with their tomahawks and take out coons, of the existence of which we had never noticed the slightest trace. In winter, after the first snow, we frequently saw three or four Indians hunting deer in company, running like hounds on the fresh, exciting tracks. The escape of the deer from these noiseless, tireless hunters was said to be well-nigh impossible; they were followed to the death.

Most of our neighbors brought some sort of gun from the old country, but seldom took time to hunt, even after the first hard work of fencing and clearing was over, except to shoot a duck or prairie chicken now and then that happened to come in their way. It was only the less industrious American settlers who left their work to go far a-hunting. Two or three of our most enterprising American neighbors went off every fall with their teams to the pine regions and cranberry marshes in the northern part of the State to hunt and gather berries. I well remember seeing their wagons loaded with game when they returned from a successful hunt. Their loads consisted usually of half a dozen deer or more, one or two black bears, and fifteen or twenty bushels of cranberries; all solidly frozen. Part of both the berries and meat was usually sold in Portage; the balance furnished their families with abundance of venison, bear grease, and pies.

Winter wheat is sown in the fall, and when it is a month or so old the deer, like the wild geese, are very fond of it, especially since other kinds of food are then becoming scarce. One of our neighbors across the Fox River killed a large number, some thirty or forty, on a small patch of wheat, simply by lying in wait for them every night. Our wheat-field was the first that was sown in the neighborhood. The deer soon found it and came in every night to feast, but it was eight or nine years before we ever disturbed them. David then killed one deer, the only one killed by any of our family. He went out shortly after sundown at the time of full moon to one of our wheat-fields, carrying a double-barreled shotgun loaded with buckshot. After lying in wait an hour or so, he saw a doe and her fawn jump the fence and come cautiously into the wheat. After they were within sixty or seventy yards of him, he was surprised when he tried to take aim that about half of the moon's disc was mysteriously darkened as if covered by the edge of a dense cloud. This proved to be an eclipse. Nevertheless, he fired at the mother, and she immediately ran off, jumped the fence, and took to the woods by the way she came. The fawn danced about bewildered, wondering what had become of its mother, but finally fled to the woods. David fired at the poor deserted thing as it ran past him but happily missed it. Hearing the shots, I joined David to learn his luck. He said he thought he must have wounded the mother, and when we were strolling about in the woods in search of her we saw three or four deer on their way to the wheat-field, led by a fine buck. They were walking rapidly, but cautiously halted at intervals of a few rods to listen and look ahead and scent the air. They failed to notice us, though by this time the moon was out of the eclipse shadow and we were standing only about fifty yards from them. I was carrying the gun. David had fired both barrels but when he was reloading one of them he happened to put the wad intended to cover the shot into the empty barrel, and so when we were climbing over the fence the buckshot had rolled out, and when I fired at the big buck I knew by the report that there was nothing but powder in the charge. The startled deer danced about in confusion for a few seconds, uncertain which way to run until they caught sight of us, when they bounded off through the woods. Next morning we found the poor mother lying about three hundred yards from the place where she was shot. She had run this distance and jumped a high fence after one of the buckshot had passed through her heart.

Excepting Sundays we boys had only two days of the year to ourselves, the 4th of July and the 1st of January. Sundays were less than half our own, on account of Bible lessons, Sunday-school lessons and church services; all the others were labor days, rain or shine, cold or warm. No wonder, then, that our two holidays were precious and that it was not easy to decide what to do with them. They were usually spent on the highest rocky hill in the neighborhood, called the Observatory; in visiting our boy friends on adjacent farms to hunt, fish, wrestle, and play games; in reading some new favorite book we had managed to borrow or buy; or in making models of machines I had invented.

One of our July days was spent with two Scotch boys of our own age hunting redwing blackbirds then busy in the corn-fields. Our party had only one single-barreled shotgun, which, as the oldest and perhaps because I was thought to be the best shot, I had the honor of carrying. We marched through the corn without getting sight of a single redwing, but just as we reached the far side of the field, a red-headed woodpecker flew up, and the Lawson boys cried: "Shoot him! Shoot him! he is just as bad as a blackbird. He eats corn!" This memorable woodpecker alighted in the top of a white oak tree about fifty feet high. I fired from a position almost immediately beneath him, and he fell straight down at my feet. When I picked him up and was admiring his plumage, he moved his legs slightly, and I said, "Poor bird, he's no deed yet and we'll hae to kill him to put him oot o' pain,"—sincerely pitying him, after we had taken pleasure in shooting him. I had seen servant girls wringing chicken necks, so with desperate humanity I took the limp unfortunate by the head, swung him around three or four times thinking I was wringing his neck, and then threw him hard on the ground to quench the last possible spark of life and make quick death doubly sure. But to our astonishment the moment he struck the ground he gave a cry of alarm and flew right straight up like a rejoicing lark into the top of the same tree, and perhaps to the same branch he had fallen from, and began to adjust his ruffled feathers, nodding and chirping and looking down at us as if wondering what in the bird world we had been doing to him. This of course banished all thought of killing, as far as that revived woodpecker was concerned, no matter how many ears of corn he might spoil, and we all heartily congratulated him on his wonderful, triumphant resurrection from three kinds of death,—shooting, neck-wringing, and destructive concussion. I suppose only one pellet had touched him, glancing on his head.

Another extraordinary shooting-affair happened one summer morning shortly after daybreak. When I went to the stable to feed the horses I noticed a big white-breasted hawk on a tall oak in front of the chicken-house, evidently waiting for a chicken breakfast. I ran to the house for the gun, and when I fired he fell about halfway down the tree, caught a branch with his claws, hung back downward and fluttered a few seconds, then managed to stand erect. I fired again to put him out of pain, and to my surprise the second shot seemed to restore his strength instead of killing him, for he flew out of the tree and over the meadow with strong and regular wing-beats for thirty or forty rods apparently as well as ever, but died suddenly in the air and dropped like a stone.

We hunted muskrats whenever we had time to run down to the lake. They are brown bunchy animals about twenty-three inches long, the tail being about nine inches in length, black in color and flattened vertically for sculling, and the hind feet are half-webbed. They look like little beavers, usually have from ten to a dozen young, are easily tamed and make interesting pets. We liked to watch them at their work and at their meals. In the spring when the snow vanishes and the lake ice begins to melt, the first open spot is always used as a feeding-place, where they dive from the edge of the ice and in a minute or less reappear with a mussel or a mouthful of pontederia or water-lily leaves, climb back on to the ice and sit up to nibble their food, handling it very much like squirrels or marmots. It is then that they are most easily shot, a solitary hunter oftentimes shooting thirty or forty in a single day. Their nests on the rushy margins of lakes and streams, far from being hidden like those of most birds, are conspicuously large, and conical in shape like Indian wigwams. They are built of plants—rushes, sedges, mosses, etc.—and ornamented around the base with mussel-shells. It was always pleasant and interesting to see them in the fall as soon as the nights began to be frosty, hard at work cutting sedges on the edge of the meadow or swimming out through the rushes, making long glittering ripples as they sculled themselves along, diving where the water is perhaps six or eight feet deep and reappearing in a minute or so with large mouthfuls of the weedy tangled plants gathered from the bottom, returning to their big wigwams, climbing up and depositing their loads where most needed to make them yet larger and firmer and warmer, foreseeing the freezing weather just like ourselves when we banked up our house to keep out the frost.

They lie snug and invisible all winter but do not hibernate. Through a channel carefully kept open they swim out under the ice for mussels, and the roots and stems of water-lilies, etc., on which they feed just as they do in summer. Sometimes the oldest and most enterprising of them venture to orchards near the water in search of fallen apples; very seldom, however, do they interfere with anything belonging to their mortal enemy man. Notwithstanding they are so well hidden and protected during the winter, many of them are killed by Indian hunters, who creep up softly and spear them through the thick walls of their cabins. Indians are fond of their flesh, and so are some of the wildest of the white trappers. They are easily caught in steel traps, and after vainly trying to drag their feet from the cruel crushing jaws, they sometimes in their agony gnaw them off. Even after having gnawed off a leg they are so guileless that they never seem to learn to know and fear traps, for some are occasionally found that have been caught twice and have gnawed off a second foot. Many other animals suffering excruciating pain in these cruel traps gnaw off their legs. Crabs and lobsters are so fortunate as to be able to shed their limbs when caught or merely frightened, apparently without suffering any pain, simply by giving themselves a little shivery shake.

The muskrat is one of the most notable and widely distributed of American animals, and millions of the gentle, industrious, beaver-like creatures are shot and trapped and speared every season for their skins, worth a dime or so,—like shooting boys and girls for their garments.

Surely a better time must be drawing nigh when godlike human beings will become truly humane, and learn to put their animal fellow mortals in their hearts instead of on their backs or in their dinners. In the mean time we may just as well as not learn to live clean, innocent lives instead of slimy, bloody ones. All hale, red-blooded boys are savage, the best and boldest the savagest, fond of hunting and fishing. But when thoughtless childhood is past, the best rise the highest above all this bloody flesh and sport business, the wild foundational animal dying out day by day, as divine uplifting, transfiguring charity grows in.

Hares and rabbits were seldom seen when we first settled in the Wisconsin woods, but they multiplied rapidly after the animals that preyed upon them had been thinned out or exterminated, and food and shelter supplied in grain-fields and log fences and the thickets of young oaks that grew up in pastures after the annual grass fires were kept out. Catching hares in the winter-time, when they were hidden in hollow fence-logs, was a favorite pastime with many of the boys whose fathers allowed them time to enjoy the sport. Occasionally a stout, lithe hare was carried out into an open snow-covered field, set free, and given a chance for its life in a race with a dog. When the snow was not too soft and deep, it usually made good its escape, for our dogs were only fat, short-legged mongrels. We sometimes discovered hares in standing hollow trees, crouching on decayed punky wood at the bottom, as far back as possible from the opening, but when alarmed they managed to climb to a considerable height if the hollow was not too wide, by bracing themselves against the sides.

Foxes, though not uncommon, we boys held steadily to work seldom saw, and as they found plenty of prairie chickens for themselves and families, they did not often come near the farmer's hen-roosts. Nevertheless the discovery of their dens was considered important. No matter how deep the den might be, it was thoroughly explored with pick and shovel by sport-loving settlers at a time when they judged the fox was likely to be at home, but I cannot remember any case in our neighborhood where the fox was actually captured. In one of the dens a mile or two from our farm a lot of prairie chickens were found and some smaller birds.

Badger dens were far more common than fox dens. One of our fields was named Badger Hill from the number of badger holes in a hill at the end of it, but I cannot remember seeing a single one of the inhabitants.

On a stormy day in the middle of an unusually severe winter, a black bear, hungry, no doubt, and seeking something to eat, came strolling down through our neighborhood from the northern pine woods. None had been seen here before, and it caused no little excitement and alarm, for the European settlers imagined that these poor, timid, bashful bears were as dangerous as man-eating lions and tigers, and that they would pursue any human being that came in their way. This species is common in the north part of the State, and few of our enterprising Yankee hunters who went to the pineries in the fall failed to shoot at least one of them.

We saw very little of the owlish, serious-looking coons, and no wonder, since they lie hidden nearly all day in hollow trees and we never had time to hunt them. We often heard their curious, quavering, whinnying cries on still evenings, but only once succeeded in tracing an unfortunate family through our corn-field to their den in a big oak and catching them all. One of our neighbors, Mr. McRath, a Highland Scotchman, caught one and made a pet of it. It became very tame and had perfect confidence in the good intentions of its kind friend and master. He always addressed it in speaking to it as a "little man." When it came running to him and jumped on his lap or climbed up his trousers, he would say, while patting its head as if it were a dog or a child, "Coonie, ma mannie, Coonie, ma mannie, hoo are ye the day? I think you're hungry,"—as the comical pet began to examine his pockets for nuts and bits of bread,—"Na, na, there's nathing in my pooch for ye the day, my wee mannie, but I'll get ye something." He would then fetch something it liked,—bread, nuts, a carrot, or perhaps a piece of fresh meat. Anything scattered for it on the floor it felt with its paw instead of looking at it, judging of its worth more by touch than sight.

The outlet of our Fountain Lake flowed past Mr. McRath's door, and the coon was very fond of swimming in it and searching for frogs and mussels. It seemed perfectly satisfied to stay about the house without being confined, occupied a comfortable bed in a section of a hollow tree, and never wandered far. How long it lived after the death of its kind master I don't know.

I suppose that almost any wild animal may be made a pet, simply by sympathizing with it and entering as much as possible into its life. In Alaska I saw one of the common gray mountain marmots kept as a pet in an Indian family. When its master entered the house it always seemed glad, almost like a dog, and when cold or tired it snuggled up in a fold of his blanket with the utmost confidence.

We have all heard of ferocious animals, lions and tigers, etc., that were fed and spoken to only by their masters, becoming perfectly tame; and, as is well known, the faithful dog that follows man and serves him, and looks up to him and loves him as if he were a god, is a descendant of the blood-thirsty wolf or jackal. Even frogs and toads and fishes may be tamed, provided they have the uniform sympathy of one person, with whom they become intimately acquainted without the distracting and varying attentions of strangers. And surely all God's people, however serious and savage, great or small, like to play. Whales and elephants, dancing, humming gnats, and invisibly small mischievous microbes,—all are warm with divine radium and must have lots of fun in them.

As far as I know, all wild creatures keep themselves clean. Birds, it seems to me, take more pains to bathe and dress themselves than any other animals. Even ducks, though living so much in water, dip and scatter cleansing showers over their backs, and shake and preen their feathers as carefully as land-birds. Watching small singers taking their morning baths is very interesting, particularly when the weather is cold. Alighting in a shallow pool, they oftentimes show a sort of dread of dipping into it, like children hesitating about taking a plunge, as if they felt the same kind of shock, and this makes it easy for us to sympathize with the little feathered people.

Occasionally I have seen from my study-window red-headed linnets bathing in dew when water elsewhere was scarce. A large Monterey cypress with broad branches and innumerable leaves on which the dew lodges in still nights made favorite bathing-places. Alighting gently, as if afraid to waste the dew, they would pause and fidget as they do before beginning to plash in pools, then dip and scatter the drops in showers and get as thorough a bath as they would in a pool. I have also seen the same kind of baths taken by birds on the boughs of silver firs on the edge of a glacier meadow, but nowhere have I seen the dew-drops so abundant as on the Monterey cypress; and the picture made by the quivering wings and irised dew was memorably beautiful. Children, too, make fine pictures plashing and crowing in their little tubs. How widely different from wallowing pigs, bathing with great show of comfort and rubbing themselves dry against rough-barked trees!

Some of our own species seem fairly to dread the touch of water. When the necessity of absolute cleanliness by means of frequent baths was being preached by a friend who had been reading Combe's Physiology, in which he had learned something of the wonders of the skin with its millions of pores that had to be kept open for health, one of our neighbors remarked: "Oh! that's unnatural. It's well enough to wash in a tub maybe once or twice a year, but not to be paddling in the water all the time like a frog in a spring-hole." Another neighbor, who prided himself on his knowledge of big words, said with great solemnity: "I never can believe that man is amphibious!"

Natives of tropic islands pass a large part of their lives in water, and seem as much at home in the sea as on the land; swim and dive, pursue fishes, play in the waves like surf-ducks and seals, and explore the coral gardens and groves and seaweed meadows as if truly amphibious. Even the natives of the far north bathe at times. I once saw a lot of Eskimo boys ducking and plashing right merrily in the Arctic Ocean.

It seemed very wonderful to us that the wild animals could keep themselves warm and strong in winter when the temperature was far below zero. Feeble-looking rabbits scud away over the snow, lithe and elastic, as if glorying in the frosty, sparkling weather and sure of their dinners. I have seen gray squirrels dragging ears of corn about as heavy as themselves out of our field through loose snow and up a tree, balancing them on limbs and eating in comfort with their dry, electric tails spread airily over their backs. Once I saw a fine hardy fellow go into a knot-hole. Thrusting in my hand I caught him and pulled him out. As soon as he guessed what I was up to, he took the end of my thumb in his mouth and sank his teeth right through it, but I gripped him hard by the neck, carried him home, and shut him up in a box that contained about half a bushel of hazel- and hickory-nuts, hoping that he would not be too much frightened and discouraged to eat while thus imprisoned after the rough handling he had suffered. I soon learned, however, that sympathy in this direction was wasted, for no sooner did I pop him in than he fell to with right hearty appetite, gnawing and munching the nuts as if he had gathered them himself and was very hungry that day. Therefore, after allowing time enough for a good square meal, I made haste to get him out of the nut-box and shut him up in a spare bedroom, in which father had hung a lot of selected ears of Indian corn for seed. They were hung up by the husks on cords stretched across from side to side of the room. The squirrel managed to jump from the top of one of the bed-posts to the cord, cut off an ear, and let it drop to the floor. He then jumped down, got a good grip of the heavy ear, carried it to the top of one of the slippery, polished bed-posts, seated himself comfortably, and, holding it well balanced, deliberately pried out one kernel at a time with his long chisel teeth, ate the soft, sweet germ, and dropped the hard part of the kernel. In this masterly way, working at high speed, he demolished several ears a day, and with a good warm bed in a box made himself at home and grew fat. Then naturally, I suppose, free romping in the snow and tree-tops with companions came to mind. Anyhow he began to look for a way of escape. Of course he first tried the window, but found that his teeth made no impression on the glass. Next he tried the sash and gnawed the wood off level with the glass; then father happened to come upstairs and discovered the mischief that was being done to his seed corn and window and immediately ordered him out of the house.

The flying squirrel was one of the most interesting of the little animals we found in the woods, a beautiful brown creature, with fine eyes and smooth, soft fur like that of a mole or field mouse. He is about half as long as the gray squirrel, but his wide-spread tail and the folds of skin along his sides that form the wings make him look broad and flat, something like a kite. In the evenings our cat often brought them to her kittens at the shanty, and later we saw them fly during the day from the trees we were chopping. They jumped and glided off smoothly and apparently without effort, like birds, as soon as they heard and felt the breaking shock of the strained fibres at the stump, when the trees they were in began to totter and groan. They can fly, or rather glide, twenty or thirty yards from the top of a tree twenty or thirty feet high to the foot of another, gliding upward as they reach the trunk, or if the distance is too great they alight comfortably on the ground and make haste to the nearest tree, and climb just like the wingless squirrels.

Every boy and girl loves the little fairy, airy striped chipmunk, half squirrel, half spermophile. He is about the size of a field mouse, and often made us think of linnets and song sparrows as he frisked about gathering nuts and berries. He likes almost all kinds of grain, berries, and nuts,—hazel-nuts, hickory-nuts, strawberries, huckleberries, wheat, oats, corn,—he is fond of them all and thrives on them. Most of the hazel bushes on our farm grew along the fences as if they had been planted for the chipmunks alone, for the rail fences were their favorite highways. We never wearied watching them, especially when the hazel-nuts were ripe and the little fellows were sitting on the rails nibbling and handling them like tree-squirrels. We used to notice too that, although they are very neat animals, their lips and fingers were dyed red like our own, when the strawberries and huckleberries were ripe. We could always tell when the wheat and oats were in the milk by seeing the chipmunks feeding on the ears. They kept nibbling at the wheat until it was harvested and then gleaned in the stubble, keeping up a careful watch for their enemies,—dogs, hawks, and shrikes. They are as widely distributed over the continent as the squirrels, various species inhabiting different regions on the mountains and lowlands, but all the different kinds have the same general characteristics of light, airy cheerfulness and good nature.

Before the arrival of farmers in the Wisconsin woods the small ground squirrels, called "gophers," lived chiefly on the seeds of wild grasses and weeds, but after the country was cleared and ploughed no feasting animal fell to more heartily on the farmer's wheat and corn. Increasing rapidly in numbers and knowledge, they became very destructive, especially in the spring when the corn was planted, for they learned to trace the rows and dig up and eat the three or four seeds in each hill about as fast as the poor farmers could cover them. And unless great pains were taken to diminish the numbers of the cunning little robbers, the fields had to be planted two or three times over, and even then large gaps in the rows would be found. The loss of the grain they consumed after it was ripe, together with the winter stores laid up in their burrows, amounted to little as compared with the loss of the seed on which the whole crop depended.

One evening about sundown, when my father sent me out with the shotgun to hunt them in a stubble field, I learned something curious and interesting in connection with these mischievous gophers, though just then they were doing no harm. As I strolled through the stubble watching for a chance for a shot, a shrike flew past me and alighted on an open spot at the mouth of a burrow about thirty yards ahead of me. Curious to see what he was up to, I stood still to watch him. He looked down the gopher hole in a listening attitude, then looked back at me to see if I was coming, looked down again and listened, and looked back at me. I stood perfectly still, and he kept twitching his tail, seeming uneasy and doubtful about venturing to do the savage job that I soon learned he had in his mind. Finally, encouraged by my keeping so still, to my astonishment he suddenly vanished in the gopher hole.

A bird going down a deep narrow hole in the ground like a ferret or a weasel seemed very strange, and I thought it would be a fine thing to run forward, clap my hand over the hole, and have the fun of imprisoning him and seeing what he would do when he tried to get out. So I ran forward but stopped when I got within a dozen or fifteen yards of the hole, thinking it might perhaps be more interesting to wait and see what would naturally happen without my interference. While I stood there looking and listening, I heard a great disturbance going on in the burrow, a mixed lot of keen squeaking, shrieking, distressful cries, telling that down in the dark something terrible was being done. Then suddenly out popped a half-grown gopher, four and a half or five inches long, and, without stopping a single moment to choose a way of escape, ran screaming through the stubble straight away from its home, quickly followed by another and another, until some half-dozen were driven out, all of them crying and running in different directions as if at this dreadful time home, sweet home, was the most dangerous and least desirable of any place in the wide world. Then out came the shrike, flew above the runaway gopher children, and, diving on them, killed them one after another with blows at the back of the skull. He then seized one of them, dragged it to the top of a small clod so as to be able to get a start, and laboriously made out to fly with it about ten or fifteen yards, when he alighted to rest. Then he dragged it to the top of another clod and flew with it about the same distance, repeating this hard work over and over again until he managed to get one of the gophers on to the top of a log fence. How much he ate of his hard-won prey, or what he did with the others, I can't tell, for by this time the sun was down and I had to hurry home to my chores.

VI

THE PLOUGHBOY

The Crops—Doing Chores—The Sights and Sounds of Winter—Road-Making—The Spirit-Rapping Craze—Tuberculosis among the Settlers—A Cruel Brother—The Rights of the Indians—Put to the Plough at the Age of Twelve—In the Harvest-Field—Over-Industry among the Settlers—Running the Breaking-Plough—Digging a Well—Choke-Damp—Lining Bees.

AT FIRST, WHEAT, corn, and potatoes were the principal crops we raised; wheat especially. But in four or five years the soil was so exhausted that only five or six bushels an acre, even in the better fields, was obtained, although when first ploughed twenty and twenty-five bushels was about the ordinary yield. More attention was then paid to corn, but without fertilizers the corn-crop also became very meagre. At last it was discovered that English clover would grow on even the exhausted fields, and that when ploughed under and planted with corn, or even wheat, wonderful crops were raised. This caused a complete change in farming methods; the farmers raised fertilizing clover, planted corn, and fed the crop to cattle and hogs.

But no crop raised in our wilderness was so surprisingly rich and sweet and purely generous to us boys and, indeed, to everybody as the watermelons and muskmelons. We planted a large patch on a sunny hill-slope the very first spring, and it seemed miraculous that a few handfuls of little flat seeds should in a few months send up a hundred wagon-loads of crisp, sumptuous, red-hearted and yellow-hearted fruits covering all the hill. We soon learned to know when they were in their prime, and when over-ripe and mealy. Also that if a second crop was taken from the same ground without fertilizing it, the melons would be small and what we called soapy; that is, soft and smooth, utterly uncrisp, and without a trace of the lively freshness and sweetness of those raised on virgin soil. Coming in from the farm work at noon, the half-dozen or so of melons we had placed in our cold spring were a glorious luxury that only weary barefooted farm boys can ever know.

Spring was not very trying as to temperature, and refreshing rains fell at short intervals. The work of ploughing commenced as soon as the frost was out of the ground. Corn- and potato-planting and the sowing of spring wheat was comparatively light work, while the nesting birds sang cheerily, grass and flowers covered the marshes and meadows and all the wild, uncleared parts of the farm, and the trees put forth their new leaves, those of the oaks forming beautiful purple masses as if every leaf were a petal; and with all this we enjoyed the mild soothing winds, the humming of innumerable small insects and hylas, and the freshness and fragrance of everything. Then, too, came the wonderful passenger pigeons streaming from the south, and flocks of geese and cranes, filling all the sky with whistling wings.

The summer work, on the contrary, was deadly heavy, especially harvesting and corn-hoeing. All the ground had to be hoed over for the first few years, before father bought cultivators or small weed-covering ploughs, and we were not allowed a moment's rest. The hoes had to be kept working up and down as steadily as if they were moved by machinery. Ploughing for winter wheat was comparatively easy, when we walked barefooted in the furrows, while the fine autumn tints kindled in the woods, and the hillsides were covered with golden pumpkins.

In summer the chores were grinding scythes, feeding the animals, chopping stove-wood, and carrying water up the hill from the spring on the edge of the meadow, etc. Then breakfast, and to the harvest or hay-field. I was foolishly ambitious to be first in mowing and cradling, and by the time I was sixteen led all the hired men. An hour was allowed at noon for dinner and more chores. We stayed in the field until dark, then supper, and still more chores, family worship, and to bed; making altogether a hard, sweaty day of about sixteen or seventeen hours. Think of that, ye blessed eight-hour-day laborers!

In winter father came to the foot of the stairs and called us at six o'clock to feed the horses and cattle, grind axes, bring in wood, and do any other chores required, then breakfast, and out to work in the mealy, frosty snow by daybreak, chopping, fencing, etc. So in general our winter work was about as restless and trying as that of the long-day summer. No matter what the weather, there was always something to do. During heavy rains or snow-storms we worked in the barn, shelling corn, fanning wheat, thrashing with the flail, making ax-handles or ox-yokes, mending things, or sprouting and sorting potatoes in the cellar.

No pains were taken to diminish or in any way soften the natural hardships of this pioneer farm life; nor did any of the Europeans seem to know how to find reasonable ease and comfort if they would. The very best oak and hickory fuel was embarrassingly abundant and cost nothing but cutting and common sense; but instead of hauling great heart-cheering loads of it for wide, open, all-welcoming, climate-changing, beauty-making, Godlike ingle-fires, it was hauled with weary heart-breaking industry into fences and waste places to get it out of the way of the plough, and out of the way of doing good. The only fire for the whole house was the kitchen stove, with a fire-box about eighteen inches long and eight inches wide and deep,—scant space for three or four small sticks, around which in hard zero weather all the family of ten persons shivered, and beneath which in the morning we found our socks and coarse, soggy boots frozen solid. We were not allowed to start even this despicable little fire in its black box to thaw them. No, we had to squeeze our throbbing, aching, chilblained feet into them, causing greater pain than toothache, and hurry out to chores. Fortunately the miserable chilblain pain began to abate as soon as the temperature of our feet approached the freezing-point, enabling us in spite of hard work and hard frost to enjoy the winter beauty,—the wonderful radiance of the snow when it was starry with crystals, and the dawns and the sunsets and white noons, and the cheery, enlivening company of the brave chickadees and nut-hatches.

The winter stars far surpassed those of our stormy Scotland in brightness, and we gazed and gazed as though we had never seen stars before. Oftentimes the heavens were made still more glorious by auroras, the long lance rays, called "Merry Dancers" in Scotland, streaming with startling tremulous motion to the zenith. Usually the electric auroral light is white or pale yellow, but in the third or fourth of our Wisconsin winters there was a magnificently colored aurora that was seen and admired over nearly all the continent. The whole sky was draped in graceful purple and crimson folds glorious beyond description. Father called us out into the yard in front of the house where we had a wide view, crying, "Come! Come, mother! Come, bairns! and see the glory of God. All the sky is clad in a robe of red light. Look straight up to the crown where the folds are gathered. Hush and wonder and adore, for surely this is the clothing of the Lord Himself, and perhaps He will even now appear looking down from his high heaven." This celestial show was far more glorious than anything we had ever yet beheld, and throughout that wonderful winter hardly anything else was spoken of.

We even enjoyed the snow-storms; the thronging crystals, like daisies, coming down separate and distinct, were very different from the tufted flakes we enjoyed so much in Scotland, when we ran into the midst of the slow-falling feathery throng shouting with enthusiasm: "Jennie's plucking her doos! Jennie's plucking her doos (doves)!"

Nature has many ways of thinning and pruning and trimming her forests,—lightning-strokes, heavy snow, and storm-winds to shatter and blow down whole trees here and there or break off branches as required. The results of these methods I have observed in different forests, but only once have I seen pruning by rain. The rain froze on the trees as it fell and grew so thick and heavy that many of them lost a third or more of their branches. The view of the woods after the storm had passed and the sun shone forth was something never to be forgotten. Every twig and branch and rugged trunk was encased in pure crystal ice, and each oak and hickory and willow became a fairy crystal palace. Such dazzling brilliance, such effects of white light and irised light glowing and flashing I had never seen before, nor have I since. This sudden change of the leafless woods to glowing silver was, like the great aurora, spoken of for years, and is one of the most beautiful of the many pictures that enriches my life. And besides the great shows there were thousands of others even in the coldest weather manifesting the utmost fineness and tenderness of beauty and affording noble compensation for hardship and pain.

One of the most striking of the winter sounds was the loud roaring and rumbling of the ice on our lake, from its shrinking and expanding with the changes of the weather. The fishermen who were catching pickerel said that they had no luck when this roaring was going on above the fish. I remember how frightened we boys were when on one of our New Year holidays we were taking a walk on the ice and heard for the first time the sudden rumbling roar beneath our feet and running on ahead of us, creaking and whooping as if all the ice eighteen or twenty inches thick was breaking.

In the neighborhood of our Wisconsin farm there were extensive swamps consisting in great part of a thick sod of very tough carex roots covering thin, watery lakes of mud. They originated in glacier lakes that were gradually overgrown. This sod was so tough that oxen with loaded wagons could be driven over it without cutting down through it, although it was afloat. The carpenters who came to build our frame house, noticing how the sedges sunk beneath their feet, said that if they should break through, they would probably be well on their way to California before touching bottom. On the contrary, all these lake-basins are shallow as compared with their width. When we went into the Wisconsin woods there was not a single wheel-track or cattle-track. The only man-made road was an Indian trail along the Fox River between Portage and Packwauckee Lake. Of course the deer, foxes, badgers, coons, skunks, and even the squirrels had well-beaten tracks from their dens and hiding-places in thickets, hollow trees, and the ground, but they did not reach far, and but little noise was made by the soft-footed travelers in passing over them, only a slight rustling and swishing among fallen leaves and grass.

Corduroying the swamps formed the principal part of road-making among the early settlers for many a day. At these annual road-making gatherings opportunity was offered for discussion of the news, politics, religion, war, the state of the crops, comparative advantages of the new country over the old, and so forth, but the principal opportunities, recurring every week, were the hours after Sunday church services. I remember hearing long talks on the wonderful beauty of the Indian corn; the wonderful melons, so wondrous fine for "sloken a body on hot days"; their contempt for tomatoes, so fine to look at with their sunny colors and so disappointing in taste; the miserable cucumbers the "Yankee bodies" ate, though tasteless as rushes; the character of the Yankees, etcetera. Then there were long discussions about the Russian war, news of which was eagerly gleaned from Greeley's "New York Tribune"; the great battles of the Alma, the charges at Balaklava and Inkerman; the siege of Sebastopol; the military genius of Todleben; the character of Nicholas; the character of the Russian soldier, his stubborn bravery, who for the first time in history withstood the British bayonet charges; the probable outcome of the terrible war; the fate of Turkey, and so forth.

Very few of our old-country neighbors gave much heed to what are called spirit-rappings. On the contrary, they were regarded as a sort of sleight-of-hand humbug. Some of these spirits seem to be stout able-bodied fellows, judging by the weights they lift and the heavy furniture they bang about. But they do no good work that I know of; never saw wood, grind corn, cook, feed the hungry, or go to the help of poor anxious mothers at the bedsides of their sick children. I noticed when I was a boy that it was not the strongest characters who followed so-called mediums. When a rapping-storm was at its height in Wisconsin, one of our neighbors, an old Scotchman, remarked, "Thay puir silly medium-bodies may gang to the deil wi' their rappin' speerits, for they dae nae gude, and I think the deil's their faything."

Although in the spring of 1849 there was no other settler within a radius of four miles of our Fountain Lake farm, in three or four years almost every quarter-section of government land was taken up, mostly by enthusiastic home-seekers from Great Britain, with only here and there Yankee families from adjacent states, who had come drifting indefinitely westward in covered wagons, seeking their fortunes like winged seeds; all alike striking root and gripping the glacial drift soil as naturally as oak and hickory trees; happy and hopeful, establishing homes and making wider and wider fields in the hospitable wilderness. The ax and plough were kept very busy; cattle, horses, sheep, and pigs multiplied; barns and corn-cribs were filled up, and man and beast were well fed; a schoolhouse was built, which was used also for a church; and in a very short time the new country began to look like an old one.

Comparatively few of the first settlers suffered from serious accidents. One of our neighbors had a finger shot off, and on a bitter, frosty night had to be taken to a surgeon in Portage, in a sled drawn by slow, plodding oxen, to have the shattered stump dressed. Another fell from his wagon and was killed by the wheel passing over his body. An acre of ground was reserved and fenced for graves, and soon consumption came to fill it. One of the saddest instances was that of a Scotch family from Edinburgh, consisting of a father, son, and daughter, who settled on eighty acres of land within half a mile of our place. The daughter died of consumption the third year after their arrival, the son one or two years later, and at last the father followed his two children. Thus sadly ended bright hopes and dreams of a happy home in rich and free America.

Another neighbor, I remember, after a lingering illness died of the same disease in mid-winter, and his funeral was attended by the neighbors in sleighs during a driving snow-storm when the thermometer was fifteen or twenty degrees below zero. The great white plague carried off another of our near neighbors, a fine Scotchman, the father of eight promising boys, when he was only about forty-five years of age. Most of those who suffered from this disease seemed hopeful and cheerful up to a very short time before their death, but Mr. Reid, I remember, on one of his last visits to our house, said with brave resignation: "I know that never more in this world can I be well, but I must just submit. I must just submit."

One of the saddest deaths from other causes than consumption was that of a poor feeble-minded man whose brother, a sturdy blacksmith and preacher, etc., was a very hard taskmaster. Poor half-witted Charlie was kept steadily at work,—although he was not able to do much, for his body was about as feeble as his mind. He never could be taught the right use of an ax, and when he was set to chopping down trees for firewood he feebly hacked and chipped round and round them, sometimes spending several days in nibbling down a tree that a beaver might have gnawed down in half the time. Occasionally when he had an extra large tree to chop, he would go home and report that the tree was too tough and strong for him and that he could never make it fall. Then his brother, calling him a useless creature, would fell it with a few well-directed strokes, and leave Charlie to nibble away at it for weeks trying to make it into stove-wood.

The brawny blacksmith minister punished his feeble brother without any show of mercy for every trivial offense or mistake or pathetic little shortcoming. All the neighbors pitied him, especially the women, who never missed an opportunity to give him kind words, cookies, and pie; above all, they bestowed natural sympathy on the poor imbecile as if he were an unfortunate motherless child. In particular, his nearest neighbors, Scotch Highlanders, warmly welcomed him to their home and never wearied in doing everything that tender sympathy could suggest. To those friends he ran away at every opportunity. But after years of suffering from overwork and punishment his feeble health failed, and he told his Scotch friends one day that he was not able to work any more or do anything that his brother wanted him to do, that he was beaten every day, and that he had come to thank them for their kindness and to bid them good-bye, for he was going to drown himself in Muir's lake. "Oh, Charlie! Charlie!" they cried, "you mustn't talk that way. Cheer up! You will soon be stronger. We all love you. Cheer up! Cheer up! And always come here whenever you need anything."

"Oh, no!" he pathetically replied, "I know you love me, but I can't cheer up any more. My heart's gone, and I want to die."

Next day, when Mr. Anderson, a carpenter whose house was on the west shore of our lake, was going to a spring he saw a man wade out through the rushes and lily-pads and throw himself forward into deep water. This was poor Charlie. Fortunately, Mr. Anderson had a skiff close by, and as the distance was not great he reached the broken-hearted imbecile in time to save his life, and after trying to cheer him took him home to his brother. But even this terrible proof of despair failed to soften the latter. He seemed to regard the attempt at suicide simply as a crime calculated to bring the reproach of the neighbors upon him. One morning, after receiving another beating, Charlie was set to work chopping firewood in front of the house, and after feebly swinging his ax a few times he pitched forward on his face and died on the wood-pile. The unnatural brother then walked over to the neighbor who had saved Charlie from drowning, and after talking on ordinary affairs, crops, the weather, etc., said in a careless tone: "I have a little job of carpenter work for you, Mr. Anderson." "What is it, Mr.—?" "I want you to make a coffin." "A coffin!" said the startled carpenter. "Who is dead?" "Charlie," he coolly replied. All the neighbors were in tears over the poor child man's fate. But, strange to say, in all that excessively law-abiding neighborhood none was bold enough or kind enough to break the blacksmith's jaw.

The mixed lot of settlers around us offered a favorable field for observation of the different kinds of people of our own race. We were swift to note the way they behaved, the differences in their religion and morals, and in their ways of drawing a living from the same kind of soil under the same general conditions; how they protected themselves from the weather; how they were influenced by new doctrines and old ones seen in new lights in preaching, lecturing, debating, bringing up their children, etc., and how they regarded the Indians, those first settlers and owners of the ground that was being made into farms.

I well remember my father's discussing with a Scotch neighbor, a Mr. George Mair, the Indian question as to the rightful ownership of the soil. Mr. Mair remarked one day that it was pitiful to see how the unfortunate Indians, children of Nature, living on the natural products of the soil, hunting, fishing, and even cultivating small corn-fields on the most fertile spots, were now being robbed of their lands and pushed ruthlessly back into narrower and narrower limits by alien races who were cutting off their means of livelihood. Father replied that surely it could never have been the intention of God to allow Indians to rove and hunt over so fertile a country and hold it forever in unproductive wildness, while Scotch and Irish and English farmers could put it to so much better use. Where an Indian required thousands of acres for his family, these acres in the hands of industrious, God-fearing farmers would support ten or a hundred times more people in a far worthier manner, while at the same time helping to spread the gospel.

Mr. Mair urged that such farming as our first immigrants were practicing was in many ways rude and full of the mistakes of ignorance, yet, rude as it was, and ill-tilled as were most of our Wisconsin farms by unskillful, inexperienced settlers who had been merchants and mechanics and servants in the old countries, how should we like to have specially trained and educated farmers drive us out of our homes and farms, such as they were, making use of the same argument, that God could never have intended such ignorant, unprofitable, devastating farmers as we were to occupy land upon which scientific farmers could raise five or ten times as much on each acre as we did? And I well remember thinking that Mr. Mair had the better side of the argument. It then seemed to me that, whatever the final outcome might be, it was at this stage of the fight only an example of the rule of might with but little or no thought for the right or welfare of the other fellow if he were the weaker; that "they should take who had the power, and they should keep who can," as Wordsworth makes the marauding Scottish Highlanders say.

Many of our old neighbors toiled and sweated and grubbed themselves into their graves years before their natural dying days, in getting a living on a quarter-section of land and vaguely trying to get rich, while bread and raiment might have been serenely won on less than a fourth of this land, and time gained to get better acquainted with God.

I was put to the plough at the age of twelve, when my head reached but little above the handles, and for many years I had to do the greater part of the ploughing. It was hard work for so small a boy; nevertheless, as good ploughing was exacted from me as if I were a man, and very soon I had to become a good ploughman, or rather ploughboy. None could draw a straighter furrow. For the first few years the work was particularly hard on account of the tree-stumps that had to be dodged. Later the stumps were all dug and chopped out to make way for the McCormick reaper, and because I proved to be the best chopper and stump-digger I had nearly all of it to myself. It was dull, hard work leaning over on my knees all day, chopping out those tough oak and hickory stumps, deep down below the crowns of the big roots. Some, though fortunately not many, were two feet or more in diameter.

And as I was the eldest boy, the greater part of all the other hard work of the farm quite naturally fell on me. I had to split rails for long lines of zigzag fences. The trees that were tall enough and straight enough to afford one or two logs ten feet long were used for rails, the others, too knotty or cross-grained, were disposed of in log and cordwood fences. Making rails was hard work and required no little skill. I used to cut and split a hundred a day from our short, knotty oak timber, swinging the ax and heavy mallet, often with sore hands, from early morning to night. Father was not successful as a rail-splitter. After trying the work with me a day or two, he in despair left it all to me. I rather liked it, for I was proud of my skill, and tried to believe that I was as tough as the timber I mauled, though this and other heavy jobs stopped my growth and earned for me the title "Runt of the family."

In those early days, long before the great labor-saving machines came to our help, almost everything connected with wheat-raising abounded in trying work,—cradling in the long, sweaty dog-days, raking and binding, stacking, thrashing,—and it often seemed to me that our fierce, over-industrious way of getting the grain from the ground was too closely connected with grave-digging. The staff of life, naturally beautiful, oftentimes suggested the grave-digger's spade. Men and boys, and in those days even women and girls, were cut down while cutting the wheat. The fat folk grew lean and the lean leaner, while the rosy cheeks brought from Scotland and other cool countries across the sea faded to yellow like the wheat. We were all made slaves through the vice of over-industry. The same was in great part true in making hay to keep the cattle and horses through the long winters. We were called in the morning at four o'clock and seldom got to bed before nine, making a broiling, seething day seventeen hours long loaded with heavy work, while I was only a small stunted boy; and a few years later my brothers David and Daniel and my older sisters had to endure about as much as I did. In the harvest dog-days and dog-nights and dog-mornings, when we arose from our clammy beds, our cotton shirts clung to our backs as wet with sweat as the bathing-suits of swimmers, and remained so all the long, sweltering days. In mowing and cradling, the most exhausting of all the farm work, I made matters worse by foolish ambition in keeping ahead of the hired men. Never a warning word was spoken of the dangers of over-work. On the contrary, even when sick we were held to our tasks as long as we could stand. Once in harvest-time I had the mumps and was unable to swallow any food except milk, but this was not allowed to make any difference, while I staggered with weakness and sometimes fell headlong among the sheaves. Only once was I allowed to leave the harvest-field—when I was stricken down with pneumonia. I lay gasping for weeks, but the Scotch are hard to kill and I pulled through. No physician was called, for father was an enthusiast, and always said and believed that God and hard work were by far the best doctors.

None of our neighbors were so excessively industrious as father; though nearly all of the Scotch, English, and Irish worked too hard, trying to make good homes and to lay up money enough for comfortable independence. Excepting small garden-patches, few of them had owned land in the old country. Here their craving land-hunger was satisfied, and they were naturally proud of their farms and tried to keep them as neat and clean and well-tilled as gardens. To accomplish this without the means for hiring help was impossible. Flowers were planted about the neatly kept log or frame houses; barnyards, granaries, etc., were kept in about as neat order as the homes, and the fences and corn-rows were rigidly straight. But every uncut weed distressed them; so also did every ungathered ear of grain, and all that was lost by birds and gophers; and this over-carefulness bred endless work and worry.

As for money, for many a year there was precious little of it in the country for anybody. Eggs sold at six cents a dozen in trade, and five-cent calico was exchanged at twenty-five cents a yard. Wheat brought fifty cents a bushel in trade. To get cash for it before the Portage Railway was built, it had to be hauled to Milwaukee, a hundred miles away. On the other hand, food was abundant,—eggs, chickens, pigs, cattle, wheat, corn, potatoes, garden vegetables of the best, and wonderful melons as luxuries. No other wild country I have ever known extended a kinder welcome to poor immigrants. On the arrival in the spring, a log house could be built, a few acres ploughed, the virgin sod planted with corn, potatoes, etc., and enough raised to keep a family comfortably the very first year; and wild hay for cows and oxen grew in abundance on the numerous meadows. The American settlers were wisely content with smaller fields and less of everything, kept indoors during excessively hot or cold weather, rested when tired, went off fishing and hunting at the most favorable times and seasons of the day and year, gathered nuts and berries, and in general tranquilly accepted all the good things the fertile wilderness offered.

After eight years of this dreary work of clearing the Fountain Lake farm, fencing it and getting it in perfect order, building a frame house and the necessary outbuildings for the cattle and horses,—after all this had been victoriously accomplished, and we had made out to escape with life,—father bought a half-section of wild land about four or five miles to the eastward and began all over again to clear and fence and break up other fields for a new farm, doubling all the stunting, heartbreaking chopping, grubbing, stump-digging, rail-splitting, fence-building, barn-building, house-building, and so forth.

By this time I had learned to run the breaking plough. Most of these ploughs were very large, turning furrows from eighteen inches to two feet wide, and were drawn by four or five yoke of oxen. They were used only for the first ploughing, in breaking up the wild sod woven into a tough mass, chiefly by the cordlike roots of perennial grasses, reinforced by the taproots of oak and hickory bushes, called "grubs," some of which were more than a century old and four or five inches in diameter. In the hardest ploughing on the most difficult ground, the grubs were said to be as thick as the hair on a dog's back. If in good trim, the plough cut through and turned over these grubs as if the century-old wood were soft like the flesh of carrots and turnips; but if not in good trim the grubs promptly tossed the plough out of the ground. A stout Highland Scot, our neighbor, whose plough was in bad order and who did not know how to trim it, was vainly trying to keep it in the ground by main strength, while his son, who was driving and merrily whipping up the cattle, would cry encouragingly, "Haud her in, fayther! Haud her in!"

"But hoo i' the deil can I haud her in when she'll no *stop* in?" his perspiring father would reply, gasping for breath between each word. On the contrary, with the share and coulter sharp and nicely adjusted, the plough, instead of shying at every grub and jumping out, ran straight ahead without need of steering or holding, and gripped the ground so firmly that it could hardly be thrown out at the end of the furrow.

Our breaker turned a furrow two feet wide, and on our best land, where the sod was toughest, held so firm a grip that at the end of the field my brother, who was driving the oxen, had to come to my assistance in throwing it over on its side to be drawn around the end of the landing; and it was all I could do to set it up again. But I learned to keep that plough in such trim that after I got started on a new furrow I used to ride on the crossbar between the handles with my feet resting comfortably on the beam, without having to steady or steer it in any way on the whole length of the field, unless we had to go round a stump, for it sawed through the biggest grubs without flinching.

The growth of these grubs was interesting to me. When an acorn or hickory-nut had sent up its first season's sprout, a few inches long, it was burned off in the autumn grass fires; but the root continued to hold on to life, formed a callus over the wound and sent up one or more shoots the next spring. Next autumn these new shoots were burned off, but the root and calloused head, about level with the surface of the ground, continued to grow and send up more new shoots; and so on, almost every year until very old, probably far more than a century, while the tops, which would naturally have become tall broad-headed trees, were only mere sprouts seldom more than two years old. Thus the ground was kept open like a prairie, with only five or six trees to the acre, which had escaped the fire by having the good fortune to grow on a bare spot at the door of a fox or badger den, or between straggling grass-tufts wide apart on the poorest sandy soil.

The uniformly rich soil of the Illinois and Wisconsin prairies produced so close and tall a growth of grasses for fires that no tree could live on it. Had there been no fires, these fine prairies, so marked a feature of the country, would have been covered by the heaviest forests. As soon as the oak openings in our neighborhood were settled, and the farmers had prevented running grass-fires, the grubs grew up into trees and formed tall thickets so dense that it was difficult to walk through them and every trace of the sunny "openings" vanished.

We called our second farm Hickory Hill, from its many fine hickory trees and the long gentle slope leading up to it. Compared with Fountain Lake farm it lay high and dry. The land was better, but it had no living water, no spring or stream or meadow or lake. A well ninety feet deep had to be dug, all except the first ten feet or so in fine-grained sandstone. When the sandstone was struck, my father, on the advice of a man who had worked in mines, tried to blast the rock; but from lack of skill the blasting went on very slowly, and father decided to have me do all the work with mason's chisels, a long, hard job, with a good deal of danger in it. I had to sit cramped in a space about three feet in diameter, and wearily chip, chip, with heavy hammer and chisels from early morning until dark, day after day, for weeks and months. In the morning, father and David lowered me in a wooden bucket by a windlass, hauled up what chips were left from the night before, then went away to the farm work and left me until noon, when they hoisted me out for dinner. After dinner I was promptly lowered again, the forenoon's accumulation of chips hoisted out of the way, and I was left until night.

One morning, after the dreary bore was about eighty feet deep, my life was all but lost in deadly choke-damp,—carbonic acid gas that had settled at the bottom during the night. Instead of clearing away the chips as usual when I was lowered to the bottom, I swayed back and forth and began to sink under the poison. Father, alarmed that I did not make any noise, shouted, "What's keeping you so still?" to which he got no reply. Just as I was settling down against the side of the wall, I happened to catch a glimpse of a branch of a bur-oak tree which leaned out over the mouth of the shaft. This suddenly awakened me, and to father's excited shouting I feebly murmured, "Take me out." But when he began to hoist he found I was not in the bucket and in wild alarm shouted, "Get in! Get in the bucket and hold on! Hold on!" Somehow I managed to get into the bucket, and that is all I remembered until I was dragged out, violently gasping for breath.

One of our near neighbors, a stone mason and miner by the name of William Duncan, came to see me, and after hearing the particulars of the accident he solemnly said: "Weel, Johnnie, it's God's mercy that you're alive. Many a companion of mine have I seen dead with choke-damp, but none that I ever saw or heard of was so near to death in it as you were and escaped without help." Mr. Duncan taught father to throw water down the shaft to absorb the gas, and also to drop a bundle of brush or hay attached to a light rope, dropping it again and again to carry down pure air and stir up the poison. When, after a day or two, I had recovered from the shock, father lowered me again to my work, after taking the precaution to test the air with a candle and stir it up well with a brush-and-hay bundle. The weary hammer-and-chisel-chipping went on as before, only more slowly, until ninety feet down, when at last I struck a fine, hearty gush of water. Constant dropping wears away stone. So does constant chipping, while at the same time wearing away the chipper. Father never spent an hour in that well. He trusted me to sink it straight and plumb, and I did, and built a fine covered top over it, and swung two iron-bound buckets in it from which we all drank for many a day.

The honey-bee arrived in America long before we boys did, but several years passed ere we noticed any on our farm. The introduction of the honey-bee into flowery America formed a grand epoch in bee history. This sweet humming creature, companion and friend of the flowers, is now distributed over the greater part of the continent, filling countless hollows in rocks and trees with honey as well as the millions of hives prepared for them by honey-farmers, who keep and tend their flocks of sweet winged cattle, as shepherds keep sheep,—a charming employment, "like directing sunbeams," as Thoreau says. The Indians call the honey-bee the white man's fly; and though they had long been acquainted with several species of bumblebees that yielded more or less honey, how gladly surprised they must have been when they discovered that, in the hollow trees where before they had found only coons or squirrels, they found swarms of brown flies with fifty or even a hundred pounds of honey sealed up in beautiful cells. With their keen hunting senses they of course were not slow to learn the habits of the little brown immigrants and the best methods of tracing them to their sweet homes, however well hidden. During the first few years none were seen on our farm, though we sometimes heard father's hired men talking about "lining bees." None of us boys ever found a bee tree, or tried to find any until about ten years after our arrival in the woods. On the Hickory Hill farm there is a ridge of moraine material, rather dry, but flowery with goldenrods and asters of many species, upon which we saw bees feeding in the late autumn just when their hives were fullest of honey, and it occurred to me one day after I was of age and my own master that I must try to find a bee tree. I made a little box about six inches long and four inches deep and wide; bought half a pound of honey, went to the goldenrod hill, swept a bee into the box and closed it. The lid had a pane of glass in it so I could see when the bee had sucked its fill and was ready to go home. At first it groped around trying to get out, but, smelling the honey, it seemed to forget everything else, and while it was feasting I carried the box and a small sharp-pointed stake to an open spot, where I could see about me, fixed the stake in the ground, and placed the box on the flat top of it. When I thought that the little feaster must be about full, I opened the box, but it was in no hurry to fly. It slowly crawled up to the edge of the box, lingered a minute or two cleaning its legs that had become sticky with honey, and when it took wing, instead of making what is called a bee-line for home, it buzzed around the box and minutely examined it as if trying to fix a clear picture of it in its mind so as to be able to recognize it when it returned for another load, then circled around at a little distance as if looking for something to locate it by. I was the nearest object, and the thoughtful worker buzzed in front of my face and took a good stare at me, and then flew up on to the top of an oak on the side of the open spot in the centre of which the honey-box was. Keeping a keen watch, after a minute or two of rest or wing-cleaning, I saw it fly in wide circles round the tops of the trees nearest the honey-box, and, after apparently satisfying itself, make a bee-line for the hive. Looking endwise on the line of flight, I saw that what is called a bee-line is not an absolutely straight line, but a line in general straight made of many slight, wavering, lateral curves. After taking as true a bearing as I could, I waited and watched. In a few minutes, probably ten, I was surprised to see that bee arrive at the end of the outleaning limb of the oak mentioned above, as though that was the first point it had fixed in its memory to be depended on in retracing the way back to the honey-box. From the tree-top it came straight to my head, thence straight to the box, entered without the least hesitation, filled up and started off after the same preparatory dressing and taking of bearings as before. Then I took particular pains to lay down the exact course so I would be able to trace it to the hive. Before doing so, however, I made an experiment to test the worth of the impression I had that the little insect found the way back to the box by fixing telling points in its mind. While it was away, I picked up the honey-box and set it on the stake a few rods from the position it had thus far occupied, and stood there watching. In a few minutes I saw the bee arrive at its guide-mark, the overleaning branch on the tree-top, and thence came bouncing down right to the spaces in the air which had been occupied by my head and the honey-box, and when the cunning little honey-gleaner found nothing there but empty air it whirled round and round as if confused and lost; and although I was standing with the open honey-box within fifty or sixty feet of the former feasting-spot, it could not, or at least did not, find it.

Now that I had learned the general direction of the hive, I pushed on in search of it. I had gone perhaps a quarter of a mile when I caught another bee, which, after getting loaded, went through the same performance of circling round and round the honey-box, buzzing in front of me and staring me in the face to be able to recognize me; but as if the adjacent trees and bushes were sufficiently well known, it simply looked around at them and bolted off without much dressing, indicating, I thought, that the distance to the hive was not great. I followed on and very soon discovered it in the bottom log of a corn-field fence, but some lucky fellow had discovered it before me and robbed it. The robbers had chopped a large hole in the log, taken out most of the honey, and left the poor bees late in the fall, when winter was approaching, to make haste to gather all the honey they could from the latest flowers to avoid starvation in the winter.

VII

KNOWLEDGE AND INVENTIONS

Hungry for Knowledge—Borrowing Books—Paternal Opposition—Snatched Moments—Early rising proves a way out of difficulties—The Cellar Workshop—Inventions—An Early-Rising Machine—Novel Clocks—Hygrometers etc.—A Neighbor's Advice.

I LEARNED ARITHMETIC in Scotland without understanding any of it, though I had the rules by heart. But when I was about fifteen or sixteen years of age, I began to grow hungry for real knowledge, and persuaded father, who was willing enough to have me study provided my farm work was kept up, to buy me a higher arithmetic. Beginning at the beginning, in one summer I easily finished it without assistance, in the short intervals between the end of dinner and the afternoon start for the harvest- and hay-fields, accomplishing more without a teacher in a few scraps of time than in years in school before my mind was ready for such work. Then in succession I took up algebra, geometry, and trigonometry and made some little progress in each, and reviewed grammar. I was fond of reading, but father had brought only a few religious books from Scotland. Fortunately, several of our neighbors had brought a dozen or two of all sorts of books, which I borrowed and read, keeping all of them except the religious ones carefully hidden from father's eye. Among these were Scott's novels, which, like all other novels, were strictly forbidden, but devoured with glorious pleasure in secret. Father was easily persuaded to buy Josephus' "Wars of the Jews," and D'Aubigné's "History of the Reformation," and I tried hard to get him to buy Plutarch's Lives, which, as I told him, everybody, even religious people, praised as a grand good book; but he would have nothing to do with the old pagan until the graham bread and anti-flesh doctrines came suddenly into our backwoods neighborhood, making a stir something like phrenology and spirit-rappings, which were as mysterious in their attacks as influenza. He then thought it possible that Plutarch might be turned to account on the food question by revealing what those old Greeks and Romans ate to make them strong; and so at last we gained our glorious Plutarch. Dick's "Christian Philosopher," which I borrowed from a neighbor, I thought I might venture to read in the open, trusting that the word "Christian" would be proof against its cautious condemnation. But father balked at the word "Philosopher," and quoted from the Bible a verse which spoke of "philosophy falsely so-called." I then ventured to speak in defense of the book, arguing that we could not do without at least a little of the most useful kinds of philosophy.

"Yes, we can," he said with enthusiasm, "the Bible is the only book human beings can possibly require throughout all the journey from earth to heaven."

"But how," I contended, "can we find the way to heaven without the Bible, and how after we grow old can we read the Bible without a little helpful science? Just think, father, you cannot read your Bible without spectacles, and millions of others are in the same fix; and spectacles cannot be made without some knowledge of the science of optics."

"Oh!" he replied, perceiving the drift of the argument, "there will always be plenty of worldly people to make spectacles."

To this I stubbornly replied with a quotation from the Bible with reference to the time coming when "all shall know the Lord from the least even to the greatest," and then who will make the spectacles? But he still objected to my reading that book, called me a contumacious quibbler too fond of disputation, and ordered me to return it to the accommodating owner. I managed, however, to read it later.

On the food question father insisted that those who argued for a vegetable diet were in the right, because our teeth showed plainly that they were made with reference to fruit and grain and not for flesh like those of dogs and wolves and tigers. He therefore promptly adopted a vegetable diet and requested mother to make the bread from graham flour instead of bolted flour. Mother put both kinds on the table, and meat also, to let all the family take their choice, and while father was insisting on the foolishness of eating flesh, I came to her help by calling father's attention to the passage in the Bible which told the story of Elijah the prophet who, when he was pursued by enemies who wanted to take his life, was hidden by the Lord by the brook Cherith, and fed by ravens; and surely the Lord knew what was good to eat, whether bread or meat. And on what, I asked, did the Lord feed Elijah? On vegetables or graham bread? No, he directed the ravens to feed his prophet on flesh. The Bible being the sole rule, father at once acknowledged that he was mistaken. The Lord never would have sent flesh to Elijah by the ravens if graham bread were better.

I remember as a great and sudden discovery that the poetry of the Bible, Shakespeare, and Milton was a source of inspiring, exhilarating, uplifting pleasure; and I became anxious to know all the poets, and saved up small sums to buy as many of their books as possible. Within three or four years I was the proud possessor of parts of Shakespeare's, Milton's, Cowper's, Henry Kirke White's, Campbell's, and Akenside's works, and quite a number of others seldom read nowadays. I think it was in my fifteenth year that I began to relish good literature with enthusiasm, and smack my lips over favorite lines, but there was desperately little time for reading, even in the winter evenings,—only a few stolen minutes now and then. Father's strict rule was, straight to bed immediately after family worship, which in winter was usually over by eight o'clock. I was in the habit of lingering in the kitchen with a book and candle after the rest of the family had retired, and considered myself fortunate if I got five minutes' reading before father noticed the light and ordered me to bed; an order that of course I immediately obeyed. But night after night I tried to steal minutes in the same lingering way, and how keenly precious those minutes were, few nowadays can know. Father failed perhaps two or three times in a whole winter to notice my light for nearly ten minutes, magnificent golden blocks of time, long to be remembered like holidays or geological periods. One evening when I was reading Church history father was particularly irritable, and called out with hope-killing emphasis, "*John, go to bed!* Must I give you a separate order every night to get you to go to bed? Now, I will have no irregularity in the family; you *must* go when the rest go, and without my having to tell you." Then, as an afterthought, as if judging that his words and tone of voice were too severe for so pardonable an offense as reading a religious book he unwarily added: "If you *will* read, get up in the morning and read. You may get up in the morning as early as you like."

That night I went to bed wishing with all my heart and soul that somebody or something might call me out of sleep to avail myself of this wonderful indulgence; and next morning to my joyful surprise

I awoke before father called me. A boy sleeps soundly after working all day in the snowy woods, but that frosty morning I sprang out of bed as if called by a trumpet blast, rushed downstairs, scarce feeling my chilblains, enormously eager to see how much time I had won; and when I held up my candle to a little clock that stood on a bracket in the kitchen I found that it was only one o'clock. I had gained five hours, almost half a day! "Five hours to myself!" I said, "five huge, solid hours!" I can hardly think of any other event in my life, any discovery I ever made that gave birth to joy so transportingly glorious as the possession of these five frosty hours.

In the glad, tumultuous excitement of so much suddenly acquired time-wealth, I hardly knew what to do with it. I first thought of going on with my reading, but the zero weather would make a fire necessary, and it occurred to me that father might object to the cost of fire-wood that took time to chop. Therefore, I prudently decided to go down cellar, and begin work on a model of a self-setting saw-mill I had invented. Next morning I managed to get up at the same gloriously early hour, and though the temperature of the cellar was a little below the freezing point, and my light was only a tallow candle the mill work went joyfully on. There were a few tools in a corner of the cellar,—a vise, files, a hammer, chisels, etc., that father had brought from Scotland, but no saw excepting a coarse crooked one that was unfit for sawing dry hickory or oak. So I made a fine-tooth saw suitable for my work out of a strip of steel that had formed part of an old-fashioned corset, that cut the hardest wood smoothly. I also made my own bradawls, punches, and a pair of compasses, out of wire and old files.

My workshop was immediately under father's bed, and the filing and tapping in making cogwheels, journals, cams, etc., must, no doubt, have annoyed him, but with the permission he had granted in his mind, and doubtless hoping that I would soon tire of getting up at one o'clock, he impatiently waited about two weeks before saying a word. I did not vary more than five minutes from one o'clock all winter, nor did I feel any bad effects whatever, nor did I think at all about the subject as to whether so little sleep might be in any way injurious; it was a grand triumph of will-power over cold and common comfort and work-weariness in abruptly cutting down my ten hours' allowance of sleep to five. I simply felt that I was rich beyond anything I could have dreamed of or hoped for. I was far more than happy. Like Tam o'Shanter I was glorious, "O'er a' the ills o' life victorious."

Father, as was customary in Scotland, gave thanks and asked a blessing before meals, not merely as a matter of form and decent Christian manners, for he regarded food as a gift derived directly from the hands of the Father in heaven. Therefore every meal to him was a sacrament requiring conduct and attitude of mind not unlike that befitting the Lord's Supper. No idle word was allowed to be spoken at our table, much less any laughing or fun or story-telling. When we were at the breakfast-table, about two weeks after the great golden time-discovery, father cleared his throat preliminary, as we all knew, to saying something considered important. I feared that it was to be on the subject of my early rising, and dreaded the withdrawal of the permission he had granted on account of the noise I made, but still hoping that, as he had given his word that I might get up as early as I wished, he would as a Scotchman stand to it, even though it was given in an unguarded moment and taken in a sense unreasonably far-reaching. The solemn sacramental silence was broken by the dreaded question:—

"John, what time is it when you get up in the morning?"

"About one o'clock," I replied in a low, meek, guilty tone of voice.

"And what kind of a time is that, getting up in the middle of the night and disturbing the whole family?"

I simply reminded him of the permission he had freely granted me to get up as early as I wished.

"I *know* it," he said, in an almost agonized tone of voice, "I *know* I gave you that miserable permission, but I never imagined that you would get up in the middle of the night."

To this I cautiously made no reply, but continued to listen for the heavenly one-o'clock call, and it never failed.

After completing my self-setting sawmill I dammed one of the streams in the meadow and put the mill in operation. This invention was speedily followed by a lot of others,—water-wheels, curious doorlocks and latches, thermometers, hygrometers, pyrometers, clocks, a barometer, an automatic contrivance for feeding the horses at any required hour, a lamp-lighter and fire-lighter, an early-or-late-rising machine, and so forth.

After the sawmill was proved and discharged from my mind, I happened to think it would be a fine thing to make a timekeeper which would tell the day of the week and the day of the month, as well as strike like a common clock and point out the hours; also to have an attachment whereby it could be connected with a bedstead to set me on my feet at any hour in the morning; also to start fires, light lamps, etc. I had learned the time laws of the pendulum from a book, but with this exception I knew nothing of timekeepers, for I had never seen the inside of any sort of clock or watch. After long brooding, the novel clock was at length completed in my mind, and was tried and found to be durable and to work well and look well before I had begun to build it in wood. I carried small parts of it in my pocket to whittle at when I was out at work on the farm, using every spare or stolen moment within reach without father's knowing anything about it. In the middle of summer, when harvesting was in progress, the novel time-machine was nearly completed. It was hidden upstairs in a spare bedroom where some tools were kept. I did the making and mending on the farm, but one day at noon, when I happened to be away, father went upstairs for a hammer or something and discovered the mysterious machine back of the bedstead. My sister Margaret saw him on his knees examining it, and at the first opportunity whispered in my ear, "John, fayther saw that thing you're making upstairs." None of the family knew what I was doing, but they knew very well that all such work was frowned on by father, and kindly warned me of any danger that threatened my plans. The fine invention seemed doomed to destruction before its time-ticking commenced, though I thought it handsome, had so long carried it in my mind, and like the nest of Burns's wee mousie it had cost me mony a weary whittling nibble. When we were at dinner several days after the sad discovery, father began to clear his throat to speak, and I feared the doom of martyrdom was about to be pronounced on my grand clock.

"John," he inquired, "what is that thing you are making upstairs?"

I replied in desperation that I didn't know what to call it.

"What! You mean to say you don't know what you are trying to do?"

"Oh, yes," I said, "I know very well what I am doing."

"What, then, is the thing for?"

"It's for a lot of things," I replied, "but getting people up early in the morning is one of the main things it is intended for; therefore it might perhaps be called an early-rising machine."

After getting up so extravagantly early, all the last memorable winter to make a machine for getting up perhaps still earlier seemed so ridiculous that he very nearly laughed. But after controlling himself and getting command of a sufficiently solemn face and voice he said severely, "Do you not think it is very wrong to waste your time on such nonsense?"

"No," I said meekly, "I don't think I'm doing any wrong."

"Well," he replied, "I assure you I do; and if you were only half as zealous in the study of religion as you are in contriving and whittling these useless, nonsensical things, it would be infinitely better for you. I want you to be like Paul, who said that he desired to know nothing among men but Christ and Him crucified."

To this I made no reply, gloomily believing my fine machine was to be burned, but still taking what comfort I could in realizing that anyhow I had enjoyed inventing and making it.

After a few days, finding that nothing more was to be said, and that father after all had not had the heart to destroy it, all necessity for secrecy being ended, I finished it in the half-hours that we had at noon and set it in the parlor between two chairs, hung moraine boulders that had come from the direction of Lake Superior on it for weights, and set it running. We were then hauling grain into the barn. Father at this period devoted himself entirely to the Bible and did no farm work whatever. The clock had a good loud tick, and when he heard it strike, one of my sisters told me that he left his study, went to the parlor, got down on his knees and carefully examined the machinery, which was all in plain sight, not being enclosed in a case. This he did repeatedly, and evidently seemed a little proud of my ability to invent and whittle such a thing, though careful to give no encouragement for anything more of the kind in future.

But somehow it seemed impossible to stop. Inventing and whittling faster than ever, I made another hickory clock, shaped like a scythe to symbolize the scythe of Father Time. The pendulum is a bunch of arrows symbolizing the flight of time. It hangs on a leafless mossy oak snag showing the effect of time, and on the snath is written, "All flesh is grass." This, especially the inscription, rather pleased father, and, of course, mother and all my sisters and brothers admired it. Like the first it indicates the days of the week and month, starts fires and beds at any given hour and minute, and, though made more than fifty years ago, is still a good timekeeper.

My mind still running on clocks, I invented a big one like a town clock with four dials, with the time-figures so large they could be read by all our immediate neighbors as well as ourselves when at work in the fields, and on the side next the house the days of the week and month were indicated. It was to be placed on the peak of the barn roof. But just as it was all but finished, father stopped me, saying that it would bring too many people around the barn. I then asked permission to put it on the top of a black-oak tree near the house. Studying the larger main branches, I thought I could secure a sufficiently rigid foundation for it, while the trimmed sprays and leaves would conceal the angles of the cabin required to shelter the works from the weather, and the two-second pendulum, fourteen feet long, could be snugly encased on the side of the trunk. Nothing about the grand, useful timekeeper, I argued, would disfigure the tree, for it would look something like a big hawk's nest. "But that," he objected, "would draw still bigger bothersome trampling crowds about the place, for who ever heard of anything so queer as a big clock on the top of a tree?" So I had to lay aside its big wheels and cams and rest content with the pleasure of inventing it, and looking at it in my mind and listening to the deep solemn throbbing of its long two-second pendulum with its two old axes back to back for the bob.

One of my inventions was a large thermometer made of an iron rod, about three feet long and five eighths of an inch in diameter, that had formed part of a wagon-box. The expansion and contraction of this rod was multiplied by a series of levers made of strips of hoop iron. The pressure of the rod against the levers was kept constant by a small counterweight, so that the slightest change in the length of the rod was instantly shown on a dial about three feet wide multiplied about thirty-two thousand times. The zero-point was gained by packing the rod in wet snow. The scale was so large that the big black hand on the white-painted dial could be seen distinctly and the temperature read while we were ploughing in the field below the house. The extremes of heat and cold caused the hand to make several revolutions. The number of these revolutions was indicated on a small dial marked on the larger one. This thermometer was fastened on the side of the house, and was so sensitive that when any one approached it within four or five feet the heat radiated from the observer's body caused the hand of the dial to move so fast that the motion was plainly visible, and when he stepped back, the hand moved slowly back to its normal position. It was regarded as a great wonder by the neighbors and even by my own all-Bible father.

Boys are fond of the books of travelers, and I remember that one day, after I had been reading Mungo Park's travels in Africa, mother said: "Weel, John, maybe you will travel like Park and Humboldt some day." Father overheard her and cried out in solemn deprecation, "Oh, Anne! dinna put sic notions in the laddie's heed." But at this time there was precious little need of such prayers. My brothers left the farm when they came of age, but I stayed a year longer, loath to leave home. Mother hoped I might be a minister some day; my sisters that I would be a great inventor. I often thought I should like to be a physician, but I saw no way of making money and getting the necessary education, excepting as an inventor. So, as a beginning, I decided to try to get into a big shop or factory and live a while among machines. But I was naturally extremely shy and had been taught to have a poor opinion of myself, as of no account, though all our neighbors encouragingly called me a genius, sure to rise in the world. When I was talking over plans one day with a friendly neighbor, he said: "Now, John, if you wish to get into a machine-shop, just take some of your inventions to the State Fair, and you may be sure that as soon as they are seen they will open the door of any shop in the country for you. You will be welcomed everywhere." And when doubtingly asked if people would care to look at things made of wood, he said, "Made of wood! Made of wood! What does it matter what they're made of when they are so out-and-out original. There's nothing else like them in the world. That is what will attract attention, and besides they're mighty handsome things anyway to come from the backwoods." So I was encouraged to leave home and go at his direction to the State Fair when it was being held in Madison.

VIII

THE WORLD AND THE UNIVERSITY

Leaving Home—Creating a Sensation in Pardeeville—A Ride on a Locomotive—At the State Fair in Madison—Employment in a Machine-Shop at Prairie du Chien—Back to Madison—Entering the University—Teaching School—First Lesson in Botany—More Inventions—The University of the Wilderness.

WHEN I TOLD father that I was about to leave home, and inquired whether, if I should happen to be in need of money, he would send me a little, he said, "No; depend entirely on yourself." Good advice, I suppose, but surely needlessly severe for a bashful, home-loving boy who had worked so hard. I had the gold sovereign that my grandfather had given me when I left Scotland, and a few dollars, perhaps ten, that I had made by raising a few bushels of grain on a little patch of sandy abandoned ground. So when I left home to try the world I had only about fifteen dollars in my pocket.

Strange to say, father carefully taught us to consider ourselves very poor worms of the dust, conceived in sin, etc., and devoutly believed that quenching every spark of pride and self-confidence was a sacred duty, without realizing that in so doing he might at the same time be quenching everything else. Praise he considered most venomous, and tried to assure me that when I was fairly out in the wicked world making my own way I would soon learn that although I might have thought him a hard taskmaster at times, strangers were far harder. On the contrary, I found no lack of kindness and sympathy. All the baggage I carried was a package made up of the two clocks and a small thermometer made of a piece of old washboard, all three tied together, with no covering or case of any sort, the whole looking like one very complicated machine.

The aching parting from mother and my sisters was, of course, hard to bear. Father let David drive me down to Pardeeville, a place I had never before seen, though it was only nine miles south of the Hickory Hill home. When we arrived at the village tavern, it seemed deserted. Not a single person was in sight. I set my clock baggage on the rickety platform. David said good-bye and started for home, leaving me alone in the world. The grinding noise made by the wagon in turning short brought out the landlord, and the first thing that caught his eye was my strange bundle. Then he looked at me and said, "Hello, young man, what's this?"

"Machines," I said, "for keeping time and getting up in the morning, and so forth."

"Well! Well! That's a mighty queer get-up. You must be a Down-East Yankee. Where did you get the pattern for such a thing?"

"In my head," I said.

Some one down the street happened to notice the landlord looking intently at something and came up to see what it was. Three or four people in that little village formed an attractive crowd, and in fifteen or twenty minutes the greater part of the population of Pardeeville stood gazing in a circle around my strange hickory belongings. I kept outside of the circle to avoid being seen, and had the advantage of hearing the remarks without being embarrassed. Almost every one as he came up would say, "What's that? What's it for? Who made it?" The landlord would answer them all alike, "Why, a young man that lives out in the country somewhere made it, and he says it's a thing for keeping time, getting up in the morning, and something that I didn't understand. I don't know what he meant." "Oh, no!" one of the crowd would say, "that can't be. It's for something else—something mysterious. Mark my words, you'll see all about it in the newspapers some of these days." A curious little fellow came running up the street, joined the crowd, stood on tiptoe to get sight of the wonder, quickly made up his mind, and shouted in crisp, confident, cock-crowing style, "I know what that contraption's for. It's a machine for taking the bones out of fish."

This was in the time of the great popular phrenology craze, when the fences and barns along the roads throughout the country were plastered with big skull-bump posters, headed, "Know Thyself," and advising everybody to attend schoolhouse lectures to have their heads explained and be told what they were good for and whom they ought to marry. My mechanical bundle seemed to bring a good deal of this phrenology to mind, for many of the onlookers would say, "I wish I could see that boy's head,—he must have a tremendous bump of invention." Others complimented me by saying, "I wish

I had that fellow's head. I'd rather have it than the best farm in the State."

I stayed overnight at this little tavern, waiting for a train. In the morning I went to the station, and set my bundle on the platform. Along came the thundering train, a glorious sight, the first train I had ever waited for. When the conductor saw my queer baggage, he cried, "Hello! What have we here?"

"Inventions for keeping time, early rising, and so forth. May I take them into the car with me?"

"You can take them where you like," he replied, "but you had better give them to the baggage-master. If you take them into the car they will draw a crowd and might get broken."

So I gave them to the baggage-master and made haste to ask the conductor whether I might ride on the engine. He good-naturedly said: "Yes, it's the right place for you. Run ahead, and tell the engineer what I say." But the engineer bluntly refused to let me on, saying: "It don't matter what the conductor told you. *I* say you can't ride on my engine."

By this time the conductor, standing ready to start his train, was watching to see what luck I had, and when he saw me returning came ahead to meet me.

"The engineer won't let me on," I reported.

"Won't he?" said the kind conductor. "Oh! I guess he will. You come down with me." And so he actually took the time and patience to walk the length of that long train to get me on to the engine.

"Charlie," said he, addressing the engineer, "don't you ever take a passenger?"

"Very seldom," he replied.

"Anyhow, I wish you would take this young man on. He has the strangest machines in the baggage-car I ever saw in my life. I believe he could make a locomotive. He wants to see the engine running. Let him on." Then in a low whisper he told me to jump on, which I did gladly, the engineer offering neither encouragement nor objection.

As soon as the train was started, the engineer asked what the "strange thing" the conductor spoke of really was.

"Only inventions for keeping time, getting folk up in the morning, and so forth," I hastily replied, and before he could ask any more questions I asked permission to go outside of the cab to see the machinery. This he kindly granted, adding, "Be careful not to fall off, and when you hear me whistling for a station you come back, because if it is reported against me to the superintendent that I allow boys to run all over my engine I might lose my job."

Assuring him that I would come back promptly, I went out and walked along the foot-board on the side of the boiler, watching the magnificent machine rushing through the landscapes as if glorying in its strength like a living creature. While seated on the cow-catcher platform, I seemed to be fairly flying, and the wonderful display of power and motion was enchanting. This was the first time I had ever been on a train, much less a locomotive, since I had left Scotland. When I got to Madison, I thanked the kind conductor and engineer for my glorious ride, inquired the way to the Fair, shouldered my inventions, and walked to the Fair Ground.

When I applied for an admission ticket at a window by the gate I told the agent that I had something to exhibit.

"What is it?" he inquired.

"Well, here it is. Look at it."

When he craned his neck through the window and got a glimpse of my bundle, he cried excitedly, "Oh! *you* don't need a ticket,—come right in."

When I inquired of the agent where such things as mine should be exhibited, he said, "You see that building up on the hill with a big flag on it? That's the Fine Arts Hall, and it's just the place for your wonderful invention."

So I went up to the Fine Arts Hall and looked in, wondering if they would allow wooden things in so fine a place.

I was met at the door by a dignified gentleman, who greeted me kindly and said, "Young man, what have we got here?"

"Two clocks and a thermometer," I replied.

"Did you make these? They look wonderfully beautiful and novel and must, I think, prove the most interesting feature of the fair."

"Where shall I place them?" I inquired.

"Just look around, young man, and choose the place you like best, whether it is occupied or not. You can have your pick of all the building, and a carpenter to make the necessary shelving and assist you every way possible!"

So I quickly had a shelf made large enough for all of them, went out on the hill and picked up some glacial boulders of the right size for weights, and in fifteen or twenty minutes the clocks were running. They seemed to attract more attention than anything else in the hall. I got lots of praise from the crowd and the newspaper-reporters. The local press reports were copied into the Eastern papers. It was considered wonderful that a boy on a farm had been able to invent and make such things, and almost every spectator foretold good fortune. But I had been so lectured by my father above all things to avoid praise that I was afraid to read those kind newspaper notices, and never clipped out or preserved any of them, just glanced at them and turned away my eyes from beholding vanity. They gave me a prize of ten or fifteen dollars and a diploma for wonderful things not down in the list of exhibits.

Many years later, after I had written articles and books, I received a letter from the gentleman who had charge of the Fine Arts Hall. He proved to be the Professor of English Literature in the University of Wisconsin at this Fair time, and long afterward he sent me clippings of reports of his lectures. He had a lecture on me, discussing style, etcetera, and telling how well he remembered my arrival at the Hall in my shirtsleeves with those mechanical wonders on my shoulder, and so forth, and so forth. These inventions, though of little importance, opened all doors for me and made marks that have lasted many years, simply, I suppose, because they were original and promising.

I was looking around in the mean time to find out where I should go to seek my fortune. An inventor at the Fair, by the name of Wiard, was exhibiting an iceboat he had invented to run on the upper Mississippi from Prairie du Chien to St. Paul during the winter months, explaining how useful it would be thus to make a highway of the river while it was closed to ordinary navigation by ice. After he saw my inventions he offered me a place in his foundry and machine-shop in Prairie du Chien and promised to assist me all he could. So I made up my mind to accept his offer and rode with him to Prairie du Chien in his iceboat, which was mounted on a flat car. I soon found, however, that he was seldom at home and that I was not likely to learn much at his small shop. I found a place where I could work for my board and devote my spare hours to mechanical drawing, geometry, and physics, making but little headway, however, although the Pelton family, for whom I worked, were very kind. I made up my mind after a few months' stay in Prairie du Chien to return to Madison, hoping that in some way I might be able to gain an education.

At Madison I raised a few dollars by making and selling a few of those bedsteads that set the sleepers on their feet in the morning,—inserting in the footboard the works of an ordinary clock that could be bought for a dollar. I also made a few dollars addressing circulars in an insurance office, while at the same time I was paying my board by taking care of a pair of horses and going errands. This is of no great interest except that I was thus winning my bread while hoping that something would turn up that might enable me to make money enough to enter the State University. This was my ambition, and it never wavered no matter what I was doing. No University, it seemed to me, could be more admirably situated, and as I sauntered about it, charmed with its fine lawns and trees and beautiful lakes, and saw the students going and coming with their books, and occasionally practising with a theodolite in measuring distances, I thought that if I could only join them it would be the greatest joy of life. I was desperately hungry and thirsty for knowledge and willing to endure anything to get it.

One day I chanced to meet a student who had noticed my inventions at the Fair and now recognized me. And when I said, "You are fortunate fellows to be allowed to study in this beautiful place. I wish I could join you." "Well, why don't you?" he asked. "I haven't money enough," I said. "Oh, as to money," he reassuringly explained, "very little is required. I presume you're able to enter the Freshman class, and you can board yourself as quite a number of us do at a cost of about a dollar a week. The baker and milkman come every day. You can live on bread and milk." Well, I thought, maybe I have money enough for at least one beginning term. Anyhow I couldn't help trying.

With fear and trembling, overladen with ignorance, I called on Professor Stirling, the Dean of the Faculty, who was then Acting President, presented my case, and told him how far I had got on with my studies at home, and that I hadn't been to school since leaving Scotland at the age of eleven years, excepting one short term of a couple of months at a district school, because I could not be spared from the farm work. After hearing my story, the kind professor welcomed me to the glorious University—next, it seemed to me, to the Kingdom of Heaven. After a few weeks in the preparatory department I entered the Freshman class. In Latin I found that one of the books in use I had already studied in Scotland. So, after an interruption of a dozen years, I began my Latin over again where I had left off; and, strange to say, most of it came back to me, especially the grammar which I had committed to memory at the Dunbar Grammar School.

During the four years that I was in the University, I earned enough in the harvest-fields during the long summer vacations to carry me through the balance of each year, working very hard, cutting with a cradle four acres of wheat a day, and helping to put it in the shock. But, having to buy books and paying, I think, thirty-two dollars a year for instruction, and occasionally buying acids and retorts, glass tubing, bell-glasses, flasks, etc., I had to cut down expenses for board now and then to half a dollar a week.

One winter I taught school ten miles north of Madison, earning much-needed money at the rate of twenty dollars a month, "boarding round," and keeping up my University work by studying at night. As I was not then well enough off to own a watch, I used one of my hickory clocks, not only for keeping time, but for starting the school fire in the cold mornings, and regulating class-times. I carried it out on my shoulder to the old log schoolhouse, and set it to work on a little shelf nailed to one of the knotty, bulging logs. The winter was very cold, and I had to go to the schoolhouse and start the fire about eight o'clock to warm it before the arrival of the scholars. This was a rather trying job, and one that my clock might easily be made to do. Therefore, after supper one evening I told the head of the family with whom I was boarding that if he would give me a candle I would go back to the schoolhouse and make arrangements for lighting the fire at eight o'clock, without my having to be present until time to open the school at nine. He said, "Oh! young man, you have some curious things in the school-room, but I don't think you can do that." I said, "Oh, yes! It's easy," and in hardly more than an hour the simple job was completed. I had only to place a teaspoonful of powdered chlorate of potash and sugar on the stove-hearth near a few shavings and kindling, and at the required time make the clock, through a simple arrangement, touch the inflammable mixture with a drop of sulphuric acid. Every evening after school was dismissed, I shoveled out what was left of the fire into the snow, put in a little kindling, filled up the big box stove with heavy oak wood, placed the lighting arrangement on the hearth, and set the clock to drop the acid at the hour of eight; all this requiring only a few minutes.

The first morning after I had made this simple arrangement I invited the doubting farmer to watch the old squat schoolhouse from a window that overlooked it, to see if a good smoke did not rise from the stovepipe. Sure enough, on the minute, he saw a tall column curling gracefully up through the frosty air, but instead of congratulating me on my success he solemnly shook his head and said in a hollow, lugubrious voice, "Young man, you will be setting fire to the schoolhouse." All winter long that faithful clock fire never failed, and by the time I got to the schoolhouse the stove was usually red-hot.

At the beginning of the long summer vacations I returned to the Hickory Hill farm to earn the means in the harvest-fields to continue my University course, walking all the way to save railroad fares. And although I cradled four acres of wheat a day, I made the long, hard, sweaty day's work still longer and harder by keeping up my study of plants. At the noon hour I collected a large handful, put them in water to keep them fresh, and after supper got to work on them and sat up till after midnight, analyzing and classifying, thus leaving only four hours for sleep; and by the end of the first year, after taking up botany, I knew the principal flowering plants of the region.

I received my first lesson in botany from a student by the name of Griswold, who is now County Judge of the County of Waukesha, Wisconsin. In the University he was often laughed at on account of his anxiety to instruct others, and his frequently saying with fine emphasis, "Imparting instruction is my greatest enjoyment." One memorable day in June, when I was standing on the stone steps of the north dormitory, Mr. Griswold joined me and at once began to teach. He reached up, plucked a flower from an overspreading branch of a locust tree, and, handing it to me, said, "Muir, do you know what family this tree belongs to?"

"No," I said, "I don't know anything about botany."

"Well, no matter," said he, "what is it like?"

"It's like a pea flower," I replied.

"That's right. You're right," he said, "it belongs to the Pea Family."

"But how can that be," I objected, "when the pea is a weak,

clinging, straggling herb, and the locust a big, thorny hardwood tree?"

"Yes, that is true," he replied, "as to the difference in size, but it is also true that in all their essential characters they are alike, and therefore they must belong to one and the same family. Just look at the peculiar form of the locust flower; you see that the upper petal, called the banner, is broad and erect, and so is the upper petal of the pea flower; the two lower petals, called the wings, are outspread and wing-shaped; so are those of the pea; and the two petals below the wings are united on their edges, curve upward, and form what is called the keel, and so you see are the corresponding petals of the pea flower. And now look at the stamens and pistils. You see that nine of the ten stamens have their filaments united into a sheath around the pistil, but the tenth stamen has its filament free. These are very marked characters, are they not? And, strange to say, you will find them the same in the tree and in the vine. Now look at the ovules or seeds of the locust, and you will see that they are arranged in a pod or legume like those of the pea. And look at the leaves. You see the leaf of the locust is made up of several leaflets, and so also is the leaf of the pea. Now taste the locust leaf."

I did so and found that it tasted like the leaf of the pea. Nature has used the same seasoning for both, though one is a straggling vine, the other a big tree.

"Now, surely you cannot imagine that all these similar characters are mere coincidences. Do they not rather go to show that the Creator in making the pea vine and locust tree had the same idea in mind, and that plants are not classified arbitrarily? Man has nothing to do with their classification. Nature has attended to all that, giving essential unity with boundless variety, so that the botanist has only to examine plants to learn the harmony of their relations."

This fine lesson charmed me and sent me flying to the woods and meadows in wild enthusiasm. Like everybody else I was always fond of flowers, attracted by their external beauty and purity. Now my eyes were opened to their inner beauty, all alike revealing glorious traces of the thoughts of God, and leading on and on into the infinite cosmos. I wandered away at every opportunity, making long excursions round the lakes, gathering specimens and keeping them fresh in a bucket in my room to study at night after my regular class tasks were learned; for my eyes never closed on the plant glory I had seen.

Nevertheless, I still indulged my love of mechanical inventions. I invented a desk in which the books I had to study were arranged in order at the beginning of each term. I also made a bed which set me on my feet every morning at the hour determined on, and in dark winter mornings just as the bed set me on the floor it lighted a lamp. Then, after the minutes allowed for dressing had elapsed, a click was heard and the first book to be studied was pushed up from a rack below the top of the desk, thrown open, and allowed to remain there the number of minutes required. Then the machinery closed the book and allowed it to drop back into its stall, then moved the rack forward and threw up the next in order, and so on, all the day being divided according to the times of recitation, and time required and allotted to each study. Besides this, I thought it would be a fine thing in the summer-time when the sun rose early, to dispense with the clock-controlled bed machinery, and make use of sunbeams instead. This I did simply by taking a lens out of my small spy-glass, fixing it on a frame on the sill of my bedroom window, and pointing it to the sunrise; the sunbeams focused on a thread burned it through, allowing the bed machinery to put me on my feet. When I wished to arise at any given time after sunrise, I had only to turn the pivoted frame that held the lens the requisite number of degrees or minutes. Thus I took Emerson's advice and hitched my dumping-wagon bed to a star.

I also invented a machine to make visible the growth of plants and the action of the sunlight, a very delicate contrivance, enclosed in glass. Besides this I invented a barometer and a lot of novel scientific apparatus. My room was regarded as a sort of show place by the professors, who oftentimes brought visitors to it on Saturdays and holidays. And when, some eighteen years after I had left the University, I was sauntering over the campus in time of vacation, and spoke to a man who seemed to be taking some charge of the grounds, he informed me that he was the janitor; and when I inquired what had become of Pat, the janitor in my time, and a favorite with the students, he replied that Pat was still alive and well, but now too old to do much work. And when I pointed to the dormitory room that I long ago occupied, he said: "Oh! then I know who you are," and mentioned my name. "How comes it that you know my name?" I inquired. He explained that "Pat always pointed out that room to newcomers and told long stories about the wonders that used to be in it." So long had the memory of my little inventions survived.

Although I was four years at the University, I did not take the regular course of studies, but instead picked out what I thought would be most useful to me, particularly chemistry, which opened a new world, and mathematics and physics, a little Greek and Latin, botany and geology. I was far from satisfied with what I had learned, and should have stayed longer. Anyhow I wandered away on a glorious botanical and geological excursion, which has lasted nearly fifty years and is not yet completed, always happy and free, poor and rich, without thought of a diploma or of making a name, urged on and on through endless, inspiring, Godful beauty.

From the top of a hill on the north side of Lake Mendota I gained a last wistful, lingering view of the beautiful University grounds and buildings where I had spent so many hungry and happy and hopeful days. There with streaming eyes I bade my blessed Alma Mater farewell. But I was only leaving one University for another, the Wisconsin University for the University of the Wilderness.

THE MOUNTAINS OF CALIFORNIA

CHAPTER I

THE SIERRA NEVADA

GO WHERE YOU may within the bounds of California, mountains are ever in sight, charming and glorifying every landscape. Yet so simple and massive is the topography of the State in general views, that the main central portion displays only one valley, and two chains of mountains which seem almost perfectly regular in trend and height: the Coast Range on the west side, the Sierra Nevada on the east. These two ranges coming together in curves on the north and south inclose a magnificent basin, with a level floor more than 400 miles long, and from 35 to 60 miles wide. This is the grand Central Valley of California, the waters of which have only one outlet to the sea through the Golden Gate. But with this general simplicity of features there is great complexity of hidden detail. The Coast Range, rising as a grand green barrier against the ocean, from 2000 to 8000 feet high, is composed of innumerable forest-crowned spurs, ridges, and rolling hill-waves which inclose a multitude of smaller valleys; some looking out through long, forest-lined vistas to the sea; others, with but few trees, to the Central Valley; while a thousand others yet smaller are embosomed and concealed in mild, round-browed hills, each with its own climate, soil, and productions.

Making your way through the mazes of the Coast Range to the summit of any of the inner peaks or passes opposite San Francisco, in the clear springtime, the grandest and most telling of all California landscapes is outspread before you. At your feet lies the great Central Valley glowing golden in the sunshine, extending north and south farther than the eye can reach, one smooth, flowery, lake-like bed of fertile soil. Along its eastern margin rises the mighty Sierra, miles in height, reposing like a smooth, cumulous cloud in the sunny sky, and so gloriously colored, and so luminous, it seems to be not clothed with light, but wholly composed of it, like the wall of some celestial city. Along the top, and extending a good way down, you see a pale, pearl-gray belt of snow; and below it a belt of blue and dark purple, marking the extension of the forests; and along the base of the range a broad belt of rose-purple and yellow, where lie the miner's gold-fields and the foot-hill gardens. All these colored belts blending smoothly make a wall of light ineffably fine, and as beautiful as a rainbow, yet firm as adamant.

When I first enjoyed this superb view, one glowing April day, from the summit of the Pacheco Pass, the Central Valley, but little trampled or ploughed as yet, was one furred, rich sheet of golden compositæ, and the luminous wall of the mountains shone in all its glory. Then it seemed to me the Sierra should be called not the Nevada, or Snowy Range, but the Range of Light. And after ten years spent in the heart of it, rejoicing and wondering, bathing in its glorious floods of light, seeing the sunbursts of morning among the icy peaks, the noonday radiance on the trees and rocks and snow, the flush of the alpenglow, and a thousand dashing waterfalls with their marvelous abundance of irised spray, it still seems to me above all others the Range of Light, the most divinely beautiful of all the mountain-chains I have ever seen.

The Sierra is about 500 miles long, 70 miles wide, and from 7000 to nearly 15,000 feet high. In general views no mark of man is visible on it, nor anything to suggest the richness of the life it cherishes, or the depth and grandeur of its sculpture. None of its magnificent forest-crowned ridges rises much above the general level to publish its wealth. No great valley or lake is seen, or river, or group of well-marked features of any kind, standing out in distinct pictures. Even the summit-peaks, so clear and high in the sky, seem comparatively smooth and featureless. Nevertheless, glaciers are still at work in the shadows of the peaks, and thousands of lakes and meadows shine and bloom beneath them, and the whole range is furrowed with cañons to a depth of from 2000 to 5000 feet, in which once flowed majestic glaciers, and in which now flow and sing a band of beautiful rivers.

Though of such stupendous depth, these famous cañons are not raw, gloomy, jagged-walled gorges, savage and inaccessible. With rough passages here and there they still make delightful pathways for the mountaineer, conducting from the fertile lowlands to the highest icy fountains, as a kind of mountain streets full of charming life and light, graded and sculptured by the ancient glaciers, and presenting, throughout all their courses, a rich variety of novel and attractive scenery, the most attractive that has yet been discovered in the mountain-ranges of the world.

In many places, especially in the middle region of the western flank of the range, the main cañons widen into spacious valleys or parks, diversified like artificial landscape-gardens, with charming groves and meadows, and thickets of blooming bushes, while the lofty, retiring walls, infinitely varied in form and sculpture, are fringed with ferns, flowering-plants of many species, oaks, and evergreens, which find anchorage on a thousand narrow steps and benches; while the whole is enlivened and made glorious with rejoicing streams that come dancing and foaming over the sunny brows of the cliffs to join the shining river that flows in tranquil beauty down the middle of each one of them.

The walls of these park valleys of the Yosemite kind are made up of rocks mountains in size, partly separated from each other by narrow gorges and side-cañons; and they are so sheer in front, and so compactly built together on a level floor, that, comprehensively seen, the parks they inclose look like immense halls or temples lighted from above. Every rock seems to glow with life. Some lean back in majestic repose; others, absolutely sheer, or nearly so, for thousands of feet, advance their brows in thoughtful attitudes beyond their companions, giving welcome to storms and calms alike, seemingly conscious yet heedless of everything going on about them, awful in stern majesty, types of permanence, yet associated with beauty of the frailest and most fleeting forms; their feet set in pine-groves and gay emerald meadows, their brows in the sky; bathed in light, bathed in floods of singing water, while snow-clouds, avalanches, and the winds shine and surge and wreathe about them as the years go by, as if into these mountain mansions Nature had taken pains to gather her choicest treasures to draw her lovers into close and confiding communion with her.

Here, too, in the middle region of deepest cañons are the grandest forest-trees, the Sequoia, king of conifers, the noble Sugar and Yellow Pines, Douglas Spruce, Libocedrus, and the Silver Firs, each a giant of its kind, assembled together in one and the same forest, surpassing all other coniferous forests in the world, both in the number of its species and in the size and beauty of its trees. The winds flow in melody through their colossal spires, and they are vocal everywhere with the songs of birds and running water. Miles of fragrant ceanothus and manzanita bushes bloom beneath them, and lily gardens and meadows, and damp, ferny glens in endless variety of fragrance and color, compelling the admiration of every observer.

Sweeping on over ridge and valley, these noble trees extend a continuous belt from end to end of the range, only slightly interrupted by sheer-walled cañons at intervals of about fifteen and twenty miles. Here the great burly brown bears delight to roam, harmonizing with the brown boles of the trees beneath which they feed. Deer, also, dwell here, and find food and shelter in the ceanothus tangles, with a multitude of smaller people. Above this region of giants, the trees grow smaller until the utmost limit of the timber line is reached on the stormy mountain-slopes at a height of from ten to twelve thousand feet above the sea, where the Dwarf Pine is so lowly and hard beset by storms and heavy snow, it is pressed into flat tangles, over the tops of which we may easily walk. Below the main forest belt the trees likewise diminish in size, frost and burning drouth repressing and blasting alike.

The rose-purple zone along the base of the range comprehends nearly all the famous gold region of California. And here it was that miners from every country under the sun assembled in a wild, torrent-like rush to seek their fortunes. On the banks of every river, ravine, and gully they have left their marks. Every gravel- and boulder-bed has been desperately riddled over and over again. But in this region the pick and shovel, once wielded with savage enthusiasm, have been laid away, and only quartz-mining is now being carried on to any considerable extent. The zone in general is made up of low, tawny, waving foot-hills, roughened here and there with brush and trees, and outcropping masses of slate, colored gray and red with lichens. The smaller masses of slate, rising abruptly from the dry, grassy sod in leaning slabs, look like ancient tombstones in a deserted burying-ground. In early spring, say from February to April, the whole of this foot-hill belt is a paradise of bees and flowers. Refreshing rains then fall freely, birds are busy building their nests, and the sunshine is balmy and delightful. But by the end of May the soil, plants, and sky seem to have been baked in an oven. Most of the plants crumble to dust beneath the foot, and the ground is full of cracks; while the thirsty traveler gazes with eager longing through the burning glare to the snowy summits looming like hazy clouds in the distance.

The trees, mostly *Quercus douglasii* and *Pinus sabiniana*, thirty to forty feet high, with thin, pale-green foliage, stand far apart and cast but little shade. Lizards glide about on the rocks enjoying a constitution that no drouth can dry, and ants in amazing numbers, whose tiny sparks of life seem to burn the brighter with the increasing heat, ramble industriously in long trains in search of food. Crows, ravens, magpies—friends in distress—gather on the ground beneath the best shade-trees, panting with drooping wings and bills wide open, scarce a note from any of them during the midday hours. Quails, too, seek the shade during the heat of the day about tepid pools in the channels of the larger mid-river streams. Rabbits scurry from thicket to thicket among the ceanothus bushes, and occasionally a long-eared hare is seen cantering gracefully across the wider openings. The nights are calm and dewless during the summer, and a thousand voices proclaim the abundance of life, notwithstanding the desolating effect of dry sunshine on the plants and larger animals. The hylas make a delightfully pure and tranquil music after sunset; and coyotes, the little, despised dogs of the wilderness, brave, hardy fellows, looking like withered wisps of hay, bark in chorus for hours. Mining-towns, most of them dead, and a few living ones with bright bits of cultivation about them, occur at long intervals along the belt, and cottages covered with climbing roses, in the midst of orange and peach orchards, and sweet-scented hay-fields in fertile flats where water for irrigation may be had. But they are mostly far apart, and make scarce any mark in general views.

Every winter the High Sierra and the middle forest region get snow in glorious abundance, and even the foot-hills are at times whitened. Then all the range looks like a vast beveled wall of purest marble. The rough places are then made smooth, the death and decay of the year is covered gently and kindly, and the ground seems as clean as the sky. And though silent in its flight from the clouds, and when it is taking its place on rock, or tree, or grassy meadow, how soon the gentle snow finds a voice! Slipping from the heights, gathering in avalanches, it booms and roars like thunder, and makes a glorious show as it sweeps down the mountain-side, arrayed in long, silken streamers and wreathing, swirling films of crystal dust.

The north half of the range is mostly covered with floods of lava, and dotted with volcanoes and craters, some of them recent and perfect in form, others in various stages of decay. The south half is composed of granite nearly from base to summit, while a considerable number of peaks, in the middle of the range, are capped with metamorphic slates, among which are Mounts Dana and Gibbs to the east of Yosemite Valley. Mount Whitney, the culminating point of the range near its southern extremity, lifts its helmet-shaped crest to a height of nearly 14,700 feet. Mount Shasta, a colossal volcanic cone, rises to a height of 14,440 feet at the northern extremity, and forms a noble landmark for all the surrounding region within a radius of a hundred miles. Residual masses of volcanic rocks occur throughout most of the granitic southern portion also, and a considerable number of old volcanoes on the flanks, especially along the eastern base of the range near Mono Lake and southward. But it is only to the northward that the entire range, from base to summit, is covered with lava.

From the summit of Mount Whitney only granite is seen. Innumerable peaks and spires but little lower than its own storm-beaten crags rise in groups like forest-trees, in full view, segregated by cañons of tremendous depth and ruggedness. On Shasta nearly every feature in the vast view speaks of the old volcanic fires. Far to the northward, in Oregon, the icy volcanoes of Mount Pitt and the Three Sisters rise above the dark evergreen woods. Southward innumerable smaller craters and cones are distributed along the axis of the range and on each flank. Of these, Lassen's Butte is the highest, being nearly 11,000 feet above sea-level. Miles of its flanks are reeking and bubbling with hot springs, many of them so boisterous and sulphurous they seem ever ready to become spouting geysers like those of the Yellowstone.

The Cinder Cone near marks the most recent volcanic eruption in the Sierra. It is a symmetrical truncated cone about 700 feet high, covered with gray cinders and ashes, and has a regular unchanged crater on its summit, in which a few small Two-leaved Pines are growing. These show that the age of the cone is not less than eighty years. It stands between two lakes, which a short time ago were one. Before the cone was built, a flood of rough vesicular lava was poured into the lake, cutting it in two, and, overflowing its banks, the fiery flood advanced into the pine-woods, overwhelming the trees in its way, the charred ends of some of which may still be seen projecting from beneath the snout of the lava-stream where it came to rest. Later still there was an eruption of ashes and loose obsidian cinders, probably from the same vent, which, besides forming the Cinder Cone, scattered a heavy shower over the surrounding woods for miles to a depth of from six inches to several feet.

The history of this last Sierra eruption is also preserved in the traditions of the Pitt River Indians. They tell of a fearful time of darkness, when the sky was black with ashes and smoke that threatened every living thing with death, and that when at length the sun appeared once more it was red like blood.

Less recent craters in great numbers roughen the adjacent region; some of them with lakes in their throats, others overgrown with trees and flowers, Nature in these old hearths and firesides having literally given beauty for ashes. On the northwest side of Mount Shasta there is a subordinate cone about 3000 feet below the summit, which has been active subsequent to the breaking up of the main ice-cap that once covered the mountain, as is shown by its comparatively unwasted crater and the streams of unglaciated lava radiating from it. The main summit is about a mile and a half in diameter, bounded by small crumbling peaks and ridges, among which we seek in vain for the outlines of the ancient crater.

These ruinous masses, and the deep glacial grooves that flute the sides of the mountain, show that it has been considerably lowered and wasted by ice; how much we have no sure means of knowing. Just below the extreme summit hot sulphurous gases and vapor issue from irregular fissures, mixed with spray derived from melting snow, the last feeble expression of the mighty force that built the mountain. Not in one great convulsion was Shasta given birth. The crags of the summit and the sections exposed by the glaciers down the sides display enough of its internal framework to prove that comparatively long periods of quiescence intervened between many distinct eruptions, during which the cooling lavas ceased to flow, and became permanent additions to the bulk of the growing mountain. With alternate haste and deliberation eruption succeeded eruption till the old volcano surpassed even its present sublime height.

Standing on the icy top of this, the grandest of all the fire-mountains of the Sierra, we can hardly fail to look forward to its next eruption. Gardens, vineyards, homes have been planted confidingly on the flanks of volcanoes which, after remaining steadfast for ages, have suddenly blazed into violent action, and poured forth overwhelming floods of fire. It is known that more than a thousand years of cool calm have intervened between violent eruptions. Like gigantic geysers spouting molten rock instead of water, volcanoes work and rest, and we have no sure means of knowing whether they are dead when still, or only sleeping.

Along the western base of the range a telling series of sedimentary rocks containing the early history of the Sierra are now being studied. But leaving for the present these first chapters, we see that only a very short geological time ago, just before the coming on of that winter of winters called the glacial period, a vast deluge of molten rocks poured from many a chasm and crater on the flanks and summit of the range, filling lake basins and river channels, and obliterating nearly every existing feature on the northern portion. At length these all-destroying floods ceased to flow. But while the great volcanic cones built up along the axis still burned and smoked, the whole Sierra passed under the domain of ice and snow. Then over the bald, featureless, fire-blackened mountains, glaciers began to crawl, covering them from the summits to the sea with a mantle of ice; and then with infinite deliberation the work went on of sculpturing the range anew. These mighty agents of erosion, halting never through unnumbered centuries, crushed and ground the flinty lavas and granites beneath their crystal folds, wasting and building until in the fullness of time the Sierra was born again, brought to light nearly as we behold it today, with glaciers and snow-crushed pines at the top of the range, wheat-fields and orange-groves at the foot of it.

This change from icy darkness and death to life and beauty was slow, as we count time, and is still going on, north and south, over all the world wherever glaciers exist, whether in the form of distinct rivers, as in Switzerland, Norway, the mountains of Asia, and the Pacific Coast; or in continuous mantling folds, as in portions of Alaska, Greenland, Franz-Joseph-Land, Nova Zembla, Spitzbergen, and the lands about the South Pole. But in no country, as far as I know, may these majestic changes be studied to better advantage than in the plains and mountains of California.

Toward the close of the glacial period, when the snow-clouds became less fertile and the melting waste of sunshine became greater, the lower folds of the ice-sheet in California, discharging fleets of icebergs into the sea, began to shallow and recede from the lowlands, and then move slowly up the flanks of the Sierra in compliance with the changes of climate. The great white mantle on the mountains broke up into a series of glaciers more or less distinct and river-like, with many tributaries, and these again were melted and divided into still smaller glaciers, until now only a few of the smallest residual topmost branches of the grand system exist on the cool slopes of the summit peaks.

Plants and animals, biding their time, closely followed the retiring ice, bestowing quick and joyous animation on the new-born landscapes. Pine-trees marched up the sun-warmed moraines in long, hopeful files, taking the ground and establishing themselves as soon as it was ready for them; brown-spiked sedges fringed the shores of the new-born lakes; young rivers roared in the abandoned channels of the glaciers; flowers bloomed around the feet of the great burnished domes,—while with quick fertility mellow beds of soil, settling and warming, offered food to multitudes of Nature's waiting children, great and small, animals as well as plants; mice, squirrels, marmots, deer, bears, elephants, etc. The ground burst into bloom with magical rapidity, and the young forests into bird-song: life in every form warming and sweetening and growing richer as the years passed away over the mighty Sierra so lately suggestive of death and consummate desolation only.

It is hard without long and loving study to realize the magnitude of the work done on these mountains during the last glacial period by glaciers, which are only streams of closely compacted snow-crystals. Careful study of the phenomena presented goes to show that the pre-glacial condition of the range was comparatively simple: one vast wave of stone in which a thousand mountains, domes, cañons, ridges, etc., lay concealed. And in the development of these Nature chose for a tool not the earthquake or lightning to rend and split asunder, not the stormy torrent or eroding rain, but the tender snow-flowers noiselessly falling through unnumbered centuries, the offspring of the sun and sea. Laboring harmoniously in united strength they crushed and ground and wore away the rocks in their march, making vast beds of soil, and at the same time developed and fashioned the landscapes into the delightful variety of hill and dale and lordly mountain that mortals call beauty. Perhaps more than a mile in average depth has the range been thus degraded during the last glacial period,—a quantity of mechanical work almost inconceivably great. And our admiration must be excited again and again as we toil and study and learn that this vast job of rockwork, so far-reaching in its influences, was done by agents so fragile and small as are these flowers of the mountain clouds. Strong only by force of numbers, they carried away entire mountains, particle by particle, block by block, and cast them into the sea; sculptured, fashioned, modeled all the range, and developed its predestined beauty. All these new Sierra landscapes were evidently predestined, for the physical structure of the rocks on which the features of the scenery depend was acquired while they lay at least a mile deep below the pre-glacial surface. And it was while these features were taking form in the depths of the range, the particles of the rocks marching to their appointed places in the dark with reference to the coming beauty, that the particles of icy vapor in the sky marching to the same music assembled to bring them to the light. Then, after their grand task was done, these bands of snow-flowers, these mighty glaciers, were melted and removed as if of no more importance than dew destined to last but an hour. Few, however, of Nature's agents have left monuments so noble and enduring as they. The great granite domes a mile high, the cañons as deep, the noble peaks, the Yosemite valleys, these, and indeed nearly all other features of the Sierra scenery, are glacier monuments.

Contemplating the works of these flowers of the sky, one may easily fancy them endowed with life: messengers sent down to work in the mountain mines on errands of divine love. Silently flying through the darkened air, swirling, glinting, to their appointed places, they seem to have taken counsel together, saying, "Come, we are feeble; let us help one another. We are many, and together we will be strong. Marching in close, deep ranks, let us roll away the stones from these mountain sepulchers, and set the landscapes free. Let us uncover these clustering domes. Here let us carve a lake basin; there, a Yosemite Valley; here, a channel for a river with fluted steps and brows for the plunge of songful cataracts. Yonder let us spread broad sheets of soil, that man and beast may be fed; and here pile trains of boulders for pines and giant Sequoias. Here make ground for a meadow; there, for a garden and grove, making it smooth and fine for small daisies and violets and beds of heathy bryanthus, spicing it well with crystals, garnet feldspar, and zircon." Thus and so on it has oftentimes seemed to me sang and planned and labored the hearty snow-flower crusaders; and nothing that I can write can possibly exaggerate the grandeur and beauty of their work. Like morning mist they have vanished in sunshine, all save the few small companies that still linger on the coolest mountainsides, and, as residual glaciers, are still busily at work completing the last of the lake basins, the last beds of soil, and the sculpture of some of the highest peaks.

CHAPTER II

THE GLACIERS

OF THE SMALL residual glaciers mentioned in the preceding chapter, I have found sixty-five in that portion of the range lying between latitude 36° 30′ and 39°. They occur singly or in small groups on the north sides of the peaks of the High Sierra, sheltered beneath broad frosty shadows, in amphitheaters of their own making, where the snow, shooting down from the surrounding heights in avalanches, is most abundant. Over two thirds of the entire number lie between latitude 37° and 38°, and form the highest fountains of the San Joaquin, Merced, Tuolumne, and Owen's rivers.

The glaciers of Switzerland, like those of the Sierra, are mere wasting remnants of mighty ice-floods that once filled the great valleys and poured into the sea. So, also, are those of Norway, Asia, and South America. Even the grand continuous mantles of ice that still cover Greenland, Spitzbergen, Nova Zembla, Franz-Joseph-Land, parts of Alaska, and the south polar region are shallowing and shrinking. Every glacier in the world is smaller than it once was. All the world is growing warmer, or the crop of snow-flowers is diminishing. But in contemplating the condition of the glaciers of the world, we must bear in mind while trying to account for the changes going on that the same sunshine that wastes them builds them. Every glacier records the expenditure of an enormous amount of sun-heat in lifting the vapor for the snow of which it is made from the ocean to the mountains, as Tyndall strikingly shows.

The number of glaciers in the Alps, according to the Schlagintweit brothers, is 1100, of which 100 may be regarded as primary, and the total area of ice, snow, and *névé* is estimated at 1177 square miles, or an average for each glacier of little more than one square mile. On the same authority, the average height above sea-level at which they melt is about 7414 feet. The Grindelwald glacier descends below 4000 feet, and one of the Mont Blanc glaciers reaches nearly as low a point. One of the largest of the Himalaya glaciers on the head waters of the Ganges does not, according to Captain Hodgson, descend below 12,914 feet. The largest of the Sierra glaciers on Mount Shasta descends to within 9500 feet of the level of the sea, which, as far as I have observed, is the lowest point reached by any glacier within the bounds of California, the average height of all being not far from 11,000 feet.

The changes that have taken place in the glacial conditions of the Sierra from the time of greatest extension is well illustrated by the series of glaciers of every size and form extending along the mountains of the coast to Alaska. A general exploration of this instructive region shows that to the north of California, through Oregon and Washington, groups of active glaciers still exist on all the high volcanic cones of the Cascade Range,—Mount Pitt, the Three Sisters, Mounts Jefferson, Hood, St. Helens, Adams, Rainier, Baker, and others,—some of them of considerable size, though none of them approach the sea. Of these mountains Rainier, in Washington, is the highest and iciest. Its dome-like summit, between 14,000 and 15,000 feet high, is capped with ice, and eight glaciers, seven to twelve miles long, radiate from it as a center, and form the sources of the principal streams of the State. The lowest-descending of this fine group flows through beautiful forests to within 3500 feet of the sea-level, and sends forth a river laden with glacier mud and sand. On through British Columbia and southeastern Alaska the broad, sustained mountain-chain, extending along the coast, is generally glacier-bearing. The upper branches of nearly all the main cañons and fiords are occupied by glaciers, which gradually increase in size, and descend lower until the high region between Mount Fairweather and Mount St. Elias is reached, where a considerable number discharge into the waters of the ocean. This is preëminently the ice-land of Alaska and of the entire Pacific Coast.

Northward from here the glaciers gradually diminish in size and thickness, and melt at higher levels. In Prince William Sound and Cook's Inlet many fine glaciers are displayed, pouring from the surrounding mountains; but to the north of latitude 62° few, if any, glaciers remain, the ground being mostly low and the snowfall light. Between latitude 56° and 60° there are probably more than 5000 glaciers, not counting the smallest. Hundreds of the largest size descend through the forests to the level of the sea, or near it, though as far as my own observations have reached, after a pretty thorough examination of the region, not more than twenty-five discharge icebergs into the sea. All the long high-walled fiords into which these great glaciers of the first class flow are of course crowded with icebergs of every conceivable form, which are detached with thundering noise at intervals of a few minutes from an imposing ice-wall that is thrust forward into deep water. But these Pacific Coast icebergs are small as compared with those of Greenland and the Antarctic region, and only a few of them escape from the intricate system of channels, with which this portion of the coast is fringed, into the open sea. Nearly all of them are swashed and drifted by wind and tide back and forth in the fiords until finally melted by the ocean water, the sunshine, the warm winds, and the copious rains of summer. Only one glacier on the coast, observed by Prof. Russell, discharges its bergs directly into the open sea, at Icy Cape, opposite Mount St. Elias. The southernmost of the glaciers that reach the sea occupies a narrow, picturesque fiord about twenty miles to the northwest of the mouth of the Stikeen River, in latitude 56° 50′. The fiord is called by the natives "Hutli," or Thunder Bay, from the noise made by the discharge of the icebergs. About one degree farther north there are four of these complete glaciers, discharging at the heads of the long arms of Holkam Bay. At the head of the Tahkoo Inlet, still farther north, there is one; and at the head and around the sides of Glacier Bay, trending in a general northerly direction from Cross Sound in latitude 58° to 59°, there are seven of these complete glaciers pouring bergs into the bay and its branches, and keeping up an eternal thundering. The largest of this group, the Muir, has upward of 200 tributaries, and a width below the confluence of the main tributaries of about twenty-five miles. Between the west side of this icy bay and the ocean all the ground, high and low, excepting the peaks of the Fairweather Range, is covered with a mantle of ice from 1000 to probably 3000 feet thick, which discharges by many distinct mouths.

This fragmentary ice-sheet, and the immense glaciers about Mount St. Elias, together with the multitude of separate river-like glaciers that load the slopes of the coast mountains, evidently once formed part of a continuous ice-sheet that flowed over all the region hereabouts, and only a comparatively short time ago extended as far southward as the mouth of the Strait of Juan de Fuca, probably farther. All the islands of the Alexander Archipelago, as well as the headlands and promontories of the mainland, display telling traces of this great mantle that are still fresh and unmistakable. They all have the forms of the greatest strength with reference to the action of a vast rigid press of oversweeping ice from the north and northwest, and their surfaces have a smooth, rounded, overrubbed appearance, generally free from angles. The intricate labyrinth of canals, channels, straits, passages, sounds, narrows, etc., between the islands, and extending into the mainland, of course manifest in their forms and trends and general characteristics the same subordination to the grinding action of universal glaciation as to their origin, and differ from the islands and banks of the fiords only in being portions of the pre-glacial margin of the continent more deeply eroded, and therefore covered by the ocean waters which flowed into them as the ice was melted out of them. The formation and extension of fiords in this manner is still going on, and may be witnessed in many places in Glacier Bay, Yakutat Bay, and adjacent regions. That the domain of the sea is being extended over the land by the wearing away of its shores, is well known, but in these icy regions of Alaska, and even as far south as Vancouver Island, the coast rocks have been so short a time exposed to wave-action they are but little wasted as yet. In these regions the extension of the sea effected by its own action in post-glacial time is scarcely appreciable as compared with that effected by ice-action.

Traces of the vanished glaciers made during the period of greater extension abound on the Sierra as far south as latitude 36°. Even the polished rock surfaces, the most evanescent of glacial records, are still found in a wonderfully perfect state of preservation on the upper half of the middle portion of the range, and form the most striking of all the glacial phenomena. They occur in large irregular patches in the summit and middle regions, and though they have been subjected to the action of the weather with its corroding storms for thousands of years, their mechanical excellence is such that they still reflect the sunbeams like glass, and attract the attention of every observer. The attention of the mountaineer is seldom arrested by moraines, however regular and high they may be, or by cañons, however deep, or by rocks, however noble in form and sculpture; but he stoops and rubs his hands admiringly on the shining surfaces and tries hard to account for their mysterious smoothness. He has seen the snow descending in avalanches, but concludes this cannot be the work of snow, for he finds it where no avalanches occur. Nor can water have done it, for he sees this smoothness glowing on the sides and tops of the highest domes. Only the winds of all the agents he knows seem capable of flowing in the directions indicated by the scoring. Indians, usually so little curious about geological phenomena, have come to me occasionally and asked me, "What makeum the ground so smooth at Lake Tenaya?" Even horses and dogs gaze wonderingly at the strange brightness of the ground, and smell the polished spaces and place their feet cautiously on them when they come to them for the first time, as if afraid of sinking. The most perfect of the polished pavements and walls lie at an elevation of from 7000 to 9000 feet above the sea, where the rock is compact silicious granite. Small dim patches may be found as low as 3000 feet on the driest and most enduring portions of sheer walls with a southern exposure, and on compact swelling bosses partially protected from rain by a covering of large boulders. On the north half of the range the striated and polished surfaces are less common, not only because this part of the chain is lower, but because the surface rocks are chiefly porous lavas subject to comparatively rapid waste. The ancient moraines also, though well preserved on most of the south half of the range, are nearly obliterated to the northward, but their material is found scattered and disintegrated.

A similar blurred condition of the superficial records of glacial action obtains throughout most of Oregon, Washington, British Columbia, and Alaska, due in great part to the action of excessive moisture. Even in southeastern Alaska, where the most extensive glaciers on the continent are, the more evanescent of the traces of their former greater extension, though comparatively recent, are more obscure than those of the ancient California glaciers where the climate is drier and the rocks more resisting.

These general views of the glaciers of the Pacific Coast will enable my readers to see something of the changes that have taken place in California, and will throw light on the residual glaciers of the High Sierra.

Prior to the autumn of 1871 the glaciers of the Sierra were unknown. In October of that year I discovered the Black Mountain Glacier in a shadowy amphitheater between Black and Red Mountains, two of the peaks of the Merced group. This group is the highest portion of a spur that straggles out from the main axis of the range in the direction of Yosemite Valley. At the time of this interesting discovery I was exploring the *névé* amphitheaters of the group, and tracing the courses of the ancient glaciers that once poured from its ample fountains through the Illilouette Basin and the Yosemite Valley, not expecting to find any active glaciers so far south in the land of sunshine.

Beginning on the northwestern extremity of the group, I explored the chief tributary basins in succession, their moraines, roches moutonnées, and splendid glacier pavements, taking them in regular succession without any reference to the time consumed in their study. The monuments of the tributary that poured its ice from between Red and Black Mountains I found to be the most interesting of them all; and when I saw its magnificent moraines extending in majestic curves from the spacious amphitheater between the mountains, I was exhilarated with the work that lay before me. It was one of the golden days of the Sierra Indian summer, when the rich sunshine glorifies every landscape however rocky and cold, and suggests anything rather than glaciers. The path of the vanished glacier was warm now, and shone in many places as if washed with silver. The tall pines growing on the moraines stood transfigured in the glowing light, the poplar groves on the levels of the basin were masses of orange-yellow, and the late-blooming goldenrods added gold to gold. Pushing on over my rosy glacial highway, I passed lake after lake set in solid basins of granite, and many a thicket and meadow watered by a stream that issues from the amphitheater and links the lakes together; now wading through plushy bogs knee-deep in yellow and purple sphagnum; now passing over bare rock. The main lateral moraines that bounded the view on either hand are from 100 to nearly 200 feet high, and about as regular as artificial embankments, and covered with a superb growth of Silver Fir and Pine. But this garden and forest luxuriance was speedily left behind. The trees were dwarfed as I ascended; patches of the alpine bryanthus and cassiope began to appear, and arctic willows pressed into flat carpets by the winter snow. The lakelets, which a few miles down the valley were so richly embroidered with flowery meadows, had here, at an elevation of 10,000 feet, only small brown mats of carex, leaving bare rocks around more than half their shores. Yet amid this alpine suppression the Mountain Pine bravely tossed his storm-beaten branches on the ledges and buttresses of Red Mountain, some specimens being over 100 feet high, and 24 feet in circumference, seemingly as fresh and vigorous as the giants of the lower zones.

Evening came on just as I got fairly within the portal of the main amphitheater. It is about a mile wide, and a little less than two miles long. The crumbling spurs and battlements of Red Mountain bound it on the north, the somber, rudely sculptured precipices of Black Mountain on the south, and a hacked, splintery *col*, curving around from mountain to mountain, shuts it in on the east.

I chose a camping-ground on the brink of one of the lakes where a thicket of Hemlock Spruce sheltered me from the night wind. Then, after making a tin-cupful of tea, I sat by my camp-fire reflecting on the grandeur and significance of the glacial records I had seen. As the night advanced the mighty rock walls of my mountain mansion seemed to come nearer, while the starry sky in glorious brightness stretched across like a ceiling from wall to wall, and fitted closely down into all the spiky irregularities of the summits. Then, after a long fireside rest and a glance at my note-book, I cut a few leafy branches for a bed, and fell into the clear, death-like sleep of the tired mountaineer.

Early next morning I set out to trace the grand old glacier that had done so much for the beauty of the Yosemite region back to its farthest fountains, enjoying the charm that every explorer feels in Nature's untrodden wildernesses. The voices of the mountains were still asleep. The wind scarce stirred the pine-needles. The sun was up, but it was yet too cold for the birds and the few burrowing animals that dwell here. Only the stream, cascading from pool to pool, seemed to be wholly awake. Yet the spirit of the opening day called to action. The sunbeams came streaming gloriously through the jagged openings of the *col*, glancing on the burnished pavements and lighting the silvery lakes, while every sun-touched rock burned white on its edges like melting iron in a furnace. Passing round the north shore of my camp lake I followed the central stream past many cascades from lakelet to lakelet. The scenery became more rigidly arctic, the Dwarf Pines and Hemlocks disappeared, and the stream was bordered with icicles. As the sun rose higher rocks were loosened on shattered portions of the cliffs, and came down in rattling avalanches, echoing wildly from crag to crag.

The main lateral moraines that extend from the jaws of the amphitheater into the Illilouette Basin are continued in straggling masses along the walls of the amphitheater, while separate boulders, hundreds of tons in weight, are left stranded here and there out in the middle of the channel. Here, also, I observed a series of small terminal moraines ranged along the south wall of the amphitheater, corresponding in size and form with the shadows cast by the highest portions. The meaning of this correspondence between moraines and shadows was afterward made plain. Tracing the stream back to the last of its chain of lakelets, I noticed a deposit of fine gray mud on the bottom except where the force of the entering current had prevented its settling. It looked like the mud worn from a grindstone, and I at once suspected its glacial origin, for the stream that was carrying it came gurgling out of the base of a raw moraine that seemed in process of formation. Not a plant or weather-stain was visible on its rough, unsettled surface. It is from 60 to over 100 feet high, and plunges forward at an angle of 38°. Cautiously picking my way, I gained the top of the moraine and was delighted to see a small but well characterized glacier swooping down from the gloomy precipices of Black Mountain in a finely graduated curve to the moraine on which I stood. The compact ice appeared on all the lower portions of the glacier, though gray with dirt and stones embedded in it. Farther up the ice disappeared beneath coarse granulated snow. The surface of the glacier was further characterized by dirt bands and the outcropping edges of the blue veins, showing the laminated structure of the ice. The uppermost crevasse, or "bergschrund," where the *névé* was attached to the mountain, was from 12 to 14 feet wide, and was bridged in a few places by the remains of snow avalanches. Creeping along the edge of the schrund, holding on with benumbed fingers, I discovered clear sections where the bedded structure was beautifully revealed. The surface snow, though sprinkled with stones shot down from the cliffs, was in some places almost pure, gradually becoming crystalline and changing to whitish porous ice of different shades of color, and this again changing at a depth of 20 or 30 feet to blue ice, some of the ribbon-like bands of which were nearly pure, and blended with the paler bands in the most gradual and delicate manner imaginable. A series of rugged zigzags enabled me to make my way down into the weird underworld of the crevasse. Its chambered hollows were hung with a multitude of clustered icicles, amid which pale, subdued light pulsed and shimmered with indescribable loveliness. Water dripped and tinkled overhead, and from far below came strange, solemn murmurings from currents that were feeling their way through veins and fissures in the dark. The chambers of a glacier are perfectly enchanting, notwithstanding one feels out of place in their frosty beauty. I was soon cold in my shirtsleeves, and the leaning wall threatened to engulf me; yet it was hard to leave the delicious music of the water and the lovely light. Coming again to the surface, I noticed boulders of every size on their journeys to the terminal moraine—journeys of more than a hundred years, without a single stop, night or day, winter or summer.

The sun gave birth to a network of sweet-voiced rills that ran gracefully down the glacier, curling and swirling in their shining channels, and cutting clear sections through the porous surface-ice into the solid blue, where the structure of the glacier was beautifully illustrated.

The series of small terminal moraines which I had observed in the morning, along the south wall of the amphitheater, correspond in every way with the moraine of this glacier, and their distribution with reference to shadows was now understood. When the climatic changes came on that caused the melting and retreat of the main glacier that filled the amphitheater, a series of residual glaciers were left in the cliff shadows, under the protection of which they lingered, until they formed the moraines we are studying. Then, as the snow became still less abundant, all of them vanished in succession, except the one just described; and the cause of its longer life is sufficiently apparent in the greater area of snow-basin it drains, and its more perfect protection from wasting sunshine. How much longer this little glacier will last depends, of course, on the amount of snow it receives from year to year, as compared with melting waste.

After this discovery, I made excursions over all the High Sierra, pushing my explorations summer after summer, and discovered that what at first sight in the distance looked like extensive snow-fields, were in great part glaciers, busily at work completing the sculpture of the summit-peaks so grandly blocked out by their giant predecessors.

On August 21, I set a series of stakes in the Maclure Glacier, near Mount Lyell, and found its rate of motion to be little more than an inch a day in the middle, showing a great contrast to the Muir Glacier in Alaska, which, near the front, flows at a rate of from five to ten feet in twenty-four hours.

Mount Shasta has three glaciers, but Mount Whitney, although it is the highest mountain in the range, does not now cherish a single glacier. Small patches of lasting snow and ice occur on its northern slopes, but they are shallow, and present no well-marked evidence of glacial motion. Its sides, however, are scored and polished in many places by the action of its ancient glaciers that flowed east and west as tributaries of the great glaciers that once filled the valleys of the Kern and Owen's rivers.

CHAPTER III

THE SNOW

THE FIRST SNOW that whitens the Sierra, usually falls about the end of October or early in November, to a depth of a few inches, after months of the most charming Indian summer weather imaginable. But in a few days, this light covering mostly melts from the slopes exposed to the sun and causes but little apprehension on the part of mountaineers who may be lingering among the high peaks at this time. The first general winter storm that yields snow that is to form a lasting portion of the season's supply, seldom breaks on the mountains before the end of November. Then, warned by the sky, cautious mountaineers, together with the wild sheep, deer, and most of the birds and bears, make haste to the lowlands or foot-hills; and burrowing marmots, mountain beavers, woodrats, and such people go into winter quarters, some of them not again to see the light of day until the general awakening and resurrection of the spring in June or July. The first heavy fall is usually from about two to four feet in depth. Then, with intervals of splendid sunshine, storm succeeds storm, heaping snow on snow, until thirty to fifty feet has fallen. But on account of its settling and compacting, and the almost constant waste from melting and evaporation, the average depth actually found at any time seldom exceeds ten feet in the forest region, or fifteen feet along the slopes of the summit peaks.

Even during the coldest weather evaporation never wholly ceases, and the sunshine that abounds between the storms is sufficiently powerful to melt the surface more or less through all the winter months. Waste from melting also goes on to some extent on the bottom from heat stored up in the rocks, and given off slowly to the snow in contact with them, as is shown by the rising of the streams on all the higher regions after the first snowfall, and their steady sustained flow all winter.

The greater portion of the snow deposited around the lofty summits of the range falls in small crisp flakes and broken crystals, or when accompanied by strong winds and low temperature, the crystals, instead of being locked together in their fall to form tufted flakes, are beaten and broken into meal and fine dust. But down in the forest region the greater portion comes gently to the ground, light and feathery, some of the flakes in mild weather being nearly an inch in diameter, and it is evenly distributed and kept from drifting to any great extent by the shelter afforded by the large trees. Every tree during the progress of gentle storms is loaded with fairy bloom at the coldest and darkest time of year, bending the branches, and hushing every singing needle. But as soon as the storm is over, and the sun shines, the snow at once begins to shift and settle and fall from the branches in miniature avalanches, and the white forest soon becomes green again. The snow on the ground also settles and thaws every bright day, and freezes at night, until it becomes coarsely granulated, and loses every trace of its rayed crystalline structure, and then a man may walk firmly over its frozen surface as if on ice. The forest region up to an elevation of 7000 feet is usually in great part free from snow in June, but at this time the higher regions are still heavy-laden, and are not touched by spring weather to any considerable extent before the middle or end of July.

One of the most striking effects of the snow on the mountains is the burial of the rivers and small lakes.

> As the snaw fa's in the river
> A moment white, then lost forever,

sang Burns, in illustrating the fleeting character of human pleasure. The first snowflakes that fall into the Sierra rivers vanish thus suddenly; but in great storms, when the temperature is low, the abundance of the snow at length chills the water nearly to the freezing-point, and then, of course, it ceases to melt and consume the snow so suddenly. The falling flakes and crystals form cloudlike masses of blue sludge, which are swept forward with the current and carried down to warmer climates many miles distant, while some are lodged against logs and rocks and projecting points of the banks, and last for days, piled high above the level of the water, and show white again, instead of being at once "lost forever," while the rivers themselves are at length lost for months during the snowy period. The snow is first built out from the banks in bossy, over-curling drifts, compacting and cementing until the streams are spanned. They then flow in the dark beneath a continuous covering across the snowy zone, which is about thirty miles wide. All the Sierra rivers and their tributaries in these high regions are thus lost every winter, as if another glacial period had come on. Not a drop of running water is to be seen excepting at a few points where large falls occur, though the rush and rumble of the heavier currents may still be heard. Toward spring, when the weather is warm during the day and frosty at night, repeated thawing and freezing and new layers of snow render the bridging-masses dense and firm, so that one may safely walk across the streams, or even lead a horse across them without danger of falling through. In June the thinnest parts of the winter ceiling, and those most exposed to sunshine, begin to give way, forming dark, rugged-edged, pit-like sinks, at the bottom of which the rushing water may be seen. At the end of June only here and there may the mountaineer find a secure snow-bridge. The most lasting of the winter bridges, thawing from below as well as from above, because of warm currents of air passing through the tunnels, are strikingly arched and sculptured; and by the occasional freezing of the oozing, dripping water of the ceiling they become brightly and picturesquely icy. In some of the reaches, where there is a free margin, we may walk through them. Small skylights appearing here and there, these tunnels are not very dark. The roaring river fills all the arching way with impressively loud reverberating music, which is sweetened at times by the ouzel, a bird that is not afraid to go wherever a stream may go, and to sing wherever a stream sings.

All the small alpine pools and lakelets are in like manner obliterated from the winter landscapes, either by being first frozen and then covered by snow, or by being filled in by avalanches. The first avalanche of the season shot into a lake basin may perhaps find the surface frozen. Then there is a grand crashing of breaking ice and dashing of waves mingled with the low, deep booming of the avalanche. Detached masses of the invading snow, mixed with fragments of ice, drift about in sludgy, island-like heaps, while the main body of it forms a talus with its base wholly or in part resting on the bottom of the basin, as controlled by its depth and the size of the avalanche. The next avalanche, of course, encroaches still farther, and so on with each in succession until the entire basin may be filled and its water sponged up or displaced. This huge mass of sludge, more or less mixed with sand, stones, and perhaps timber, is frozen to a considerable depth, and much sun-heat is required to thaw it. Some of these unfortunate lakelets are not clear of ice and snow until near the end of summer. Others are never quite free, opening only on the side opposite the entrance of the avalanches. Some show only a narrow crescent of water lying between the shore and sheer bluffs of icy compacted snow, masses of which breaking off float in front like icebergs in a miniature Arctic Ocean, while the avalanche heaps leaning back against the mountains look like small glaciers. The frontal cliffs are in some instances quite picturesque, and with the berg-dotted waters in front of them lighted with sunshine are exceedingly beautiful. It often happens that while one side of a lake basin is hopelessly snow-buried and frozen, the other, enjoying sunshine, is adorned with beautiful flower-gardens. Some of the smaller lakes are extinguished in an instant by a heavy avalanche either of rocks or snow. The rolling, sliding, ponderous mass entering on one side sweeps across the bottom and up the opposite side, displacing the water and even scraping the basin clean, and shoving the accumulated rocks and sediments up the farther bank and taking full possession. The dislodged water is in part absorbed, but most of it is sent around the front of the avalanche and down the channel of the outlet, roaring and hurrying as if frightened and glad to escape.

SNOW-BANNERS

The most magnificent storm phenomenon I ever saw, surpassing in showy grandeur the most imposing effects of clouds, floods, or avalanches, was the peaks of the High Sierra, back of Yosemite Valley, decorated with snow-banners. Many of the starry snow-flowers, out of which these banners are made, fall before they are ripe, while most of those that do attain perfect development as six-rayed crystals glint and chafe against one another in their fall through the frosty air, and are broken into fragments. This dry fragmentary snow is still further prepared for the formation of banners by the action of the wind. For, instead of finding rest at once, like the snow which falls into the tranquil depths of the forests, it is rolled over and over, beaten against rock-ridges, and swirled in pits and hollows, like boulders, pebbles, and sand in the pot-holes of a river, until finally the delicate angles of the crystals are worn off, and the whole mass is reduced to dust. And whenever storm-winds find this prepared snow-dust in a loose condition on exposed slopes, where there is a free upward sweep to leeward, it is tossed back into the sky, and borne onward from peak to peak in the form of banners or cloudy drifts, according to the velocity of the wind and the conformation of the slopes up or around which it is driven. While thus flying through the air, a small portion makes good its escape, and remains in the sky as vapor. But far the greater part, after being driven into the sky again and again, is at length locked fast in bossy drifts, or in the wombs of glaciers, some of it to remain silent and rigid for centuries before it is finally melted and sent singing down the mountainsides to the sea.

Yet, notwithstanding the abundance of winter snow-dust in the mountains, and the frequency of high winds, and the length of time the dust remains loose and exposed to their action, the occurrence of well-formed banners is, for causes we shall hereafter note, comparatively rare. I have seen only one display of this kind that seemed in every way perfect. This was in the winter of 1873, when the snow-laden summits were swept by a wild "norther." I happened at the time to be wintering in Yosemite Valley, that sublime Sierra temple where every day one may see the grandest sights. Yet even here the wild gala-day of the north wind seemed surpassingly glorious. I was awakened in the morning by the rocking of my cabin and the beating of pine-burs on the roof. Detached torrents and avalanches from the main wind-flood overhead were rushing wildly down the narrow side cañons, and over the precipitous walls, with loud resounding roar, rousing the pines to enthusiastic action, and making the whole valley vibrate as though it were an instrument being played.

But afar on the lofty exposed peaks of the range standing so high in the sky, the storm was expressing itself in still grander characters, which I was soon to see in all their glory. I had long been anxious to study some points in the structure of the ice-cone that is formed every winter at the foot of the upper Yosemite fall, but the blinding spray by which it is invested had hitherto prevented me from making a sufficiently near approach. This morning the entire body of the fall was torn into gauzy shreds, and blown horizontally along the face of the cliff, leaving the cone dry; and while making my way to the top of an overlooking ledge to seize so favorable an opportunity to examine the interior of the cone, the peaks of the Merced group came in sight over the shoulder of the South Dome, each waving a resplendent banner against the blue sky, as regular in form, and as firm in texture, as if woven of fine silk. So rare and splendid a phenomenon, of course, overbore all other considerations, and I at once let the ice-cone go, and began to force my way out of the valley to some dome or ridge sufficiently lofty to command a general view of the main summits, feeling assured that I should find them bannered still more gloriously; nor was I in the least disappointed. Indian Cañon, through which I climbed, was choked with snow that had been shot down in avalanches from the high cliffs on either side, rendering the ascent difficult; but inspired by the roaring storm, the tedious wallowing brought no fatigue, and in four hours I gained the top of a ridge above the valley, 8000 feet high. And there in bold relief, like a clear painting, appeared a most imposing scene. Innumerable peaks, black and sharp, rose grandly into the dark blue sky, their bases set in solid white, their sides streaked and splashed with snow, like ocean rocks with foam; and from every summit, all free and unconfused, was streaming a beautiful silky silvery banner, from half a mile to a mile in length, slender at the point of attachment, then widening gradually as it extended from the peak until it was about 1000 or 1500 feet in breadth, as near as I could estimate. The cluster of peaks called the "Crown of the Sierra," at the head of the Merced and Tuolumne rivers,—Mounts Dana, Gibbs, Conness, Lyell, Maclure, Ritter, with their nameless compeers,—each had its own refulgent banner, waving with a clearly visible motion in the sunglow, and there was not a single cloud in the sky to mar their simple grandeur. Fancy yourself standing on this Yosemite ridge looking eastward. You notice a strange garish glitter in the air. The gale drives wildly overhead with a fierce, tempestuous roar, but its violence is not felt, for you are looking through a sheltered opening in the woods as through a window. There, in the immediate foreground of your picture, rises a majestic forest of Silver Fir blooming in eternal freshness, the foliage yellow-green, and the snow beneath the trees strewn with their beautiful plumes, plucked off by the wind. Beyond, and extending over all the middle ground, are somber swaths of pine, interrupted by huge swelling ridges and domes; and just beyond the dark forest you see the monarchs of the High Sierra waving their magnificent banners. They are twenty miles away, but you would not wish them nearer, for every feature is distinct, and the whole glorious show is seen in its right proportions. After this general view, mark how sharply the dark snowless ribs and buttresses and summits of the peaks are defined, excepting the portions veiled by the banners, and how delicately their sides are streaked with snow, where it has come to rest in narrow flutings and gorges. Mark, too, how grandly the banners wave as the wind is deflected against their sides, and how trimly each is attached to the very summit of its peak, like a streamer at a masthead; how smooth and silky they are in texture, and how finely their fading fringes are penciled on the azure sky. See how dense and opaque they are at the point of attachment, and how filmy and translucent toward the end, so that the peaks back of them are seen dimly, as though you were looking through ground glass. Yet again observe how some of the longest, belonging to the loftiest summits, stream perfectly free all the way across intervening notches and passes from peak to peak, while others overlap and partly hide each other. And consider how keenly every particle of this wondrous cloth of snow is flashing out jets of light. These are the main features of the beautiful and terrible picture as seen from the forest window; and it would still be surpassingly glorious were the fore- and middle-grounds obliterated altogether, leaving only the black peaks, the white banners, and the blue sky.

Glancing now in a general way at the formation of snow-banners, we find that the main causes of the wondrous beauty and perfection of those we have been contemplating were the favorable direction and great force of the wind, the abundance of snow-dust, and the peculiar conformation of the slopes of the peaks. It is essential not only that the wind should move with great velocity and steadiness to supply a sufficiently copious and continuous stream of snow-dust, but that it should come from the north. No perfect banner is ever hung on the Sierra peaks by a south wind. Had the gale that day blown from the south, leaving other conditions unchanged, only a dull, confused, fog-like drift would have been produced; for the snow, instead of being spouted up over the tops of the peaks in concentrated currents to be drawn out as streamers, would have been shed off around the sides, and piled down into the glacier wombs. The cause of the concentrated action of the north wind is found in the peculiar form of the north sides of the peaks, where the amphitheaters of the residual glaciers are. In general the south sides are convex and irregular, while the north sides are concave both in their vertical and horizontal sections; the wind in ascending these curves converges toward the summits, carrying the snow in concentrating currents with it, shooting it almost straight up into the air above the peaks, from which it is then carried away in a horizontal direction.

This difference in form between the north and south sides of the peaks was almost wholly produced by the difference in the kind and quantity of the glaciation to which they have been subjected, the north sides having been hollowed by residual shadow-glaciers of a form that never existed on the sun-beaten sides.

It appears, therefore, that shadows in great part determine not only the forms of lofty icy mountains, but also those of the snow-banners that the wild winds hang on them.

CHAPTER IV

A NEAR VIEW OF THE HIGH SIERRA

EARLY ONE BRIGHT morning in the middle of Indian summer, while the glacier meadows were still crisp with frost crystals, I set out from the foot of Mount Lyell, on my way down to Yosemite Valley, to replenish my exhausted store of bread and tea. I had spent the past summer, as many preceding ones, exploring the glaciers that lie on the head waters of the San Joaquin, Tuolumne, Merced, and Owen's rivers; measuring and studying their movements, trends, crevasses, moraines, etc., and the part they had played during the period of their greater extension in the creation and development of the landscapes of this alpine wonderland. The time for this kind of work was nearly over for the year, and I began to look forward with delight to the approaching winter with its wondrous storms, when I would be warmly snow-bound in my Yosemite cabin with plenty of bread and books; but a tinge of regret came on when I considered that possibly I might not see this favorite region again until the next summer, excepting distant views from the heights about the Yosemite walls.

To artists, few portions of the High Sierra are, strictly speaking, picturesque. The whole massive uplift of the range is one great picture, not clearly divisible into smaller ones; differing much in this respect from the older, and what may be called, riper mountains of the Coast Range. All the landscapes of the Sierra, as we have seen, were born again, remodeled from base to summit by the developing ice-floods of the last glacial winter. But all these new landscapes were not brought forth simultaneously; some of the highest, where the ice lingered longest, are tens of centuries younger than those of the warmer regions below them. In general, the younger the mountain-landscapes,—younger, I mean, with reference to the time of their emergence from the ice of the glacial period,—the less separable are they into artistic bits capable of being made into warm, sympathetic, lovable pictures with appreciable humanity in them.

Here, however, on the head waters of the Tuolumne, is a group of wild peaks on which the geologist may say that the sun has but just begun to shine, which is yet in a high degree picturesque, and in its main features so regular and evenly balanced as almost to appear conventional—one somber cluster of snow-laden peaks with gray pine-fringed granite bosses braided around its base, the whole surging free into the sky from the head of a magnificent valley, whose lofty walls are beveled away on both sides so as to embrace it all without admitting anything not strictly belonging to it. The foreground was now aflame with autumn colors, brown and purple and gold, ripe in the mellow sunshine; contrasting brightly with the deep, cobalt blue of the sky, and the black and gray, and pure, spiritual white of the rocks and glaciers. Down through the midst, the young Tuolumne was seen pouring from its crystal fountains, now resting in glassy pools as if changing back again into ice, now leaping in white cascades as if turning to snow; gliding right and left between granite bosses, then sweeping on through the smooth, meadowy levels of the valley, swaying pensively from side to side with calm, stately gestures past dipping willows and sedges, and around groves of arrowy pine; and throughout its whole eventful course, whether flowing fast or slow, singing loud or low, ever filling the landscape with spiritual animation, and manifesting the grandeur of its sources in every movement and tone.

Pursuing my lonely way down the valley, I turned again and again to gaze on the glorious picture, throwing up my arms to inclose it as in a frame. After long ages of growth in the darkness beneath the glaciers, through sunshine and storms, it seemed now to be ready and waiting for the elected artist, like yellow wheat for the reaper; and I could not help wishing that I might carry colors and brushes with me on my travels, and learn to paint. In the mean time I had to be content with photographs on my mind and sketches in my note-books. At length, after I had rounded a precipitous headland that puts out from the west wall of the valley, every peak vanished from sight, and I pushed rapidly along the frozen meadows, over the divide between the waters of the Merced and Tuolumne, and down through the forests that clothe the slopes of Clouds' Rest, arriving in Yosemite in due time—which, with me, is *any* time. And, strange to say, among the first people I met here were two artists who, with letters of introduction, were awaiting my return. They inquired whether in the course of my explorations in the adjacent mountains

I had ever come upon a landscape suitable for a large painting; whereupon I began a description of the one that had so lately excited my admiration. Then, as I went on further and further into details, their faces began to glow, and I offered to guide them to it, while they declared that they would gladly follow, far or near, whithersoever I could spare the time to lead them.

Since storms might come breaking down through the fine weather at any time, burying the colors in snow, and cutting off the artists' retreat, I advised getting ready at once.

I led them out of the valley by the Vernal and Nevada Falls, thence over the main dividing ridge to the Big Tuolumne Meadows, by the old Mono trail, and thence along the upper Tuolumne River to its head. This was my companions' first excursion into the High Sierra, and as I was almost always alone in my mountaineering, the way that the fresh beauty was reflected in their faces made for me a novel and interesting study. They naturally were affected most of all by the colors—the intense azure of the sky, the purplish grays of the granite, the red and browns of dry meadows, and the translucent purple and crimson of huckleberry bogs; the flaming yellow of aspen groves, the silvery flashing of the streams, and the bright green and blue of the glacier lakes. But the general expression of the scenery—rocky and savage—seemed sadly disappointing; and as they threaded the forest from ridge to ridge, eagerly scanning the landscapes as they were unfolded, they said: "All this is huge and sublime, but we see nothing as yet at all available for effective pictures. Art is long, and art is limited, you know; and here are foregrounds, middlegrounds, backgrounds, all alike; bare rock-waves, woods, groves, diminutive flecks of meadow, and strips of glittering water." "Never mind," I replied, "only bide a wee, and I will show you something you will like."

At length, toward the end of the second day, the Sierra Crown began to come into view, and when we had fairly rounded the projecting headland before mentioned, the whole picture stood revealed in the flush of the alpenglow. Their enthusiasm was excited beyond bounds, and the more impulsive of the two, a young Scotchman, dashed ahead, shouting and gesticulating and tossing his arms in the air like a madman. Here, at last, was a typical alpine landscape.

After feasting awhile on the view, I proceeded to make camp in a sheltered grove a little way back from the meadow, where pine-boughs could be obtained for beds, and where there was plenty of dry wood for fires, while the artists ran here and there, along the river-bends and up the sides of the cañon, choosing foregrounds for sketches. After dark, when our tea was made and a rousing fire had been built, we began to make our plans. They decided to remain several days, at the least, while I concluded to make an excursion in the mean time to the untouched summit of Ritter.

It was now about the middle of October, the springtime of snow-flowers. The first winter-clouds had already bloomed, and the peaks were strewn with fresh crystals, without, however, affecting the climbing to any dangerous extent. And as the weather was still profoundly calm, and the distance to the foot of the mountain only a little more than a day, I felt that I was running no great risk of being storm-bound.

Mount Ritter is king of the mountains of the middle portion of the High Sierra, as Shasta of the north and Whitney of the south sections. Moreover, as far as I know, it had never been climbed. I had explored the adjacent wilderness summer after summer, but my studies thus far had never drawn me to the top of it. Its height above sea-level is about 13,300 feet, and it is fenced round by steeply inclined glaciers, and cañons of tremendous depth and ruggedness, which render it almost inaccessible. But difficulties of this kind only exhilarate the mountaineer.

Next morning, the artists went heartily to their work and I to mine. Former experiences had given good reason to know that passionate storms, invisible as yet, might be brooding in the calm sun-gold; therefore, before bidding farewell, I warned the artists not to be alarmed should I fail to appear before a week or ten days, and advised them, in case a snow-storm should set in, to keep up big fires and shelter themselves as best they could, and on no account to become frightened and attempt to seek their way back to Yosemite alone through the drifts.

My general plan was simply this: to scale the cañon wall, cross over to the eastern flank of the range, and then make my way southward to the northern spurs of Mount Ritter in compliance with the intervening topography; for to push on directly southward from camp through the innumerable peaks and pinnacles that adorn this portion of the axis of the range, however interesting, would take too much time, besides being extremely difficult and dangerous at this time of year.

All my first day was pure pleasure; simply mountaineering indulgence, crossing the dry pathways of the ancient glaciers, tracing happy streams, and learning the habits of the birds and marmots in the groves and rocks. Before I had gone a mile from camp, I came to the foot of a white cascade that beats its way down a rugged gorge in the cañon wall, from a height of about nine hundred feet, and pours its throbbing waters into the Tuolumne. I was acquainted with its fountains, which, fortunately, lay in my course. What a fine traveling companion it proved to be, what songs it sang, and how passionately it told the mountain's own joy! Gladly I climbed along its dashing border, absorbing its divine music, and bathing from time to time in waftings of irised spray. Climbing higher, higher, new beauty came streaming on the sight: painted meadows, late-blooming gardens, peaks of rare architecture, lakes here and there, shining like silver, and glimpses of the forested middle region and the yellow lowlands far in the west. Beyond the range I saw the so-called Mono Desert, lying dreamily silent in thick purple light—a desert of heavy sun-glare beheld from a desert of ice-burnished granite. Here the waters divide, shouting in glorious enthusiasm, and falling eastward to vanish in the volcanic sands and dry sky of the Great Basin, or westward to the Great Valley of California, and thence through the Bay of San Francisco and the Golden Gate to the sea.

Passing a little way down over the summit until I had reached an elevation of about 10,000 feet, I pushed on southward toward a group of savage peaks that stand guard about Ritter on the north and west, groping my way, and dealing instinctively with every obstacle as it presented itself. Here a huge gorge would be found cutting across my path, along the dizzy edge of which I scrambled until some less precipitous point was discovered where I might safely venture to the bottom and then, selecting some feasible portion of the opposite wall, reascend with the same slow caution. Massive, flat-topped spurs alternate with the gorges, plunging abruptly from the shoulders of the snowy peaks, and planting their feet in the warm desert. These were everywhere marked and adorned with characteristic sculptures of the ancient glaciers that swept over this entire region like one vast ice-wind, and the polished surfaces produced by the ponderous flood are still so perfectly preserved that in many places the sunlight reflected from them is about as trying to the eyes as sheets of snow.

God's glacial-mills grind slowly, but they have been kept in motion long enough in California to grind sufficient soil for a glorious abundance of life, though most of the grist has been carried to the lowlands, leaving these high regions comparatively lean and bare; while the post-glacial agents of erosion have not yet furnished sufficient available food over the general surface for more than a few tufts of the hardiest plants, chiefly carices and eriogonæ. And it is interesting to learn in this connection that the sparseness and repressed character of the vegetation at this height is caused more by want of soil than by harshness of climate; for, here and there, in sheltered hollows (countersunk beneath the general surface) into which a few rods of well-ground moraine chips have been dumped, we find groves of spruce and pine thirty to forty feet high, trimmed around the edges with willow and huckleberry bushes, and oftentimes still further by an outer ring of tall grasses, bright with lupines, larkspurs, and showy columbines, suggesting a climate by no means repressingly severe. All the streams, too, and the pools at this elevation are furnished with little gardens wherever soil can be made to lie, which, though making scarce any show at a distance, constitute charming surprises to the appreciative observer. In these bits of leafiness a few birds find grateful homes. Having no acquaintance with man, they fear no ill, and flock curiously about the stranger, almost allowing themselves to be taken in the hand. In so wild and so beautiful a region was spent my first day, every sight and sound inspiring, leading one far out of himself, yet feeding and building up his individuality.

Now came the solemn, silent evening. Long, blue, spiky shadows crept out across the snow-fields, while a rosy glow, at first scarce discernible, gradually deepened and suffused every mountain-top, flushing the glaciers and the harsh crags above them. This was the alpenglow, to me one of the most impressive of all the terrestrial manifestations of God. At the touch of this divine light, the mountains seemed to kindle to a rapt, religious consciousness, and stood hushed and waiting like devout worshipers. Just before the alpenglow began to fade, two crimson clouds came streaming across the summit like wings of flame, rendering the sublime scene yet more impressive; then came darkness and the stars.

Icy Ritter was still miles away, but I could proceed no farther that night. I found a good campground on the rim of a glacier basin about 11,000 feet above the sea. A small lake nestles in the bottom of it, from which I got water for my tea, and a stormbeaten thicket near by furnished abundance of resiny fire-wood. Somber peaks, hacked and shattered, circled halfway around the horizon, wearing a savage aspect in the gloaming, and a waterfall chanted solemnly across the lake on its way down from the foot of a glacier. The fall and the lake and the glacier were almost equally bare; while the scraggy pines anchored in the rock-fissures were so dwarfed and shorn by storm-winds that you might walk over their tops. In tone and aspect the scene was one of the most desolate I ever beheld. But the darkest scriptures of the mountains are illumined with bright passages of love that never fail to make themselves felt when one is alone.

I made my bed in a nook of the pine-thicket, where the branches were pressed and crinkled overhead like a roof, and bent down around the sides. These are the best bedchambers the high mountains afford—snug as squirrel-nests, well ventilated, full of spicy odors, and with plenty of wind-played needles to sing one asleep. I little expected company, but, creeping in through a low side-door, I found five or six birds nestling among the tassels. The night-wind began to blow soon after dark; at first only a gentle breathing, but increasing toward midnight to a rough gale that fell upon my leafy roof in ragged surges like a cascade, bearing wild sounds from the crags overhead. The waterfall sang in chorus, filling the old ice-fountain with its solemn roar, and seeming to increase in power as the night advanced—fit voice for such a landscape. I had to creep out many times to the fire during the night, for it was biting cold and I had no blankets. Gladly I welcomed the morning star.

The dawn in the dry, wavering air of the desert was glorious. Everything encouraged my undertaking and betokened success. There was no cloud in the sky, no storm-tone in the wind. Breakfast of bread and tea was soon made. I fastened a hard, durable crust to my belt by way of provision, in case I should be compelled to pass a night on the mountain-top; then, securing the remainder of my little stock against wolves and wood-rats, I set forth free and hopeful.

How glorious a greeting the sun gives the mountains! To behold this alone is worth the pains of any excursion a thousand times over. The highest peaks burned like islands in a sea of liquid shade. Then the lower peaks and spires caught the glow, and long lances of light, streaming through many a notch and pass, fell thick on the frozen meadows. The majestic form of Ritter was full in sight, and I pushed rapidly on over rounded rock-bosses and pavements, my iron-shod shoes making a clanking sound, suddenly hushed now and then in rugs of bryanthus, and sedgy lake-margins soft as moss. Here, too, in this so-called "land of desolation," I met cassiope, growing in fringes among the battered rocks. Her blossoms had faded long ago, but they were still clinging with happy memories to the evergreen sprays, and still so beautiful as to thrill every fiber of one's being. Winter and summer, you may hear her voice, the low, sweet melody of her purple bells. No evangel among all the mountain plants speaks Nature's love more plainly than cassiope. Where she dwells, the redemption of the coldest solitude is complete. The very rocks and glaciers seem to feel her presence, and become imbued with her own fountain sweetness. All things were warming and awakening. Frozen rills began to flow, the marmots came out of their nests in boulder-piles and climbed sunny rocks to bask, and the dun-headed sparrows were flitting about seeking their breakfasts. The lakes seen from every ridge-top were brilliantly rippled and spangled, shimmering like the thickets of the low Dwarf Pines. The rocks, too, seemed responsive to the vital heat—rock-crystals and snow-crystals thrilling alike. I strode on exhilarated, as if never more to feel fatigue, limbs moving of themselves, every sense unfolding like the thawing flowers, to take part in the new day harmony.

All along my course thus far, excepting when down in the cañons, the landscapes were mostly open to me, and expansive, at least on one side. On the left were the purple plains of Mono, reposing dreamily and warm; on the right, the near peaks springing keenly into the thin sky with more and more impressive sublimity. But these larger views were at length lost. Rugged spurs, and moraines, and huge, projecting buttresses began to shut me in. Every feature became more rigidly alpine, without, however, producing any chilling effect; for going to the mountains is like going home. We always find that the strangest objects in these fountain wilds are in some degree familiar, and we look upon them with a vague sense of having seen them before.

On the southern shore of a frozen lake, I encountered an extensive field of hard, granular snow, up which I scampered in fine tone, intending to follow it to its head, and cross the rocky spur against which it leans, hoping thus to come direct upon the base of the main Ritter peak. The surface was pitted with oval hollows, made by stones and drifted pine-needles that had melted themselves into the mass by the radiation of absorbed sun-heat. These afforded good footholds, but the surface curved more and more steeply at the head, and the pits became shallower and less abundant, until I found myself in danger of being shed off like avalanching snow. I persisted, however, creeping on all fours, and shuffling up the smoothest places on my back, as I had often done on burnished granite, until, after slipping several times, I was compelled to retrace my course to the bottom, and make my way around the west end of the lake, and thence up to the summit of the divide between the head waters of Rush Creek and the northernmost tributaries of the San Joaquin.

Arriving on the summit of this dividing crest, one of the most exciting pieces of pure wilderness was disclosed that I ever discovered in all my mountaineering. There, immediately in front, loomed the majestic mass of Mount Ritter, with a glacier swooping down its face nearly to my feet, then curving westward and pouring its frozen flood into a dark blue lake, whose shores were bound with precipices of crystalline snow; while a deep chasm drawn between the divide and the glacier separated the massive picture from everything else. I could see only the one sublime mountain, the one glacier, the one lake; the whole veiled with one blue shadow—rock, ice, and water close together without a single leaf or sign of life. After gazing spellbound, I began instinctively to scrutinize every notch and gorge and weathered buttress of the mountain, with reference to making the ascent. The entire front above the glacier appeared as one tremendous precipice, slightly receding at the top, and bristling with spires and pinnacles set above one another in formidable array. Massive lichen-stained battlements stood forward here and there, hacked at the top with angular notches, and separated by frosty gullies and recesses that have been veiled in shadow ever since their creation; while to right and left, as far as I could see, were huge, crumbling buttresses, offering no hope to the climber. The head of the glacier sends up a few finger-like branches through narrow *couloirs*; but these seemed too steep and short to be available, especially as I had no ax with which to cut steps, and the numerous narrow-throated gullies down which stones and snow are avalanched seemed hopelessly steep, besides being interrupted by vertical cliffs; while the whole front was rendered still more terribly forbidding by the chill shadow and the gloomy blackness of the rocks.

Descending the divide in a hesitating mood, I picked my way across the yawning chasm at the foot, and climbed out upon the glacier. There were no meadows now to cheer with their brave colors, nor could I hear the dun-headed sparrows, whose cheery notes so often relieve the silence of our highest mountains. The only sounds were the gurgling of small rills down in the veins and crevasses of the glacier, and now and then the rattling report of falling stones, with the echoes they shot out into the crisp air.

I could not distinctly hope to reach the summit from this side, yet I moved on across the glacier as if driven by fate. Contending with myself, the season is too far spent, I said, and even should I be successful, I might be storm-bound on the mountain; and in the cloud-darkness, with the cliffs and crevasses covered with snow, how could I escape? No; I must wait till next summer. I would only approach the mountain now, and inspect it, creep about its flanks, learn what I could of its history, holding myself ready to flee on the approach of the first storm-cloud. But we little know until tried how much of the uncontrollable there is in us, urging across glaciers and torrents, and up dangerous heights, let the judgment forbid as it may.

I succeeded in gaining the foot of the cliff on the eastern extremity of the glacier, and there discovered the mouth of a narrow avalanche gully, through which I began to climb, intending to follow it as far as possible, and at least obtain some fine wild views for my pains. Its general course is oblique to the plane of the mountain-face, and the metamorphic slates of which the mountain is built are cut by cleavage planes in such a way that they weather off in angular blocks, giving rise to irregular steps that greatly facilitate climbing on the sheer places. I thus made my way into a wilderness of crumbling spires and battlements, built together in bewildering combinations, and glazed in many places with a thin coating of ice, which I had to hammer off with stones. The situation was becoming gradually more perilous; but, having passed several dangerous spots, I dared not think of descending; for so steep was the entire ascent, one would inevitably fall to the glacier in case a single misstep were made. Knowing, therefore, the tried danger beneath, I became all the more anxious concerning the developments to be made above, and began to be conscious of a vague foreboding of what actually befell; not that I was given to fear, but rather because my instincts, usually so positive and true, seemed vitiated in some way, and were leading me astray. At length, after attaining an elevation of about 12,800 feet, I found myself at the foot of a sheer drop in the bed of the avalanche channel I was tracing, which seemed absolutely to bar further progress. It was only about forty-five or fifty feet high, and somewhat roughened by fissures and projections; but these seemed so slight and insecure, as footholds, that I tried hard to avoid the precipice altogether, by scaling the wall of the channel on either side. But, though less steep, the walls were smoother than the obstructing rock, and repeated efforts only showed that I must either go right ahead or turn back. The tried dangers beneath seemed even greater than that of the cliff in front; therefore, after scanning its face again and again, I began to scale it, picking my holds with intense caution. After gaining a point about halfway to the top, I was suddenly brought to a dead stop, with arms outspread, clinging close to the face of the rock, unable to move hand or foot either up or down. My doom appeared fixed. I *must* fall. There would be a moment of bewilderment, and then a lifeless tumble down the one general precipice to the glacier below.

When this final danger flashed upon me, I became nerve-shaken for the first time since setting foot on the mountains, and my mind seemed to fill with a stifling smoke. But this terrible eclipse lasted only a moment, when life blazed forth again with preternatural clearness. I seemed suddenly to become possessed of a new sense. The other self, bygone experiences, Instinct, or Guardian Angel,—call it what you will,—came forward and assumed control. Then my trembling muscles became firm again, every rift and flaw in the rock was seen as through a microscope, and my limbs moved with a positiveness and precision with which I seemed to have nothing at all to do. Had I been borne aloft upon wings, my deliverance could not have been more complete.

Above this memorable spot, the face of the mountain is still more savagely hacked and torn. It is a maze of yawning chasms and gullies, in the angles of which rise beetling crags and piles of detached boulders that seem to have been gotten ready to be launched below. But the strange influx of strength I had received seemed inexhaustible. I found a way without effort, and soon stood upon the topmost crag in the blessed light.

How truly glorious the landscape circled around this noble summit!—giant mountains, valleys innumerable, glaciers and meadows, rivers and lakes, with the wide blue sky bent tenderly over them all. But in my first hour of freedom from that terrible shadow, the sunlight in which I was laving seemed all in all.

Looking southward along the axis of the range, the eye is first caught by a row of exceedingly sharp and slender spires, which rise openly to a height of about a thousand feet, above a series of short, residual glaciers that lean back against their bases; their fantastic sculpture and the unrelieved sharpness with which they spring out of the ice rendering them peculiarly wild and striking. These are "The Minarets." Beyond them you behold a sublime wilderness of mountains, their snowy summits towering together in crowded abundance, peak beyond peak, swelling higher, higher as they sweep on southward, until the culminating point of the range is reached on Mount Whitney, near the head of the Kern River, at an elevation of nearly 14,700 feet above the level of the sea.

Westward, the general flank of the range is seen flowing sublimely away from the sharp summits, in smooth undulations; a sea of huge gray granite waves dotted with lakes and meadows, and fluted with stupendous cañons that grow steadily deeper as they recede in the distance. Below this gray region lies the dark forest zone, broken here and there by upswelling ridges and domes; and yet beyond lies a yellow, hazy belt, marking the broad plain of the San Joaquin, bounded on its farther side by the blue mountains of the coast.

Turning now to the northward, there in the immediate foreground is the glorious Sierra Crown, with Cathedral Peak, a temple of marvelous architecture, a few degrees to the left of it; the gray, massive form of Mammoth Mountain to the right; while Mounts Ord, Gibbs, Dana, Conness, Tower Peak, Castle Peak, Silver Mountain, and a host of noble companions, as yet nameless, make a sublime show along the axis of the range.

Eastward, the whole region seems a land of desolation covered with beautiful light. The torrid volcanic basin of Mono, with its one bare lake fourteen miles long; Owen's Valley and the broad lava table-land at its head, dotted with craters, and the massive Inyo Range, rivaling even the Sierra in height; these are spread, map-like, beneath you, with countless ranges beyond, passing and overlapping one another and fading on the glowing horizon.

At a distance of less than 3000 feet below the summit of Mount Ritter you may find tributaries of the San Joaquin and Owen's rivers, bursting forth from the ice and snow of the glaciers that load its flanks; while a little to the north of here are found the highest affluents of the Tuolumne and Merced. Thus, the fountains of four of the principal rivers of California are within a radius of four or five miles.

Lakes are seen gleaming in all sorts of places,—round, or oval, or square, like very mirrors; others narrow and sinuous, drawn close around the peaks like silver zones, the highest reflecting only rocks, snow, and the sky. But neither these nor the glaciers, nor the bits of brown meadow and moorland that occur here and there, are large enough to make any marked impression upon the mighty wilderness of mountains. The eye, rejoicing in its freedom, roves about the vast expanse, yet returns again and again to the fountain peaks. Perhaps some one of the multitude excites special attention, some gigantic castle with turret and battlement, or some Gothic cathedral more abundantly spired than Milan's. But, generally, when looking for the first time from an all-embracing standpoint like this, the inexperienced observer is oppressed by the incomprehensible grandeur, variety, and abundance of the mountains rising shoulder to shoulder beyond the reach of vision; and it is only after they have been studied one by one, long and lovingly, that their far-reaching harmonies become manifest. Then, penetrate the wilderness where you may, the main telling features, to which all the surrounding topography is subordinate, are quickly perceived, and the most complicated clusters of peaks stand revealed harmoniously correlated and fashioned like works of art—eloquent monuments of the ancient ice-rivers that brought them into relief from the general mass of the range. The cañons, too, some of them a mile deep, mazing wildly through the mighty host of mountains, however lawless and ungovernable at first sight they appear, are at length recognized as the necessary effects of causes which followed each other in harmonious sequence—Nature's poems carved on tables of stone—the simplest and most emphatic of her glacial compositions.

Could we have been here to observe during the glacial period, we should have overlooked a wrinkled ocean of ice as continuous as that now covering the landscapes of Greenland; filling every valley and cañon with only the tops of the fountain peaks rising darkly above the rock-encumbered ice-waves like islets in a stormy sea—those islets the only hints of the glorious landscapes now smiling in the sun. Standing here in the deep, brooding silence all the wilderness seems motionless, as if the work of creation were done. But in the midst of this outer steadfastness we know there is incessant motion and change. Ever and anon, avalanches are falling from yonder peaks. These cliff-bound glaciers, seemingly wedged and immovable, are flowing like water and grinding the rocks beneath them. The lakes are lapping their granite shores and wearing them away, and every one of these rills and young rivers is fretting the air into music, and carrying the mountains to the plains. Here are the roots of all the life of the valleys, and here more simply than elsewhere is the eternal flux of nature manifested. Ice changing to water, lakes to meadows, and mountains to plains. And while we thus contemplate Nature's methods of landscape creation, and, reading the records she has carved on the rocks, reconstruct, however imperfectly, the landscapes of the past, we also learn that as these we now behold have succeeded those of the pre-glacial age, so they in turn are withering and vanishing to be succeeded by others yet unborn.

But in the midst of these fine lessons and landscapes, I had to remember that the sun was wheeling far to the west, while a new way down the mountain had to be discovered to some point on the timber line where I could have a fire; for I had not even burdened myself with a coat. I first scanned the western spurs, hoping some way might appear through which I might reach the northern glacier, and cross its snout; or pass around the lake into which it flows, and thus strike my morning track. This route was soon sufficiently unfolded to show that, if practicable at all, it would require so much time that reaching camp that night would be out of the question. I therefore scrambled back eastward, descending the southern slopes obliquely at the same time. Here the crags seemed less formidable, and the head of a glacier that flows northeast came in sight, which I determined to follow as far as possible, hoping thus to make my way to the foot of the peak on the east side, and thence across the intervening cañons and ridges to camp.

The inclination of the glacier is quite moderate at the head, and, as the sun had softened the *névé*, I made safe and rapid progress, running and sliding, and keeping up a sharp outlook for crevasses. About half a mile from the head, there is an ice-cascade, where the glacier pours over a sharp declivity and is shattered into massive blocks separated by deep, blue fissures. To thread my way through the slippery mazes of this crevassed portion seemed impossible, and I endeavored to avoid it by climbing off to the shoulder of the mountain. But the slopes rapidly steepened and at length fell away in sheer precipices, compelling a return to the ice. Fortunately, the day had been warm enough to loosen the ice-crystals so as to admit of hollows being dug in the rotten portions of the blocks, thus enabling me to pick my way with far less difficulty than I had anticipated. Continuing down over the snout, and along the left lateral moraine, was only a confident saunter, showing that the ascent of the mountain by way of this glacier is easy, provided one is armed with an ax to cut steps here and there.

The lower end of the glacier was beautifully waved and barred by the outcropping edges of the bedded ice-layers which represent the annual snow-falls, and to some extent the irregularities of structure caused by the weathering of the walls of crevasses, and by separate snow-falls which have been followed by rain, hail, thawing and freezing, etc. Small rills were gliding and swirling over the melting surface with a smooth, oily appearance, in channels of pure ice—their quick, compliant movements contrasting most impressively with the rigid, invisible flow of the glacier itself, on whose back they all were riding.

Night drew near before I reached the eastern base of the mountain, and my camp lay many a rugged mile to the north; but ultimate success was assured. It was now only a matter of endurance and ordinary mountain-craft. The sunset was, if possible, yet more beautiful than that of the day before. The Mono landscape seemed to be fairly saturated with warm, purple light. The peaks marshaled along the summit were in shadow, but through every notch and pass streamed vivid sunfire, soothing and irradiating their rough, black angles, while companies of small, luminous clouds hovered above them like very angels of light.

Darkness came on, but I found my way by the trends of the cañons and the peaks projected against the sky. All excitement died with the light, and then I was weary. But the joyful sound of the waterfall across the lake was heard at last, and soon the stars were seen reflected in the lake itself. Taking my bearings from these, I discovered the little pine thicket in which my nest was, and then I had a rest such as only a tired mountaineer may enjoy. After lying loose and lost for awhile, I made a sunrise fire, went down to the lake, dashed water on my head, and dipped a cupful for tea. The revival brought about by bread and tea was as complete as the exhaustion from excessive enjoyment and toil. Then I crept beneath the pine-tassels to bed. The wind was frosty and the fire burned low, but my sleep was none the less sound, and the evening constellations had swept far to the west before I awoke.

After thawing and resting in the morning sunshine, I sauntered home,—that is, back to the Tuolumne camp,—bearing away toward a cluster of peaks that hold the fountain snows of one of the north tributaries of Rush Creek. Here I discovered a group of beautiful glacier lakes, nestled together in a grand amphitheater. Toward evening, I crossed the divide separating the Mono waters from those of the Tuolumne, and entered the glacier basin that now holds the fountain snows of the stream that forms the upper Tuolumne cascades. This stream I traced down through its many dells and gorges, meadows and bogs, reaching the brink of the main Tuolumne at dusk.

A loud whoop for the artists was answered again and again. Their camp-fire came in sight, and half an hour afterward I was with them. They seemed unreasonably glad to see me. I had been absent only three days; nevertheless, though the weather was fine, they had already been weighing chances as to whether I would ever return, and trying to decide whether they should wait longer or begin to seek their way back to the lowlands. Now their curious troubles were over. They packed their precious sketches, and next morning we set out homeward bound, and in two days entered the Yosemite Valley from the north by way of Indian Cañon.

CHAPTER V

THE PASSES

THE SUSTAINED GRANDEUR of the High Sierra is strikingly illustrated by the great height of the passes. Between latitude 36° 20′ and 38° the lowest pass, gap, gorge, or notch of any kind cutting across the axis of the range, as far as I have discovered, exceeds 9000 feet in height above the level of the sea; while the average height of all that are in use, either by Indians or whites, is perhaps not less than 11,000 feet, and not one of these is a carriage-pass.

Farther north a carriage-road has been constructed through what is known as the Sonora Pass, on the head waters of the Stanislaus and Walker's rivers, the summit of which is about 10,000 feet above the sea. Substantial wagon-roads have also been built through the Carson and Johnson passes, near the head of Lake Tahoe, over which immense quantities of freight were hauled from California to the mining regions of Nevada, before the construction of the Central Pacific Railroad.

Still farther north a considerable number of comparatively low passes occur, some of which are accessible to wheeled vehicles, and through these rugged defiles during the exciting years of the gold period long emigrant-trains with foot-sore cattle wearily toiled. After the toil-worn adventurers had escaped a thousand dangers and had crawled thousands of miles across the plains the snowy Sierra at last loomed in sight, the eastern wall of the land of gold. And as with shaded eyes they gazed through the tremulous haze of the desert, with what joy must they have descried the pass through which they were to enter the better land of their hopes and dreams!

Between the Sonora Pass and the southern extremity of the High Sierra, a distance of nearly 160 miles, there are only five passes through which trails conduct from one side of the range to the other. These are barely practicable for animals; a pass in these regions meaning simply any notch or cañon through which one may, by the exercise of unlimited patience, make out to lead a mule, or a sure-footed mustang; animals that can slide or jump as well as walk. Only three of the five passes may be said to be in use, viz.: the Kearsarge, Mono, and Virginia Creek; the tracks leading through the others being only obscure Indian trails, not graded in the least, and scarcely traceable by white men; for much of the way is over solid rock and earthquake avalanche taluses, where the unshod ponies of the Indians leave no appreciable sign. Only skilled mountaineers are able to detect the marks that serve to guide the Indians, such as slight abrasions of the looser rocks, the displacement of stones here and there, and bent bushes and weeds. A general knowledge of the topography is, then, the main guide, enabling one to determine where the trail ought to go—*must* go. One of these Indian trails crosses the range by a nameless pass between the head waters of the south and middle forks of the San Joaquin, the other between the north and middle forks of the same river, just to the south of "The Minarets"; this last being about 9000 feet high, is the lowest of the five. The Kearsarge is the highest, crossing the summit near the head of the south fork of King's River, about eight miles to the north of Mount Tyndall, through the midst of the most stupendous rock-scenery. The summit of this pass is over 12,000 feet above sea-level; nevertheless, it is one of the safest of the five, and is used every summer, from July to October or November, by hunters, prospectors, and stock-owners, and to some extent by enterprising pleasure-seekers also. For, besides the surpassing grandeur of the scenery about the summit, the trail, in ascending the western flank of the range, conducts through a grove of the giant Sequoias, and through the magnificent Yosemite Valley of the south fork of King's River. This is, perhaps, the highest traveled pass on the North American continent.

The Mono Pass lies to the east of Yosemite Valley, at the head of one of the tributaries of the south fork of the Tuolumne. This is the best known and most extensively traveled of all that exist in the High Sierra. A trail was made through it about the time of the Mono gold excitement, in the year 1858, by adventurous miners and prospectors—men who would build a trail down the throat of darkest Erebus on the way to gold. Though more than a thousand feet lower than the Kearsarge, it is scarcely less sublime in rock-scenery, while in snowy, falling water it far surpasses it. Being so favorably situated for the stream of Yosemite travel, the more adventurous tourists cross over through this glorious gateway to the volcanic region around Mono Lake. It has therefore gained a name and fame above every other pass in the range. According to the few barometrical observations made upon it, its highest point is 10,765 feet above the sea. The other pass of the five we have been considering is somewhat lower, and crosses the axis of the range a few miles to the north of the Mono Pass, at the head of the southernmost tributary of Walker's River. It is used chiefly by roaming bands of the Pah Ute Indians and "sheepmen."

But, leaving wheels and animals out of the question, the free mountaineer with a sack of bread on his shoulders and an ax to cut steps in ice and frozen snow can make his way across the range almost everywhere, and at any time of year when the weather is calm. To him nearly every notch between the peaks is a pass, though much patient step-cutting is at times required up and down steeply inclined glaciers, with cautious climbing over precipices that at first sight would seem hopelessly inaccessible.

In pursuing my studies, I have crossed from side to side of the range at intervals of a few miles all along the highest portion of the chain, with far less real danger than one would naturally count on. And what fine wildness was thus revealed—storms and avalanches, lakes and waterfalls, gardens and meadows, and interesting animals—only those will ever know who give the freest and most buoyant portion of their lives to climbing and seeing for themselves.

To the timid traveler, fresh from the sedimentary levels of the lowlands, these highways, however picturesque and grand, seem terribly forbidding—cold, dead, gloomy gashes in the bones of the mountains, and of all Nature's ways the ones to be most cautiously avoided. Yet they are full of the finest and most telling examples of Nature's love; and though hard to travel, none are safer. For they lead through regions that lie far above the ordinary haunts of the devil, and of the pestilence that walks in darkness. True, there are innumerable places where the careless step will be the last step; and a rock falling from the cliffs may crush without warning like lightning from the sky; but what then? Accidents in the mountains are less common than in the lowlands, and these mountain mansions are decent, delightful, even divine, places to die in, compared with the doleful chambers of civilization. Few places in this world are more dangerous than home. Fear not, therefore, to try the mountain-passes. They will kill care, save you from deadly apathy, set you free, and call forth every faculty into vigorous, enthusiastic action. Even the sick should try these so-called dangerous passes, because for every unfortunate they kill, they cure a thousand.

All the passes make their steepest ascents on the eastern flank. On this side the average rise is not far from a thousand feet to the mile, while on the west it is about two hundred feet. Another marked difference between the eastern and western portions of the passes is that the former begin at the very foot of the range, while the latter can hardly be said to begin lower than an elevation of from seven to ten thousand feet. Approaching the range from the gray levels of Mono and Owen's Valley on the east, the traveler sees before him the steep, short passes in full view, fenced in by rugged spurs that come plunging down from the shoulders of the peaks on either side, the courses of the more direct being disclosed from top to bottom without interruption. But from the west one sees nothing of the way he may be seeking until near the summit, after days have been spent in threading the forests growing on the main dividing ridges between the river cañons.

It is interesting to observe how surely the alp-crossing animals of every kind fall into the same trails. The more rugged and inaccessible the general character of the topography of any particular region, the more surely will the trails of white men, Indians, bears, wild sheep, etc., be found converging into the best passes. The Indians of the western slope venture cautiously over the passes in settled weather to attend dances, and obtain loads of pine-nuts and the larvæ of a small fly that breeds in Mono and Owen's lakes, which, when dried, forms an important article of food; while the Pah Utes cross over from the east to hunt the deer and obtain supplies of acorns; and it is truly astonishing to see what immense loads the haggard old squaws make out to carry bare-footed through these rough passes, oftentimes for a distance of sixty or seventy miles. They are always accompanied by the men, who stride on, unburdened and erect, a little in advance, kindly stooping at difficult places to pile stepping-stones for their patient, pack-animal wives, just as they would prepare the way for their ponies.

Bears evince great sagacity as mountaineers, but although they are tireless and enterprising travelers they seldom cross the range. I have several times tracked them through the Mono Pass, but only in late years, after cattle and sheep had passed that way, when they doubtless were following to feed on the stragglers and on those that had been killed by falling over the rocks. Even the wild sheep, the best mountaineers of all, choose regular passes in making journeys across the summits. Deer seldom cross the range in either direction. I have never yet observed a single specimen of the mule-deer of the Great Basin west of the summit, and rarely one of the black-tailed species on the eastern slope, notwithstanding many of the latter ascend the range nearly to the summit every summer, to feed in the wild gardens and bring forth their young.

The glaciers are the pass-makers, and it is by them that the courses of all mountaineers are predestined. Without exception every pass in the Sierra was created by them without the slightest aid or predetermining guidance from any of the cataclysmic agents. I have seen elaborate statements of the amount of drilling and blasting accomplished in the construction of the railroad across the Sierra, above Donner Lake; but for every pound of rock moved in this way, the glaciers which descended east and west through this same pass, crushed and carried away more than a hundred tons.

The so-called practicable road-passes are simply those portions of the range more degraded by glacial action than the adjacent portions, and degraded in such a way as to leave the summits rounded, instead of sharp; while the peaks, from the superior strength and hardness of their rocks, or from more favorable position, having suffered less degradation, are left towering above the passes as if they had been heaved into the sky by some force acting from beneath.

The scenery of all the passes, especially at the head, is of the wildest and grandest description,—lofty peaks massed together and laden around their bases with ice and snow; chains of glacier lakes; cascading streams in endless variety, with glorious views, westward over a sea of rocks and woods, and eastward over strange ashy plains, volcanoes, and the dry, dead-looking ranges of the Great Basin. Every pass, however, possesses treasures of beauty all its own.

Having thus in a general way indicated the height, leading features, and distribution of the principal passes, I will now endeavor to describe the Mono Pass in particular, which may, I think, be regarded as a fair example of the higher alpine passes in general.

The main portion of the Mono Pass is formed by Bloody Cañon, which begins at the summit of the range, and runs in a general east-northeasterly direction to the edge of the Mono Plain.

The first white men who forced a way through its somber depths were, as we have seen, eager gold-seekers. But the cañon was known and traveled as a pass by the Indians and mountain animals long before its discovery by white men, as is shown by the numerous tributary trails which come into it from every direction. Its name accords well with the character of the "early times" in California, and may perhaps have been suggested by the predominant color of the metamorphic slates in which it is in great part eroded; or more probably by blood-stains made by the unfortunate animals which were compelled to slip and shuffle awkwardly over its rough, cutting rocks. I have never known an animal, either mule or horse, to make its way through the cañon, either in going up or down, without losing more or less blood from wounds on the legs. Occasionally one is killed outright—falling headlong and rolling over precipices like a boulder. But such accidents are rarer than from the terrible appearance of the trail one would be led to expect; the more experienced when driven loose find their way over the dangerous places with a caution and sagacity that is truly wonderful. During the gold excitement it was at times a matter of considerable pecuniary importance to force a way through the cañon with pack-trains early in the spring while it was yet heavily blocked with snow; and then the mules with their loads had sometimes to be let down over the steepest drifts and avalanche beds by means of ropes.

A good bridle-path leads from Yosemite through many a grove and meadow up to the head of the cañon, a distance of about thirty miles. Here the scenery undergoes a sudden and startling condensation. Mountains, red, gray, and black, rise close at hand on the right, whitened around their bases with banks of enduring snow; on the left swells the huge red mass of Mount Gibbs, while in front the eye wanders down the shadowy cañon, and out on the warm plain of Mono, where the lake is seen gleaming like a burnished metallic disk, with clusters of lofty volcanic cones to the south of it.

When at length we enter the mountain gateway, the somber rocks seem aware of our presence, and seem to come thronging closer about us. Happily the ouzel and the old familiar robin are here to sing us welcome, and azure daisies beam with trustfulness and sympathy, enabling us to feel something of Nature's love even here, beneath the gaze of her coldest rocks.

The effect of this expressive outspokenness on the part of the cañon-rocks is greatly enhanced by the quiet aspect of the alpine meadows through which we pass just before entering the narrow gateway. The forests in which they lie, and the mountain-tops rising beyond them, seem quiet and tranquil. We catch their restful spirit, yield to the soothing influences of the sunshine, and saunter dreamily on through flowers and bees, scarce touched by a definite thought; then suddenly we find ourselves in the shadowy cañon, closeted with Nature in one of her wildest strongholds.

After the first bewildering impression begins to wear off, we perceive that it is not altogether terrible; for besides the reassuring birds and flowers we discover a chain of shining lakelets hanging down from the very summit of the pass, and linked together by a silvery stream. The highest are set in bleak, rough bowls, scantily fringed with brown and yellow sedges. Winter storms blow snow through the cañon in blinding drifts, and avalanches shoot from the heights. Then are these sparkling tarns filled and buried, leaving not a hint of their existence. In June and July they begin to blink and thaw out like sleepy eyes, the carices thrust up their short brown spikes, the daisies bloom in turn, and the most profoundly buried of them all is at length warmed and summered as if winter were only a dream.

Red Lake is the lowest of the chain, and also the largest. It seems rather dull and forbidding at first sight, lying motionless in its deep, dark bed. The cañon wall rises sheer from the water's edge on the south, but on the opposite side there is sufficient space and sunshine for a sedgy daisy garden, the center of which is brilliantly lighted with lilies, castilleias, larkspurs, and columbines, sheltered from the wind by leafy willows, and forming a most joyful outburst of plant-life keenly emphasized by the chill baldness of the onlooking cliffs.

After indulging here in a dozing, shimmering lake-rest, the happy stream sets forth again, warbling and trilling like an ouzel, ever delightfully confiding, no matter how dark the way; leaping, gliding, hither, thither, clear or foaming: manifesting the beauty of its wildness in every sound and gesture.

One of its most beautiful developments is the Diamond Cascade, situated a short distance below Red Lake. Here the tense, crystalline water is first dashed into coarse, granular spray mixed with dusty foam, and then divided into a diamond pattern by following the diagonal cleavage-joints that intersect the face of the precipice over which it pours. Viewed in front, it resembles a strip of embroidery of definite pattern, varying through the seasons with the temperature and the volume of water. Scarce a flower may be seen along its snowy border. A few bent pines look on from a distance, and small fringes of cassiope and rock-ferns are growing in fissures near the head, but these are so lowly and undemonstrative that only the attentive observer will be likely to notice them.

On the north wall of the cañon, a little below the Diamond Cascade, a glittering side stream makes its appearance, seeming to leap directly out of the sky. It first resembles a crinkled ribbon of silver hanging loosely down the wall, but grows wider as it descends, and dashes the dull rock with foam. A long rough talus curves up against this part of the cliff, overgrown with snow-pressed willows, in which the fall disappears with many an eager surge and swirl and plashing leap, finally beating its way down to its confluence with the main cañon stream.

Below this point the climate is no longer arctic. Butterflies become larger and more abundant, grasses with imposing spread of panicle wave above your shoulders, and the summery drone of the bumblebee thickens the air. The Dwarf Pine, the tree-mountaineer that climbs highest and braves the coldest blasts, is found scattered in stormbeaten clumps from the summit of the pass about halfway down the cañon. Here it is succeeded by the hardy Two-leaved Pine, which is speedily joined by the taller Yellow and Mountain Pines. These, with the burly juniper, and shimmering aspen, rapidly grow larger as the sunshine becomes richer, forming groves that block the view; or they stand more apart here and there in picturesque groups, that make beautiful and obvious harmony with the rocks and with one another. Blooming underbrush becomes abundant,—azalea, spiræa, and the brier-rose weaving fringes for the streams, and shaggy rugs to relieve the stern, unflinching rock-bosses.

Through this delightful wilderness, Cañon Creek roves without any constraining channel, throbbing and wavering; now in sunshine, now in thoughtful shade; falling, swirling, flashing from side to side in weariless exuberance of energy. A glorious milky way of cascades is thus developed, of which Bower Cascade, though one of the smallest, is perhaps the most beautiful of them all. It is situated in the lower region of the pass, just where the sunshine begins to mellow between the cold and warm climates. Here the glad creek, grown strong with tribute gathered from many a snowy fountain on the heights, sings richer strains, and becomes more human and lovable at every step. Now you may by its side find the rose and homely yarrow, and small meadows full of bees and clover. At the head of a low-browed rock, luxuriant dogwood bushes and willows arch over from bank to bank, embowering the stream with their leafy branches; and drooping plumes, kept in motion by the current, fringe the brow of the cascade in front. From this leafy covert the stream leaps out into the light in a fluted curve thick sown with sparkling crystals, and falls into a pool filled with brown boulders, out of which it creeps gray with foam-bells and disappears in a tangle of verdure like that from which it came.

Hence, to the foot of the cañon, the metamorphic slates give place to granite, whose nobler sculpture calls forth expressions of corresponding beauty from the stream in passing over it,—bright trills of rapids, booming notes of falls, solemn hushes of smooth-gliding sheets, all chanting and blending in glorious harmony. When, at length, its impetuous alpine life is done, it slips through a meadow with scarce an audible whisper, and falls asleep in Moraine Lake.

This water-bed is one of the finest I ever saw. Evergreens wave soothingly about it, and the breath of flowers floats over it like incense. Here our blessed stream rests from its rocky wanderings, all its mountaineering done,—no more foaming rock-leaping, no more wild, exulting song. It falls into a smooth, glassy sleep, stirred only by the night-wind, which, coming down the cañon, makes it croon and mutter in ripples along its broidered shores.

Leaving the lake, it glides quietly through the rushes, destined never more to touch the living rock. Henceforth its path lies through ancient moraines and reaches of ashy sage-plain, which nowhere afford rocks suitable for the development of cascades or sheer falls. Yet this beauty of maturity, though less striking, is of a still higher order, enticing us lovingly on through gentian meadows and groves of rustling aspen to Lake Mono, where, spirit-like, our happy stream vanishes in vapor, and floats free again in the sky.

Bloody Cañon, like every other in the Sierra, was recently occupied by a glacier, which derived its fountain snows from the adjacent summits, and descended into Mono Lake, at a time when its waters stood at a much higher level than now. The principal characters in which the history of the ancient glaciers is preserved are displayed here in marvelous freshness and simplicity, furnishing the student with extraordinary advantages for the acquisition of knowledge of this sort. The most striking passages are polished and striated surfaces, which in many places reflect the rays of the sun like smooth water. The dam of Red Lake is an elegantly modeled rib of metamorphic slate, brought into relief because of its superior strength, and because of the greater intensity of the glacial erosion of the rock immediately above it, caused by a steeply inclined tributary glacier, which entered the main trunk with a heavy down-thrust at the head of the lake.

Moraine Lake furnishes an equally interesting example of a basin formed wholly, or in part, by a terminal moraine dam curved across the path of a stream between two lateral moraines.

At Moraine Lake the cañon proper terminates, although apparently continued by the two lateral moraines of the vanished glacier. These moraines are about 300 feet high, and extend unbrokenly from the sides of the cañon into the plain, a distance of about five miles, curving and tapering in beautiful lines. Their sunward sides are gardens, their shady sides are groves; the former devoted chiefly to eriogonæ, compositæ, and graminæ; a square rod containing five or six profusely flowered eriogonums of several species, about the same number of bahia and linosyris, and a few grass tufts; each species being planted trimly apart, with bare gravel between, as if cultivated artificially.

My first visit to Bloody Cañon was made in the summer of 1869, under circumstances well calculated to heighten the impressions that are the peculiar offspring of mountains. I came from the blooming tangles of Florida, and waded out into the plant-gold of the great valley of California, when its flora was as yet untrodden. Never before had I beheld congregations of social flowers half so extensive or half so glorious. Golden compositæ covered all the ground from the Coast Range to the Sierra like a stratum of curdled sunshine, in which I reveled for weeks, watching the rising and setting of their innumerable suns; then I gave myself up to be borne forward on the crest of the summer wave that sweeps annually up the Sierra and spends itself on the snowy summits.

At the Big Tuolumne Meadows I remained more than a month, sketching, botanizing, and climbing among the surrounding mountains. The mountaineer with whom I then happened to be camping was one of those remarkable men one so frequently meets in California, the hard angles and bosses of whose characters have been brought into relief by the grinding excitements of the gold period, until they resemble glacial landscapes. But at this late day, my friend's activities had subsided, and his craving for rest caused him to become a gentle shepherd and literally to lie down with the lamb.

Recognizing the unsatisfiable longings of my Scotch Highland instincts, he threw out some hints concerning Bloody Cañon, and advised me to explore it. "I have never seen it myself," he said, "for I never was so unfortunate as to pass that way. But I have heard many a strange story about it, and I warrant you will at least find it wild enough."

Then of course I made haste to see it. Early next morning I made up a bundle of bread, tied my note-book to my belt, and strode away in the bracing air, full of eager, indefinite hope. The plushy lawns that lay in my path served to soothe my morning haste. The sod in many places was starred with daisies and blue gentians, over which I lingered. I traced the paths of the ancient glaciers over many a shining pavement, and marked the gaps in the upper forests that told the power of the winter avalanches. Climbing higher, I saw for the first time the gradual dwarfing of the pines in compliance with climate, and on the summit discovered creeping mats of the arctic willow overgrown with silky catkins, and patches of the dwarf vaccinium with its round flowers sprinkled in the grass like purple hail; while in every direction the landscape stretched sublimely away in fresh wildness—a manuscript written by the hand of Nature alone.

At length, as I entered the pass, the huge rocks began to close around in all their wild, mysterious impressiveness, when suddenly, as I was gazing eagerly about me, a drove of gray hairy beings came in sight, lumbering toward me with a kind of boneless, wallowing motion like bears.

I never turn back, though often so inclined, and in this particular instance, amid such surroundings, everything seemed singularly unfavorable for the calm acceptance of so grim a company. Suppressing my fears, I soon discovered that although as hairy as bears and as crooked as summit pines, the strange creatures were sufficiently erect to belong to our own species. They proved to be nothing more formidable than Mono Indians dressed in the skins of sage-rabbits. Both the men and the women begged persistently for whiskey and tobacco, and seemed so accustomed to denials that I found it impossible to convince them that I had none to give. Excepting the names of these two products of civilization, they seemed to understand not a word of English; but I afterward learned that they were on their way to Yosemite Valley to feast awhile on trout and procure a load of acorns to carry back through the pass to their huts on the shore of Mono Lake.

Occasionally a good countenance may be seen among the Mono Indians, but these, the first specimens I had seen, were mostly ugly, and some of them altogether hideous. The dirt on their faces was fairly stratified, and seemed so ancient and so undisturbed it might almost possess a geological significance. The older faces were, moreover, strangely blurred and divided into sections by furrows that looked like the cleavage-joints of rocks, suggesting exposure on the mountains in a cast-away condition for ages. Somehow they seemed to have no right place in the landscape, and I was glad to see them fading out of sight down the pass.

Then came evening, and the somber cliffs were inspired with the ineffable beauty of the alpenglow. A solemn calm fell upon everything. All the lower portion of the cañon was in gloaming shadow, and I crept into a hollow near one of the upper lakelets to smooth the ground in a sheltered nook for a bed. When the short twilight faded, I kindled a sunny fire, made a cup of tea, and lay down to rest and look at the stars. Soon the night-wind began to flow and pour in torrents among the jagged peaks, mingling strange tones with those of the waterfalls sounding far below; and as I drifted toward sleep I began to experience an uncomfortable feeling of nearness to the furred Monos. Then the full moon looked down over the edge of the cañon wall, her countenance seemingly filled with intense concern, and apparently so near as to produce a startling effect as if she had entered my bedroom, forgetting all the world, to gaze on me alone.

The night was full of strange sounds, and I gladly welcomed the morning. Breakfast was soon done, and I set forth in the exhilarating freshness of the new day, rejoicing in the abundance of pure wildness so close about me. The stupendous rocks, hacked and scarred with centuries of storms, stood sharply out in the thin early light, while down in the bottom of the cañon grooved and polished bosses heaved and glistened like swelling sea-waves, telling a grand old story of the ancient glacier that poured its crushing floods above them.

Here for the first time I met the arctic daisies in all their perfection of purity and spirituality,—gentle mountaineers face to face with the stormy sky, kept safe and warm by a thousand miracles. I leaped lightly from rock to rock, glorying in the eternal freshness and sufficiency of Nature, and in the ineffable tenderness with which she nurtures her mountain darlings in the very fountains of storms. Fresh beauty appeared at every step, delicate rock-ferns, and groups of the fairest flowers. Now another lake came to view, now a waterfall. Never fell light in brighter spangles, never fell water in whiter foam. I seemed to float through the cañon enchanted, feeling nothing of its roughness, and was out in the Mono levels before I was aware.

Looking back from the shore of Moraine Lake, my morning ramble seemed all a dream. There curved Bloody Cañon, a mere glacial furrow 2000 feet deep, with smooth rocks projecting from the sides and braided together in the middle, like bulging, swelling muscles. Here the lilies were higher than my head, and the sunshine was warm enough for palms. Yet the snow around the arctic willows was plainly visible only four miles away, and between were narrow specimen zones of all the principal climates of the globe.

On the bank of a small brook that comes gurgling down the side of the left lateral moraine, I found a camp-fire still burning, which no doubt belonged to the gray Indians I had met on the summit, and I listened instinctively and moved cautiously forward, half expecting to see some of their grim faces peering out of the bushes.

Passing on toward the open plain, I noticed three well-defined terminal moraines curved gracefully across the cañon stream, and joined by long splices to the two noble laterals. These mark the halting-places of the vanished glacier when it was retreating into its summit shadows on the breaking-up of the glacial winter.

Five miles below the foot of Moraine Lake, just where the lateral moraines lose themselves in the plain, there was a field of wild rye, growing in magnificent waving bunches six to eight feet high, bearing heads from six to twelve inches long. Rubbing out some of the grains, I found them about five eighths of an inch long, dark-colored, and sweet. Indian women were gathering it in baskets, bending down large handfuls, beating it out, and fanning it in the wind. They were quite picturesque, coming through the rye, as one caught glimpses of them here and there, in winding lanes and openings, with splendid tufts arching above their heads, while their incessant chat and laughter showed their heedless joy.

Like the rye-field, I found the so-called desert of Mono blooming in a high state of natural cultivation with the wild rose, cherry, aster, and the delicate abronia; also innumerable gilias, phloxes, poppies, and bush-compositæ. I observed their gestures and the various expressions of their corollas, inquiring how they could be so fresh and beautiful out in this volcanic desert. They told as happy a life as any plant-company I ever met, and seemed to enjoy even the hot sand and the wind.

But the vegetation of the pass has been in great part destroyed, and the same may be said of all the more accessible passes throughout the range. Immense numbers of starving sheep and cattle have been driven through them into Nevada, trampling the wild gardens and meadows almost out of existence. The lofty walls are untouched by any foot, and the falls sing on unchanged; but the sight of crushed flowers and stripped, bitten bushes goes far toward destroying the charm of wildness.

The cañon should be seen in winter. A good, strong traveler, who knows the way and the weather, might easily make a safe excursion through it from Yosemite Valley on snow-shoes during some tranquil time, when the storms are hushed. The lakes and falls would be buried then; but so, also, would be the traces of destructive feet, while the views of the mountains in their winter garb, and the ride at lightning speed down the pass between the snowy walls, would be truly glorious.

CHAPTER VI

THE GLACIER LAKES

AMONG THE MANY unlooked-for treasures that are bound up and hidden away in the depths of Sierra solitudes, none more surely charm and surprise all kinds of travelers than the glacier lakes. The forests and the glaciers and the snowy fountains of the streams advertise their wealth in a more or less telling manner even in the distance, but nothing is seen of the lakes until we have climbed above them. All the upper branches of the rivers are fairly laden with lakes, like orchard trees with fruit. They lie embosomed in the deep woods, down in the grovy bottoms of cañons, high on bald table-lands, and around the feet of the icy peaks, mirroring back their wild beauty over and over again. Some conception of their lavish abundance may be made from the fact that, from one standpoint on the summit of Red Mountain, a day's journey to the east of Yosemite Valley, no fewer than forty-two are displayed within a radius of ten miles. The whole number in the Sierra can hardly be less than fifteen hundred, not counting the smaller pools and tarns, which are innumerable. Perhaps two thirds or more lie on the western flank of the range, and all are restricted to the alpine and sub-alpine regions. At the close of the last glacial period, the middle and foot-hill regions also abounded in lakes, all of which have long since vanished as completely as the magnificent ancient glaciers that brought them into existence.

Though the eastern flank of the range is excessively steep, we find lakes pretty regularly distributed throughout even the most precipitous portions. They are mostly found in the upper branches of the cañons, and in the glacial amphitheaters around the peaks.

Occasionally long, narrow specimens occur upon the steep sides of dividing ridges, their basins swung lengthwise like hammocks, and very rarely one is found lying so exactly on the summit of the range at the head of some pass that its waters are discharged down both flanks when the snow is melting fast. But, however situated, they soon cease to form surprises to the studious mountaineer; for, like all the love-work of Nature, they are harmoniously related to one another, and to all the other features of the mountains. It is easy, therefore, to find the bright lake-eyes in the roughest and most ungovernable-looking topography of any landscape countenance. Even in the lower regions, where they have been closed for many a century, their rocky orbits are still discernible, filled in with the detritus of flood and avalanche. A beautiful system of grouping in correspondence with the glacial fountains is soon perceived; also their extension in the direction of the trends of the ancient glaciers; and in general their dependence as to form, size, and position upon the character of the rocks in which their basins have been eroded, and the quantity and direction of application of the glacial force expended upon each basin.

In the upper cañons we usually find them in pretty regular succession, strung together like beads on the bright ribbons of their feeding-streams, which pour, white and gray with foam and spray, from one to the other, their perfect mirror stillness making impressive contrasts with the grand blare and glare of the connecting cataracts. In Lake Hollow, on the north side of the Hoffmann spur, immediately above the great Tuolumne cañon, there are ten lovely lakelets lying near together in one general hollow, like eggs in a nest. Seen from above, in a general view, feathered with Hemlock Spruce, and fringed with sedge, they seem to me the most singularly beautiful and interestingly located lake-cluster I have ever yet discovered.

Lake Tahoe, 22 miles long by about 10 wide, and from 500 to over 1600 feet in depth, is the largest of all the Sierra lakes. It lies just beyond the northern limit of the higher portion of the range between the main axis and a spur that puts out on the east side from near the head of the Carson River. Its forested shores go curving in and out around many an emerald bay and pine-crowned promontory, and its waters are everywhere as keenly pure as any to be found among the highest mountains.

Donner Lake, rendered memorable by the terrible fate of the Donner party, is about three miles long, and lies about ten miles to the north of Tahoe, at the head of one of the tributaries of the Truckee. A few miles farther north lies Lake Independence, about the same size as Donner. But far the greater number of the lakes lie much higher and are quite small, few of them exceeding a mile in length, most of them less than half a mile.

Along the lower edge of the lake-belt, the smallest have disappeared by the filling-in of their basins, leaving only those of considerable size. But all along the upper freshly glaciated margin of the lake-bearing zone, every hollow, however small, lying within reach of any portion of the close network of streams, contains a bright, brimming pool; so that the landscape viewed from the mountain-tops seems to be sown broadcast with them. Many of the larger lakes are encircled with smaller ones like central gems girdled with sparkling brilliants. In general, however, there is no marked dividing line as to size. In order, therefore, to prevent confusion, I would state here that in giving numbers, I include none less than 500 yards in circumference.

In the basin of the Merced River, I counted 131, of which 111 are upon the tributaries that fall so grandly into Yosemite Valley. Pohono Creek, which forms the fall of that name, takes its rise in a beautiful lake, lying beneath the shadow of a lofty granite spur that puts out from Buena Vista peak. This is now the only lake left in the whole Pohono Basin. The Illilouette has sixteen, the Nevada no fewer than sixty-seven, the Tenaya eight, Hoffmann Creek five, and Yosemite Creek fourteen. There are but two other lake-bearing affluents of the Merced, viz., the South Fork with fifteen, and Cascade Creek with five, both of which unite with the main trunk below Yosemite.

The Merced River, as a whole, is remarkably like an elm-tree, and it requires but little effort on the part of the imagination to picture it standing upright, with all its lakes hanging upon its spreading branches, the topmost eighty miles in height. Now add all the other lake-bearing rivers of the Sierra, each in its place, and you will have a truly glorious spectacle,—an avenue the length and width of the range; the long, slender, gray shafts of the main trunks, the milky way of arching branches, and the silvery lakes, all clearly defined and shining on the sky. How excitedly such an addition to the scenery would be gazed at! Yet these lakeful rivers are still more excitingly beautiful and impressive in their natural positions to those who have the eyes to see them as they lie imbedded in their meadows and forests and glacier-sculptured rocks.

When a mountain lake is born,—when, like a young eye, it first opens to the light,—it is an irregular, expressionless crescent, inclosed in banks of rock and ice,—bare, glaciated rock on the lower side, the rugged snout of a glacier on the upper. In this condition it remains for many a year, until at length, toward the end of some auspicious cluster of seasons, the glacier recedes beyond the upper margin of the basin, leaving it open from shore to shore for the first time, thousands of years after its conception beneath the glacier that excavated its basin. The landscape, cold and bare, is reflected in its pure depths; the winds ruffle its glassy surface, and the sun fills it with throbbing spangles, while its waves begin to lap and murmur around its leafless shores,—sun-spangles during the day and reflected stars at night its only flowers, the winds and the snow its only visitors. Meanwhile, the glacier continues to recede, and numerous rills, still younger than the lake itself, bring down glacier-mud, sand-grains, and pebbles, giving rise to margin-rings and plats of soil. To these fresh soil-beds come many a waiting plant. First, a hardy carex with arching leaves and a spike of brown flowers; then, as the seasons grow warmer, and the soil-beds deeper and wider, other sedges take their appointed places, and these are joined by blue gentians, daisies, dodecatheons, violets, honeyworts, and many a lowly moss. Shrubs also hasten in time to the new gardens,—kalmia with its glossy leaves and purple flowers, the arctic willow, making soft woven carpets, together with the heathy bryanthus and cassiope, the fairest and dearest of them all. Insects now enrich the air, frogs pipe cheerily in the shallows, soon followed by the ouzel, which is the first bird to visit a glacier lake, as the sedge is the first of plants.

So the young lake grows in beauty, becoming more and more humanly lovable from century to century. Groves of aspen spring up, and hardy pines, and the Hemlock Spruce, until it is richly overshadowed and embowered. But while its shores are being enriched, the soil-beds creep out with incessant growth, contracting its area, while the lighter mud-particles deposited on the bottom cause it to grow constantly shallower, until at length the last remnant of the lake vanishes,—closed forever in ripe and natural old age. And now its feeding-stream goes winding on without halting through the new gardens and groves that have taken its place.

The length of the life of any lake depends ordinarily upon the capacity of its basin, as compared with the carrying power of the streams that flow into it, the character of the rocks over which these streams flow, and the relative position of the lake toward other lakes. In a series whose basins lie in the same cañon, and are fed by one and the same main stream, the uppermost will, of course, vanish first unless some other lake-filling agent comes in to modify the result;

because at first it receives nearly all of the sediments that the stream brings down, only the finest of the mud-particles being carried through the highest of the series to the next below. Then the next higher, and the next would be successively filled, and the lowest would be the last to vanish. But this simplicity as to duration is broken in upon in various ways, chiefly through the action of side-streams that enter the lower lakes direct. For, notwithstanding many of these side tributaries are quite short, and, during late summer, feeble, they all become powerful torrents in springtime when the snow is melting, and carry not only sand and pine-needles, but large trunks and boulders tons in weight, sweeping them down their steeply inclined channels and into the lake basins with astounding energy. Many of these side affluents also have the advantage of access to the main lateral moraines of the vanished glacier that occupied the cañon, and upon these they draw for lake-filling material, while the main trunk stream flows mostly over clean glacier pavements, where but little moraine matter is ever left for them to carry. Thus a small rapid stream with abundance of loose transportable material within its reach may fill up an extensive basin in a few centuries, while a large perennial trunk stream, flowing over clean, enduring pavements, though ordinarily a hundred times larger, may not fill a smaller basin in thousands of years.

The comparative influence of great and small streams as lake-fillers is strikingly illustrated in Yosemite Valley, through which the Merced flows. The bottom of the valley is now composed of level meadow-lands and dry, sloping soil-beds planted with oak and pine, but it was once a lake stretching from wall to wall and nearly from one end of the valley to the other, forming one of the most beautiful cliff-bound sheets of water that ever existed in the Sierra. And though never perhaps seen by human eye, it was but yesterday, geologically speaking, since it disappeared, and the traces of its existence are still so fresh, it may easily be restored to the eye of imagination and viewed in all its grandeur, about as truly and vividly as if actually before us. Now we find that the detritus which fills this magnificent basin was not brought down from the distant mountains by the main streams that converge here to form the river, however powerful and available for the purpose at first sight they appear; but almost wholly by the small local tributaries, such as those of Indian Cañon, the Sentinel, and the Three Brothers, and by a few small residual glaciers which lingered in the shadows of the walls long after the main trunk glacier had receded beyond the head of the valley.

Had the glaciers that once covered the range been melted at once, leaving the entire surface bare from top to bottom simultaneously, then of course all the lakes would have come into existence at the same time, and the highest, other circumstances being equal, would, as we have seen, be the first to vanish. But because they melted gradually from the foot of the range upward, the lower lakes were the first to see the light and the first to be obliterated. Therefore, instead of finding the lakes of the present day at the foot of the range, we find them at the top. Most of the lower lakes vanished thousands of years before those now brightening the alpine landscapes were born. And in general, owing to the deliberation of the upward retreat of the glaciers, the lowest of the existing lakes are also the oldest, a gradual transition being apparent throughout the entire belt, from the older, forested, meadow-rimmed and contracted forms all the way up to those that are new born, lying bare and meadowless among the highest peaks.

A few small lakes unfortunately situated are extinguished suddenly by a single swoop of an avalanche, carrying down immense numbers of trees, together with the soil they were growing upon. Others are obliterated by land-slips, earthquake taluses, etc., but these lake-deaths compared with those resulting from the deliberate and incessant deposition of sediments, may be termed accidental. Their fate is like that of trees struck by lightning.

The lake-line is of course still rising, its present elevation being about 8000 feet above sea-level; somewhat higher than this toward the southern extremity of the range, lower toward the northern, on account of the difference in time of the withdrawal of the glaciers, due to difference in climate. Specimens occur here and there considerably below this limit, in basins specially protected from inwashing detritus, or exceptional in size. These, however, are not sufficiently numerous to make any marked irregularity in the line. The highest I have yet found lies at an elevation of about 12,000 feet, in a glacier womb, at the foot of one of the highest of the summit peaks, a few miles to the north of Mount Ritter. The basins of perhaps twenty-five or thirty are still in process of formation beneath the few lingering glaciers, but by the time they are born, an equal or greater number will probably have died. Since the beginning of the close of the ice-period the whole number in the range has perhaps never been greater than at present.

A rough approximation to the average duration of these mountain lakes may be made from data already suggested, but I cannot stop here to present the subject in detail. I must also forgo, in the mean time, the pleasure of a full discussion of the interesting question of lake-basin formation, for which fine, clear, demonstrative material abounds in these mountains. In addition to what has been already given on the subject, I will only make this one statement. Every lake in the Sierra is a glacier lake. Their basins were not merely remodeled and scoured out by this mighty agent, but in the first place were eroded from the solid.

I must now make haste to give some nearer views of representative specimens lying at different elevations on the main lake-belt, confining myself to descriptions of the features most characteristic of each.

SHADOW LAKE

This is a fine specimen of the oldest and lowest of the existing lakes. It lies about eight miles above Yosemite Valley, on the main branch of the Merced, at an elevation of about 7350 feet above the sea; and is everywhere so securely cliff-bound that without artificial trails only wild animals can get down to its rocky shores from any direction. Its original length was about a mile and a half; now it is only half a mile in length by about a fourth of a mile in width, and over the lowest portion of the basin ninety-eight feet deep. Its crystal waters are clasped around on the north and south by majestic granite walls sculptured in true Yosemitic style into domes, gables, and battlemented headlands, which on the south come plunging down sheer into deep water, from a height of from 1500 to 2000 feet. The South Lyell glacier eroded this magnificent basin out of solid porphyritic granite while forcing its way westward from the summit fountains toward Yosemite, and the exposed rocks around the shores, and the projecting bosses of the walls, ground and burnished beneath the vast ice-flood, still glow with silvery radiance, notwithstanding the innumerable corroding storms that have fallen upon them. The general conformation of the basin, as well as the moraines laid along the top of the walls, and the grooves and scratches on the bottom and sides, indicate in the most unmistakable manner the direction pursued by this mighty ice-river, its great depth, and the tremendous energy it exerted in thrusting itself into and out of the basin; bearing down with superior pressure upon this portion of its channel, because of the greater declivity, consequently eroding it deeper than the other portions about it, and producing the lake-bowl as the necessary result.

With these magnificent ice-characters so vividly before us it is not easy to realize that the old glacier that made them vanished tens of centuries ago; for, excepting the vegetation that has sprung up, and the changes effected by an earthquake that hurled rock-avalanches from the weaker headlands, the basin as a whole presents the same appearance that it did when first brought to light. The lake itself, however, has undergone marked changes; one sees at a glance that it is growing old. More than two thirds of its original area is now dry land, covered with meadow-grasses and groves of pine and fir, and the level bed of alluvium stretching across from wall to wall at the head is evidently growing out all along its lakeward margin, and will at length close the lake forever.

Every lover of fine wildness would delight to saunter on a summer day through the flowery groves now occupying the filled-up portion of the basin. The curving shore is clearly traced by a ribbon of white sand upon which the ripples play; then comes a belt of broad-leafed sedges, interrupted here and there by impenetrable tangles of willows; beyond this there are groves of trembling aspen; then a dark, shadowy belt of Two-leaved Pine, with here and there a round carex meadow ensconced nest-like in its midst; and lastly, a narrow outer margin of majestic Silver Fir 200 feet high. The ground beneath the trees is covered with a luxuriant crop of grasses, chiefly triticum, bromus, and calamagrostis, with purple spikes and panicles arching to one's shoulders; while the open meadow patches glow throughout the summer with showy flowers,—heleniums, goldenrods, erigerons, lupines, castilleias, and lilies, and form favorite hiding- and feeding-grounds for bears and deer.

The rugged south wall is feathered darkly along the top with an imposing array of spirey Silver Firs, while the rifted precipices all the way down to the water's edge are adorned with picturesque old junipers, their cinnamon-colored bark showing finely upon the neutral gray of the granite. These, with a few venturesome Dwarf Pines and Spruces, lean out over fissured ribs and tablets, or stand erect back in shadowy niches, in an indescribably wild and fearless manner. Moreover, the white-flowered Douglas spiræa and dwarf evergreen oak form graceful fringes along the narrower seams, wherever the slightest hold can be effected. Rock-ferns, too, are here, such as allosorus, pellæa, and cheilanthes, making handsome rosettes on the drier fissures; and the delicate maidenhair, cistoperis, and woodsia hide back in mossy grottoes, moistened by some trickling rill; and then the orange wall-flower holds up its showy panicles here and there in the sunshine, and bahia makes bosses of gold. But, notwithstanding all this plant beauty, the general impression in looking across the lake is of stern, unflinching rockiness; the ferns and flowers are scarcely seen, and not one fiftieth of the whole surface is screened with plant life.

The sunnier north wall is more varied in sculpture, but the general tone is the same. A few headlands, flat-topped and soil-covered, support clumps of cedar and pine; and up-curving tangles of chinquapin and live-oak, growing on rough earthquake taluses, girdle their bases. Small streams come cascading down between them, their foaming margins brightened with gay primulas, gilias, and mimuluses. And close along the shore on this side there is a strip of rocky meadow enameled with buttercups, daisies, and white violets, and the purple-topped grasses out on its beveled border dip their leaves into the water.

The lower edge of the basin is a dam-like swell of solid granite, heavily abraded by the old glacier, but scarce at all cut into as yet by the outflowing stream, though it has flowed on unceasingly since the lake came into existence.

As soon as the stream is fairly over the lake-lip it breaks into cascades, never for a moment halting, and scarce abating one jot of its glad energy, until it reaches the next filled-up basin, a mile below. Then swirling and curving drowsily through meadow and grove, it breaks forth anew into gray rapids and falls, leaping and gliding in glorious exuberance of wild bound and dance down into another and yet another filled-up lake basin. Then, after a long rest in the levels of Little Yosemite, it makes its grandest display in the famous Nevada Fall. Out of the clouds of spray at the foot of the fall the battered, roaring river gropes its way, makes another mile of cascades and rapids, rests a moment in Emerald Pool, then plunges over the grand cliff of the Vernal Fall, and goes thundering and chafing down a boulder-choked gorge of tremendous depth and wildness into the tranquil reaches of the old Yosemite lake basin.

The color-beauty about Shadow Lake during the Indian summer is much richer than one could hope to find in so young and so glacial a wilderness. Almost every leaf is tinted then, and the goldenrods are in bloom; but most of the color is given by the ripe grasses, willows, and aspens. At the foot of the lake you stand in a trembling aspen grove, every leaf painted like a butterfly, and away to right and left round the shores sweeps a curving ribbon of meadow, red and brown dotted with pale yellow, shading off here and there into hazy purple. The walls, too, are dashed with bits of bright color that gleam out on the neutral granite gray. But neither the walls, nor the margin meadow, nor yet the gay, fluttering grove in which you stand, nor the lake itself, flashing with spangles, can long hold your attention; for at the head of the lake there is a gorgeous mass of orange-yellow, belonging to the main aspen belt of the basin, which seems the very fountain whence all the color below it had flowed, and here your eye is filled and fixed. This glorious mass is about thirty feet high, and extends across the basin nearly from wall to wall. Rich bosses of willow flame in front of it, and from the base of these the brown meadow comes forward to the water's edge, the whole being relieved against the unyielding green of the coniferæ, while thick sun-gold is poured over all.

During these blessed color-days no cloud darkens the sky, the winds are gentle, and the landscape rests, hushed everywhere, and indescribably impressive. A few ducks are usually seen sailing on the lake, apparently more for pleasure than anything else, and the ouzels at the head of the rapids sing always; while robins, grosbeaks, and the Douglas squirrels are busy in the groves, making delightful company, and intensifying the feeling of grateful sequestration without ruffling the deep, hushed calm and peace.

This autumnal mellowness usually lasts until the end of November. Then come days of quite another kind. The winter clouds grow, and bloom, and shed their starry crystals on every leaf and rock, and all the colors vanish like a sunset. The deer gather and hasten down their well-known trails, fearful of being snowbound. Storm succeeds storm, heaping snow on the cliffs and meadows, and bending the slender pines to the ground in wide arches, one over the other, clustering and interlacing like lodged wheat. Avalanches rush and boom from the shelving heights, piling immense heaps upon the frozen lake, and all the summer glory is buried and lost. Yet in the midst of this hearty winter the sun shines warm at times, calling the Douglas squirrel to frisk in the snowy pines and seek out his hidden stores; and the weather is never so severe as to drive away the grouse and little nut-hatches and chickadees.

Toward May, the lake begins to open. The hot sun sends down innumerable streams over the cliffs, streaking them round and round with foam. The snow slowly vanishes, and the meadows show tintings of green. Then spring comes on apace; flowers and flies enrich the air and the sod, and the deer come back to the upper groves like birds to an old nest.

I first discovered this charming lake in the autumn of 1872, while on my way to the glaciers at the head of the river. It was rejoicing then in its gayest colors, untrodden, hidden in the glorious wildness like unmined gold. Year after year I walked its shores without discovering any other trace of humanity than the remains of an Indian camp-fire, and the thigh-bones of a deer that had been broken to get at the marrow. It lies out of the regular ways of Indians, who love to hunt in more accessible fields adjacent to trails. Their knowledge of deer-haunts had probably enticed them here some hunger-time when they wished to make sure of a feast; for hunting in this lake-hollow is like hunting in a fenced park. I had told the beauty of Shadow Lake only to a few friends, fearing it might come to be trampled and "improved" like Yosemite. On my last visit, as I was sauntering along the shore on the strip of sand between the water and sod, reading the tracks of the wild animals that live here, I was startled by a human track, which I at once saw belonged to some shepherd; for each step was turned out 35° or 40° from the general course pursued, and was also run over in an uncertain sprawling fashion at the heel, while a row of round dots on the right indicated the staff that shepherds carry. None but a shepherd could make such a track, and after tracing it a few minutes I began to fear that he might be seeking pasturage; for what else could he be seeking? Returning from the glaciers shortly afterward, my worst fears were realized. A trail had been made down the mountainside from the north, and all the gardens and meadows were destroyed by a horde of hoofed locusts, as if swept by a fire. The money-changers were in the temple.

ORANGE LAKE

Besides these larger cañon lakes, fed by the main cañon streams, there are many smaller ones lying aloft on the top of rock benches, entirely independent of the general drainage channels, and of course drawing their supplies from a very limited area. Notwithstanding they are mostly small and shallow, owing to their immunity from avalanche detritus and the inwashings of powerful streams, they often endure longer than others many times larger but less favorably situated. When very shallow they become dry toward the end of summer; but because their basins are ground out of seamless stone they suffer no loss save from evaporation alone; and the great depth of snow that falls, lasting into June, makes their dry season short in any case.

Orange Lake is a fair illustration of this bench form. It lies in the middle of a beautiful glacial pavement near the lower margin of the lake-line, about a mile and a half to the northwest of Shadow Lake. It is only about 100 yards in circumference. Next the water there is a girdle of carices with wide overarching leaves, then in regular order a shaggy ruff of huckleberry bushes, a zone of willows with here and there a bush of the Mountain Ash, then a zone of aspens with a few pines around the outside. These zones are of course concentric, and together form a wall beyond which the naked ice-burnished granite stretches away in every direction, leaving it conspicuously relieved, like a bunch of palms in a desert.

In autumn, when the colors are ripe, the whole circular grove, at a little distance, looks like a big handful of flowers set in a cup to be kept fresh—a tuft of goldenrods. Its feeding-streams are exceedingly beautiful, notwithstanding their inconstancy and extreme shallowness. They have no channel whatever, and consequently are left free to spread in thin sheets upon the shining granite and wander at will. In many places the current is less than a fourth of an inch deep, and flows with so little friction it is scarcely visible. Sometimes there is not a single foam-bell, or drifting pine-needle, or irregularity of any sort to manifest its motion. Yet when observed narrowly it is seen to form a web of gliding lacework exquisitely woven, giving beautiful reflections from its minute curving ripples and eddies, and differing from the water-laces of large cascades in being everywhere transparent. In spring, when the snow is melting, the lake-bowl is brimming full, and sends forth quite a large stream that slips glassily for 200 yards or so, until it comes to an almost vertical precipice 800 feet high, down which it plunges in a fine cataract; then it gathers its scattered waters and goes smoothly over folds of gently dipping granite to its confluence with the main cañon stream. During the greater portion of the year, however, not a single water sound will you hear either at head or foot of the lake, not even the whispered lappings of ripple-waves along the shore; for the winds are fenced out. But the deep mountain silence is sweetened now and then by birds that stop here to rest and drink on their way across the cañon.

LAKE STARR KING

A beautiful variety of the bench-top lakes occurs just where the great lateral moraines of the main glaciers have been shoved forward in outswelling concentric rings by small residual tributary glaciers. Instead of being encompassed by a narrow ring of trees like Orange Lake, these lie embosomed in dense moraine woods, so dense that in seeking them you may pass them by again and again, although you may know nearly where they lie concealed.

Lake Starr King, lying to the north of the cone of that name, above the Little Yosemite Valley, is a fine specimen of this variety. The ouzels pass it by, and so do the ducks; they could hardly get into it if they would, without plumping straight down inside the circling trees.

Yet these isolated gems, lying like fallen fruit detached from the branches, are not altogether without inhabitants and joyous, animating visitors. Of course fishes cannot get into them, and this is generally true of nearly every glacier lake in the range, but they are all well stocked with happy frogs. How did the frogs get into them in the first place? Perhaps their sticky spawn was carried in on the feet of ducks or other birds, else their progenitors must have made some exciting excursions through the woods and up the sides of the cañons. Down in the still, pure depths of these hidden lakelets you may also find the larvæ of innumerable insects and a great variety of beetles, while the air above them is thick with humming wings, through the midst of which fly-catchers are constantly darting. And in autumn, when the huckleberries are ripe, bands of robins and grosbeaks come to feast, forming altogether delightful little byworlds for the naturalist.

Pushing our way upward toward the axis of the range, we find lakes in greater and greater abundance; and more youthful in aspect. At an elevation of about 9000 feet above sea-level they seem to have arrived at middle age,—that is, their basins seem to be about half filled with alluvium. Broad sheets of meadow-land are seen extending into them, imperfect and boggy in many places and more nearly level than those of the older lakes below them, and the vegetation of their shores is of course more alpine. Kalmia, ledum, and cassiope fringe the meadow rocks, while the luxuriant, waving groves, so characteristic of the lower lakes, are represented only by clumps of the Dwarf Pine and Hemlock Spruce. These, however, are oftentimes very picturesquely grouped on rocky headlands around the outer rim of the meadows, or with still more striking effect crown some rocky islet.

Moreover, from causes that we cannot stop here to explain, the cliffs about these middle-aged lakes are seldom of the massive Yosemite type, but are more broken, and less sheer, and they usually stand back, leaving the shores comparatively free; while the few precipitous rocks that do come forward and plunge directly into deep water are seldom more than three or four hundred feet high.

I have never yet met ducks in any of the lakes of this kind, but the ouzel is never wanting where the feeding-streams are perennial. Wild sheep and deer may occasionally be seen on the meadows, and very rarely a bear. One might camp on the rugged shores of these bright fountains for weeks, without meeting any animal larger than the marmots that burrow beneath glacier boulders along the edges of the meadows.

The highest and youngest of all the lakes lie nestled in glacier wombs. At first sight, they seem pictures of pure bloodless desolation, miniature arctic seas, bound in perpetual ice and snow, and overshadowed by harsh, gloomy, crumbling precipices. Their waters are keen ultramarine blue in the deepest parts, lively grass-green toward the shore shallows and around the edges of the small bergs usually floating about in them. A few hardy sedges, frost-pinched every night, are occasionally found making soft sods along the sun-touched portions of their shores, and when their northern banks slope openly to the south, and are soil-covered, no matter how coarsely, they are sure to be brightened with flowers. One lake in particular now comes to mind which illustrates the floweriness of the sun-touched banks of these icy gems. Close up under the shadow of the Sierra Matterhorn, on the eastern slope of the range, lies one of the iciest of these glacier lakes at an elevation of about 12,000 feet. A short, ragged-edged glacier crawls into it from the south, and on the opposite side it is embanked and dammed by a series of concentric terminal moraines, made by the glacier when it entirely filled the basin. Half a mile below lies a second lake, at a height of 11,500 feet, about as cold and as pure as a snow-crystal. The waters of the first come gurgling down into it over and through the moraine dam, while a second stream pours into it direct from a glacier that lies to the southeast. Sheer precipices of crystalline snow rise out of deep water on the south, keeping perpetual winter on that side, but there is a fine summery spot on the other, notwithstanding the lake is only about 300 yards wide. Here, on August 25, 1873, I found a charming company of flowers, not pinched, crouching dwarfs, scarce able to look up, but warm and juicy, standing erect in rich cheery color and bloom. On a narrow strip of shingle, close to the water's edge, there were a few tufts of carex gone to seed; and a little way back up the rocky bank at the foot of a crumbling wall so inclined as to absorb and radiate as well as reflect a considerable quantity of sun-heat, was the garden, containing a thrifty thicket of Cowania covered with large yellow flowers; several bushes of the alpine ribes with berries nearly ripe and wildly acid; a few handsome grasses belonging to two distinct species, and one goldenrod; a few hairy lupines and radiant spragueas, whose blue and rose-colored flowers were set off to fine advantage amid green carices; and along a narrow seam in the very warmest angle of the wall a perfectly gorgeous fringe of *Epilobium obcordatum* with flowers an inch wide, crowded together in lavish profusion, and colored as royal a purple as ever was worn by any high-bred plant of the tropics; and best of all, and greatest of all, a noble thistle in full bloom, standing erect, head and shoulders above his companions, and thrusting out his lances in sturdy vigor as if growing on a Scottish brae. All this brave warm bloom among the raw stones; right in the face of the onlooking glaciers.

As far as I have been able to find out, these upper lakes are snow-buried in winter to a depth of about thirty-five or forty feet, and those most exposed to avalanches, to a depth of even a hundred feet or more. These last are, of course, nearly lost to the landscape. Some remain buried for years, when the snowfall is exceptionally great, and many open only on one side late in the season. The snow of the closed side is composed of coarse granules compacted and frozen into a firm, faintly stratified mass, like the *névé* of a glacier. The lapping waves of the open portion gradually undermine and cause it to break off in large masses like icebergs, which gives rise to a precipitous front like the discharging wall of a glacier entering the sea. The play of the lights among the crystal angles of these snow-cliffs, the pearly white of the outswelling bosses, the bergs drifting in front, aglow in the sun and edged with green water, and the deep blue disk of the lake itself extending to your feet,—this forms a picture that enriches all your afterlife, and is never forgotten. But however perfect the season and the day, the cold incompleteness of these young lakes is always keenly felt. We approach them with a kind of mean caution, and steal unconfidingly around their crystal shores, dashed and ill at ease, as if expecting to hear some forbidding voice. But the love-songs of the ouzels and the love-looks of the daisies gradually reassure us, and manifest the warm fountain humanity that pervades the coldest and most solitary of them all.

CHAPTER VII

THE GLACIER MEADOWS

AFTER THE LAKES on the High Sierra come the glacier meadows. They are smooth, level, silky lawns, lying embedded in the upper forests, on the floors of the valleys, and along the broad backs of the main dividing ridges, at a height of about 8000 to 9500 feet above the sea.

They are nearly as level as the lakes whose places they have taken, and present a dry, even surface free from rock-heaps, mossy bogginess, and the frowsy roughness of rank, coarse-leaved, weedy, and shrubby vegetation. The sod is close and fine, and so complete that you cannot see the ground; and at the same time so brightly enameled with flowers and butterflies that it may well be called a garden-meadow, or meadow-garden; for the plushy sod is in many places so crowded with gentians, daisies, ivesias, and various species of orthocarpus that the grass is scarcely noticeable, while in others the flowers are only pricked in here and there singly, or in small ornamental rosettes.

The most influential of the grasses composing the sod is a delicate calamagrostis with fine filiform leaves, and loose, airy panicles that seem to float above the flowery lawn like a purple mist. But, write as I may, I cannot give anything like an adequate idea of the exquisite beauty of these mountain carpets as they lie smoothly outspread in the savage wilderness. What words are fine enough to picture them? to what shall we liken them? The flowery levels of the prairies of the old West, the luxuriant savannahs of the South, and the finest of cultivated meadows are coarse in comparison. One may at first sight compare them with the carefully tended lawns of pleasure-grounds; for they are as free from weeds as they, and as smooth, but here the likeness ends; for these wild lawns, with all their exquisite fineness, have no trace of that painful, licked, snipped, repressed appearance that pleasure-ground lawns are apt to have even when viewed at a distance. And, not to mention the flowers with which they are brightened, their grasses are very much finer both in color and texture, and instead of lying flat and motionless, matted together like a dead green cloth, they respond to the touches of every breeze, rejoicing in pure wildness, blooming and fruiting in the vital light.

Glacier meadows abound throughout all the alpine and subalpine regions of the Sierra in still greater numbers than the lakes. Probably from 2500 to 3000 exist between latitude 36° 30′ and 39°, distributed, of course, like the lakes, in concordance with all the other glacial features of the landscape.

On the head waters of the rivers there are what are called "Big Meadows," usually about from five to ten miles long. These occupy the basins of the ancient ice-seas, where many tributary glaciers came together to form the grand trunks. Most, however, are quite small, averaging perhaps but little more than three fourths of a mile in length.

One of the very finest of the thousands I have enjoyed lies hidden in an extensive forest of the Two-leaved Pine, on the edge of the basin of the ancient Tuolumne Mer de Glace, about eight miles to the west of Mount Dana.

Imagine yourself at the Tuolumne Soda Springs on the bank of the river, a day's journey above Yosemite Valley. You set off northward through a forest that stretches away indefinitely before you, seemingly unbroken by openings of any kind. As soon as you are fairly into the woods, the gray mountain-peaks, with their snowy gorges and hollows, are lost to view. The ground is littered with fallen trunks that lie crossed and recrossed like storm-lodged wheat; and besides this close forest of pines, the rich moraine soil supports a luxuriant growth of ribbon-leaved grasses—bromus, triticum, calamagrostis, agrostis, etc., which rear their handsome spikes and panicles above your waist. Making your way through the fertile wilderness,—finding lively bits of interest now and then in the squirrels and Clark crows, and perchance in a deer or bear,—after the lapse of an hour or two vertical bars of sunshine are seen ahead between the brown shafts of the pines, showing that you are approaching an open space, and then you suddenly emerge from the forest shadows upon a delightful purple lawn lying smooth and free in the light like a lake. This is a glacier meadow. It is about a mile and a half long by a quarter of a mile wide. The trees come pressing forward all around in close serried ranks, planting their feet exactly on its margin, and holding themselves erect, strict and orderly like soldiers on parade; thus bounding the meadow with exquisite precision, yet with free curving lines such as Nature alone can draw. With inexpressible delight you wade out into the grassy sun-lake, feeling yourself contained in one of Nature's most sacred chambers, withdrawn from the sterner influences of the mountains, secure from all intrusion, secure from yourself, free in the universal beauty. And notwithstanding the scene is so impressively spiritual, and you seem dissolved in it, yet everything about you is beating with warm, terrestrial, human love and life delightfully substantial and familiar. The resiny pines are types of health and steadfastness; the robins feeding on the sod belong to the same species you have known since childhood; and surely these daisies, larkspurs, and goldenrods are the very friend-flowers of the old home garden. Bees hum as in a harvest noon, butterflies waver above the flowers, and like them you lave in the vital sunshine, too richly and homogeneously joy-filled to be capable of partial thought. You are all eye, sifted through and through with light and beauty. Sauntering along the brook that meanders silently through the meadow from the east, special flowers call you back to discriminating consciousness. The sod comes curving down to the water's edge, forming bossy outswelling banks, and in some places overlapping countersunk boulders and forming bridges. Here you find mats of the curious dwarf willow scarce an inch high, yet sending up a multitude of gray silky catkins, illumined here and there with the purple cups and bells of bryanthus and vaccinium.

Go where you may, you everywhere find the lawn divinely beautiful, as if Nature had fingered and adjusted every plant this very day. The floating grass panicles are scarcely felt in brushing through their midst, so fine are they, and none of the flowers have tall or rigid stalks. In the brightest places you find three species of gentians with different shades of blue, daisies pure as the sky, silky leaved ivesias with warm yellow flowers, several species of orthocarpus with blunt, bossy spikes, red and purple and yellow; the alpine goldenrod, pentstemon, and clover, fragrant and honeyful, with their colors massed and blended. Parting the grasses and looking more closely you may trace the branching of their shining stems, and note the marvelous beauty of their mist of flowers, the glumes and pales exquisitely penciled, the yellow dangling stamens, and feathery pistils. Beneath the lowest leaves you discover a fairy realm of mosses,—hypnum, dicranum, polytrichum, and many others,—their precious spore-cups poised daintily on polished shafts, curiously hooded, or open, showing the richly ornate peristomas worn like royal crowns. Creeping liverworts are here also in abundance, and several rare species of fungi, exceedingly small, and frail, and delicate, as if made only for beauty. Caterpillars, black beetles, and ants roam the wilds of this lower world, making their way through miniature groves and thickets like bears in a thick wood.

And how rich, too, is the life of the sunny air! Every leaf and flower seems to have its winged representative overhead. Dragonflies shoot in vigorous zigzags through the dancing swarms, and a rich profusion of butterflies—the leguminosæ of insects—make a fine addition to the general show. Many of these last are comparatively small at this elevation, and as yet almost unknown to science; but every now and then a familiar vanessa or papilio comes sailing past. Humming-birds, too, are quite common here, and the robin is always found along the margin of the stream, or out in the shallowest portions of the sod, and sometimes the grouse and mountain quail, with their broods of precious fluffy chickens. Swallows skim the grassy lake from end to end, fly-catchers come and go in fitful flights from the tops of dead spars, while woodpeckers swing across from side to side in graceful festoon curves,—birds, insects, and flowers all in their own way telling a deep summer joy.

The influences of pure nature seem to be so little known as yet, that it is generally supposed that complete pleasure of this kind, permeating one's very flesh and bones, unfits the student for scientific pursuits in which cool judgment and observation are required. But the effect is just the opposite. Instead of producing a dissipated condition, the mind is fertilized and stimulated and developed like sun-fed plants. All that we have seen here enables us to see with surer vision the fountains among the summit-peaks to the east whence flowed the glaciers that ground soil for the surrounding forest; and down at the foot of the meadow the moraine which formed the dam which gave rise to the lake that occupied this basin before the meadow was made; and around the margin the stones that were shoved back and piled up into a rude wall by the expansion of the lake ice during long bygone winters; and along the sides of the streams the slight hollows of the meadow which mark those portions of the old lake that were the last to vanish.

I would fain ask my readers to linger awhile in this fertile wilderness, to trace its history from its earliest glacial beginnings, and learn what we may of its wild inhabitants and visitors. How happy the birds are all summer and some of them all winter; how the pouched marmots drive tunnels under the snow, and how fine and brave a life the slandered coyote lives here, and the deer and bears! But, knowing well the difference between reading and seeing, I will only ask attention to some brief sketches of its varying aspects as they are presented throughout the more marked seasons of the year.

The summer life we have been depicting lasts with but little abatement until October, when the night frosts begin to sting, bronzing the grasses, and ripening the leaves of the creeping heathworts along the banks of the stream to reddish purple and crimson; while the flowers disappear, all save the goldenrods and a few daisies, that continue to bloom on unscathed until the beginning of snowy winter. In still nights the grass panicles and every leaf and stalk are laden with frost crystals, through which the morning sunbeams sift in ravishing splendor, transforming each to a precious diamond radiating the colors of the rainbow. The brook shallows are plaited across and across with slender lances of ice, but both these and the grass crystals are melted before midday, and, notwithstanding the great elevation of the meadow, the afternoons are still warm enough to revive the chilled butterflies and call them out to enjoy the late-flowering goldenrods. The divine alpenglow flushes the surrounding forest every evening, followed by a crystal night with hosts of lily stars, whose size and brilliancy cannot be conceived by those who have never risen above the lowlands.

Thus come and go the bright sun-days of autumn, not a cloud in the sky, week after week until near December. Then comes a sudden change. Clouds of a peculiar aspect with a slow, crawling gait gather and grow in the azure, throwing out satiny fringes, and becoming gradually darker until every lake-like rift and opening is closed and the whole bent firmament is obscured in equal structureless gloom. Then comes the snow, for the clouds are ripe, the meadows of the sky are in bloom, and shed their radiant blossoms like an orchard in the spring. Lightly, lightly they lodge in the brown grasses and in the tasseled needles of the pines, falling hour after hour, day after day, silently, lovingly,—all the winds hushed,—glancing and circling hither, thither, glinting against one another, rays interlocking in flakes as large as daisies; and then the dry grasses, and the trees, and the stones are all equally abloom again. Thunder-showers occur here during the summer months, and impressive it is to watch the coming of the big transparent drops, each a small world in itself,—one unbroken ocean without islands hurling free through the air like planets through space. But still more impressive to me is the coming of the snow-flowers,—falling stars, winter daisies,—giving bloom to all the ground alike. Raindrops blossom brilliantly in the rainbow, and change to flowers in the sod, but snow comes in full flower direct from the dark, frozen sky.

The later snow-storms are oftentimes accompanied by winds that break up the crystals, when the temperature is low, into single petals and irregular dusty fragments; but there is comparatively little drifting on the meadow, so securely is it embosomed in the woods. From December to May, storm succeeds storm, until the snow is about fifteen or twenty feet deep, but the surface is always as smooth as the breast of a bird.

Hushed now is the life that so late was beating warmly. Most of the birds have gone down below the snow-line, the plants sleep, and all the fly-wings are folded. Yet the sun beams gloriously many a cloudless day in midwinter, casting long lance shadows athwart the dazzling expanse. In June small flecks of the dead, decaying sod begin to appear, gradually widening and uniting with one another, covered with creeping rags of water during the day, and ice by night, looking as hopeless and unvital as crushed rocks just emerging from the darkness of the glacial period. Walk the meadow now! Scarce the memory of a flower will you find. The ground seems twice dead. Nevertheless, the annual resurrection is drawing near. The life-giving sun pours his floods, the last snow-wreath melts, myriads of growing points push eagerly through the steaming mold, the birds come back, new wings fill the air, and fervid summer life comes surging on, seemingly yet more glorious than before.

This is a perfect meadow, and under favorable circumstances exists without manifesting any marked changes for centuries. Nevertheless, soon or late it must inevitably grow old and vanish. During the calm Indian summer, scarce a sand-grain moves around its banks, but in flood-times and storm-times, soil is washed forward upon it and laid in successive sheets around its gently sloping rim, and is gradually extended to the center, making it dryer. Through a considerable period the meadow vegetation is not greatly affected thereby, for it gradually rises with the rising ground, keeping on the surface like water-plants rising on the swell of waves. But at length the elevation of the meadow-land goes on so far as to produce too dry a soil for the specific meadow-plants, when, of course, they have to give up their places to others fitted for the new conditions. The most characteristic of the newcomers at this elevation above the sea are principally sun-loving gilias, eriogonæ, and compositæ, and finally forest-trees. Henceforward the obscuring changes are so manifold that the original lake-meadow can be unveiled and seen only by the geologist.

Generally speaking, glacier lakes vanish more slowly than the meadows that succeed them, because, unless very shallow, a greater quantity of material is required to fill up their basins and obliterate them than is required to render the surface of the meadow too high and dry for meadow vegetation. Furthermore, owing to the weathering to which the adjacent rocks are subjected, material of the finer sort, susceptible of transportation by rains and ordinary floods, is more abundant during the meadow period than during the lake period. Yet doubtless many a fine meadow favorably situated exists in almost prime beauty for thousands of years, the process of extinction being exceedingly slow, as we reckon time. This is especially the case with meadows circumstanced like the one we have described—embosomed in deep woods, with the ground rising gently away from it all around, the network of tree-roots in which all the ground is clasped preventing any rapid torrential washing. But, in exceptional cases, beautiful lawns formed with great deliberation are overwhelmed and obliterated at once by the action of landslips, earthquake avalanches, or extraordinary floods, just as lakes are.

In those glacier meadows that take the places of shallow lakes which have been fed by feeble streams, glacier mud and fine vegetable humus enter largely into the composition of the soil; and on account of the shallowness of this soil, and the seamless, watertight, undrained condition of the rock-basins, they are usually wet, and therefore occupied by tall grasses and sedges, whose coarse appearance offers a striking contrast to that of the delicate lawn-making kind described above. These shallow-soiled meadows are oftentimes still further roughened and diversified by partially buried moraines and swelling bosses of the bed-rock, which, with the trees and shrubs growing upon them, produce a striking effect as they stand in relief like islands in the grassy level, or sweep across in rugged curves from one forest wall to the other.

Throughout the upper meadow region, wherever water is sufficiently abundant and low in temperature, in basins secure from flood-washing, handsome bogs are formed with a deep growth of brown and yellow sphagnum picturesquely ruffled with patches of kalmia and ledum which ripen masses of beautiful color in the autumn. Between these cool, spongy bogs and the dry, flowery meadows there are many interesting varieties which are graduated into one another by the varied conditions already alluded to, forming a series of delightful studies.

HANGING MEADOWS

Another very well-marked and interesting kind of meadow, differing greatly both in origin and appearance from the lake-meadows, is found lying aslant upon moraine-covered hillsides trending in the direction of greatest declivity, waving up and down over rock heaps and ledges, like rich green ribbons brilliantly illumined with tall flowers. They occur both in the alpine and sub-alpine regions in considerable numbers, and never fail to make telling features in the landscape. They are often a mile or more in length, but never very wide—usually from thirty to fifty yards. When the mountain or cañon side on which they lie dips at the required angle, and other conditions are at the same time favorable, they extend from above the timber line to the bottom of a cañon or lake basin, descending in fine, fluent lines like cascades, breaking here and there into a kind of spray on large boulders, or dividing and flowing around on either side of some projecting islet. Sometimes a noisy stream goes brawling down through them, and again, scarcely a drop of water is in sight. They owe their existence, however, to streams, whether visible or invisible, the wildest specimens being found where some perennial fountain, as a glacier or snowbank or moraine spring sends down its waters across a rough sheet of soil in a dissipated web of feeble, oozing rivulets. These conditions give rise to a meadowy vegetation, whose extending roots still more obstruct the free flow of the waters, and tend to dissipate them out over a yet wider area. Thus the moraine soil and the necessary moisture requisite for the better class of meadow plants are at times combined about as perfectly as if smoothly outspread on a level surface. Where the soil happens to be composed of the finer qualities of glacial detritus and the water is not in excess, the nearest approach is made by the vegetation to that of the lake-meadow. But where, as is more commonly the case, the soil is coarse and bouldery, the vegetation is correspondingly rank. Tall, wide-leaved grasses take their places along the sides, and rushes and nodding carices in the wetter portions, mingled with the most beautiful and imposing flowers,—orange lilies and larkspurs seven or eight feet high, lupines, senecios, alliums, painted-cups, many species of mimulus and pentstemon, the ample boat-leaved *Veratrum alba*, and the magnificent alpine columbine, with spurs an inch and a half long. At an elevation of from seven to nine thousand feet showy flowers frequently form the bulk of the vegetation; then the hanging meadows become hanging gardens.

In rare instances we find an alpine basin the bottom of which is a perfect meadow, and the sides nearly all the way round, rising in gentle curves, are covered with moraine soil, which, being saturated with melting snow from encircling fountains, gives rise to an almost continuous girdle of down-curving meadow vegetation that blends gracefully into the level meadow at the bottom, thus forming a grand, smooth, soft, meadow-lined mountain nest. It is in meadows of this sort that the mountain beaver (*Haplodon*) loves to make his home, excavating snug chambers beneath the sod, digging canals, turning the underground waters from channel to channel to suit his convenience, and feeding the vegetation.

Another kind of meadow or bog occurs on densely timbered hillsides where small perennial streams have been dammed at short intervals by fallen trees. Still another kind is found hanging down smooth, flat precipices, while corresponding leaning meadows rise to meet them.

There are also three kinds of small pot-hole meadows one of which is found along the banks of the main streams, another on the summits of rocky ridges, and the third on glacier pavements, all of them interesting in origin and brimful of plant beauty.

CHAPTER VIII

THE FORESTS

THE CONIFEROUS FORESTS of the Sierra are the grandest and most beautiful in the world, and grow in a delightful climate on the most interesting and accessible of mountain-ranges, yet strange to say they are not well known. More than sixty years ago David Douglas, an enthusiastic botanist and tree lover, wandered alone through fine sections of the Sugar Pine and Silver Fir woods wild with delight. A few years later, other botanists made short journeys from the coast into the lower woods. Then came the wonderful multitude of miners into the foot-hill zone, mostly blind with gold-dust, soon followed by "sheepmen," who, with wool over their eyes, chased their flocks through all the forest belts from one end of the range to the other. Then the Yosemite Valley was discovered, and thousands of admiring tourists passed through sections of the lower and middle zones on their way to that wonderful park, and gained fine glimpses of the Sugar Pines and Silver Firs along the edges of dusty trails and roads. But few indeed, strong and free with eyes undimmed with care, have gone far enough and lived long enough with the trees to gain anything like a loving conception of their grandeur and significance as manifested in the harmonies of their distribution and varying aspects throughout the seasons, as they stand arrayed in their winter garb rejoicing in storms, putting forth their fresh leaves in the spring while steaming with resiny fragrance, receiving the thunder-showers of summer, or reposing heavy-laden with ripe cones in the rich sungold of autumn. For knowledge of this kind one must dwell with the trees and grow with them, without any reference to time in the almanac sense.

The distribution of the general forest in belts is readily perceived. These, as we have seen, extend in regular order from one extremity of the range to the other; and however dense and somber they may appear in general views, neither on the rocky heights nor down in the leafiest hollows will you find anything to remind you of the dank, malarial selvas of the Amazon and Orinoco, with their "boundless contiguity of shade," the monotonous uniformity of the Deodar forests of the Himalaya, the Black Forest of Europe, or the dense dark woods of Douglas Spruce where rolls the Oregon. The giant pines, and firs, and Sequoias hold their arms open to the sunlight, rising above one another on the mountain benches, marshaled in glorious array, giving forth the utmost expression of grandeur and beauty with inexhaustible variety and harmony.

The inviting openness of the Sierra woods is one of their most distinguishing characteristics. The trees of all the species stand more or less apart in groves, or in small, irregular groups, enabling one to find a way nearly everywhere, along sunny colonnades and through openings that have a smooth, park-like surface, strewn with brown needles and burs. Now you cross a wild garden, now a meadow, now a ferny, willowy stream; and ever and anon you emerge from all the groves and flowers upon some granite pavement or high, bare ridge commanding superb views above the waving sea of evergreens far and near.

One would experience but little difficulty in riding on horseback through the successive belts all the way up to the storm-beaten fringes of the icy peaks. The deep cañons, however, that extend from the axis of the range, cut the belts more or less completely into sections, and prevent the mounted traveler from tracing them lengthwise.

This simple arrangement in zones and sections brings the forest, as a whole, within the comprehension of every observer. The different species are ever found occupying the same relative positions to one another, as controlled by soil, climate, and the comparative vigor of each species in taking and holding the ground; and so appreciable are these relations, one need never be at a loss in determining, within a few hundred feet, the elevation above sea-level by the trees alone; for, notwithstanding some of the species range upward for several thousand feet, and all pass one another more or less, yet even those possessing the greatest vertical range are available in this connection, in as much as they take on new forms corresponding with the variations in altitude.

Crossing the treeless plains of the Sacramento and San Joaquin from the west and reaching the Sierra foot-hills, you enter the lower fringe of the forest, composed of small oaks and pines, growing so far apart that not one twentieth of the surface of the ground is in shade at clear noonday. After advancing fifteen or twenty miles, and making an ascent of from two to three thousand feet, you reach the lower margin of the main pine belt, composed of the gigantic Sugar Pine, Yellow Pine, Incense Cedar, and Sequoia. Next you come to the magnificent Silver Fir belt, and lastly to the upper pine belt, which sweeps up the rocky acclivities of the summit peaks in a dwarfed, wavering fringe to a height of from ten to twelve thousand feet.

This general order of distribution, with reference to climate dependent on elevation, is perceived at once, but there are other harmonies, as far-reaching in this connection, that become manifest only after patient observation and study. Perhaps the most interesting of these is the arrangement of the forests in long, curving bands, braided together into lace-like patterns, and outspread in charming variety. The key to this beautiful harmony is the ancient glaciers; where they flowed the trees followed, tracing their wavering courses along cañons, over ridges, and over high, rolling plateaus. The Cedars of Lebanon, says Hooker, are growing upon one of the moraines of an ancient glacier. All the forests of the Sierra are growing upon moraines. But moraines vanish like the glaciers that make them. Every storm that falls upon them wastes them, cutting gaps, disintegrating boulders, and carrying away their decaying material into new formations, until at length they are no longer recognizable by any save students, who trace their transitional forms down from the fresh moraines still in process of formation, through those that are more and more ancient, and more and more obscured by vegetation and all kinds of post-glacial weathering.

Had the ice-sheet that once covered all the range been melted simultaneously from the foot-hills to the summits, the flanks would, of course, have been left almost bare of soil, and these noble forests would be wanting. Many groves and thickets would undoubtedly have grown up on lake and avalanche beds, and many a fair flower and shrub would have found food and a dwelling-place in weathered nooks and crevices, but the Sierra as a whole would have been a bare, rocky desert.

It appears, therefore, that the Sierra forests in general indicate the extent and positions of the ancient moraines as well as they do lines of climate. For forests, properly speaking, cannot exist without soil; and, since the moraines have been deposited upon the solid rock, and only upon elected places, leaving a considerable portion of the old glacial surface bare, we find luxuriant forests of pine and fir abruptly terminated by scored and polished pavements on which not even a moss is growing, though soil alone is required to fit them for the growth of trees 200 feet in height.

THE NUT PINE
(*Pinus sabiniana*)

The Nut Pine, the first conifer met in ascending the range from the west, grows only on the torrid foot-hills, seeming to delight in the most ardent sun-heat, like a palm; springing up here and there singly, or in scattered groups of five or six, among scrubby White Oaks and thickets of ceanothus and manzanita; its extreme upper limit being about 4000 feet above the sea, its lower about from 500 to 800 feet.

This tree is remarkable for its airy, widespread, tropical appearance, which suggests a region of palms, rather than cool, resiny pine woods. No one would take it at first sight to be a conifer of any kind, it is so loose in habit and so widely branched, and its foliage is so thin and gray. Full-grown specimens are from forty to fifty feet in height, and from two to three feet in diameter. The trunk usually divides into three or four main branches, about fifteen and twenty feet from the ground, which, after bearing away from one another, shoot straight up and form separate summits; while the crooked subordinate branches aspire, and radiate, and droop in ornamental sprays. The slender, grayish-green needles are from eight to twelve inches long, loosely tasseled, and inclined to droop in handsome curves, contrasting with the stiff, dark-colored trunk and branches in a very striking manner. No other tree of my acquaintance, so substantial in body, is in its foliage so thin and so pervious to the light. The sunbeams sift through even the leafiest trees with scarcely any interruption, and the weary, heated traveler finds but little protection in their shade.

The generous crop of nutritious nuts which the Nut Pine yields makes it a favorite with Indians, bears, and squirrels. The cones are most beautiful, measuring from five to eight inches in length, and not much less in thickness, rich chocolate-brown in color, and protected by strong, down-curving hooks which terminate the scales. Nevertheless, the little Douglas squirrel can open them. Indians gathering the ripe nuts make a striking picture. The men climb the trees like bears and beat off the cones with sticks, or recklessly cut off the more fruitful branches with hatchets, while the squaws gather the big, generous cones, and roast them until the scales open sufficiently to allow the hard-shelled seeds to be beaten out. Then, in the cool evenings, men, women, and children, with their capacity for dirt greatly increased by the soft resin with which they are all bedraggled, form circles around camp-fires, on the bank of the nearest stream, and lie in easy independence cracking nuts and laughing and chattering, as heedless of the future as the squirrels.

Pinus tuberculata

This curious little pine is found at an elevation of from 1500 to 3000 feet, growing in close, willowy groves. It is exceedingly slender and graceful in habit, although trees that chance to stand alone outside the groves sweep forth long, curved branches, producing a striking contrast to the ordinary grove form. The foliage is of the same peculiar gray-green color as that of the Nut Pine, and is worn about as loosely, so that the body of the tree is scarcely obscured by it.

At the age of seven or eight years it begins to bear cones, not on branches, but on the main axis, and, as they never fall off, the trunk is soon picturesquely dotted with them. The branches also become fruitful after they attain sufficient size. The average size of the older trees is about thirty or forty feet in height, and twelve to fourteen inches in diameter. The cones are about four inches long, exceedingly hard, and covered with a sort of silicious varnish and gum, rendering them impervious to moisture, evidently with a view to the careful preservation of the seeds.

No other conifer in the range is so closely restricted to special localities. It is usually found apart, standing deep in chaparral on sunny hill- and cañon-sides where there is but little depth of soil, and, where found at all, it is quite plentiful; but the ordinary traveler, following carriage-roads and trails, may ascend the range many times without meeting it.

While exploring the lower portion of the Merced Cañon I found a lonely miner seeking his fortune in a quartz vein on a wild mountain-side planted with this singular tree. He told me that he called it the Hickory Pine, because of the whiteness and toughness of the wood. It is so little known, however, that it can hardly be said to have a common name. Most mountaineers refer to it as "that queer little pine-tree covered all over with burs." In my studies of this species I found a very interesting and significant group of facts, whose relations will be seen almost as soon as stated:

1st. All the trees in the groves I examined, however unequal in size, are of the same age.

2d. Those groves are all planted on dry hillsides covered with chaparral, and therefore are liable to be swept by fire.

3d. There are no seedlings or saplings in or about the living groves, but there is always a fine, hopeful crop springing up on the ground once occupied by any grove that has been destroyed by the burning of the chaparral.

4th. The cones never fall off and never discharge their seeds until the tree or branch to which they belong dies.

A full discussion of the bearing of these facts upon one another would perhaps be out of place here, but I may at least call attention to the admirable adaptation of the tree to the fire-swept regions where alone it is found. After a grove has been destroyed, the ground is at once sown lavishly with all the seeds ripened during its whole life, which seem to have been carefully held in store with reference to such a calamity. Then a young grove immediately springs up, giving beauty for ashes.

SUGAR PINE
(*Pinus lambertiana*)

This is the noblest pine yet discovered, surpassing all others not merely in size but also in kingly beauty and majesty.

It towers sublimely from every ridge and cañon of the range, at an elevation of from three to seven thousand feet above the sea, attaining most perfect development at a height of about 5000 feet.

Full-grown specimens are commonly about 220 feet high, and from six to eight feet in diameter near the ground, though some grand old patriarch is occasionally met that has enjoyed five or six centuries of storms, and attained a thickness of ten or even twelve feet, living on undecayed, sweet and fresh in every fiber.

In southern Oregon, where it was first discovered by David Douglas, on the head waters of the Umpqua, it attains still grander dimensions, one specimen having been measured that was 245 feet high, and over eighteen feet in diameter three feet from the ground. The discoverer was the Douglas for whom the noble Douglas Spruce is named, and many other plants which will keep his memory sweet and fresh as long as trees and flowers are loved. His first visit to the Pacific Coast was made in the year 1825. The Oregon Indians watched him with curiosity as he wandered in the woods collecting specimens, and, unlike the fur-gathering strangers they had hitherto known, caring nothing about trade. And when at length they came to know him better, and saw that from year to year the growing things of the woods and prairies were his only objects of pursuit, they called him "The Man of Grass," a title of which he was proud. During his first summer on the waters of the Columbia he made Fort Vancouver his headquarters, making excursions from this Hudson Bay post in every direction. On one of his long trips he saw in an Indian's pouch some of the seeds of a new species of pine which he learned were obtained from a very large tree far to the southward of the Columbia. At the end of the next summer, returning to Fort Vancouver after the setting in of the winter rains, bearing in mind the big pine he had heard of, he set out on an excursion up the Willamette Valley in search of it; and how he fared, and what dangers and hardships he endured, are best told in his own journal, from which I quote as follows:

> OCTOBER 26, 1826. Weather dull. Cold and cloudy. When my friends in England are made acquainted with my travels I fear they will think I have told them nothing but my miseries. . . . I quitted my camp early in the morning to survey the neighboring country, leaving my guide to take charge of the horses until my return in the evening. About an hour's walk from the camp I met an Indian, who on perceiving me instantly strung his bow, placed on his left arm a sleeve of raccoon skin and stood on the defensive. Being quite sure that conduct was prompted by fear and not by hostile intentions, the poor fellow having probably never seen such a being as myself before, I laid my gun at my feet on the ground and waved my hand for him to come to me, which he did

slowly and with great caution. I then made him place his bow and quiver of arrows beside my gun, and striking a light gave him a smoke out of my own pipe and a present of a few beads. With my pencil I made a rough sketch of the cone and pine tree which I wanted to obtain, and drew his attention to it, when he instantly pointed with his hand to the hills fifteen or twenty miles distant towards the south; and when I expressed my intention of going thither, cheerfully set out to accompany me. At midday I reached my long-wished-for pines, and lost no time in examining them and endeavoring to collect specimens and seeds. New and strange things seldom fail to make strong impressions, and are therefore frequently over-rated; so that, lest I should never see my friends in England to inform them verbally of this most beautiful and immensely grand tree, I shall here state the dimensions of the largest I could find among several that had been blown down by the wind. At 3 feet from the ground its circumference is 57 feet 9 inches; at 134 feet, 17 feet 5 inches; the extreme length 245 feet. . . . As it was impossible either to climb the tree or hew it down, I endeavored to knock off the cones by firing at them with ball, when the report of my gun brought eight Indians, all of them painted with red earth, armed with bows, arrows, bone-tipped spears, and flint-knives. They appeared anything but friendly. I explained to them what I wanted, and they seemed satisfied and sat down to smoke; but presently I saw one of them string his bow, and another sharpen his flint knife with a pair of wooden pincers and suspend it on the wrist of his right hand. Further testimony of their intentions was unnecessary. To save myself by flight was impossible, so without hesitation I stepped back about five paces, cocked my gun, drew one of the pistols out of my belt, and holding it in my left hand and the gun in my right, showed myself determined to fight for my life. As much as possible I endeavored to preserve my coolness, and thus we stood looking at one another without making any movement or uttering a word for perhaps ten minutes, when one at last, who seemed to be the leader, gave a sign that they wished for some tobacco; this I signified that they should have if they fetched a quantity of cones. They went off immediately in

> search of them, and no sooner were they all out of sight than I picked up my three cones and some twigs of the trees and made the quickest possible retreat, hurrying back to the camp, which I reached before dusk. . . . I now write lying on the grass with my gun cocked beside me, and penning these lines by the light of my Columbian candle, namely, an ignited piece of rosin-wood.

This grand pine discovered under such exciting circumstances Douglas named in honor of his friend Dr. Lambert of London.

The trunk is a smooth, round, delicately tapered shaft, mostly without limbs, and colored rich purplish-brown, usually enlivened with tufts of yellow lichen. At the top of this magnificent bole, long, curving branches sweep gracefully outward and downward, sometimes forming a palm-like crown, but far more nobly impressive than any palm crown I ever beheld. The needles are about three inches long, finely tempered and arranged in rather close tassels at the ends of slender branchlets that clothe the long, outsweeping limbs. How well they sing in the wind, and how strikingly harmonious an effect is made by the immense cylindrical cones that depend loosely from the ends of the main branches! No one knows what Nature can do in the way of pine-burs until he has seen those of the Sugar Pine. They are commonly from fifteen to eighteen inches long, and three in diameter; green, shaded with dark purple on their sunward sides. They are ripe in September and October. Then the flat scales open and the seeds take wing, but the empty cones become still more beautiful and effective, for their diameter is nearly doubled by the spreading of the scales, and their color changes to a warm yellowish-brown; while they remain swinging on the tree all the following winter and summer, and continue effectively beautiful even on the ground many years after they fall. The wood is deliciously fragrant, and fine in grain and texture; it is of a rich cream-yellow, as if formed of condensed sunbeams. *Retinospora obtusa, Siebold*, the glory of Eastern forests, is called "Fu-si-no-ki" (tree of the sun) by the Japanese; the Sugar Pine is the sun-tree of the Sierra. Unfortunately it is greatly prized by the lumbermen, and in accessible places is always the first tree in the woods to feel their steel. But the regular lumbermen, with their saw-mills, have been less generally destructive thus far than the shingle-makers. The wood splits freely, and there is a constant demand for the shingles. And because an ax, and saw, and frow are all the capital required for the business, many of that drifting, unsteady class of men so large in California engage in it for a few months in the year. When prospectors, hunters, ranch hands, etc., touch their "bottom dollar" and find themselves out of employment, they say, "Well, I can at least go to the Sugar Pines and make shingles." A few posts are set in the ground, and a single length cut from the first tree felled produces boards enough for the walls and roof of a cabin; all the rest the lumberman makes is for sale, and he is speedily independent. No gardener or haymaker is more sweetly perfumed than these rough mountaineers while engaged in this business, but the havoc they make is most deplorable.

The sugar, from which the common name is derived, is to my taste the best of sweets—better than maple sugar. It exudes from the heart-wood, where wounds have been made, either by forest fires, or the ax, in the shape of irregular, crisp, candy-like kernels, which are crowded together in masses of considerable size, like clusters of resin-beads. When fresh, it is perfectly white and delicious, but, because most of the wounds on which it is found have been made by fire, the exuding sap is stained on the charred surface, and the hardened sugar becomes brown. Indians are fond of it, but on account of its laxative properties only small quantities may be eaten. Bears, so fond of sweet things in general, seem never to taste it; at least I have failed to find any trace of their teeth in this connection.

No lover of trees will ever forget his first meeting with the Sugar Pine, nor will he afterward need a poet to call him to "listen what the pine-tree saith." In most pine-trees there is a sameness of expression, which, to most people, is apt to become monotonous; for the typical spiry form, however beautiful, affords but little scope for appreciable individual character. The Sugar Pine is as free from conventionalities of form and motion as any oak. No two are alike, even to the most inattentive observer; and, notwithstanding they are ever tossing out their immense arms in what might seem most extravagant gestures, there is a majesty and repose about them that precludes all possibility of the grotesque, or even picturesque, in their general expression. They are the priests of pines, and seem ever to be addressing the surrounding forest. The Yellow Pine is found growing with them on warm hillsides, and the White Silver Fir on cool northern slopes; but, noble as these are, the Sugar Pine is easily king, and spreads his arms above them in blessing while they rock and wave in sign of recognition. The main branches are sometimes found to be forty feet in length, yet persistently simple, seldom dividing at all, excepting near the end; but anything like a bare cable appearance is prevented by the small, tasseled branchlets that extend all around them; and when these superb limbs sweep out symmetrically on all sides, a crown sixty or seventy feet wide is formed, which, gracefully poised on the summit of the noble shaft, and filled with sunshine, is one of the most glorious forest objects conceivable. Commonly, however, there is a great preponderance of limbs toward the east, away from the direction of the prevailing winds.

No other pine seems to me so unfamiliar and self-contained. In approaching it, we feel as if in the presence of a superior being, and begin to walk with a light step, holding our breath. Then, perchance, while we gaze awe-stricken, along comes a merry squirrel, chattering and laughing, to break the spell, running up the trunk with no ceremony, and gnawing off the cones as if they were made only for him; while the carpenter-woodpecker hammers away at the bark, drilling holes in which to store his winter supply of acorns.

Although so wild and unconventional when full-grown, the Sugar Pine is a remarkably proper tree in youth. The old is the most original and independent in appearance of all the Sierra evergreens; the young is the most regular,—a strict follower of coniferous fashions,—slim, erect, with leafy, supple branches kept exactly in place, each tapering in outline and terminating in a spirey point. The successive transitional forms presented between the cautious neatness of youth and bold freedom of maturity offer a delightful study. At the age of fifty or sixty years, the shy, fashionable form begins to be broken up. Specialized branches push out in the most unthought-of places, and bend with the great cones, at once marking individual character, and this being constantly augmented from year to year by the varying action of the sunlight, winds, snowstorms, etc., the individuality of the tree is never again lost in the general forest.

The most constant companion of this species is the Yellow Pine, and a worthy companion it is. The Douglas Spruce, Libocedrus, Sequoia, and the White Silver Fir are also more or less associated with it; but on many deep-soiled mountain-sides, at an elevation of about 5000 feet above the sea, it forms the bulk of the forest, filling every swell and hollow and down-plunging ravine. The majestic crowns, approaching each other in bold curves, make a glorious canopy through which the tempered sunbeams pour, silvering the needles, and gilding the massive boles, and flowery, park-like ground, into a scene of enchantment.

On the most sunny slopes the white-flowered fragrant chamoebatia is spread like a carpet, brightened during early summer with the crimson Sarcodes, the wild rose, and innumerable violets and gilias. Not even in the shadiest nooks will you find any rank, untidy weeds or unwholesome darkness. On the north sides of ridges the boles are more slender, and the ground is mostly occupied by an underbrush of hazel, ceanothus, and flowering dogwood, but never so densely as to prevent the traveler from sauntering where he will; while the crowning branches are never impenetrable to the rays of the sun, and never so interblended as to lose their individuality.

View the forest from beneath or from some commanding ridge-top; each tree presents a study in itself, and proclaims the surpassing grandeur of the species.

YELLOW, OR SILVER PINE
(*Pinus ponderosa*)

The Silver, or Yellow, Pine, as it is commonly called, ranks second among the pines of the Sierra as a lumber tree, and almost rivals the Sugar Pine in stature and nobleness of port. Because of its superior powers of enduring variations of climate and soil, it has a more extensive range than any other conifer growing on the Sierra. On the western slope it is first met at an elevation of about 2000 feet, and extends nearly to the upper limit of the timber line. Thence, crossing the range by the lowest passes, it descends to the eastern base, and pushes out for a considerable distance into the hot volcanic plains, growing bravely upon well-watered moraines, gravelly lake basins, arctic ridges, and torrid lava-beds; planting itself upon the lips of craters, flourishing vigorously even there, and tossing ripe cones among the ashes and cinders of Nature's hearths.

The average size of full-grown trees on the western slope, where it is associated with the Sugar Pine, is a little less than 200 feet in height and from five to six feet in diameter, though specimens may easily be found that are considerably larger. I measured one, growing at an elevation of 4000 feet in the valley of the Merced, that is a few inches over eight feet in diameter, and 220 feet high.

Where there is plenty of free sunshine and other conditions are favorable, it presents a striking contrast in form to the Sugar Pine, being a symmetrical spire, formed of a straight round trunk, clad with innumerable branches that are divided over and over again. About one half of the trunk is commonly branchless, but where it grows at all close, three fourths or more become naked; the tree presenting then a more slender and elegant shaft than any other tree in the woods. The bark is mostly arranged in massive plates, some of them measuring four or five feet in length by eighteen inches in width, with a thickness of three or four inches, forming a quite marked and distinguishing feature. The needles are of a fine, warm, yellow-green color, six to eight inches long, firm and elastic, and crowded in handsome, radiant tassels on the upturning ends of the branches. The cones are about three or four inches long, and two and a half wide, growing in close, sessile clusters among the leaves.

The species attains its noblest form in filled-up lake basins, especially in those of the older yosemites, and so prominent a part does it form of their groves that it may well be called the Yosemite Pine. Ripe specimens favorably situated are almost always 200 feet or more in height, and the branches clothe the trunk nearly to the ground.

The Jeffrey variety attains its finest development in the northern portion of the range, in the wide basins of the McCloud and Pitt rivers, where it forms magnificent forests scarcely invaded by any other tree. It differs from the ordinary form in size, being only about half as tall, and in its redder and more closely furrowed bark, grayish-green foliage, less divided branches, and larger cones; but intermediate forms come in which make a clear separation impossible, although some botanists regard it as a distinct species. It is this variety that climbs storm-swept ridges, and wanders out among the volcanoes of the Great Basin. Whether exposed to extremes of heat or cold, it is dwarfed like every other tree, and becomes all knots and angles, wholly unlike the majestic forms we have been sketching. Old specimens, bearing cones about as big as pineapples, may sometimes be found clinging to rifted rocks at an elevation of seven or eight thousand feet, whose highest branches scarce reach above one's shoulders.

I have oftentimes feasted on the beauty of these noble trees when they were towering in all their winter grandeur, laden with snow—one mass of bloom; in summer, too, when the brown, staminate clusters hang thick among the shimmering needles, and the big purple burs are ripening in the mellow light; but it is during cloudless wind-storms that these colossal pines are most impressively beautiful. Then they bow like willows, their leaves streaming forward all in one direction, and, when the sun shines upon them at the required angle, entire groves glow as if every leaf were burnished silver. The fall of tropic light on the royal crown of a palm is a truly glorious spectacle, the fervid sun-flood breaking upon the glossy leaves in long lance-rays, like mountain water among boulders. But to me there is something more impressive in the fall of light upon these Silver Pines. It seems beaten to the finest dust, and is shed off in myriads of minute sparkles that seem to come from the very heart of the trees, as if, like rain falling upon fertile soil, it had been absorbed, to reappear in flowers of light.

This species also gives forth the finest music to the wind. After listening to it in all kinds of winds, night and day, season after season, I think I could approximate to my position on the mountains by this pine-music alone. If you would catch the tones of separate needles, climb a tree. They are well tempered, and give forth no uncertain sound, each standing out, with no interference excepting during heavy gales; then you may detect the click of one needle upon another, readily distinguishable from their free, wing-like hum. Some idea of their temper may be drawn from the fact that, notwithstanding they are so long, the vibrations that give rise to the peculiar shimmering of the light are made at the rate of about two hundred and fifty per minute.

When a Sugar Pine and one of this species equal in size are observed together, the latter is seen to be far more simple in manners, more lithely graceful, and its beauty is of a kind more easily appreciated; but then, it is, on the other hand, much less dignified and original in demeanor. The Silver Pine seems eager to shoot aloft. Even while it is drowsing in autumn sun-gold, you may still detect a skyward aspiration. But the Sugar Pine seems too unconsciously noble, and too complete in every way, to leave room for even a heavenward care.

DOUGLAS SPRUCE
(*Pseudotsuga douglasii*)

This tree is the king of the spruces, as the Sugar Pine is king of pines. It is by far the most majestic spruce I ever beheld in any forest, and one of the largest and longest lived of the giants that flourish throughout the main pine belt, often attaining a height of nearly 200 feet, and a diameter of six or seven. Where the growth is not too close, the strong, spreading branches come more than halfway down the trunk, and these are hung with innumerable slender, swaying sprays, that are handsomely feathered with the short leaves which radiate at right angles all around them. This vigorous spruce is ever beautiful, welcoming the mountain winds and the snow as well as the mellow summer light, and maintaining its youthful freshness undiminished from century to century through a thousand storms.

It makes its finest appearance in the months of June and July. The rich brown buds with which its sprays are tipped swell and break about this time, revealing the young leaves, which at first are bright yellow, making the tree appear as if covered with gay blossoms; while the pendulous bracted cones with their shell-like scales are a constant adornment.

The young trees are mostly gathered into beautiful family groups, each sapling exquisitely symmetrical. The primary branches are whorled regularly around the axis, generally in fives, while each is draped with long, feathery sprays, that descend in curves as free and as finely drawn as those of falling water.

In Oregon and Washington it grows in dense forests, growing tall and mast-like to a height of 300 feet, and is greatly prized as a lumber tree. But in the Sierra it is scattered among other trees, or forms small groves, seldom ascending higher than 5500 feet, and never making what would be called a forest. It is not particular in its choice of soil—wet or dry, smooth or rocky, it makes out to live well on them all. Two of the largest specimens I have measured are in Yosemite Valley, one of which is more than eight feet in diameter, and is growing upon the terminal moraine of the residual glacier that occupied the South Fork Cañon; the other is nearly as large, growing upon angular blocks of granite that have been shaken from the precipitous front of the Liberty Cap near the Nevada Fall. No other tree seems so capable of adapting itself to earthquake taluses, and many of these rough boulder-slopes are occupied by it almost exclusively, especially in yosemite gorges moistened by the spray of waterfalls.

INCENSE CEDAR

(*Libocedrus decurrens*)

The Incense Cedar is another of the giants quite generally distributed throughout this portion of the forest, without exclusively occupying any considerable area, or even making extensive groves. It ascends to about 5000 feet on the warmer hillsides, and reaches the climate most congenial to it at about from 3000 to 4000 feet, growing vigorously at this elevation on all kinds of soil, and in particular it is capable of enduring more moisture about its roots than any of its companions, excepting only the Sequoia.

The largest specimens are about 150 feet high, and seven feet in diameter. The bark is brown, of a singularly rich tone very attractive to artists, and the foliage is tinted with a warmer yellow than that of any other evergreen in the woods. Casting your eye over the general forest from some ridge-top, the color alone of its spiry summits is sufficient to identify it in any company.

In youth, say up to the age of seventy or eighty years, no other tree forms so strictly tapered a cone from top to bottom. The branches swoop outward and downward in bold curves, excepting the younger ones near the top, which aspire, while the lowest droop to the ground, and all spread out in flat, ferny plumes, beautifully fronded, and imbricated upon one another. As it becomes older, it grows strikingly irregular and picturesque. Large special branches put out at right angles from the trunk, form big, stubborn elbows, and then shoot up parallel with the axis. Very old trees are usually dead at the top, the main axis protruding above ample masses of green plumes, gray and lichen-covered, and drilled full of acorn holes by the woodpeckers. The plumes are exceedingly beautiful; no waving fern-frond in shady dell is more unreservedly beautiful in form and texture, or half so inspiring in color and spicy fragrance. In its prime, the whole tree is thatched with them, so that they shed off rain and snow like a roof, making fine mansions for storm-bound birds and mountaineers. But if you would see the *Libocedrus* in all its glory, you must go to the woods in winter. Then it is laden with myriads of four-sided staminate cones about the size of wheat grains,—winter wheat,—producing a golden tinge, and forming a noble illustration of Nature's immortal vigor and virility. The fertile cones are about three fourths of an inch long, borne on the outside of the plumy branchlets, where they serve to enrich still more the surpassing beauty of this grand winter-blooming goldenrod.

WHITE SILVER FIR
(*Abies concolor*)

We come now to the most regularly planted of all the main forest belts, composed almost exclusively of two noble firs—*A. concolor* and *A. magnifica*. It extends with no marked interruption for 450 miles, at an elevation of from 5000 to nearly 9000 feet above the sea. In its youth *A. concolor* is a charmingly symmetrical tree with branches regularly whorled in level collars around its whitish-gray axis, which terminates in a strong, hopeful shoot. The leaves are in two horizontal rows, along branchlets that commonly are less than eight years old, forming handsome plumes, pinnated like the fronds of ferns. The cones are grayish-green when ripe, cylindrical, about from three to four inches long by one and a half to two inches wide, and stand upright on the upper branches.

Full-grown trees, favorably situated as to soil and exposure, are about 200 feet high, and five or six feet in diameter near the ground, though larger specimens are by no means rare.

As old age creeps on, the bark becomes rougher and grayer, the branches lose their exact regularity, many are snow-bent or broken off, and the main axis often becomes double or otherwise irregular from accidents to the terminal bud or shoot; but throughout all the vicissitudes of its life on the mountains, come what may, the noble grandeur of the species is patent to every eye.

MAGNIFICENT SILVER FIR, OR RED FIR
(*Abies magnifica*)

This is the most charmingly symmetrical of all the giants of the Sierra woods, far surpassing its companion species in this respect, and easily distinguished from it by the purplish-red bark, which is also more closely furrowed than that of the white, and by its larger cones, more regularly whorled and fronded branches, and by its leaves, which are shorter, and grow all around the branchlets and point upward.

In size, these two Silver Firs are about equal, the *magnifica* perhaps a little the taller. Specimens from 200 to 250 feet high are not rare on well-ground moraine soil, at an elevation of from 7500 to 8500 feet above sea-level. The largest that I measured stands back three miles from the brink of the north wall of Yosemite Valley. Fifteen years ago it was 240 feet high, with a diameter of a little more than five feet.

Happy the man with the freedom and the love to climb one of these superb trees in full flower and fruit. How admirable the forest-work of Nature is then seen to be, as one makes his way up through the midst of the broad, fronded branches, all arranged in exquisite order around the trunk, like the whorled leaves of lilies, and each branch and branchlet about as strictly pinnate as the most symmetrical fern-frond. The staminate cones are seen growing straight downward from the under side of the young branches in lavish profusion, making fine purple clusters amid the grayish-green foliage. On the topmost branches the fertile cones are set firmly on end like small casks. They are about six inches long, three wide, covered with a fine gray down, and streaked with crystal balsam that seems to have been poured upon each cone from above.

Both the Silver Firs live 250 years or more when the conditions about them are at all favorable. Some venerable patriarch may often be seen, heavily storm-marked, towering in severe majesty above the rising generation, with a protecting grove of saplings pressing close around his feet, each dressed with such loving care that not a leaf seems wanting. Other companies are made up of trees near the prime of life, exquisitely harmonized to one another in form and gesture, as if Nature had culled them one by one with nice discrimination from all the rest of the woods.

It is from this tree, called Red Fir by the lumberman, that mountaineers always cut boughs to sleep on when they are so fortunate as to be within its limits. Two rows of the plushy branches overlapping along the middle, and a crescent of smaller plumes mixed with ferns and flowers for a pillow, form the very best bed imaginable. The essences of the pressed leaves seem to fill every pore of one's body, the sounds of falling water make a soothing hush, while the spaces between the grand spires afford noble openings through which to gaze dreamily into the starry sky. Even in the matter of sensuous ease, any combination of cloth, steel springs, and feathers seems vulgar in comparison.

The fir woods are delightful sauntering-grounds at any time of year, but most so in autumn. Then the noble trees are hushed in the hazy light, and drip with balsam; the cones are ripe, and the seeds, with their ample purple wings, mottle the air like flocks of butterflies; while deer feeding in the flowery openings between the groves, and birds and squirrels in the branches, make a pleasant stir which enriches the deep, brooding calm of the wilderness, and gives a peculiar impressiveness to every tree. No wonder the enthusiastic Douglas went wild with joy when he first discovered this species. Even in the Sierra, where so many noble evergreens challenge admiration, we linger among these colossal firs with fresh love, and extol their beauty again and again, as if no other in the world could henceforth claim our regard.

It is in these woods the great granite domes rise that are so striking and characteristic a feature of the Sierra. And here too we find the best of the garden meadows. They lie level on the tops of the dividing ridges, or sloping on the sides of them, embedded in the magnificent forest. Some of these meadows are in great part occupied by *Veratrum alba*, which here grows rank and tall, with boat-shaped leaves thirteen inches long and twelve inches wide, ribbed like those of cypripedium. Columbine grows on the drier margins with tall larkspurs and lupines waist-deep in grasses and sedges; several species of castilleia also make a bright show in beds of blue and white violets and daisies. But the glory of these forest meadows is a lily—*L. parvum*. The flowers are orange-colored and quite small, the smallest I ever saw of the true lilies; but it is showy nevertheless, for it is seven to eight feet high and waves magnificent racemes of ten to twenty flowers or more over one's head, while it stands out in the open ground with just enough of grass and other plants about it to make a fringe for its feet and show it off to best advantage.

A dry spot a little way back from the margin of a Silver Fir lily garden makes a glorious campground, especially where the slope is toward the east and opens a view of the distant peaks along the summit of the range. The tall lilies are brought forward in all their glory by the light of your blazing camp-fire, relieved against the outer darkness, and the nearest of the trees with their whorled branches tower above you like larger lilies, and the sky seen through the garden opening seems one vast meadow of white lily stars.

In the morning everything is joyous and bright, the delicious purple of the dawn changes softly to daffodil yellow and white; while the sunbeams pouring through the passes between the peaks give a margin of gold to each of them. Then the spires of the firs in the hollows of the middle region catch the glow, and your camp grove is filled with light. The birds begin to stir, seeking sunny branches on the edge of the meadow for sun-baths after the cold night, and looking for their breakfasts, every one of them as fresh as a lily and as charmingly arrayed. Innumerable insects begin to dance, the deer withdraw from the open glades and ridge-tops to their leafy hiding-places in the chaparral, the flowers open and straighten their petals as the dew vanishes, every pulse beats high, every life-cell rejoices, the very rocks seem to tingle with life, and God is felt brooding over everything great and small.

BIG TREE

(*Sequoia gigantea*)

Between the heavy pine and Silver Fir belts we find the Big Tree, the king of all the conifers in the world, "the noblest of a noble race." It extends in a widely interrupted belt from a small grove on the middle fork of the American River to the head of Deer Creek, a distance of about 260 miles, the northern limit being near the thirty-ninth parallel, the southern a little below the thirty-sixth, and the elevation of the belt above the sea varies from about 5000 to 8000 feet. From the American River grove to the forest on King's River the species occurs only in small isolated groups so sparsely distributed along the belt that three of the gaps in it are from forty to sixty miles wide. But from King's River southward the Sequoia is not restricted to mere groves, but extends across the broad rugged basins of the Kaweah and Tule rivers in noble forests, a distance of nearly seventy miles, the continuity of this part of the belt being broken only by deep cañons. The Fresno, the largest of the northern groves, occupies an area of three or four square miles, a short distance to the southward of the famous Mariposa Grove. Along the beveled rim of the cañon of the south fork of King's River there is a majestic forest of Sequoia about six miles long by two wide. This is the northernmost assemblage of Big Trees that may fairly be called a forest. Descending the precipitous divide between the King's River and Kaweah you enter the grand forests that form the main continuous portion of the belt. Advancing southward the giants become more and more irrepressibly exuberant, heaving their massive crowns into the sky from every ridge and slope, and waving onward in graceful compliance with the complicated topography of the region. The finest of the Kaweah section of the belt is on the broad ridge between Marble Creek and the middle fork, and extends from the granite headlands overlooking the hot plains to within a few miles of the cool glacial fountains of the summit peaks. The extreme upper limit of the belt is reached between the middle and south forks of the Kaweah at an elevation of 8400 feet. But the finest block of Big Tree forest in the entire belt is on the north fork of Tule River. In the northern groves there are comparatively few young trees or saplings. But here for every old, storm-stricken giant there are many in all the glory of prime vigor, and for each of these a crowd of eager, hopeful young trees and saplings growing heartily on moraines, rocky ledges, along watercourses, and in the moist alluvium of meadows, seemingly in hot pursuit of eternal life.

But though the area occupied by the species increases so much from north to south there is no marked increase in the size of the trees. A height of 275 feet and a diameter near the ground of about 20 feet is perhaps about the average size of full-grown trees favorably situated; specimens 25 feet in diameter are not very rare, and a few are nearly 300 feet high. In the Calaveras Grove there are four trees over 300 feet in height, the tallest of which by careful measurement is 325 feet. The largest I have yet met in the course of my explorations is a majestic old scarred monument in the King's River forest. It is 35 feet 8 inches in diameter inside the bark four feet from the ground. Under the most favorable conditions these giants probably live 5000 years or more, though few of even the larger trees are more than half as old. I never saw a Big Tree that had died a natural death; barring accidents they seem to be immortal, being exempt from all the diseases that afflict and kill other trees. Unless destroyed by man, they live on indefinitely until burned, smashed by lightning, or cast down by storms, or by the giving way of the ground on which they stand. The age of one that was felled in the Calaveras Grove, for the sake of having its stump for a dancing-floor, was about 1300 years, and its diameter, measured across the stump, 24 feet inside the bark. Another that was cut down in the King's River forest was about the same size, but nearly a thousand years older (2200 years), though not a very old-looking tree. It was felled to procure a section for exhibition, and thus an opportunity was given to count its annual rings of growth. The colossal scarred monument in the King's River forest mentioned above is burned half through, and I spent a day in making an estimate of its age, clearing away the charred surface with an ax and carefully counting the annual rings with the aid of a pocket-lens. The wood-rings in the section I laid bare were so involved and contorted in some places that I was not able to determine its age exactly, but I counted over 4000 rings, which showed that this tree was in its prime, swaying in the Sierra winds, when Christ walked the earth. No other tree in the world, as far as I know, has looked down on so many centuries as the Sequoia, or opens such impressive and suggestive views into history.

So exquisitely harmonious and finely balanced are even the very mightiest of these monarchs of the woods in all their proportions and circumstances there never is anything overgrown or monstrous-looking about them. On coming in sight of them for the first time, you are likely to say, "Oh, see what beautiful, noble-looking trees are towering there among the firs and pines!"—their grandeur being in the mean time in great part invisible, but to the living eye it will be manifested sooner or later, stealing slowly on the senses, like the grandeur of Niagara, or the lofty Yosemite domes. Their great size is hidden from the inexperienced observer as long as they are seen at a distance in one harmonious view. When, however, you approach them and walk round them, you begin to wonder at their colossal size and seek a measuring-rod. These giants bulge considerably at the base, but not more than is required for beauty and safety; and the only reason that this bulging seems in some cases excessive is that only a comparatively small section of the shaft is seen at once in near views. One that I measured in the King's River forest was 25 feet in diameter at the ground, and 10 feet in diameter 200 feet above the ground, showing that the taper of the trunk as a whole is charmingly fine. And when you stand back far enough to see the massive columns from the swelling instep to the lofty summit dissolving in a dome of verdure, you rejoice in the unrivaled display of combined grandeur and beauty. About a hundred feet or more of the trunk is usually branchless, but its massive simplicity is relieved by the bark furrows, which instead of making an irregular network run evenly parallel, like the fluting of an architectural column, and to some extent by tufts of slender sprays that wave lightly in the winds and cast flecks of shade, seeming to have been pinned on here and there for the sake of beauty only. The young trees have slender simple branches down to the ground, put on with strict regularity, sharply aspiring at the top, horizontal about halfway down, and drooping in handsome curves at the base. By the time the sapling is five or six hundred years old this spirey, feathery, juvenile habit merges into the firm, rounded dome form of middle age, which in turn takes on the eccentric picturesqueness of old age. No other tree in the Sierra forest has foliage so densely massed or presents outlines so firmly drawn and so steadily subordinate to a special type. A knotty ungovernable-looking branch five to eight feet thick may be seen pushing out abruptly from the smooth trunk, as if sure to throw the regular curve into confusion, but as soon as the general outline is reached it stops short and dissolves in spreading bosses of law-abiding sprays, just as if every tree were growing beneath some huge, invisible bell-glass, against whose sides every branch was being pressed and molded, yet somehow indulging in so many small departures from the regular form that there is still an appearance of freedom.

The foliage of the saplings is dark bluish-green in color, while the older trees ripen to a warm brownish-yellow tint like Libocedrus. The bark is rich cinnamon-brown, purplish in young trees and in shady portions of the old, while the ground is covered with brown leaves and burs forming color-masses of extraordinary richness, not to mention the flowers and underbrush that rejoice about them in their seasons. Walk the Sequoia woods at any time of year and you will say they are the most beautiful and majestic on earth. Beautiful and impressive contrasts meet you everywhere: the colors of tree and flower, rock and sky, light and shade, strength and frailty, endurance and evanescence, tangles of supple hazel-bushes, tree-pillars about as rigid as granite domes, roses and violets, the smallest of their kind, blooming around the feet of the giants, and rugs of the lowly chamæbatia where the sunbeams fall. Then in winter the trees themselves break forth in bloom, myriads of small four-sided staminate cones crowd the ends of the slender sprays, coloring the whole tree, and when ripe dusting the air and the ground with golden pollen. The fertile cones are bright grass-green, measuring about two inches in length by one and a half in thickness, and are made up of about forty firm rhomboidal scales densely packed, with from five to eight seeds at the base of each. A single cone, therefore, contains from two to three hundred seeds, which are about a fourth of an inch long by three sixteenths wide, including a thin, flat margin that makes them go glancing and wavering in their fall like a boy's kite. The fruitfulness of Sequoia may be illustrated by two specimen branches one and a half and two inches in diameter on which I counted 480 cones. No other Sierra conifer produces nearly so many seeds. Millions are ripened annually by a single tree, and in a fruitful year the product of one of the northern groves would be enough to plant all the mountain-ranges of the world. Nature takes care, however, that not one seed in a million shall germinate at all, and of those that do perhaps not one in ten thousand is suffered to live through the many vicissitudes of storm, drought, fire, and snow-crushing that beset their youth.

The Douglas squirrel is the happy harvester of most of the Sequoia cones. Out of every hundred perhaps ninety fall to his share, and unless cut off by his ivory sickle they shake out their seeds and remain on the tree for many years. Watching the squirrels at their harvest work in the Indian summer is one of the most delightful diversions imaginable. The woods are calm and the ripe colors are blazing in all their glory; the cone-laden trees stand motionless in the warm, hazy air, and you may see the crimson-crested woodcock, the prince of Sierra woodpeckers, drilling some dead limb or fallen trunk with his bill, and ever and anon filling the glens with his happy cackle. The humming-bird, too, dwells in these noble woods, and may oftentimes be seen glancing among the flowers or resting wing-weary on some leafless twig; here also are the familiar robin of the orchards, and the brown and grizzly bears so obviously fitted for these majestic solitudes; and the Douglas squirrel, making more hilarious, exuberant, vital stir than all the bears, birds, and humming wings together.

As soon as any accident happens to the crown of these Sequoias, such as being stricken off by lightning or broken by storms, then the branches beneath the wound, no matter how situated, seem to be excited like a colony of bees that have lost their queen, and become anxious to repair the damage. Limbs that have grown outward for centuries at right angles to the trunk begin to turn upward to assist in making a new crown, each speedily assuming the special form of true summits. Even in the case of mere stumps, burned half through, some mere ornamental tuft will try to go aloft and do its best as a leader in forming a new head.

Groups of two or three of these grand trees are often found standing close together, the seeds from which they sprang having probably grown on ground cleared for their reception by the fall of a large tree of a former generation. These patches of fresh, mellow soil beside the upturned roots of the fallen giant may be from forty to sixty feet wide, and they are speedily occupied by seedlings. Out of these seedling-thickets perhaps two or three may become trees, forming those close groups called "three graces," "loving couples," etc. For even supposing that the trees should stand twenty or thirty feet apart while young, by the time they are full-grown their trunks will touch and crowd against each other and even appear as one in some cases.

It is generally believed that this grand Sequoia was once far more widely distributed over the Sierra; but after long and careful study I have come to the conclusion that it never was, at least since the close of the glacial period, because a diligent search along the margins of the groves, and in the gaps between, fails to reveal a single trace of its previous existence beyond its present bounds. Notwithstanding, I feel confident that if every Sequoia in the range were to die to-day, numerous monuments of their existence would remain, of so imperishable a nature as to be available for the student more than ten thousand years hence.

In the first place we might notice that no species of coniferous tree in the range keeps its individuals so well together as Sequoia; a mile is perhaps the greatest distance of any straggler from the main body, and all of those stragglers that have come under my observation are young, instead of old monumental trees, relics of a more extended growth.

Again, Sequoia trunks frequently endure for centuries after they fall. I have a specimen block, cut from a fallen trunk, which is hardly distinguishable from specimens cut from living trees, although the old trunk-fragment from which it was derived has lain in the damp forest more than 380 years, probably thrice as long. The time measure in the case is simply this: when the ponderous trunk to which the old vestige belonged fell, it sunk itself into the ground, thus making a long, straight ditch, and in the middle of this ditch a Silver Fir is growing that is now four feet in diameter and 380 years old, as determined by cutting it half through and counting the rings, thus demonstrating that the remnant of the trunk that made the ditch has lain on the ground *more* than 380 years. For it is evident that to find the whole time, we must add to the 380 years the time that the vanished portion of the trunk lay in the ditch before being burned out of the way, plus the time that passed before the seed from which the monumental fir sprang fell into the prepared soil and took root. Now, because Sequoia trunks are never wholly consumed in one forest fire, and those fires recur only at considerable intervals, and because Sequoia ditches after being cleared are often left unplanted for centuries, it becomes evident that the trunk remnant in question may probably have lain a thousand years or more. And this instance is by no means a rare one.

But admitting that upon those areas supposed to have been once covered with Sequoia every tree may have fallen, and every trunk may have been burned or buried, leaving not a remnant, many of the ditches made by the fall of the ponderous trunks, and the bowls made by their upturning roots, would remain patent for thousands of years after the last vestige of the trunks that made them had vanished. Much of this ditch-writing would no doubt be quickly effaced by the flood-action of overflowing streams and rain-washing; but no inconsiderable portion would remain enduringly engraved on ridge-tops beyond such destructive action; for, where all the conditions are favorable, it is almost imperishable. *Now these historic ditches and root bowls occur in all the present Sequoia groves and forests, but as far as I have observed, not the faintest vestige of one presents itself outside of them.*

We therefore conclude that the area covered by Sequoia has not been diminished during the last eight or ten thousand years, and probably not at all in post-glacial times.

Is the species verging to extinction? What are its relations to climate, soil, and associated trees?

All the phenomena bearing on these questions also throw light, as we shall endeavor to show, upon the peculiar distribution of the species, and sustain the conclusion already arrived at on the question of extension.

In the northern groups, as we have seen, there are few young trees or saplings growing up around the failing old ones to perpetuate the race, and in as much as those aged Sequoias, so nearly childless, are the only ones commonly known, the species, to most observers, seems doomed to speedy extinction, as being nothing more than an expiring remnant, vanquished in the so-called struggle for life by pines and firs that have driven it into its last strongholds in moist glens where climate is exceptionally favorable. But the language of the majestic continuous forests of the south creates a very different impression. No tree of all the forest is more enduringly established in concordance with climate and soil. It grows heartily everywhere—on moraines, rocky ledges, along watercourses, and in the deep, moist alluvium of meadows, with a multitude of seedlings and saplings crowding up around the aged, seemingly abundantly able to maintain the forest in prime vigor. For every old storm-stricken tree, there is one or more in all the glory of prime; and for each of these many young trees and crowds of exuberant saplings. So that if all the trees of any section of the main Sequoia forest were ranged together according to age, a very promising curve would be presented, all the way up from last year's seedlings to giants, and with the young and middle-aged portion of the curve many times longer than the old portion. Even as far north as the Fresno, I counted 536 saplings and seedlings growing promisingly upon a piece of rough avalanche soil not exceeding two acres in area. This soil bed is about seven years old, and has been seeded almost simultaneously by pines, firs, Libocedrus, and Sequoia, presenting a simple and instructive illustration of the struggle for life among the rival species; and it was interesting to note that the conditions thus far affecting them have enabled the young Sequoias to gain a marked advantage.

In every instance like the above I have observed that the seedling Sequoia is capable of growing on both drier and wetter soil than its rivals, but requires more sunshine than they; the latter fact being clearly shown wherever a Sugar Pine or fir is growing in close contact with a Sequoia of about equal age and size, and equally exposed to the sun; the branches of the latter in such cases are always less leafy. Toward the south, however, where the Sequoia becomes *more* exuberant and numerous, the rival trees become *less* so; and where they mix with Sequoias, they mostly grow up beneath them, like slender grasses among stalks of Indian corn. Upon a bed of sandy flood-soil I counted ninety-four Sequoias, from one to twelve feet high, on a patch of ground once occupied by four large Sugar Pines which lay crumbling beneath them,—an instance of conditions which have enabled Sequoias to crowd out the pines.

I also noted eighty-six vigorous saplings upon a piece of fresh ground prepared for their reception by fire. Thus fire, the great destroyer of Sequoia, also furnishes bare virgin ground, one of the conditions essential for its growth from the seed. Fresh ground is, however, furnished in sufficient quantities for the constant renewal of the forests without fire, viz., by the fall of old trees. The soil is thus upturned and mellowed, and many trees are planted for every one that falls. Land-slips and floods also give rise to bare virgin ground; and a tree now and then owes its existence to a burrowing wolf or squirrel, but the most regular supply of fresh soil is furnished by the fall of aged trees.

The climatic changes in progress in the Sierra, bearing on the tenure of tree life, are entirely misapprehended, especially as to the time and the means employed by Nature in effecting them. It is constantly asserted in a vague way that the Sierra was vastly wetter than now, and that the increasing drought will of itself extinguish Sequoia, leaving its ground to other trees supposed capable of flourishing in a drier climate. But that Sequoia can and does grow on as dry ground as any of its present rivals, is manifest in a thousand places. "Why, then," it will be asked, "are Sequoias always found in greatest abundance in well-watered places where streams are exceptionally abundant?" Simply because a growth of Sequoias creates those streams. The thirsty mountaineer knows well that in every Sequoia grove he will find running water, but it is a mistake to suppose that the water is the cause of the grove being there; on the contrary, the grove is the cause of the water being there. Drain off the water and the trees will remain, but cut off the trees, and the streams will vanish. Never was cause more completely mistaken for effect than in the case of these related phenomena of Sequoia woods and perennial streams, and I confess that at first I shared in the blunder.

When attention is called to the method of Sequoia stream-making, it will be apprehended at once. The roots of this immense tree fill the ground, forming a thick sponge that absorbs and holds back the rains and melting snows, only allowing them to ooze and flow gently. Indeed, every fallen leaf and rootlet, as well as long clasping root, and prostrate trunk, may be regarded as a dam hoarding the bounty of storm-clouds, and dispensing it as blessings all through the summer, instead of allowing it to go headlong in short-lived floods. Evaporation is also checked by the dense foliage to a greater extent than by any other Sierra tree, and the air is entangled in masses and broad sheets that are quickly saturated; while thirsty winds are not allowed to go sponging and licking along the ground.

So great is the retention of water in many places in the main belt, that bogs and meadows are created by the killing of the trees. A single trunk falling across a stream in the woods forms a dam 200 feet long, and from ten to thirty feet high, giving rise to a pond which kills the trees within its reach. These dead trees fall in turn, thus making a clearing, while sediments gradually accumulate changing the pond into a bog, or meadow, for a growth of carices and sphagnum. In some instances a series of small bogs or meadows rise above one another on a hillside, which are gradually merged into one another, forming sloping bogs, or meadows, which make striking features of Sequoia woods, and since all the trees that have fallen into them have been preserved, they contain records of the generations that have passed since they began to form.

Since, then, it is a fact that thousands of Sequoias are growing thriftily on what is termed dry ground, and even clinging like mountain pines to rifts in granite precipices; and since it has also been shown that the extra moisture found in connection with the denser growths is an effect of their presence, instead of a cause of their presence, then the notions as to the former extension of the species and its near approach to extinction, based upon its supposed dependence on greater moisture, are seen to be erroneous.

The decrease in the rain- and snowfall since the close of the glacial period in the Sierra is much less than is commonly guessed. The highest post-glacial watermarks are well preserved in all the upper river channels, and they are not greatly higher than the spring floodmarks of the present; showing conclusively that no extraordinary decrease has taken place in the volume of the upper tributaries of post-glacial Sierra streams since they came into existence. But in the mean time, eliminating all this complicated question of climatic change, the plain fact remains that *the present rain- and snowfall is abundantly sufficient for the luxuriant growth of Sequoia forests*. Indeed, all my observations tend to show that in a prolonged drought the Sugar Pines and firs would perish before the Sequoia, not alone because of the greater longevity of individual trees, but because the species can endure more drought, and make the most of whatever moisture falls.

Again, if the restriction and irregular distribution of the species be interpreted as a result of the desiccation of the range, then instead of increasing as it does in individuals toward the south where the rainfall is less, it should diminish.

If, then, the peculiar distribution of Sequoia has not been governed by superior conditions of soil as to fertility or moisture, by what has it been governed?

In the course of my studies I observed that the northern groves, the only ones I was at first acquainted with, were located on just those portions of the general forest soil-belt that were first laid bare toward the close of the glacial period when the ice-sheet began to break up into individual glaciers. And while searching the wide basin of the San Joaquin, and trying to account for the absence of Sequoia where every condition seemed favorable for its growth, it occurred to me that this remarkable gap in the Sequoia belt is located exactly in the basin of the vast ancient *mer de glace* of the San Joaquin and King's River basins, which poured its frozen floods to the plain, fed by the snows that fell on more than fifty miles of the summit. I then perceived that the next great gap in the belt to the northward, forty miles wide, extending between the Calaveras and Tuolumne groves, occurs in the basin of the great ancient *mer de glace* of the Tuolumne and Stanislaus basins, and that the smaller gap between the Merced and Mariposa groves occurs in the basin of the smaller glacier of the Merced. *The wider the ancient glacier, the wider the corresponding gap in the Sequoia belt.*

Finally, pursuing my investigations across the basins of the Kaweah and Tule, I discovered that the Sequoia belt attained its greatest development just where, owing to the topographical peculiarities of the region, the ground had been most perfectly protected from the main ice-rivers that continued to pour past from the summit fountains long after the smaller local glaciers had been melted.

Taking now a general view of the belt, beginning at the south, we see that the majestic ancient glaciers were shed off right and left down the valleys of Kern and King's rivers by the lofty protective spurs outspread embracingly above the warm Sequoia-filled basins of the Kaweah and Tule. Then, next northward, occurs the wide Sequoia-less channel, or basin, of the ancient San Joaquin and King's River *mer de glace*; then the warm, protected spots of Fresno and Mariposa groves; then the Sequoia-less channel of the ancient Merced glacier; next the warm, sheltered ground of the Merced and Tuolumne groves; then the Sequoia-less channel of the grand ancient *mer de glace* of the Tuolumne and Stanislaus; then the warm old ground of the Calaveras and Stanislaus groves. It appears, therefore, that just where, at a certain period in the history of the Sierra, the glaciers were not, there the Sequoia is, and just where the glaciers were, there the Sequoia is not.

What the other conditions may have been that enabled Sequoia to establish itself upon these oldest and warmest portions of the main glacial soil-belt, I cannot say. I might venture to state, however, in this connection, that since the Sequoia forests present a more and more ancient aspect as they extend southward, I am inclined to think that the species was distributed from the south, while the Sugar Pine, its great rival in the northern groves, seems to have come around the head of the Sacramento valley and down the Sierra from the north; consequently, when the Sierra soil-beds were first thrown open to preëmption on the melting of the ice-sheet, the Sequoia may have established itself along the available portions of the south half of the range prior to the arrival of the Sugar Pine, while the Sugar Pine took possession of the north half prior to the arrival of Sequoia.

But however much uncertainty may attach to this branch of the question, there are no obscuring shadows upon the grand general relationship we have pointed out between the present distribution of Sequoia and the ancient glaciers of the Sierra. And when we bear in mind that all the present forests of the Sierra are young, growing on moraine soil recently deposited, and that the flank of the range itself, with all its landscapes, is new-born, recently sculptured, and brought to the light of day from beneath the ice mantle of the glacial winter, then a thousand lawless mysteries disappear, and broad harmonies take their places.

But although all the observed phenomena bearing on the post-glacial history of this colossal tree point to the conclusion that it never was more widely distributed on the Sierra since the close of the glacial epoch; that its present forests are scarcely past prime, if, indeed, they have reached prime; that the post-glacial day of the species is probably not half done; yet, when from a wider outlook the vast antiquity of the genus is considered, and its ancient richness in species and individuals; comparing our Sierra Giant and *Sequoia sempervirens* of the Coast Range, the only other living species of Sequoia, with the twelve fossil species already discovered and described by Heer and Lesquereux, some of which seem to have flourished over vast areas in the Arctic regions and in Europe and our own territories, during tertiary and cretaceous times,—then indeed it becomes plain that our two surviving species, restricted to narrow belts within the limits of California, are mere remnants of the genus, both as to species and individuals, and that they probably are verging to extinction. But the verge of a period beginning in cretaceous times may have a breadth of tens of thousands of years, not to mention the possible existence of conditions calculated to multiply and reëxtend both species and individuals. This, however, is a branch of the question into which I do not now purpose to enter.

In studying the fate of our forest king, we have thus far considered the action of purely natural causes only; but, unfortunately, *man* is in the woods, and waste and pure destruction are making rapid head-way. If the importance of forests were at all understood, even from an economic standpoint, their preservation would call forth the most watchful attention of government. Only of late years by means of forest reservations has the simplest groundwork for available legislation been laid, while in many of the finest groves every species of destruction is still moving on with accelerated speed.

In the course of my explorations I found no fewer than five mills located on or near the lower edge of the Sequoia belt, all of which were cutting considerable quantities of Big Tree lumber. Most of the Fresno group are doomed to feed the mills recently erected near them, and a company of lumbermen are now cutting the magnificent forest on King's River. In these milling operations waste far exceeds use, for after the choice young manageable trees on any given spot have been felled, the woods are fired to clear the ground of limbs and refuse with reference to further operations, and, of course, most of the seedlings and saplings are destroyed.

These mill ravages, however, are small as compared with the comprehensive destruction caused by "sheepmen." Incredible numbers of sheep are driven to the mountain pastures every summer, and their course is ever marked by desolation. Every wild garden is trodden down, the shrubs are stripped of leaves as if devoured by locusts, and the woods are burned. Running fires are set everywhere, with a view to clearing the ground of prostrate trunks, to facilitate the movements of the flocks and improve the pastures. The entire forest belt is thus swept and devastated from one extremity of the range to the other, and, with the exception of the resinous *Pinus contorta*, Sequoia suffers most of all. Indians burn off the underbrush in certain localities to facilitate deer-hunting, mountaineers and lumbermen carelessly allow their camp-fires to run; but the fires of the sheepmen, or *muttoneers*, form more than ninety per cent. of all destructive fires that range the Sierra forests.

It appears, therefore, that notwithstanding our forest king might live on gloriously in Nature's keeping, it is rapidly vanishing before the fire and steel of man; and unless protective measures be speedily invented and applied, in a few decades, at the farthest, all that will be left of *Sequoia gigantea* will be a few hacked and scarred monuments.

TWO-LEAVED, OR TAMARACK, PINE
(*Pinus contorta*, var. *murrayana*)

This species forms the bulk of the alpine forests, extending along the range, above the fir zone, up to a height of from 8000 to 9500 feet above the sea, growing in beautiful order upon moraines that are scarcely changed as yet by post-glacial weathering. Compared with the giants of the lower zones, this is a small tree, seldom attaining a height of a hundred feet. The largest specimen I ever measured was ninety feet in height, and a little over six in diameter four feet from the ground. The average height of mature trees throughout the entire belt is probably not far from fifty or sixty feet, with a diameter of two feet. It is a well-proportioned, rather handsome little pine, with grayish-brown bark, and crooked, much-divided branches, which cover the greater portion of the trunk, not so densely, however, as to prevent its being seen. The lower limbs curve downward, gradually take a horizontal position about halfway up the trunk, then aspire more and more toward the summit, thus forming a sharp, conical top. The foliage is short and rigid, two leaves in a fascicle, arranged in comparatively long, cylindrical tassels at the ends of the tough, up-curving branchlets. The cones are about two inches long, growing in stiff clusters among the needles, without making any striking effect, except while very young, when they are of a vivid crimson color, and the whole tree appears to be dotted with brilliant flowers. The sterile cones are still more showy, on account of their great abundance, often giving a reddish-yellow tinge to the whole mass of the foliage, and filling the air with pollen.

No other pine on the range is so regularly planted as this one. Moraine forests sweep along the sides of the high, rocky valleys for miles without interruption; still, strictly speaking, they are not dense, for flecks of sunshine and flowers find their way into the darkest places, where the trees grow tallest and thickest. Tall, nutritious grasses are specially abundant beneath them, growing over all the ground, in sunshine and shade, over extensive areas like a farmer's crop, and serving as pasture for the multitude of sheep that are driven from the arid plains every summer as soon as the snow is melted.

The Two-leaved Pine, more than any other, is subject to destruction by fire. The thin bark is streaked and sprinkled with resin, as though it had been showered down upon it like rain, so that even the green trees catch fire readily, and during strong winds whole forests are destroyed, the flames leaping from tree to tree, forming one continuous belt of roaring fire that goes surging and racing onward above the bending woods, like the grass-fires of a prairie. During the calm, dry season of Indian summer, the fire creeps quietly along the ground, feeding on the dry needles and burs; then, arriving at the foot of a tree, the resiny bark is ignited, and the heated air ascends in a powerful current, increasing in velocity, and dragging the flames swiftly upward; then the leaves catch fire, and an immense column of flame, beautifully spired on the edges, and tinted a rose-purple hue, rushes aloft thirty or forty feet above the top of the tree, forming a grand spectacle, especially on a dark night. It lasts, however, only a few seconds, vanishing with magical rapidity, to be succeeded by others along the fire-line at irregular intervals for weeks at a time—tree after tree flashing and darkening, leaving the trunks and branches hardly scarred. The heat, however, is sufficient to kill the trees, and in a few years the bark shrivels and falls off. Belts miles in extent are thus killed and left standing with the branches on, peeled and rigid, appearing gray in the distance, like misty clouds. Later the branches drop off, leaving a forest of bleached spars. At length the roots decay, and the forlorn trunks are blown down during some storm, and piled one upon another encumbering the ground until they are consumed by the next fire, and leave it ready for a fresh crop.

The endurance of the species is shown by its wandering occasionally out over the lava plains with the Yellow Pine, and climbing moraineless mountainsides with the Dwarf Pine, clinging to any chance support in rifts and crevices of storm-beaten rocks—always, however, showing the effects of such hardships in every feature.

Down in sheltered lake hollows, on beds of rich alluvium, it varies so far from the common form as frequently to be taken for a distinct species. Here it grows in dense sods, like grasses, from forty to eighty feet high, bending all together to the breeze and whirling in eddying gusts more lithely than any other tree in the woods. I have frequently found specimens fifty feet high less than five inches in diameter. Being thus slender, and at the same time well clad with leafy boughs, it is oftentimes bent to the ground when laden with soft snow, forming beautiful arches in endless variety, some of which last until the melting of the snow in spring.

MOUNTAIN PINE
(*Pinus monticola*)

The Mountain Pine is king of the alpine woods, brave, hardy, and long-lived, towering grandly above its companions, and becoming stronger and more imposing just where other species begin to crouch and disappear. At its best it is usually about ninety feet high and five or six in diameter, though a specimen is often met considerably larger than this. The trunk is as massive and as suggestive of enduring strength as that of an oak. About two thirds of the trunk is commonly free of limbs, but close, fringy tufts of sprays occur all the way down, like those which adorn the colossal shafts of Sequoia. The bark is deep reddish-brown upon trees that occupy exposed situations near its upper limit, and furrowed rather deeply, the main furrows running nearly parallel with each other, and connected by conspicuous cross furrows, which, with one exception, are, as far as I have noticed, peculiar to this species.

The cones are from four to eight inches long, slender, cylindrical, and somewhat curved, resembling those of the common White Pine of the Atlantic coast. They grow in clusters of about from three to six or seven, becoming pendulous as they increase in weight, chiefly by the bending of the branches.

This species is nearly related to the Sugar Pine, and, though not half so tall, it constantly suggests its noble relative in the way that it extends its long arms and in general habit. The Mountain Pine is first met on the upper margin of the fir zone, growing singly in a subdued, inconspicuous form, in what appear as chance situations, without making much impression on the general forest. Continuing up through the Two-leaved Pines in the same scattered growth, it begins to show its character, and at an elevation of about 10,000 feet attains its noblest development near the middle of the range, tossing its tough arms in the frosty air, welcoming storms and feeding on them, and reaching the grand old age of 1000 years.

JUNIPER, OR RED CEDAR
(*Juniperus occidentalis*)

The Juniper is preëminently a rock tree, occupying the baldest domes and pavements, where there is scarcely a handful of soil, at a height of from 7000 to 9500 feet. In such situations the trunk is frequently over eight feet in diameter, and not much more in height. The top is almost always dead in old trees, and great stubborn limbs push out horizontally that are mostly broken and bare at the ends, but densely covered and embedded here and there with bossy mounds of gray foliage. Some are mere weathered stumps, as broad as long, decorated with a few leafy sprays, reminding one of the crumbling towers of some ancient castle scantily draped with ivy. Only upon the head waters of the Carson have I found this species established on good moraine soil. Here it flourishes with the Silver and Two-leaved Pines, in great beauty and luxuriance, attaining a height of from forty to sixty feet, and manifesting but little of that rocky angularity so characteristic a feature throughout the greater portion of its range. Two of the largest, growing at the head of Hope Valley, measured twenty-nine feet three inches and twenty-five feet six inches in circumference, respectively, four feet from the ground. The bark is of a bright cinnamon color, and, in thrifty trees, beautifully braided and reticulated, flaking off in thin, lustrous ribbons that are sometimes used by Indians for tent-matting. Its fine color and odd picturesqueness always catch an artist's eye, but to me the Juniper seems a singularly dull and taciturn tree, never speaking to one's heart. I have spent many a day and night in its company, in all kinds of weather, and have ever found it silent, cold, and rigid, like a column of ice. Its broad stumpiness, of course, precludes all possibility of waving, or even shaking; but it is not this rocky steadfastness that constitutes its silence. In calm, sun-days the Sugar Pine preaches the grandeur of the mountains like an apostle without moving a leaf.

On level rocks it dies standing, and wastes insensibly out of existence like granite, the wind exerting about as little control over it alive or dead as it does over a glacier boulder. Some are undoubtedly over 2000 years old. All the trees of the alpine woods suffer, more or less, from avalanches, the Two-leaved Pine most of all. Gaps two or three hundred yards wide, extending from the upper limit of the tree-line to the bottoms of valleys and lake basins, are of common occurrence in all the upper forests, resembling the clearings of settlers in the old backwoods. Scarcely a tree is spared, even the soil is scraped away, while the thousands of uprooted pines and spruces are piled upon one another heads downward, and nicked snugly in along the sides of the clearing in two windrows, like lateral moraines. The pines lie with branches wilted and drooping like weeds. Not so the burly junipers. After braving in silence the storms of perhaps a dozen or twenty centuries, they seem in this, their last calamity, to become somewhat communicative, making sign of a very unwilling acceptance of their fate, holding themselves well up from the ground on knees and elbows, seemingly ill at ease, and anxious, like stubborn wrestlers, to rise again.

HEMLOCK SPRUCE
(*Tsuga pattoniana*)

The Hemlock Spruce is the most singularly beautiful of all the California coniferæ. So slender is its axis at the top, that it bends over and droops like the stalk of a nodding lily. The branches droop also, and divide into innumerable slender, waving sprays, which are arranged in a varied, eloquent harmony that is wholly indescribable. Its cones are purple, and hang free, in the form of little tassels two inches long from all the sprays from top to bottom. Though exquisitely delicate and feminine in expression, it grows best where the snow lies deepest, far up in the region of storms, at an elevation of from 9000 to 9500 feet, on frosty northern slopes; but it is capable of growing considerably higher, say 10,500 feet. The tallest specimens, growing in sheltered hollows somewhat beneath the heaviest wind-currents, are from eighty to a hundred feet high, and from two to four feet in diameter. The very largest specimen I ever found was nineteen feet seven inches in circumference four feet from the ground, growing on the edge of Lake Hollow, at an elevation of 9250 feet above the level of the sea. At the age of twenty or thirty years it becomes fruitful, and hangs out its beautiful purple cones at the ends of the slender sprays, where they swing free in the breeze, and contrast delightfully with the cool green foliage. They are translucent when young, and their beauty is delicious. After they are fully ripe, they spread their shell-like scales and allow the brown-winged seeds to fly in the mellow air, while the empty cones remain to beautify the tree until the coming of a fresh crop.

The staminate cones of all the conifer are beautiful, growing in bright clusters, yellow, and rose, and crimson. Those of the

Hemlock Spruce are the most beautiful of all, forming little conelets of blue flowers, each on a slender stem.

Under all conditions, sheltered or storm-beaten, well-fed or ill-fed, this tree is singularly graceful in habit. Even at its highest limit upon exposed ridge-tops, though compelled to crouch in dense thickets, huddled close together, as if for mutual protection, it still manages to throw out its sprays in irrepressible loveliness; while on well-ground moraine soil it develops a perfectly tropical luxuriance of foliage and fruit, and is the very loveliest tree in the forest; poised in thin white sunshine, clad with branches from head to foot, yet not in the faintest degree heavy or bunchy, it towers in unassuming majesty, drooping as if unaffected with the aspiring tendencies of its race, loving the ground while transparently conscious of heaven and joyously receptive of its blessings, reaching out its branches like sensitive tentacles, feeling the light and reveling in it. No other of our alpine conifers so finely veils its strength. Its delicate branches yield to the mountains' gentlest breath; yet is it strong to meet the wildest onsets of the gale,—strong not in resistance, but compliance, bowing, snow-laden, to the ground, gracefully accepting burial month after month in the darkness beneath the heavy mantle of winter.

When the first soft snow begins to fall, the flakes lodge in the leaves, weighing down the branches against the trunk. Then the axis bends yet lower and lower, until the slender top touches the ground, thus forming a fine ornamental arch. The snow still falls lavishly, and the whole tree is at length buried, to sleep and rest in its beautiful grave as though dead. Entire groves of young trees, from ten to forty feet high, are thus buried every winter like slender grasses. But, like the violets and daisies which the heaviest snows crush not, they are safe. It is as though this were only Nature's method of putting her darlings to sleep instead of leaving them exposed to the biting storms of winter.

Thus warmly wrapped they await the summer resurrection. The snow becomes soft in the sunshine, and freezes at night, making the mass hard and compact, like ice, so that during the months of April and May you can ride a horse over the prostrate groves without catching sight of a single leaf. At length the down-pouring sunshine sets them free. First the elastic tops of the arches begin to appear, then one branch after another, each springing loose with a gentle rustling sound, and at length the whole tree, with the assistance of the winds, gradually unbends and rises and settles back into its place in the warm air, as dry and feathery and fresh as young ferns just out of the coil.

Some of the finest groves I have yet found are on the southern slopes of Lassen's Butte. There are also many charming companies on the head waters of the Tuolumne, Merced, and San Joaquin, and, in general, the species is so far from being rare that you can scarcely fail to find groves of considerable extent in crossing the range, choose what pass you may. The Mountain Pine grows beside it, and more frequently the two-leaved species; but there are many beautiful groups, numbering 1000 individuals, or more, without a single intruder.

I wish I had space to write more of the surpassing beauty of this favorite spruce. Every tree-lover is sure to regard it with special admiration; apathetic mountaineers, even, seeking only game or gold, stop to gaze on first meeting it, and mutter to themselves: "That's a mighty pretty tree," some of them adding, "d——d pretty!" In autumn, when its cones are ripe, the little striped tamias, and the Douglas squirrel, and the Clark crow make a happy stir in its groves. The deer love to lie down beneath its spreading branches; bright streams from the snow that is always near ripple through its groves, and bryanthus spreads precious carpets in its shade. But the best words only hint its charms. Come to the mountains and see.

DWARF PINE
(*Pinus albicaulis*)

This species forms the extreme edge of the timber line throughout nearly the whole extent of the range on both flanks. It is first met growing in company with *Pinus contorta*, var. *murrayana*, on the upper margin of the belt, as an erect tree from fifteen to thirty feet high and from one to two feet in thickness; thence it goes straggling up the flanks of the summit peaks, upon moraines or crumbling ledges, wherever it can obtain a foothold, to an elevation of from 10,000 to 12,000 feet, where it dwarfs to a mass of crumpled, prostrate branches, covered with slender, upright shoots, each tipped with a short, close-packed tassel of leaves. The bark is smooth and purplish, in some places almost white. The fertile cones grow in rigid clusters upon the upper branches, dark chocolate in color while young, and bear beautiful pearly seeds about the size of peas, most of which are eaten by two species of tamias and the notable Clark crow. The staminate cones occur in clusters, about an inch wide, down among the leaves, and, as they are colored bright rose-purple, they give rise to a lively, flowery appearance little looked for in such a tree.

Pines are commonly regarded as sky-loving trees that must necessarily aspire or die. This species forms a marked exception, creeping lowly, in compliance with the most rigorous demands of climate, yet enduring bravely to a more advanced age than many of its lofty relatives in the sun-lands below. Seen from a distance, it would never be taken for a tree of any kind. Yonder, for example, is Cathedral Peak, some three miles away, with a scattered growth of this pine creeping like mosses over the roof and around the beveled edges of the north gable, nowhere giving any hint of an ascending axis. When approached quite near it still appears matted and healthy, and is so low that one experiences no great difficulty in walking over the top of it. Yet it is seldom absolutely prostrate, at its lowest usually attaining a height of three or four feet, with a main trunk, and branches outspread and intertangled above it, as if in ascending they had been checked by a ceiling, against which they had grown and been compelled to spread horizontally. The winter snow is indeed such a ceiling, lasting half the year; while the pressed, shorn surface is made yet smoother by violent winds, armed with cutting sand-grains, that beat down any shoot that offers to rise much above the general level, and carve the dead trunks and branches in beautiful patterns.

During stormy nights I have often camped snugly beneath the interlacing arches of this little pine. The needles, which have accumulated for centuries, make fine beds, a fact well known to other mountaineers, such as deer and wild sheep, who paw out oval hollows and lie beneath the larger trees in safe and comfortable concealment.

The longevity of this lowly dwarf is far greater than would be guessed. Here, for example, is a specimen, growing at an elevation of 10,700 feet, which seems as though it might be plucked up by the roots, for it is only three and a half inches in diameter, and its topmost tassel is hardly three feet above the ground. Cutting it half through and counting the annual rings with the aid of a lens, we find its age to be no less than 255 years. Here is another telling specimen about the same height, 426 years old, whose trunk is only six inches in diameter; and one of its supple branchlets, hardly an eighth of an inch in diameter inside the bark, is seventy-five years old, and so filled with oily balsam, and so well seasoned by storms, that we may tie it in knots like a whip-cord.

WHITE PINE
(*Pinus flexilis*)

This species is widely distributed throughout the Rocky Mountains, and over all the higher of the many ranges of the Great Basin, between the Wahsatch Mountains and the Sierra, where it is known as White Pine. In the Sierra it is sparsely scattered along the eastern flank, from Bloody Cañon southward nearly to the extremity of the range, opposite the village of Lone Pine, nowhere forming any appreciable portion of the general forest. From its peculiar position, in loose, straggling parties, it seems to have been derived from the Basin ranges to the eastward, where it is abundant.

It is a larger tree than the Dwarf Pine. At an elevation of about 9000 feet above the sea, it often attains a height of forty or fifty feet, and a diameter of from three to five feet. The cones open freely when ripe, and are twice as large as those of the *albicaulis*, and the foliage and branches are more open, having a tendency to sweep out in free, wild curves, like those of the Mountain Pine, to which it is closely allied. It is seldom found lower than 9000 feet above sea-level, but from this elevation it pushes upward over the roughest ledges to the extreme limit of tree-growth, where, in its dwarfed, storm-crushed condition, it is more like the white-barked species.

Throughout Utah and Nevada it is one of the principal timber-trees, great quantities being cut every year for the mines. The famous White Pine Mining District, White Pine City, and the White Pine Mountains have derived their names from it.

NEEDLE PINE

(*Pinus aristata*)

This species is restricted in the Sierra to the southern portion of the range, about the head waters of King's and Kern rivers, where it forms extensive forests, and in some places accompanies the Dwarf Pine to the extreme limit of tree-growth.

It is first met at an elevation of between 9000 and 10,000 feet, and runs up to 11,000 without seeming to suffer greatly from the climate or the leanness of the soil. It is a much finer tree than the Dwarf Pine. Instead of growing in clumps and low, heathy mats, it manages in some way to maintain an erect position, and usually stands single. Wherever the young trees are at all sheltered, they grow up straight and arrowy, with delicately tapered bole, and ascending branches terminated with glossy, bottle-brush tassels. At middle age, certain limbs are specialized and pushed far out for the bearing of cones, after the manner of the Sugar Pine; and in old age these branches droop and cast about in every direction, giving rise to very picturesque effects. The trunk becomes deep brown and rough, like that of the Mountain Pine, while the young cones are of a strange, dull, blackish-blue color, clustered on the upper branches. When ripe they are from three to four inches long, yellowish brown, resembling in every way those of the Mountain Pine. Excepting the Sugar Pine, no tree on the mountains is so capable of individual expression, while in grace of form and movement it constantly reminds one of the Hemlock Spruce.

The largest specimen I measured was a little over five feet in diameter and ninety feet in height, but this is more than twice the ordinary size.

This species is common throughout the Rocky Mountains and most of the short ranges of the Great Basin, where it is called the Fox-tail Pine, from its long dense leaf-tassels. On the Hot Creek, White Pine, and Golden Gate ranges it is quite abundant. About a foot or eighteen inches of the ends of the branches is densely packed with stiff outstanding needles which radiate like an electric fox or squirrel's tail. The needles have a glossy polish, and the sunshine sifting through them makes them burn with silvery luster, while their number and elastic temper tell delightfully in the winds. This tree is here still more original and picturesque than in the Sierra, far surpassing not only its companion conifers in this respect, but also the most noted of the lowland oaks. Some stand firmly erect, feathered with radiant tassels down to the ground, forming slender tapering towers of shining verdure; others, with two or three specialized branches pushed out at right angles to the trunk and densely clad with tasseled sprays, take the form of beautiful ornamental crosses. Again in the same woods you find trees that are made up of several boles united near the ground, spreading at the sides in a plane parallel to the axis of the mountain, with the elegant tassels hung in charming order between them, making a harp held against the main wind lines where they are most effective in playing the grand storm harmonies. And besides these there are many variable arching forms, alone or in groups, with innumerable tassels drooping beneath the arches or radiant above them, and many lowly giants of no particular form that have braved the storms of a thousand years. But whether old or young, sheltered or exposed to the wildest gales, this tree is ever found irrepressibly and extravagantly picturesque, and offers a richer and more varied series of forms to the artist than any other conifer I know of.

NUT PINE

(*Pinus monophylla*)

The Nut Pine covers or rather dots the eastern flank of the Sierra, to which it is mostly restricted, in grayish, bush-like patches, from the margin of the sage-plains to an elevation of from 7000 to 8000 feet.

A more contentedly fruitful and unaspiring conifer could not be conceived. All the species we have been sketching make departures more or less distant from the typical spire form, but none goes so far as this. Without any apparent exigency of climate or soil, it remains near the ground, throwing out crooked, divergent branches like an orchard apple-tree, and seldom pushes a single shoot higher than fifteen or twenty feet above the ground.

The average thickness of the trunk is, perhaps, about ten or twelve inches. The leaves are mostly undivided, like round awls, instead of being separated, like those of other pines, into twos and threes and fives. The cones are green while growing, and are usually found over all the tree, forming quite a marked feature as seen against the bluish-gray foliage. They are quite small, only about two inches in length, and give no promise of edible nuts; but when we come to open them, we find that about half the entire bulk of the cone is made up of sweet, nutritious seeds, the kernels of which are nearly as large as those of hazel-nuts.

This is undoubtedly the most important food-tree on the Sierra, and furnishes the Mono, Carson, and Walker River Indians with more and better nuts than all the other species taken together. It is the Indians' own tree, and many a white man have they killed for cutting it down.

In its development Nature seems to have aimed at the formation of as great a fruit-bearing surface as possible. Being so low and accessible, the cones are readily beaten off with poles, and the nuts procured by roasting them until the scales open. In bountiful seasons a single Indian will gather thirty or forty bushels of them—a fine squirrelish employment.

Of all the conifers along the eastern base of the Sierra, and on all the many mountain groups and short ranges of the Great Basin, this foodful little pine is the commonest tree, and the most important. Nearly every mountain is planted with it to a height of from 8000 to 9000 feet above the sea. Some are covered from base to summit by this one species, with only a sparse growth of juniper on the lower slopes to break the continuity of its curious woods, which, though dark-looking at a distance, are almost shadeless, and have none of the damp, leafy glens and hollows so characteristic of other pine woods. Tens of thousands of acres occur in continuous belts. Indeed, viewed comprehensively the entire Basin seems to be pretty evenly divided into level plains dotted with sage-bushes and mountain-chains covered with Nut Pines. No slope is too rough, none too dry, for these bountiful orchards of the red man.

The value of this species to Nevada is not easily overestimated. It furnishes charcoal and timber for the mines, and, with the juniper, supplies the ranches with fuel and rough fencing. In fruitful seasons the nut crop is perhaps greater than the California wheat crop, which exerts so much influence throughout the food markets of the world. When the crop is ripe, the Indians make ready the long beating-poles; bags, baskets, mats, and sacks are collected; the women out at service among the settlers, washing or drudging, assemble at the family huts; the men leave their ranch work; old and young, all are mounted on ponies and start in great glee to the nut-lands, forming curiously picturesque cavalcades; flaming scarfs and calico skirts stream loosely over the knotty ponies, two squaws usually astride of each, with baby midgets bandaged in baskets slung on their backs or balanced on the saddlebow; while nut-baskets and water-jars project from each side, and the long beating-poles make angles in every direction. Arriving at some well-known central point where grass and water are found, the squaws with baskets, the men with poles ascend the ridges to the laden trees, followed by the children. Then the beating begins right merrily, the burs fly in every direction, rolling down the slopes, lodging here and there against rocks and sage-bushes, chased and gathered by the women and children with fine natural gladness. Smoke-columns speedily mark the joyful scene of their labors as the roasting-fires are kindled, and, at night, assembled in gay circles garrulous as jays, they begin the first nut feast of the season.

The nuts are about half an inch long and a quarter of an inch in diameter, pointed at the top, round at the base, light brown in general color, and, like many other pine seeds, handsomely dotted with purple, like birds' eggs. The shells are thin and may be crushed between the thumb and finger. The kernels are white, becoming brown by roasting, and are sweet to every palate, being eaten by birds, squirrels, dogs, horses, and men. Perhaps less than one bushel in a thousand of the whole crop is ever gathered. Still, besides supplying their own wants, in times of plenty the Indians bring large quantities to market; then they are eaten around nearly every fireside in the State, and are even fed to horses occasionally instead of barley.

Of other trees growing on the Sierra, but forming a very small part of the general forest, we may briefly notice the following:

Chamæcyparis lawsoniana is a magnificent tree in the coast ranges, but small in the Sierra. It is found only well to the northward along the banks of cool streams on the upper Sacramento toward Mount Shasta. Only a few trees of this species, as far as I have seen, have as yet gained a place in the Sierra woods. It has evidently been derived from the coast range by way of the tangle of connecting mountains at the head of the Sacramento Valley.

In shady dells and on cool stream banks of the northern Sierra we also find the Yew (*Taxus brevifolia*).

The interesting Nutmeg Tree (*Torreya californica*) is sparsely distributed along the western flank of the range at an elevation of about 4000 feet, mostly in gulches and cañons. It is a small, prickly leaved, glossy evergreen, like a conifer, from twenty to fifty feet high, and one to two feet in diameter. The fruit resembles a greengage plum, and contains one seed, about the size of an acorn, and like a nutmeg, hence the common name. The wood is fine-grained and of a beautiful, creamy yellow color like box, sweet-scented when dry, though the green leaves emit a disagreeable odor.

Betula occidentalis, the only birch, is a small, slender tree restricted to the eastern flank of the range along stream-sides below the pine-belt, especially in Owen's Valley.

Alder, Maple, and Nuttall's Flowering Dogwood make beautiful bowers over swift, cool streams at an elevation of from 3000 to 5000 feet, mixed more or less with willows and cottonwood; and above these in lake basins the aspen forms fine ornamental groves, and lets its light shine gloriously in the autumn months.

The Chestnut Oak (*Quercus densiflora*) seems to have come from the coast range around the head of the Sacramento Valley, like the *Chamæcyparis*, but as it extends southward along the lower edge of the main pine-belt it grows smaller until it finally dwarfs to a mere chaparral bush. In the coast mountains it is a fine, tall, rather slender tree, about from sixty to seventy-five feet high, growing with the grand *Sequoia sempervirens*, or Redwood. But unfortunately it is too good to live, and is now being rapidly destroyed for tan-bark.

Besides the common Douglas Oak and the grand *Quercus wislizeni* of the foot-hills, and several small ones that make dense growths of chaparral, there are two mountain-oaks that grow with the pines up to an elevation of about 5000 feet above the sea, and greatly enhance the beauty of the yosemite parks. These are the Mountain Live Oak and the Kellogg Oak, named in honor of the admirable botanical pioneer of California. Kellogg's Oak (*Quercus kelloggi*) is a firm, bright, beautiful tree, reaching a height of sixty feet, four to seven feet in diameter, with wide-spreading branches, and growing at an elevation of from 3000 to 5000 feet in sunny valleys and flats among the evergreens, and higher in a dwarfed state. In the cliff-bound parks about 4000 feet above the sea it is so abundant and effective it might fairly be called the Yosemite Oak. The leaves make beautiful masses of purple in the spring, and yellow in ripe autumn; while its acorns are eagerly gathered by Indians, squirrels, and woodpeckers. The Mountain Live Oak (*Q. chrysolepis*) is a tough, rugged mountaineer of a tree, growing bravely and attaining noble dimensions on the roughest earthquake taluses in deep cañons and yosemite valleys. The trunk is usually short, dividing near the ground into great, wide-spreading limbs, and these again into a multitude of slender sprays, many of them cord-like and drooping to the ground, like those of the Great White Oak of the lowlands (*Q. lobata*). The top of the tree where there is plenty of space is broad and bossy, with a dense covering of shining leaves, making delightful canopies, the complicated system of gray, interlacing, arching branches as seen from beneath being exceedingly rich and picturesque. No other tree that I know dwarfs so regularly and completely as this under changes of climate due to changes in elevation. At the foot of a cañon 4000 feet above the sea you may find magnificent specimens of this oak fifty feet high, with craggy, bulging trunks, five to seven feet in diameter, and at the head of the cañon, 2500 feet higher, a dense, soft, low, shrubby growth of the same species, while all the way up the cañon between these extremes of size and habit a perfect gradation may be traced. The largest I have seen was fifty feet high, eight feet in diameter, and about seventy-five feet in spread. The trunk was all knots and buttresses, gray like granite, and about as angular and irregular as the boulders on which it was growing—a type of steadfast, unwedgeable strength.

CHAPTER IX

THE DOUGLAS SQUIRREL

(*Sciurus douglasii*)

THE DOUGLAS SQUIRREL is by far the most interesting and influential of the California sciuridæ, surpassing every other species in force of character, numbers, and extent of range, and in the amount of influence he brings to bear upon the health and distribution of the vast forests he inhabits.

Go where you will throughout the noble woods of the Sierra Nevada, among the giant pines and spruces of the lower zones, up through the towering Silver Firs to the storm-bent thickets of the summit peaks, you everywhere find this little squirrel the master-existence. Though only a few inches long, so intense is his fiery vigor and restlessness, he stirs every grove with wild life, and makes himself more important than even the huge bears that shuffle through the tangled underbrush beneath him. Every wind is fretted by his voice, almost every bole and branch feels the sting of his sharp feet. How much the growth of the trees is stimulated by this means it is not easy to learn, but his action in manipulating their seeds is more appreciable. Nature has made him master forester and committed most of her coniferous crops to his paws. Probably over fifty per cent. of all the cones ripened on the Sierra are cut off and handled by the Douglas alone, and of those of the Big Trees perhaps ninety per cent. pass through his hands: the greater portion is of course stored away for food to last during the winter and spring, but some of them are tucked separately into loosely covered holes, where some of the seeds germinate and become trees. But the Sierra is only one of the many provinces over which he holds sway, for his dominion extends over all the Redwood Belt of the Coast Mountains, and far northward throughout the majestic forests of Oregon, Washington, and British Columbia. I make haste to mention these facts, to show upon how substantial a foundation the importance I ascribe to him rests.

The Douglas is closely allied to the Red Squirrel or Chickaree of the eastern woods. Ours may be a lineal descendant of this species, distributed westward to the Pacific by way of the Great Lakes and the Rocky Mountains, and thence southward along our forested ranges. This view is suggested by the fact that our species becomes redder and more Chickaree-like in general, the farther it is traced back along the course indicated above. But whatever their relationship, and the evolutionary forces that have acted upon them, the Douglas is now the larger and more beautiful animal.

From the nose to the root of the tail he measures about eight inches; and his tail, which he so effectively uses in interpreting his feelings, is about six inches in length. He wears dark bluish-gray over the back and halfway down the sides, bright buff on the belly, with a stripe of dark gray, nearly black, separating the upper and under colors; this dividing stripe, however, is not very sharply defined. He has long black whiskers, which gives him a rather fierce look when observed closely, strong claws, sharp as fish-hooks, and the brightest of bright eyes, full of telling speculation.

A King's River Indian told me that they call him "Pillillooeet," which, rapidly pronounced with the first syllable heavily accented, is not unlike the lusty exclamation he utters on his way up a tree when excited. Most mountaineers in California call him the Pine Squirrel; and when I asked an old trapper whether he knew our little forester, he replied with brightening countenance: "Oh, yes, of course I know him; everybody knows him. When I'm huntin' in the woods, I often find out where the deer are by his barkin' at 'em. I call 'em Lightnin' Squirrels, because they're so mighty quick and peert."

All the true squirrels are more or less birdlike in speech and movements; but the Douglas is preëminently so, possessing, as he does, every attribute peculiarly squirrelish enthusiastically concentrated. He is the squirrel of squirrels, flashing from branch to branch of his favorite evergreens crisp and glossy and undiseased as a sunbeam. Give him wings and he would outfly any bird in the woods. His big gray cousin is a looser animal, seemingly light enough to float on the wind; yet when leaping from limb to limb, or out of one tree-top to another, he sometimes halts to gather strength, as if making efforts concerning the up-shot of which he does not always feel exactly confident. But the Douglas, with his denser body, leaps and glides in hidden strength, seemingly as independent of common muscles as a mountain stream. He threads the tasseled branches of the pines, stirring their needles like a rustling breeze; now shooting across openings in arrowy lines; now launching in curves, glinting deftly from side to side in sudden zigzags, and swirling in giddy loops and spirals around the knotty trunks; getting into what seem to be the most impossible situations without sense of danger; now on his haunches, now on his head; yet ever graceful, and punctuating his most irrepressible outbursts of energy with little dots and dashes of perfect repose. He is, without exception, the wildest animal I ever saw,—a fiery, sputtering little bolt of life, luxuriating in quick oxygen and the woods' best juices. One can hardly think of such a creature being dependent, like the rest of us, on climate and food. But, after all, it requires no long acquaintance to learn he is human, for he works for a living. His busiest time is in the Indian summer. Then he gathers burs and hazel-nuts like a plodding farmer, working continuously every day for hours; saying not a word; cutting off the ripe cones at the top of his speed, as if employed by the job, and examining every branch in regular order, as if careful that not one should escape him; then, descending, he stores them away beneath logs and stumps, in anticipation of the pinching hunger days of winter. He seems himself a kind of coniferous fruit,—both fruit and flower. The resiny essences of the pines pervade every pore of his body, and eating his flesh is like chewing gum.

One never tires of this bright chip of nature,—this brave little voice crying in the wilderness,—of observing his many works and ways, and listening to his curious language. His musical, piny gossip is as savory to the ear as balsam to the palate; and, though he has not exactly the gift of song, some of his notes are as sweet as those of a linnet—almost flute-like in softness, while others prick and tingle like thistles. He is the mocking-bird of squirrels, pouring forth mixed chatter and song like a perennial fountain; barking like a dog, screaming like a hawk, chirping like a blackbird or a sparrow; while in bluff, audacious noisiness he is a very jay.

In descending the trunk of a tree with the intention of alighting on the ground, he preserves a cautious silence, mindful, perhaps, of foxes and wildcats; but while rocking safely at home in the pine-tops there is no end to his capers and noise; and woe to the gray squirrel or chipmunk that ventures to set foot on his favorite tree! No matter how slyly they trace the furrows of the bark, they are speedily discovered, and kicked down-stairs with comic vehemence, while a torrent of angry notes comes rushing from his whiskered lips that sounds remarkably like swearing. He will even attempt at times to drive away dogs and men, especially if he has had no previous knowledge of them. Seeing a man for the first time, he approaches nearer and nearer, until within a few feet; then, with an angry outburst, he makes a sudden rush, all teeth and eyes, as if about to eat you up. But, finding that the big, forked animal doesn't scare, he prudently beats a retreat, and sets himself up to reconnoiter on some overhanging branch, scrutinizing every movement you make with ludicrous solemnity. Gathering courage, he ventures down the trunk again, churring and chirping, and jerking nervously up and down in curious loops, eyeing you all the time, as if showing off and demanding your admiration. Finally, growing calmer, he settles down in a comfortable posture on some horizontal branch commanding a good view, and beats time with his tail to a steady "Chee-up! chee-up!" or, when somewhat less excited, "Pee-ah!" with the first syllable keenly accented, and the second drawn out like the scream of a hawk,—repeating this slowly and more emphatically at first, then gradually faster, until a rate of about 150 words a minute is reached; usually sitting all the time on his haunches, with paws resting on his breast, which pulses visibly with each word. It is remarkable, too, that, though articulating distinctly, he keeps his mouth shut most of the time, and speaks through his nose. I have occasionally observed him even eating Sequoia seeds and nibbling a troublesome flea, without ceasing or in any way confusing his "Pee-ah! pee-ah!" for a single moment.

While ascending trees all his claws come into play, but in descending the weight of his body is sustained chiefly by those of the hind feet; still in neither case do his movements suggest effort, though if you are near enough you may see the bulging strength of his short, bear-like arms, and note his sinewy fists clinched in the bark.

Whether going up or down, he carries his tail extended at full length in line with his body, unless it be required for gestures. But while running along horizontal limbs or fallen trunks, it is frequently folded forward over the back, with the airy tip daintily upcurled. In cool weather it keeps him warm. Then, after he has finished his meal, you may see him crouched close on some level limb with his tail-robe neatly spread and reaching forward to his ears, the electric, outstanding hairs quivering in the breeze like pine-needles. But in wet or very cold weather he stays in his nest, and while curled up there his comforter is long enough to come forward around his nose. It is seldom so cold, however, as to prevent his going out to his stores when hungry.

Once as I lay storm-bound on the upper edge of the timber line on Mount Shasta, the thermometer nearly at zero and the sky thick with driving snow, a Douglas came bravely out several times from one of the lower hollows of a Dwarf Pine near my camp, faced the wind without seeming to feel it much, frisked lightly about over the mealy snow, and dug his way down to some hidden seeds with wonderful precision, as if to his eyes the thick snow-covering were glass.

No other of the Sierra animals of my acquaintance is better fed, not even the deer, amid abundance of sweet herbs and shrubs, or the mountain sheep, or omnivorous bears. His food consists of grass-seeds, berries, hazel-nuts, chinquapins, and the nuts and seeds of all the coniferous trees without exception,—Pine, Fir, Spruce, Libocedrus, Juniper, and Sequoia,—he is fond of them all, and they all agree with him, green or ripe. No cone is too large for him to manage, none so small as to be beneath his notice. The smaller ones, such as those of the Hemlock, and the Douglas Spruce, and the Two-leaved Pine, he cuts off and eats on a branch of the tree, without allowing them to fall; beginning at the bottom of the cone and cutting away the scales to expose the seeds; not gnawing by guess, like a bear, but turning them round and round in regular order, in compliance with their spiral arrangement.

When thus employed, his location in the tree is betrayed by a dribble of scales, shells, and seed-wings, and, every few minutes, by the fall of the stripped axis of the cone. Then of course he is ready for another, and if you are watching you may catch a glimpse of him as he glides silently out to the end of a branch and see him examining the cone-clusters until he finds one to his mind; then, leaning over, pull back the springy needles out of his way, grasp the cone with his paws to prevent its falling, snip it off in an incredibly short time, seize it with jaws grotesquely stretched, and return to his chosen seat near the trunk. But the immense size of the cones of the Sugar Pine—from fifteen to twenty inches in length—and those of the Jeffrey variety of the Yellow Pine compel him to adopt a quite different method. He cuts them off without attempting to hold them, then goes down and drags them from where they have chanced to fall up to the bare, swelling ground around the instep of the tree, where he demolishes them in the same methodical way, beginning at the bottom and following the scale-spirals to the top.

From a single Sugar Pine cone he gets from two to four hundred seeds about half the size of a hazel-nut, so that in a few minutes he can procure enough to last a week. He seems, however, to prefer those of the two Silver Firs above all others; perhaps because they are most easily obtained, as the scales drop off when ripe without needing to be cut. Both species are filled with an exceedingly pungent, aromatic oil, which spices all his flesh, and is of itself sufficient to account for his lightning energy.

You may easily know this little workman by his chips. On sunny hillsides around the principal trees they lie in big piles,—bushels and basketfuls of them, all fresh and clean, making the most beautiful kitchen-middens imaginable. The brown and yellow scales and nut-shells are as abundant and as delicately penciled and tinted as the shells along the sea-shore; while the beautiful red and purple seed-wings mingled with them would lead one to fancy that innumerable butterflies had there met their fate.

He feasts on all the species long before they are ripe, but is wise enough to wait until they are matured before he gathers them into his barns. This is in October and November, which with him are the two busiest months of the year. All kinds of burs, big and little, are now cut off and showered down alike, and the ground is speedily covered with them. A constant thudding and bumping is kept up; some of the larger cones chancing to fall on old logs make the forest reëcho with the sound. Other nut-eaters less industrious know well what is going on, and hasten to carry away the cones as they fall. But however busy the harvester may be, he is not slow to descry the pilferers below, and instantly leaves his work to drive them away. The little striped tamias is a thorn in his flesh, stealing persistently, punish him as he may. The large Gray Squirrel gives trouble also, although the Douglas has been accused of stealing from him. Generally, however, just the opposite is the case.

The excellence of the Sierra evergreens is well known to nurserymen throughout the world, consequently there is considerable demand for the seeds. The greater portion of the supply has hitherto been procured by chopping down the trees in the more accessible sections of the forest alongside of bridle-paths that cross the range.

Sequoia seeds at first brought from twenty to thirty dollars per pound, and therefore were eagerly sought after. Some of the smaller fruitful trees were cut down in the groves not protected by government, especially those of Fresno and King's River. Most of the Sequoias, however, are of so gigantic a size that the seedsmen have to look for the greater portion of their supplies to the Douglas, who soon learns he is no match for these freebooters. He is wise enough, however, to cease working the instant he perceives them, and never fails to embrace every opportunity to recover his burs whenever they happen to be stored in any place accessible to him, and the busy seedsman often finds on returning to camp that the little Douglas has exhaustively spoiled the spoiler. I know one seed-gatherer who, whenever he robs the squirrels, scatters wheat or barley beneath the trees as conscience-money.

The want of appreciable life remarked by so many travelers in the Sierra forests is never felt at this time of year. Banish all the humming insects and the birds and quadrupeds, leaving only Sir Douglas, and the most solitary of our so-called solitudes would still throb with ardent life. But if you should go impatiently even into the most populous of the groves on purpose to meet him, and walk about looking up among the branches, you would see very little of him. But lie down at the foot of one of the trees and straightway he will come. For, in the midst of the ordinary forest sounds, the falling of burs, piping of quails, the screaming of the Clark Crow, and the rustling of deer and bears among the chaparral, he is quick to detect your strange footsteps, and will hasten to make a good, close inspection of you as soon as you are still. First, you may hear him sounding a few notes of curious inquiry, but more likely the first intimation of his approach will be the prickly sounds of his feet as he descends the tree overhead, just before he makes his savage onrush to frighten you and proclaim your presence to every squirrel and bird in the neighborhood. If you remain perfectly motionless, he will come nearer and nearer, and probably set your flesh a-tingle by frisking across your body. Once, while I was seated at the foot of a Hemlock Spruce in one of the most inaccessible of the San Joaquin yosemites engaged in sketching, a reckless fellow came up behind me, passed under my bended arm, and jumped on my paper. And one warm afternoon, while an old friend of mine was reading out in the shade of his cabin, one of his Douglas neighbors jumped from the gable upon his head, and then with admirable assurance ran down over his shoulder and on to the book he held in his hand.

Our Douglas enjoys a large social circle; for, besides his numerous relatives, *Sciurus fossor*, *Tamias quadrivitatus*, *T. townsendii*, *Spermophilus beecheyi*, *S. douglasii*, he maintains intimate relations with the nut-eating birds, particularly the Clark Crow (*Picicorvus columbianus*) and the numerous woodpeckers and jays. The two spermophiles are astonishingly abundant in the lowlands and lower foot-hills, but more and more sparingly distributed up through the Douglas domains,—seldom venturing higher than six or seven thousand feet above the level of the sea. The gray sciurus ranges but little higher than this. The little striped tamias alone is associated with him everywhere. In the lower and middle zones, where they all meet, they are tolerably harmonious—a happy family, though very amusing skirmishes may occasionally be witnessed. Wherever the ancient glaciers have spread forest soil there you find our wee hero, most abundant where depth of soil and genial climate have given rise to a corresponding luxuriance in the trees, but following every kind of growth up the curving moraines to the highest glacial fountains.

Though I cannot of course expect all my readers to sympathize fully in my admiration of this little animal, few, I hope, will think this sketch of his life too long. I cannot begin to tell here how much he has cheered my lonely wanderings during all the years I have been pursuing my studies in these glorious wilds; or how much unmistakable humanity I have found in him. Take this for example: One calm, creamy Indian summer morning, when the nuts were ripe, I was camped in the upper pine-woods of the south fork of the San Joaquin, where the squirrels seemed to be about as plentiful as the ripe burs. They were taking an early breakfast before going to their regular harvest-work. While I was busy with my own breakfast I heard the thudding fall of two or three heavy cones from a Yellow Pine near me. I stole noiselessly forward within about twenty feet of the base of it to observe. In a few moments down came the Douglas. The breakfast-burs he had cut off had rolled on the gently sloping ground into a clump of ceanothus bushes, but he seemed to know exactly where they were, for he found them at once, apparently without searching for them. They were more than twice as heavy as himself, but after turning them into the right position for getting a good hold with his long sickle-teeth he managed to drag them up to the foot of the tree from which he had cut them, moving backward. Then seating himself comfortably, he held them on end, bottom up, and demolished them at his ease. A good deal of nibbling had to be done before he got anything to eat, because the lower scales are barren, but when he had patiently worked his way up to the fertile ones he found two sweet nuts at the base of each, shaped like trimmed hams, and spotted purple like birds' eggs. And notwithstanding these cones were dripping with soft balsam, and covered with prickles, and so strongly put together that a boy would be puzzled to cut them open with a jack-knife, he accomplished his meal with easy dignity and cleanliness, making less effort apparently than a man would in eating soft cookery from a plate.

Breakfast done, I whistled a tune for him before he went to work, curious to see how he would be affected by it. He had not seen me all this while; but the instant I began to whistle he darted up the tree nearest to him, and came out on a small dead limb opposite me, and composed himself to listen. I sang and whistled more than a dozen airs, and as the music changed his eyes sparkled, and he turned his head quickly from side to side, but made no other response. Other squirrels, hearing the strange sounds, came around on all sides, also chipmunks and birds. One of the birds, a handsome, speckle-breasted thrush, seemed even more interested than the squirrels. After listening for awhile on one of the lower dead sprays of a pine, he came swooping forward within a few feet of my face, and remained fluttering in the air for half a minute or so, sustaining himself with whirring wing-beats, like a humming-bird in front of a flower, while I could look into his eyes and see his innocent wonder.

By this time my performance must have lasted nearly half an hour. I sang or whistled "Bonnie Doon," "Lass o' Gowrie," "O'er the Water to Charlie," "Bonnie Woods o' Cragie Lee," etc., all of which seemed to be listened to with bright interest, my first Douglas sitting patiently through it all, with his telling eyes fixed upon me until I ventured to give the "Old Hundredth," when he screamed his Indian name, Pillillooeet, turned tail, and darted with ludicrous haste up the tree out of sight, his voice and actions in the case leaving a somewhat profane impression, as if he had said, "I'll be hanged if you get me to hear anything so solemn and unpiny." This acted as a signal for the general dispersal of the whole hairy tribe, though the birds seemed willing to wait further developments, music being naturally more in their line.

What there can be in that grand old church-tune that is so offensive to birds and squirrels I can't imagine. A year or two after this High Sierra concert, I was sitting one fine day on a hill in the Coast Range where the common Ground Squirrels were abundant. They were very shy on account of being hunted so much; but after I had been silent and motionless for half an hour or so they began to venture out of their holes and to feed on the seeds of the grasses and thistles around me as if I were no more to be feared than a tree-stump. Then it occurred to me that this was a good opportunity to find out whether they also disliked "Old Hundredth." Therefore I began to whistle as nearly as I could remember the same familiar airs that had pleased the mountaineers of the Sierra. They at once stopped eating, stood erect, and listened patiently until I came to "Old Hundredth," when with ludicrous haste every one of them rushed to their holes and bolted in, their feet twinkling in the air for a moment as they vanished.

No one who makes the acquaintance of our forester will fail to admire him; but he is far too self-reliant and warlike ever to be taken for a darling.

How long the life of a Douglas Squirrel may be, I don't know. The young seem to sprout from knot-holes, perfect from the first, and as enduring as their own trees. It is difficult, indeed, to realize that so condensed a piece of sun-fire should ever become dim or die at all. He is seldom killed by hunters, for he is too small to encourage much of their attention, and when pursued in settled regions becomes excessively shy, and keeps close in the furrows of the highest trunks, many of which are of the same color as himself. Indian boys, however, lie in wait with unbounded patience to shoot them with arrows. In the lower and middle zones a few fall a prey to rattlesnakes. Occasionally he is pursued by hawks and wildcats, etc. But, upon the whole, he dwells safely in the deep bosom of the woods, the most highly favored of all his happy tribe. May his tribe increase!

CHAPTER X

A WIND-STORM IN THE FORESTS

THE MOUNTAIN WINDS, like the dew and rain, sunshine and snow, are measured and bestowed with love on the forests to develop their strength and beauty. However restricted the scope of other forest influences, that of the winds is universal. The snow bends and trims the upper forests every winter, the lightning strikes a single tree here and there, while avalanches mow down thousands at a swoop as a gardener trims out a bed of flowers. But the winds go to every tree, fingering every leaf and branch and furrowed bole; not one is forgotten; the Mountain Pine towering with outstretched arms on the rugged buttresses of the icy peaks, the lowliest and most retiring tenant of the dells; they seek and find them all, caressing them tenderly, bending them in lusty exercise, stimulating their growth, plucking off a leaf or limb as required, or removing an entire tree or grove, now whispering and cooing through the branches like a sleepy child, now roaring like the ocean; the winds blessing the forests, the forests the winds, with ineffable beauty and harmony as the sure result.

After one has seen pines six feet in diameter bending like grasses before a mountain gale, and ever and anon some giant falling with a crash that shakes the hills, it seems astonishing that any, save the lowest thickset trees, could ever have found a period sufficiently stormless to establish themselves; or, once established, that they should not, sooner or later, have been blown down. But when the storm is over, and we behold the same forests tranquil again, towering fresh and unscathed in erect majesty, and consider what centuries of storms have fallen upon them since they were first planted,—hail, to break the tender seedlings; lightning, to scorch and shatter; snow, winds, and avalanches, to crush and overwhelm,—while the manifest result of all this wild storm-culture is the glorious perfection we behold; then faith in Nature's forestry is established, and we cease to deplore the violence of her most destructive gales, or of any other storm-implement whatsoever.

There are two trees in the Sierra forests that are never blown down, so long as they continue in sound health. These are the Juniper and the Dwarf Pine of the summit peaks. Their stiff, crooked roots grip the storm-beaten ledges like eagles' claws, while their lithe, cord-like branches bend round compliantly, offering but slight holds for winds, however violent. The other alpine conifers—the Needle Pine, Mountain Pine, Two-leaved Pine, and Hemlock Spruce—are never thinned out by this agent to any destructive extent, on account of their admirable toughness and the closeness of their growth. In general the same is true of the giants of the lower zones. The kingly Sugar Pine, towering aloft to a height of more than 200 feet, offers a fine mark to storm-winds; but it is not densely foliaged, and its long, horizontal arms swing round compliantly in the blast, like tresses of green, fluent alga in a brook; while the Silver Firs in most places keep their ranks well together in united strength. The Yellow or Silver Pine is more frequently overturned than any other tree on the Sierra, because its leaves and branches form a larger mass in proportion to its height, while in many places it is planted sparsely, leaving open lanes through which storms may enter with full force. Furthermore, because it is distributed along the lower portion of the range, which was the first to be left bare on the breaking up of the ice-sheet at the close of the glacial winter, the soil it is growing upon has been longer exposed to post-glacial weathering, and consequently is in a more crumbling, decayed condition than the fresher soils farther up the range, and therefore offers a less secure anchorage for the roots.

While exploring the forest zones of Mount Shasta, I discovered the path of a hurricane strewn with thousands of pines of this species. Great and small had been uprooted or wrenched off by sheer force, making a clean gap, like that made by a snow avalanche. But hurricanes capable of doing this class of work are rare in the Sierra, and when we have explored the forests from one extremity of the range to the other, we are compelled to believe that they are the most beautiful on the face of the earth, however we may regard the agents that have made them so.

There is always something deeply exciting, not only in the sounds of winds in the woods, which exert more or less influence over every mind, but in their varied waterlike flow as manifested by the movements of the trees, especially those of the conifers. By no other trees are they rendered so extensively and impressively visible, not even by the lordly tropic palms or tree-ferns responsive to the gentlest breeze. The waving of a forest of the giant Sequoias is indescribably impressive and sublime, but the pines seem to me the best interpreters of winds. They are mighty waving goldenrods, ever in tune, singing and writing wind-music all their long century lives. Little, however, of this noble tree-waving and tree-music will you see or hear in the strictly alpine portion of the forests. The burly Juniper, whose girth sometimes more than equals its height, is about as rigid as the rocks on which it grows. The slender lash-like sprays of the Dwarf Pine stream out in wavering ripples, but the tallest and slenderest are far too unyielding to wave even in the heaviest gales. They only shake in quick, short vibrations. The Hemlock Spruce, however, and the Mountain Pine, and some of the tallest thickets of the Two-leaved species bow in storms with considerable scope and gracefulness. But it is only in the lower and middle zones that the meeting of winds and woods is to be seen in all its grandeur.

One of the most beautiful and exhilarating storms I ever enjoyed in the Sierra occurred in December, 1874, when I happened to be exploring one of the tributary valleys of the Yuba River. The sky and the ground and the trees had been thoroughly rain-washed and were dry again. The day was intensely pure, one of those incomparable bits of California winter, warm and balmy and full of white sparkling sunshine, redolent of all the purest influences of the spring, and at the same time enlivened with one of the most bracing wind-storms conceivable. Instead of camping out, as I usually do, I then chanced to be stopping at the house of a friend. But when the storm began to sound, I lost no time in pushing out into the woods to enjoy it. For on such occasions Nature has always something rare to show us, and the danger to life and limb is hardly greater than one would experience crouching deprecatingly beneath a roof.

It was still early morning when I found myself fairly adrift. Delicious sunshine came pouring over the hills, lighting the tops of the pines, and setting free a steam of summery fragrance that contrasted strangely with the wild tones of the storm. The air was mottled with pine-tassels and bright green plumes, that went flashing past in the sunlight like birds pursued. But there was not the slightest dustiness, nothing less pure than leaves, and ripe pollen, and flecks of withered bracken and moss. I heard trees falling for hours at the rate of one every two or three minutes; some uprooted, partly on account of the loose, water-soaked condition of the ground; others broken straight across, where some weakness caused by fire had determined the spot. The gestures of the various trees made a delightful study. Young Sugar Pines, light and feathery as squirrel-tails, were bowing almost to the ground; while the grand old patriarchs, whose massive boles had been tried in a hundred storms, waved solemnly above them, their long, arching branches streaming fluently on the gale, and every needle thrilling and ringing and shedding off keen lances of light like a diamond. The Douglas Spruces, with long sprays drawn out in level tresses, and needles massed in a gray, shimmering glow, presented a most striking appearance as they stood in bold relief along the hilltops. The madroños in the dells, with their red bark and large glossy leaves tilted every way, reflected the sunshine in throbbing spangles like those one so often sees on the rippled surface of a glacier lake. But the Silver Pines were now the most impressively beautiful of all. Colossal spires 200 feet in height waved like supple goldenrods chanting and bowing low as if in worship, while the whole mass of their long, tremulous foliage was kindled into one continuous blaze of white sun-fire. The force of the gale was such that the most steadfast monarch of them all rocked down to its roots with a motion plainly perceptible when one leaned against it. Nature was holding high festival, and every fiber of the most rigid giants thrilled with glad excitement.

I drifted on through the midst of this passionate music and motion, across many a glen, from ridge to ridge; often halting in the lee of a rock for shelter, or to gaze and listen. Even when the grand anthem had swelled to its highest pitch, I could distinctly hear the varying tones of individual trees,—Spruce, and Fir, and Pine, and leafless Oak,—and even the infinitely gentle rustle of the withered grasses at my feet. Each was expressing itself in its own way,—singing its own song, and making its own peculiar gestures,—manifesting a richness of variety to be found in no other forest I have yet seen. The coniferous woods of Canada, and the Carolinas, and Florida, are made up of trees that resemble one another about as nearly as blades of grass, and grow close together in much the same way. Coniferous trees, in general, seldom possess individual character, such as is manifest among Oaks and Elms. But the California forests are made up of a greater number of distinct species than any other in the world. And in them we find, not only a marked differentiation into special groups, but also a marked individuality in almost every tree, giving rise to storm effects indescribably glorious.

Toward midday, after a long, tingling scramble through copses of hazel and ceanothus, I gained the summit of the highest ridge in the neighborhood; and then it occurred to me that it would be a fine thing to climb one of the trees to obtain a wider outlook and get my ear close to the Æolian music of its topmost needles. But under the circumstances the choice of a tree was a serious matter. One whose instep was not very strong seemed in danger of being blown down, or of being struck by others in case they should fall; another was branchless to a considerable height above the ground, and at the same time too large to be grasped with arms and legs in climbing; while others were not favorably situated for clear views. After cautiously casting about, I made choice of the tallest of a group of Douglas Spruces that were growing close together like a tuft of grass, no one of which seemed likely to fall unless all the rest fell with it. Though comparatively young, they were about 100 feet high, and their lithe, brushy tops were rocking and swirling in wild ecstasy. Being accustomed to climb trees in making botanical studies, I experienced no difficulty in reaching the top of this one, and never before did I enjoy so noble an exhilaration of motion. The slender tops fairly flapped and swished in the passionate torrent, bending and swirling backward and forward, round and round, tracing indescribable combinations of vertical and horizontal curves, while I clung with muscles firm braced, like a bobolink on a reed.

In its widest sweeps my tree-top described an arc of from twenty to thirty degrees, but I felt sure of its elastic temper, having seen others of the same species still more severely tried—bent almost to the ground indeed, in heavy snows—without breaking a fiber. I was therefore safe, and free to take the wind into my pulses and enjoy the excited forest from my superb outlook. The view from here must be extremely beautiful in any weather. Now my eye roved over the piny hills and dales as over fields of waving grain, and felt the light running in ripples and broad swelling undulations across the valleys from ridge to ridge, as the shining foliage was stirred by corresponding waves of air. Oftentimes these waves of reflected light would break up suddenly into a kind of beaten foam, and again, after chasing one another in regular order, they would seem to bend forward in concentric curves, and disappear on some hillside, like sea-waves on a shelving shore. The quantity of light reflected from the bent needles was so great as to make whole groves appear as if covered with snow, while the black shadows beneath the trees greatly enhanced the effect of the silvery splendor.

Excepting only the shadows there was nothing somber in all this wild sea of pines. On the contrary, notwithstanding this was the winter season, the colors were remarkably beautiful. The shafts of the pine and libocedrus were brown and purple, and most of the foliage was well tinged with yellow; the laurel groves, with the pale undersides of their leaves turned upward, made masses of gray; and then there was many a dash of chocolate color from clumps of manzanita, and jet of vivid crimson from the bark of the madroños, while the ground on the hillsides, appearing here and there through openings between the groves, displayed masses of pale purple and brown.

The sounds of the storm corresponded gloriously with this wild exuberance of light and motion. The profound bass of the naked branches and boles booming like waterfalls; the quick, tense vibrations of the pine-needles, now rising to a shrill, whistling hiss, now falling to a silky murmur; the rustling of laurel groves in the dells, and the keen metallic click of leaf on leaf—all this was heard in easy analysis when the attention was calmly bent.

The varied gestures of the multitude were seen to fine advantage, so that one could recognize the different species at a distance of several miles by this means alone, as well as by their forms and colors, and the way they reflected the light. All seemed strong and comfortable, as if really enjoying the storm, while responding to its most enthusiastic greetings. We hear much nowadays concerning the universal struggle for existence, but no struggle in the common meaning of the word was manifest here; no recognition of danger by any tree; no deprecation; but rather an invincible gladness as remote from exultation as from fear.

I kept my lofty perch for hours, frequently closing my eyes to enjoy the music by itself, or to feast quietly on the delicious fragrance that was streaming past. The fragrance of the woods was less marked than that produced during warm rain, when so many balsamic buds and leaves are steeped like tea; but, from the chafing of resiny branches against each other, and the incessant attrition of myriads of needles, the gale was spiced to a very tonic degree. And besides the fragrance from these local sources there were traces of scents brought from afar. For this wind came first from the sea, rubbing against its fresh, briny waves, then distilled through the redwoods, threading rich ferny gulches, and spreading itself in broad undulating currents over many a flower-enameled ridge of the coast mountains, then across the golden plains, up the purple foot-hills, and into these piny woods with the varied incense gathered by the way.

Winds are advertisements of all they touch, however much or little we may be able to read them; telling their wanderings even by their scents alone. Mariners detect the flowery perfume of land-winds far at sea, and sea-winds carry the fragrance of dulse and tangle far inland, where it is quickly recognized, though mingled with the scents of a thousand land-flowers. As an illustration of this, I may tell here that I breathed sea-air on the Firth of Forth, in Scotland, while a boy; then was taken to Wisconsin, where I remained nineteen years; then, without in all this time having breathed one breath of the sea, I walked quietly, alone, from the middle of the Mississippi Valley to the Gulf of Mexico, on a botanical excursion, and while in Florida, far from the coast, my attention wholly bent on the splendid tropical vegetation about me, I suddenly recognized a sea-breeze, as it came sifting through the palmettos and blooming vine-tangles, which at once awakened and set free a thousand dormant associations, and made me a boy again in Scotland, as if all the intervening years had been annihilated.

Most people like to look at mountain rivers, and bear them in mind; but few care to look at the winds, though far more beautiful and sublime, and though they become at times about as visible as flowing water. When the north winds in winter are making upward sweeps over the curving summits of the High Sierra, the fact is sometimes published with flying snow-banners a mile long. Those portions of the winds thus embodied can scarce be wholly invisible, even to the darkest imagination. And when we look around over an agitated forest, we may see something of the wind that stirs it, by its effects upon the trees. Yonder it descends in a rush of water-like ripples, and sweeps over the bending pines from hill to hill. Nearer, we see detached plumes and leaves, now speeding by on level currents, now whirling in eddies, or, escaping over the edges of the whirls, soaring aloft on grand, upswelling domes of air, or tossing on flame-like crests. Smooth, deep currents, cascades, falls, and swirling eddies, sing around every tree and leaf, and over all the varied topography of the region with telling changes of form, like mountain rivers conforming to the features of their channels.

After tracing the Sierra streams from their fountains to the plains, marking where they bloom white in falls, glide in crystal plumes, surge gray and foam-filled in boulder-choked gorges, and slip through the woods in long, tranquil reaches—after thus learning their language and forms in detail, we may at length hear them chanting all together in one grand anthem, and comprehend them all in clear inner vision, covering the range like lace. But even this spectacle is far less sublime and not a whit more substantial than what we may behold of these storm-streams of air in the mountain woods.

We all travel the milky way together, trees and men; but it never occurred to me until this storm-day, while swinging in the wind, that trees are travelers, in the ordinary sense. They make many journeys, not extensive ones, it is true; but our own little journeys, away and back again, are only little more than tree-wavings—many of them not so much.

When the storm began to abate, I dismounted and sauntered down through the calming woods. The storm-tones died away, and, turning toward the east, I beheld the countless hosts of the forests hushed and tranquil, towering above one another on the slopes of the hills like a devout audience. The setting sun filled them with amber light, and seemed to say, while they listened, "My peace I give unto you."

As I gazed on the impressive scene, all the so-called ruin of the storm was forgotten, and never before did these noble woods appear so fresh, so joyous, so immortal.

CHAPTER XI

THE RIVER FLOODS

THE SIERRA RIVERS are flooded every spring by the melting of the snow as regularly as the famous old Nile. They begin to rise in May, and in June high-water mark is reached. But because the melting does not go on rapidly over all the fountains, high and low, simultaneously, and the melted snow is not reinforced at this time of year by rain, the spring floods are seldom very violent or destructive. The thousand falls, however, and the cascades in the cañons are then in full bloom, and sing songs from one end of the range to the other. Of course the snow on the lower tributaries of the rivers is first melted, then that on the higher fountains most exposed to sunshine, and about a month later the cooler, shadowy fountains send down their treasures, thus allowing the main trunk streams nearly six weeks to get their waters hurried through the foot-hills and across the lowlands to the sea. Therefore very violent spring floods are avoided, and will be as long as the shading, restraining forests last. The rivers of the north half of the range are still less subject to sudden floods, because their upper fountains in great part lie protected from the changes of the weather beneath thick folds of lava, just as many of the rivers of Alaska lie beneath folds of ice, coming to the light farther down the range in large springs, while those of the high Sierra lie on the surface of solid granite, exposed to every change of temperature. More than ninety per cent. of the water derived from the snow and ice of Mount Shasta is at once absorbed and drained away beneath the porous lava folds of the mountain, where mumbling and groping in the dark they at length find larger fissures and tunnel-like caves from which they emerge, filtered and cool, in the form of large springs, some of them so large they give birth to rivers that set out on their journeys beneath the sun without any visible intermediate period of childhood. Thus the Shasta River issues from a large lake-like spring in Shasta Valley, and about two thirds of the volume of the McCloud River gushes forth suddenly from the face of a lava bluff in a roaring spring seventy-five yards wide.

These spring rivers of the north are of course shorter than those of the south whose tributaries extend up to the tops of the mountains. Fall River, an important tributary of the Pitt or Upper Sacramento, is only about ten miles long, and is all falls, cascades, and springs from its head to its confluence with the Pitt. Bountiful springs, charmingly embowered, issue from the rocks at one end of it, a snowy fall a hundred and eighty feet high thunders at the other, and a rush of crystal rapids sing and dance between. Of course such streams are but little affected by the weather. Sheltered from evaporation their flow is nearly as full in the autumn as in the time of general spring floods. While those of the high Sierra diminish to less than the hundredth part of their springtime prime, shallowing in autumn to a series of silent pools among the rocks and hollows of their channels, connected by feeble, creeping threads of water, like the sluggish sentences of a tired writer, connected by a drizzle of "ands" and "buts." Strange to say, the greatest floods occur in winter, when one would suppose all the wild waters would be muffled and chained in frost and snow. The same long, all-day storms of the so-called Rainy Season in California, that give rain to the lowlands, give dry frosty snow to the mountains. But at rare intervals warm rains and warm winds invade the mountains and push back the snow line from 2000 feet to 8000, or even higher, and then come the big floods.

I was usually driven down out of the High Sierra about the end of November, but the winter of 1874 and 1875 was so warm and calm that I was tempted to seek general views of the geology and topography of the basin of Feather River in January. And I had just completed a hasty survey of the region, and made my way down to winter quarters, when one of the grandest flood-storms that I ever saw broke on the mountains. I was then in the edge of the main forest belt at a small foot-hill town called Knoxville, on the divide between the waters of the Feather and Yuba rivers. The cause of this notable flood was simply a sudden and copious fall of warm wind and rain on the basins of these rivers at a time when they contained a considerable quantity of snow. The rain was so heavy and long-sustained that it was, of itself, sufficient to make a good wild flood, while the snow which the warm wind and rain melted on the upper and middle regions of the basins was sufficient to make another flood equal to that of the rain. Now these two distinct harvests of flood waters were gathered simultaneously and poured out on the plain in one magnificent avalanche. The basins of the Yuba and Feather, like many others of the Sierra, are admirably adapted to the growth of floods of this kind. Their many tributaries radiate far and wide, comprehending extensive areas, and the tributaries are steeply inclined, while the trunks are comparatively level. While the flood-storm was in progress the thermometer at Knoxville ranged between 44° and 50°; and when warm wind and warm rain fall simultaneously on snow contained in basins like these, both the rain and that portion of the snow which the rain and wind melt are at first sponged up and held back until the combined mass becomes sludge, which at length, suddenly dissolving, slips and descends all together to the trunk channel; and since the deeper the stream the faster it flows, the flooded portion of the current above overtakes the slower foot-hill portion below it, and all sweeping forward together with a high, overcurling front, debouches on the open plain with a violence and suddenness that at first seem wholly unaccountable. The destructiveness of the lower portion of this particular flood was somewhat augmented by mining gravel in the river channels, and by levees which gave way after having at first restrained and held back the accumulating waters. These exaggerating conditions did not, however, greatly influence the general result, the main effect having been caused by the rare combination of flood factors indicated above. It is a pity that but few people meet and enjoy storms so noble as this in their homes in the mountains, for spending themselves in the open levels of the plains, they are likely to be remembered more by the bridges and houses they carry away than by their beauty or the thousand blessings they bring to the fields and gardens of Nature.

On the morning of the flood, January 19th, all the Feather and Yuba landscapes were covered with running water, muddy torrents filled every gulch and ravine, and the sky was thick with rain. The pines had long been sleeping in sunshine; they were now awake, roaring and waving with the beating storm, and the winds sweeping along the curves of hill and dale, streaming through the woods, surging and gurgling on the tops of rocky ridges, made the wildest of wild storm melody.

It was easy to see that only a small part of the rain reached the ground in the form of drops. Most of it was thrashed into dusty spray like that into which small waterfalls are divided when they dash on shelving rocks. Never have I seen water coming from the sky in denser or more passionate streams. The wind chased the spray forward in choking drifts, and compelled me again and again to seek shelter in the dell copses and back of large trees to rest and catch my breath. Wherever I went, on ridges or in hollows, enthusiastic water still flashed and gurgled about my ankles, recalling a wild winter flood in Yosemite when a hundred waterfalls came booming and chanting together and filled the grand valley with a sea-like roar.

After drifting an hour or two in the lower woods, I set out for the summit of a hill 900 feet high, with a view to getting as near the heart of the storm as possible. In order to reach it I had to cross Dry Creek, a tributary of the Yuba that goes crawling along the base of the hill on the northwest. It was now a booming river as large as the Tuolumne at ordinary stages, its current brown with mining-mud washed down from many a "claim," and mottled with sluice-boxes, fence-rails, and logs that had long lain above its reach. A slim foot-bridge stretched across it, now scarcely above the swollen current. Here I was glad to linger, gazing and listening, while the storm was in its richest mood—the gray rain-flood above, the brown river-flood beneath. The language of the river was scarcely less enchanting than that of the wind and rain; the sublime overboom of the main bouncing, exulting current, the swash and gurgle of the eddies, the keen dash and clash of heavy waves breaking against rocks, and the smooth, downy hush of shallow currents feeling their way through the willow thickets of the margin. And amid all this varied throng of sounds I heard the smothered bumping and rumbling of boulders on the bottom as they were shoving and rolling forward against one another in a wild rush, after having lain still for probably 100 years or more.

The glad creek rose high above its banks and wandered from its channel out over many a briery sand-flat and meadow. Alders and willows waist-deep were bearing up against the current with nervous trembling gestures, as if afraid of being carried away, while supple branches bending confidingly, dipped lightly and rose again, as if stroking the wild waters in play. Leaving the bridge and passing on through the storm-thrashed woods, all the ground seemed to be moving. Pine-tassels, flakes of bark, soil, leaves, and broken branches were being swept forward, and many a rock-fragment, weathered from exposed ledges, was now receiving its first rounding and polishing in the wild streams of the storm. On they rushed through every gulch and hollow, leaping, gliding, working with a will, and rejoicing like living creatures.

Nor was the flood confined to the ground. Every tree had a water system of its own spreading far and wide like miniature Amazons and Mississippis.

Toward midday, cloud, wind, and rain reached their highest development. The storm was in full bloom, and formed, from my commanding outlook on the hilltop, one of the most glorious views I ever beheld. As far as the eye could reach, above, beneath, around, wind-driven rain filled the air like one vast waterfall. Detached clouds swept imposingly up the valley, as if they were endowed with independent motion and had special work to do in replenishing the mountain wells, now rising above the pine-tops, now descending into their midst, fondling their arrowy spires and soothing every branch and leaf with gentleness in the midst of all the savage sound and motion. Others keeping near the ground glided behind separate groves, and brought them forward into relief with admirable distinctness; or, passing in front, eclipsed whole groves in succession, pine after pine melting in their gray fringes and bursting forth again seemingly clearer than before.

The forms of storms are in great part measured, and controlled by the topography of the regions where they rise and over which they pass. When, therefore, we attempt to study them from the valleys, or from gaps and openings of the forest, we are confounded by a multitude of separate and apparently antagonistic impressions. The bottom of the storm is broken up into innumerable waves and currents that surge against the hillsides like sea-waves against a shore, and these, reacting on the nether surface of the storm, erode immense cavernous hollows and cañons, and sweep forward the resulting detritus in long trains, like the moraines of glaciers. But, as we ascend, these partial, confusing effects disappear and the phenomena are beheld united and harmonious.

The longer I gazed into the storm, the more plainly visible it became. The drifting cloud detritus gave it a kind of visible body, which explained many perplexing phenomena, and published its movements in plain terms, while the texture of the falling mass of rain rounded it out and rendered it more complete. Because raindrops differ in size they fall at different velocities and overtake and clash against one another, producing mist and spray. They also, of course, yield unequal compliance to the force of the wind, which gives rise to a still greater degree of interference, and passionate gusts sweep off clouds of spray from the groves like that torn from wave-tops in a gale. All these factors of irregularity in density, color, and texture of the general rain mass tend to make it the more appreciable and telling. It is then seen as one grand flood rushing over bank and brae, bending the pines like weeds, curving this way and that, whirling in huge eddies in hollows and dells, while the main current pours grandly over all, like ocean currents over the landscapes that lie hidden at the bottom of the sea.

I watched the gestures of the pines while the storm was at its height, and it was easy to see that they were not distressed. Several large Sugar Pines stood near the thicket in which I was sheltered, bowing solemnly and tossing their long arms as if interpreting the very words of the storm while accepting its wildest onsets with passionate exhilaration. The lions were feeding. Those who have observed sunflowers feasting on sunshine during the golden days of Indian summer know that none of their gestures express thankfulness. Their celestial food is too heartily given, too heartily taken to leave room for thanks. The pines were evidently accepting the benefactions of the storm in the same whole-souled manner; and when I looked down among the budding hazels, and still lower to the young violets and fern-tufts on the rocks, I noticed the same divine methods of giving and taking, and the same exquisite adaptations of what seems an outbreak of violent and uncontrollable force to the purposes of beautiful and delicate life. Calms like sleep come upon landscapes, just as they do on people and trees, and storms awaken them in the same way. In the dry midsummer of the lower portion of the range the withered hills and valleys seem to lie as empty and expressionless as dead shells on a shore. Even the highest mountains may be found occasionally dull and uncommunicative as if in some way they had lost countenance and shrunk to less than half their real stature. But when the lightnings crash and echo in the cañons, and the clouds come down wreathing and crowning their bald snowy heads, every feature beams with expression and they rise again in all their imposing majesty.

Storms are fine speakers, and tell all they know, but their voices of lightning, torrent, and rushing wind are much less numerous than the nameless still, small voices too low for human ears; and because we are poor listeners we fail to catch much that is fairly within reach. Our best rains are heard mostly on roofs, and winds in chimneys; and when by choice or compulsion we are pushed into the heart of a storm, the confusion made by cumbersome equipments and nervous haste and mean fear, prevent our hearing any other than the loudest expressions. Yet we may draw enjoyment from storm sounds that are beyond hearing, and storm movements we cannot see. The sublime whirl of planets around their suns is as silent as raindrops oozing in the dark among the roots of plants. In this great storm, as in every other, there were tones and gestures inexpressibly gentle manifested in the midst of what is called violence and fury, but easily recognized by all who look and listen for them. The rain brought out the colors of the woods with delightful freshness, the rich brown of the bark of the trees and the fallen burs and leaves and dead ferns; the grays of rocks and lichens; the light purple of swelling buds, and the warm yellow greens of the libocedrus and mosses. The air was steaming with delightful fragrance, not rising and wafting past in separate masses, but diffused through all the atmosphere. Pine woods are always fragrant, but most so in spring when the young tassels are opening and in warm weather when the various gums and balsams are softened by the sun. The wind was now chafing their innumerable needles and the warm rain was steeping them. Monardella grows here in large beds in the openings, and there is plenty of laurel in dells and manzanita on the hillsides, and the rosy, fragrant chamœbatia carpets the ground almost everywhere. These, with the gums and balsams of the woods, form the main local fragrance-fountains of the storm. The ascending clouds of aroma wind-rolled and rain-washed became pure like light and traveled with the wind as part of it. Toward the middle of the afternoon the main flood cloud lifted along its western border revealing a beautiful section of the Sacramento Valley some twenty or thirty miles away, brilliantly sun-lighted and glistering with rain-sheets as if paved with silver. Soon afterward a jagged bluff-like cloud with a sheer face appeared over the valley of the Yuba, dark-colored and roughened with numerous furrows like some huge lava-table. The blue Coast Range was seen stretching along the sky like a beveled wall, and the somber, craggy Marysville Buttes rose impressively out of the flooded plain like islands out of the sea. Then the rain began to abate and I sauntered down through the dripping bushes reveling in the universal vigor and freshness that inspired all the life about me. How clean and unworn and immortal the woods seemed to be!—the lofty cedars in full bloom laden with golden pollen and their washed plumes shining; the pines rocking gently and settling back into rest, and the evening sunbeams spangling on the broad leaves of the madroños, their tracery of yellow boughs relieved against dusky thickets of Chestnut Oak; liverworts, lycopodiums, ferns were exulting in glorious revival, and every moss that had ever lived seemed to be coming crowding back from the dead to clothe each trunk and stone in living green. The steaming ground seemed fairly to throb and tingle with life; smilax, fritillaria, saxifrage, and young violets were pushing up as if already conscious of the summer glory, and innumerable green and yellow buds were peeping and smiling everywhere.

As for the birds and squirrels, not a wing or tail of them was to be seen while the storm was blowing. Squirrels dislike wet weather more than cats do; therefore they were at home rocking in their dry nests. The birds were hiding in the dells out of the wind, some of the strongest of them pecking at acorns and manzanita berries, but most were perched on low twigs, their breast feathers puffed out and keeping one another company through the hard time as best they could.

When I arrived at the village about sundown, the good people bestirred themselves, pitying my bedraggled condition as if I were some benumbed castaway snatched from the sea, while I, in turn, warm with excitement and reeking like the ground, pitied them for being dry and defrauded of all the glory that Nature had spread round about them that day.

CHAPTER XII

SIERRA THUNDER-STORMS

THE WEATHER OF spring and summer in the middle region of the Sierra is usually well flecked with rains and light dustings of snow, most of which are far too obviously joyful and life-giving to be regarded as storms; and in the picturesque beauty and clearness of outlines of their clouds they offer striking contrasts to those boundless, all-embracing cloud-mantles of the storms of winter. The smallest and most perfectly individualized specimens present a richly modeled cumulous cloud rising above the dark woods, about 11 A.M., swelling with a visible motion straight up into the calm, sunny sky to a height of 12,000 to 14,000 feet above the sea, its white, pearly bosses relieved by gray and pale purple shadows in the hollows, and showing outlines as keenly defined as those of the glacier-polished domes. In less than an hour it attains full development and stands poised in the blazing sunshine like some colossal mountain, as beautiful in form and finish as if it were to become a permanent addition to the landscape. Presently a thunderbolt crashes through the crisp air, ringing like steel on steel, sharp and clear, its startling detonation breaking into a spray of echoes against the cliffs and cañon walls. Then down comes a cataract of rain. The big drops sift through the pine-needles, plash and patter on the granite pavements, and pour down the sides of ridges and domes in a network of gray, bubbling rills. In a few minutes the cloud withers to a mesh of dim filaments and disappears, leaving the sky perfectly clear and bright, every dust-particle wiped and washed out of it. Everything is refreshed and invigorated, a steam of fragrance rises, and the storm is finished—one cloud, one lightning-stroke, and one dash of rain. This is the Sierra midsummer thunder-storm reduced to its lowest terms. But some of them attain much larger proportions, and assume a grandeur and energy of expression hardly surpassed by those bred in the depths of winter, producing those sudden floods called "cloud-bursts," which are local, and to a considerable extent periodical, for they appear nearly every day about the same time for weeks, usually about eleven o'clock, and lasting from five minutes to an hour or two. One soon becomes so accustomed to see them that the noon sky seems empty and abandoned without them, as if Nature were forgetting something. When the glorious pearl and alabaster clouds of these noonday storms are being built I never give attention to anything else. No mountain or mountain-range, however divinely clothed with light, has a more enduring charm than those fleeting mountains of the sky—floating fountains bearing water for every well, the angels of the streams and lakes; brooding in the deep azure, or sweeping softly along the ground over ridge and dome, over meadow, over forest, over garden and grove; lingering with cooling shadows, refreshing every flower, and soothing rugged rock-brows with a gentleness of touch and gesture wholly divine.

The most beautiful and imposing of the summer storms rise just above the upper edge of the Silver Fir zone, and all are so beautiful that it is not easy to choose any one for particular description. The one that I remember best fell on the mountains near Yosemite Valley, July 19, 1869, while I was encamped in the Silver Fir woods. A range of bossy cumuli took possession of the sky, huge domes and peaks rising one beyond another with deep cañons between them, bending this way and that in long curves and reaches, interrupted here and there with white upboiling masses that looked like the spray of waterfalls. Zigzag lances of lightning followed each other in quick succession, and the thunder was so gloriously loud and massive it seemed as if surely an entire mountain was being shattered at every stroke. Only the trees were touched, however, so far as I could see,—a few firs 200 feet high, perhaps, and five to six feet in diameter, were split into long rails and slivers from top to bottom and scattered to all points of the compass. Then came the rain in a hearty flood, covering the ground and making it shine with a continuous sheet of water that, like a transparent film or skin, fitted closely down over all the rugged anatomy of the landscape.

It is not long, geologically speaking, since the first raindrop fell on the present landscapes of the Sierra; and in the few tens of thousands of years of stormy cultivation they have been blest with, how beautiful they have become! The first rains fell on raw, crumbling moraines and rocks without a plant. Now scarcely a drop can fail to find a beautiful mark: on the tops of the peaks, on the smooth glacier pavements, on the curves of the domes, on moraines full of crystals, on the thousand forms of yosemitic sculpture with their tender beauty of balmy, flowery vegetation, laving, plashing, glinting, pattering; some falling softly on meadows, creeping out of sight, seeking and finding every thirsty rootlet, some through the spires of the woods, sifting in dust through the needles, and whispering good cheer to each of them; some falling with blunt tapping sounds, drumming on the broad leaves of veratrum, cypripedium, saxifrage; some falling straight into fragrant corollas, kissing the lips of lilies, glinting on the sides of crystals, on shining grains of gold; some falling into the fountains of snow to swell their well-saved stores; some into the lakes and rivers, patting the smooth glassy levels, making dimples and bells and spray, washing the mountain windows, washing the wandering winds; some plashing into the heart of snowy falls and cascades as if eager to join in the dance and the song and beat the foam yet finer. Good work and happy work for the merry mountain raindrops, each one of them a brave fall in itself, rushing from the cliffs and hollows of the clouds into the cliffs and hollows of the mountains; away from the thunder of the sky into the thunder of the roaring rivers. And how far they have to go, and how many cups to fill—cassiope-cups, holding half a drop, and lake basins between the hills, each replenished with equal care—every drop God's messenger sent on its way with glorious pomp and display of power—silvery new-born stars with lake and river, mountain and valley—all that the landscape holds—reflected in their crystal depths.

CHAPTER XIII

THE WATER-OUZEL

THE WATERFALLS OF the Sierra are frequented by only one bird,—the Ouzel or Water Thrush (*Cinclus mexicanus*, Sw.). He is a singularly joyous and lovable little fellow, about the size of a robin, clad in a plain waterproof suit of bluish gray, with a tinge of chocolate on the head and shoulders. In form he is about as smoothly plump and compact as a pebble that has been whirled in a pot-hole, the flowing contour of his body being interrupted only by his strong feet and bill, the crisp wing-tips, and the up-slanted wren-like tail.

Among all the countless waterfalls I have met in the course of ten years' exploration in the Sierra, whether among the icy peaks, or warm foot-hills, or in the profound yosemitic cañons of the middle region, not one was found without its Ouzel. No cañon is too cold for this little bird, none too lonely, provided it be rich in falling water. Find a fall, or cascade, or rushing rapid, anywhere upon a clear stream, and there you will surely find its complementary Ouzel, flitting about in the spray, diving in foaming eddies, whirling like a leaf among beaten foam-bells; ever vigorous and enthusiastic, yet self-contained, and neither seeking nor shunning your company.

If disturbed while dipping about in the margin shallows, he either sets off with a rapid whir to some other feeding-ground up or down the stream, or alights on some half-submerged rock or snag out in the current, and immediately begins to nod and courtesy like a wren, turning his head from side to side with many other odd dainty movements that never fail to fix the attention of the observer.

He is the mountain streams' own darling, the humming-bird of blooming waters, loving rocky ripple-slopes and sheets of foam as a bee loves flowers, as a lark loves sunshine and meadows. Among all the mountain birds, none has cheered me so much in my lonely wanderings,—none so unfailingly. For both in winter and summer he sings, sweetly, cheerily, independent alike of sunshine and of love, requiring no other inspiration than the stream on which he dwells. While water sings, so must he, in heat or cold, calm or storm, ever attuning his voice in sure accord; low in the drought of summer and the drought of winter, but never silent.

During the golden days of Indian summer, after most of the snow has been melted, and the mountain streams have become feeble,—a succession of silent pools, linked together by shallow, transparent currents and strips of silvery lacework,—then the song of the Ouzel is at its lowest ebb. But as soon as the winter clouds have bloomed, and the mountain treasuries are once more replenished with snow, the voices of the streams and ouzels increase in strength and richness until the flood season of early summer. Then the torrents chant their noblest anthems, and then is the flood-time of our songster's melody. As for weather, dark days and sun days are the same to him. The voices of most song-birds, however joyous, suffer a long winter eclipse; but the Ouzel sings on through all the seasons and every kind of storm. Indeed no storm can be more violent than those of the waterfalls in the midst of which he delights to dwell. However dark and boisterous the weather, snowing, blowing, or cloudy, all the same he sings, and with never a note of sadness. No need of spring sunshine to thaw *his* song, for it never freezes. Never shall you hear anything wintry from *his* warm breast; no pinched cheeping, no wavering notes between sorrow and joy; his mellow, fluty voice is ever tuned to downright gladness, as free from dejection as cock-crowing.

It is pitiful to see wee frost-pinched sparrows on cold mornings in the mountain groves shaking the snow from their feathers, and hopping about as if anxious to be cheery, then hastening back to their hidings out of the wind, puffing out their breast-feathers over their toes, and subsiding among the leaves, cold and breakfastless, while the snow continues to fall, and there is no sign of clearing. But the Ouzel never calls forth a single touch of pity; not because he is strong to endure, but rather because he seems to live a charmed life beyond the reach of every influence that makes endurance necessary.

One wild winter morning, when Yosemite Valley was swept its length from west to east by a cordial snow-storm, I sallied forth to see what I might learn and enjoy. A sort of gray, gloaming-like darkness filled the valley, the huge walls were out of sight, all ordinary sounds were smothered, and even the loudest booming of the falls was at times buried beneath the roar of the heavy-laden blast. The loose snow was already over five feet deep on the meadows, making extended walks impossible without the aid of snow-shoes. I found no great difficulty, however, in making my way to a certain ripple on the river where one of my ouzels lived. He was at home, busily gleaning his breakfast among the pebbles of a shallow portion of the margin, apparently unaware of anything extraordinary in the weather. Presently he flew out to a stone against which the icy current was beating, and turning his back to the wind, sang as delightfully as a lark in springtime.

After spending an hour or two with my favorite, I made my way across the valley, boring and wallowing through the drifts, to learn as definitely as possible how the other birds were spending their time. The Yosemite birds are easily found during the winter because all of them excepting the Ouzel are restricted to the sunny north side of the valley, the south side being constantly eclipsed by the great frosty shadow of the wall. And because the Indian Cañon groves, from their peculiar exposure, are the warmest, the birds congregate there, more especially in severe weather.

I found most of the robins cowering on the lee side of the larger branches where the snow could not fall upon them, while two or three of the more enterprising were making desperate efforts to reach the mistletoe berries by clinging nervously to the under side of the snow-crowned masses, back downward, like woodpeckers. Every now and then they would dislodge some of the loose fringes of the snow-crown, which would come sifting down on them and send them screaming back to camp, where they would subside among their companions with a shiver, muttering in low, querulous chatter like hungry children.

Some of the sparrows were busy at the feet of the larger trees gleaning seeds and benumbed insects, joined now and then by a robin weary of his unsuccessful attempts upon the snow-covered berries. The brave woodpeckers were clinging to the snowless sides of the larger boles and overarching branches of the camp trees, making short flights from side to side of the grove, pecking now and then at the acorns they had stored in the bark, and chattering aimlessly as if unable to keep still, yet evidently putting in the time in a very dull way, like storm-bound travelers at a country tavern. The hardy nut-hatches were threading the open furrows of the trunks in their usual industrious manner, and uttering their quaint notes, evidently less distressed than their neighbors. The Steller jays were of course making more noisy stir than all the other birds combined; ever coming and going with loud bluster, screaming as if each had a lump of melting sludge in his throat, and taking good care to improve the favorable opportunity afforded by the storm to steal from the acorn stores of the woodpeckers. I also noticed one solitary gray eagle braving the storm on the top of a tall pine-stump just outside the main grove. He was standing bolt upright with his back to the wind, a tuft of snow piled on his square shoulders, a monument of passive endurance. Thus every snow-bound bird seemed more or less uncomfortable if not in positive distress. The storm was reflected in every gesture, and not one cheerful note, not to say song, came from a single bill; their cowering, joyless endurance offering a striking contrast to the spontaneous, irrepressible gladness of the Ouzel, who could no more help exhaling sweet song than a rose sweet fragrance. He *must* sing though the heavens fall. I remember noticing the distress of a pair of robins during the violent earthquake of the year 1872, when the pines of the Valley, with strange movements, flapped and waved their branches, and beetling rock-brows came thundering down to the meadows in tremendous avalanches. It did not occur to me in the midst of the excitement of other observations to look for the ouzels, but I doubt not they were singing straight on through it all, regarding the terrible rock-thunder as fearlessly as they do the booming of the waterfalls.

What may be regarded as the separate songs of the Ouzel are exceedingly difficult of description, because they are so variable and at the same time so confluent. Though I have been acquainted with my favorite ten years, and during most of this time have heard him sing nearly every day, I still detect notes and strains that seem new to me. Nearly all of his music is sweet and tender, lapsing from his round breast like water over the smooth lip of a pool, then breaking farther on into a sparkling foam of melodious notes, which glow with subdued enthusiasm, yet without expressing much of the strong, gushing ecstasy of the bobolink or skylark.

The more striking strains are perfect arabesques of melody, composed of a few full, round, mellow notes, embroidered with delicate trills which fade and melt in long slender cadences. In a general way his music is that of the streams refined and spiritualized. The deep booming notes of the falls are in it, the trills of rapids, the gurgling of margin eddies, the low whispering of level reaches, and the sweet tinkle of separate drops oozing from the ends of mosses and falling into tranquil pools.

The Ouzel never sings in chorus with other birds, nor with his kind, but only with the streams. And like flowers that bloom beneath the surface of the ground, some of our favorite's best song-blossoms never rise above the surface of the heavier music of the water. I have often observed him singing in the midst of beaten spray, his music completely buried beneath the water's roar; yet I knew he was surely singing by his gestures and the movements of his bill.

His food, as far as I have noticed, consists of all kinds of water insects, which in summer are chiefly procured along shallow margins. Here he wades about ducking his head under water and deftly turning over pebbles and fallen leaves with his bill, seldom choosing to go into deep water where he has to use his wings in diving.

He seems to be especially fond of the larvæ of mosquitos, found in abundance attached to the bottom of smooth rock channels where the current is shallow. When feeding in such places he wades upstream, and often while his head is under water the swift current is deflected upward along the glossy curves of his neck and shoulders, in the form of a clear, crystalline shell, which fairly incloses him like a bell-glass, the shell being broken and re-formed as he lifts and dips his head; while ever and anon he sidles out to where the too powerful current carries him off his feet; then he dexterously rises on the wing and goes gleaning again in shallower places.

But during the winter, when the stream-banks are embossed in snow, and the streams themselves are chilled nearly to the freezing-point, so that the snow falling into them in stormy weather is not wholly dissolved, but forms a thin, blue sludge, thus rendering the current opaque—then he seeks the deeper portions of the main rivers, where he may dive to clear water beneath the sludge. Or he repairs to some open lake or millpond, at the bottom of which he feeds in safety.

When thus compelled to betake himself to a lake, he does not plunge into it at once like a duck, but always alights in the first place upon some rock or fallen pine along the shore. Then flying out thirty or forty yards, more or less, according to the character of the bottom, he alights with a dainty glint on the surface, swims about, looks down, finally makes up his mind, and disappears with a sharp stroke of his wings. After feeding for two or three minutes he suddenly reappears, showers the water from his wings with one vigorous shake, and rises abruptly into the air as if pushed up from beneath, comes back to his perch, sings a few minutes, and goes out to dive again; thus coming and going, singing and diving at the same place for hours.

The Ouzel is usually found singly; rarely in pairs, excepting during the breeding season, and *very* rarely in threes or fours. I once observed three thus spending a winter morning in company, upon a small glacier lake, on the Upper Merced, about 7500 feet above the level of the sea. A storm had occurred during the night, but the morning sun shone unclouded, and the shadowy lake, gleaming darkly in its setting of fresh snow, lay smooth and motionless as a mirror. My camp chanced to be within a few feet of the water's edge, opposite a fallen pine, some of the branches of which leaned out over the lake. Here my three dearly welcome visitors took up their station, and at once began to embroider the frosty air with their delicious melody, doubly delightful to me that particular morning, as I had been somewhat apprehensive of danger in breaking my way down through the snow-choked cañons to the lowlands.

The portion of the lake bottom selected for a feeding-ground lies at a depth of fifteen or twenty feet below the surface, and is covered with a short growth of algæ and other aquatic plants,—facts I had previously determined while sailing over it on a raft. After alighting on the glassy surface, they occasionally indulged in a little play, chasing one another round about in small circles; then all three would suddenly dive together, and then come ashore and sing.

The Ouzel seldom swims more than a few yards on the surface, for, not being web-footed, he makes rather slow progress, but by means of his strong, crisp wings he swims, or rather flies, with celerity under the surface, often to considerable distances. But it is in withstanding the force of heavy rapids that his strength of wing in this respect is most strikingly manifested. The following may be regarded as a fair illustration of his power of sub-aquatic flight. One stormy morning in winter when the Merced River was blue and green with unmelted snow, I observed one of my ouzels perched on a snag out in the midst of a swift-rushing rapid, singing cheerily, as if everything was just to his mind; and while I stood on the bank admiring him, he suddenly plunged into the sludgy current, leaving his song abruptly broken off. After feeding a minute or two at the bottom, and when one would suppose that he must inevitably be swept far down-stream, he emerged just where he went down, alighted on the same snag, showered the water-beads from his feathers, and continued his unfinished song, seemingly in tranquil ease as if it had suffered no interruption.

The Ouzel alone of all birds dares to enter a white torrent. And though strictly terrestrial in structure, no other is so inseparably related to water, not even the duck, or the bold ocean albatross, or the stormy-petrel. For ducks go ashore as soon as they finish feeding in undisturbed places, and very often make long flights overland from lake to lake or field to field. The same is true of most other aquatic birds. But the Ouzel, born on the brink of a stream, or on a snag or boulder in the midst of it, seldom leaves it for a single moment. For, notwithstanding he is often on the wing, he never flies overland, but whirs with rapid, quail-like beat above the stream, tracing all its windings. Even when the stream is quite small, say from five to ten feet wide, he seldom shortens his flight by crossing a bend, however abrupt it may be; and even when disturbed by meeting some one on the bank, he prefers to fly over one's head, to dodging out over the ground. When, therefore, his flight along a crooked stream is viewed endwise, it appears most strikingly wavered—a description on the air of every curve with lightning-like rapidity.

The vertical curves and angles of the most precipitous torrents he traces with the same rigid fidelity, swooping down the inclines of cascades, dropping sheer over dizzy falls amid the spray, and ascending with the same fearlessness and ease, seldom seeking to lessen the steepness of the acclivity by beginning to ascend before reaching the base of the fall. No matter though it may be several hundred feet in height he holds straight on, as if about to dash headlong into the throng of booming rockets, then darts abruptly upward, and, after alighting at the top of the precipice to rest a moment, proceeds to feed and sing. His flight is solid and impetuous, without any intermission of wing-beats,—one homogeneous buzz like that of a laden bee on its way home. And while thus buzzing freely from fall to fall, he is frequently heard giving utterance to a long outdrawn train of unmodulated notes, in no way connected with his song, but corresponding closely with his flight in sustained vigor.

Were the flights of all the ouzels in the Sierra traced on a chart, they would indicate the direction of the flow of the entire system of ancient glaciers, from about the period of the breaking up of the ice-sheet until near the close of the glacial winter; because the streams which the ouzels so rigidly follow are, with the unimportant exceptions of a few side tributaries, all flowing in channels eroded for them out of the solid flank of the range by the vanished glaciers,—the streams tracing the ancient glaciers, the ouzels tracing the streams. Nor do we find so complete compliance to glacial conditions in the life of any other mountain bird, or animal of any kind. Bears frequently accept the pathways laid down by glaciers as the easiest to travel; but they often leave them and cross over from cañon to cañon. So also, most of the birds trace the moraines to some extent, because the forests are growing on them. But they wander far, crossing the cañons from grove to grove, and draw exceedingly angular and complicated courses.

The Ouzel's nest is one of the most extraordinary pieces of bird architecture I ever saw, odd and novel in design, perfectly fresh and beautiful, and in every way worthy of the genius of the little builder. It is about a foot in diameter, round and bossy in outline, with a neatly arched opening near the bottom, somewhat like an old-fashioned brick oven, or Hottentot's hut. It is built almost exclusively of green and yellow mosses, chiefly the beautiful fronded hypnum that covers the rocks and old drift-logs in the vicinity of waterfalls. These are deftly interwoven, and felted together into a charming little hut; and so situated that many of the outer mosses continue to flourish as if they had not been plucked. A few fine, silky-stemmed grasses are occasionally found interwoven with the mosses, but, with the exception of a thin layer lining the floor, their presence seems accidental, as they are of a species found growing with the mosses and are probably plucked with them. The site chosen for this curious mansion is usually some little rock-shelf within reach of the lighter particles of the spray of a waterfall, so that its walls are kept green and growing, at least during the time of high water.

No harsh lines are presented by any portion of the nest as seen in place, but when removed from its shelf, the back and bottom, and sometimes a portion of the top, is found quite sharply angular, because it is made to conform to the surface of the rock upon which and against which it is built, the little architect always taking advantage of slight crevices and protuberances that may chance to offer, to render his structure stable by means of a kind of gripping and dovetailing.

In choosing a building-spot, concealment does not seem to be taken into consideration; yet notwithstanding the nest is large and guilelessly exposed to view, it is far from being easily detected, chiefly because it swells forward like any other bulging moss-cushion growing naturally in such situations. This is more especially the case where the nest is kept fresh by being well sprinkled. Sometimes these romantic little huts have their beauty enhanced by rock-ferns and grasses that spring up around the mossy walls, or in front of the door-sill, dripping with crystal beads.

Furthermore, at certain hours of the day, when the sunshine is poured down at the required angle, the whole mass of the spray enveloping the fairy establishment is brilliantly irised; and it is through so glorious a rainbow atmosphere as this that some of our blessed ouzels obtain their first peep at the world.

Ouzels seem so completely part and parcel of the streams they inhabit, they scarce suggest any other origin than the streams themselves; and one might almost be pardoned in fancying they come direct from the living waters, like flowers from the ground. At least, from whatever cause, it never occurred to me to look for their nests until more than a year after I had made the acquaintance of the birds themselves, although I found one the very day on which I began the search. In making my way from Yosemite to the glaciers at the heads of the Merced and Tuolumne rivers, I camped in a particularly wild and romantic portion of the Nevada cañon where in previous excursions I had never failed to enjoy the company of my favorites, who were attracted here, no doubt, by the safe nesting-places in the shelving rocks, and by the abundance of food and falling water. The river, for miles above and below, consists of a succession of small falls from ten to sixty feet in height, connected by flat, plume-like cascades that go flashing from fall to fall, free and almost channelless, over waving folds of glacier-polished granite.

On the south side of one of the falls, that portion of the precipice which is bathed by the spray presents a series of little shelves and tablets caused by the development of planes of cleavage in the granite, and by the consequent fall of masses through the action of the water. "Now here," said I, "of all places, is the most charming spot for an Ouzel's nest." Then carefully scanning the fretted face of the precipice through the spray, I at length noticed a yellowish moss-cushion, growing on the edge of a level tablet within five or six feet of the outer folds of the fall. But apart from the fact of its being situated where one acquainted with the lives of ouzels would fancy an Ouzel's nest ought to be, there was nothing in its appearance visible at first sight, to distinguish it from other bosses of rock-moss similarly situated with reference to perennial spray; and it was not until I had scrutinized it again and again, and had removed my shoes and stockings and crept along the face of the rock within eight or ten feet of it, that I could decide certainly whether it was a nest or a natural growth.

In these moss huts three or four eggs are laid, white like foam-bubbles; and well may the little birds hatched from them sing water songs, for they hear them all their lives, and even before they are born.

I have often observed the young just out of the nest making their odd gestures, and seeming in every way as much at home as their experienced parents, like young bees on their first excursions to the flower fields. No amount of familiarity with people and their ways seems to change them in the least. To all appearance their behavior is just the same on seeing a man for the first time, as when they have seen him frequently.

On the lower reaches of the rivers where mills are built, they sing on through the din of the machinery, and all the noisy confusion of dogs, cattle, and workmen. On one occasion, while a wood-chopper was at work on the river-bank, I observed one cheerily singing within reach of the flying chips. Nor does any kind of unwonted disturbance put him in bad humor, or frighten him out of calm self-possession. In passing through a narrow gorge, I once drove one ahead of me from rapid to rapid, disturbing him four times in quick succession where he could not very well fly past me on account of the narrowness of the channel. Most birds under similar circumstances fancy themselves pursued, and become suspiciously uneasy; but, instead of growing nervous about it, he made his usual dippings, and sang one of his most tranquil strains. When observed within a few yards their eyes are seen to express remarkable gentleness and intelligence; but they seldom allow so near a view unless one wears clothing of about the same color as the rocks and trees, and knows how to sit still. On one occasion, while rambling along the shore of a mountain lake, where the birds, at least those born that season, had never seen a man, I sat down to rest on a large stone close to the water's edge, upon which it seemed the ouzels and sandpipers were in the habit of alighting when they came to feed on that part of the shore, and some of the other birds also, when they came down to wash or drink. In a few minutes, along came a whirring Ouzel and alighted on the stone beside me, within reach of my hand. Then suddenly observing me, he stooped nervously as if about to fly on the instant, but as I remained as motionless as the stone, he gained confidence, and looked me steadily in the face for about a minute, then flew quietly to the outlet and began to sing. Next came a sandpiper and gazed at me with much the same guileless expression of eye as the Ouzel. Lastly, down with a swoop came a Steller's jay out of a fir-tree, probably with the intention of moistening his noisy throat. But instead of sitting confidingly as my other visitors had done, he rushed off at once, nearly tumbling heels over head into the lake in his suspicious confusion, and with loud screams roused the neighborhood.

Love for song-birds, with their sweet human voices, appears to be more common and unfailing than love for flowers. Every one loves flowers to some extent, at least in life's fresh morning, attracted by them as instinctively as humming-birds and bees. Even the young Digger Indians have sufficient love for the brightest of those found growing on the mountains to gather them and braid them as decorations for the hair. And I was glad to discover, through the few Indians that could be induced to talk on the subject, that they have names for the wild rose and the lily, and other conspicuous flowers, whether available as food or otherwise. Most men, however, whether savage or civilized, become apathetic toward all plants that have no other apparent use than the use of beauty. But fortunately one's first instinctive love of song-birds is never wholly obliterated, no matter what the influences upon our lives may be. I have often been delighted to see a pure, spiritual glow come into the countenances of hard business-men and old miners, when a song-bird chanced to alight near them. Nevertheless, the little mouthful of meat that swells out the breasts of some song-birds is too often the cause of their death. Larks and robins in particular are brought to market in hundreds. But fortunately the Ouzel has no enemy so eager to eat his little body as to follow him into the mountain solitudes. I never knew him to be chased even by hawks.

An acquaintance of mine, a sort of foot-hill mountaineer, had a pet cat, a great, dozy, overgrown creature, about as broad-shouldered as a lynx. During the winter, while the snow lay deep, the mountaineer sat in his lonely cabin among the pines smoking his pipe and wearing the dull time away. Tom was his sole companion, sharing his bed, and sitting beside him on a stool with much the same drowsy expression of eye as his master. The good-natured bachelor was content with his hard fare of soda-bread and bacon, but Tom, the only creature in the world acknowledging dependence on him, must needs be provided with fresh meat. Accordingly he bestirred himself to contrive squirrel-traps, and waded the snowy woods with his gun, making sad havoc among the few winter birds, sparing neither robin, sparrow, nor tiny nut-hatch, and the pleasure of seeing Tom eat and grow fat was his great reward.

One cold afternoon, while hunting along the river-bank, he noticed a plain-feathered little bird skipping about in the shallows, and immediately raised his gun. But just then the confiding songster began to sing, and after listening to his summery melody the charmed hunter turned away, saying, "Bless your little heart, I can't shoot you, not even for Tom."

Even so far north as icy Alaska, I have found my glad singer. When I was exploring the glaciers between Mount Fairweather and the Stikeen River, one cold day in November, after trying in vain to force a way through the innumerable icebergs of Sum Dum Bay to the great glaciers at the head of it, I was weary and baffled and sat resting in my canoe convinced at last that I would have to leave this part of my work for another year. Then I began to plan my escape to open water before the young ice which was beginning to form should shut me in. While I thus lingered drifting with the bergs, in the midst of these gloomy forebodings and all the terrible glacial desolation and grandeur, I suddenly heard the well-known whir of an Ouzel's wings, and, looking up, saw my little comforter coming straight across the ice from the shore. In a second or two he was with me, flying three times round my head with a happy salute, as if saying, "Cheer up, old friend; you see I'm here, and all's well." Then he flew back to the shore, alighted on the topmost jag of a stranded iceberg, and began to nod and bow as though he were on one of his favorite boulders in the midst of a sunny Sierra cascade.

The species is distributed all along the mountain-ranges of the Pacific Coast from Alaska to Mexico, and east to the Rocky Mountains. Nevertheless, it is as yet comparatively little known. Audubon and Wilson did not meet it. Swainson was, I believe, the first naturalist to describe a specimen from Mexico. Specimens were shortly afterward procured by Drummond near the sources of the Athabasca River, between the fifty-fourth and fifty-sixth parallels; and it has been collected by nearly all of the numerous exploring expeditions undertaken of late through our Western States and Territories; for it never fails to engage the attention of naturalists in a very particular manner.

Such, then, is our little cinclus, beloved of every one who is so fortunate as to know him. Tracing on strong wing every curve of the most precipitous torrents from one extremity of the Sierra to the other; not fearing to follow them through their darkest gorges and coldest snow-tunnels; acquainted with every waterfall, echoing their divine music; and throughout the whole of their beautiful lives interpreting all that we in our unbelief call terrible in the utterances of torrents and storms, as only varied expressions of God's eternal love.

CHAPTER XIV

THE WILD SHEEP

(*Ovis montana*)

THE WILD SHEEP ranks highest among the animal mountaineers of the Sierra. Possessed of keen sight and scent, and strong limbs, he dwells secure amid the loftiest summits, leaping unscathed from crag to crag, up and down the fronts of giddy precipices, crossing foaming torrents and slopes of frozen snow, exposed to the wildest storms, yet maintaining a brave, warm life, and developing from generation to generation in perfect strength and beauty.

Nearly all the lofty mountain-chains of the globe are inhabited by wild sheep, most of which, on account of the remote and all but inaccessible regions where they dwell, are imperfectly known as yet. They are classified by different naturalists under from five to ten distinct species or varieties, the best known being the burrhel of the Himalaya (*Ovis burrhel*, Blyth); the argali, the large wild sheep of central and northeastern Asia (*O. ammon*, Linn., or *Caprovis argali*); the Corsican mouflon (*O. musimon*, Pal.); the aoudad of the mountains of northern Africa (*Ammotragus tragelaphus*); and the Rocky Mountain bighorn (*O. montana*, Cuv.). To this last-named species belongs the wild sheep of the Sierra. Its range, according to the late Professor Baird of the Smithsonian Institution, extends "from the region of the upper Missouri and Yellowstone to the Rocky Mountains and the high grounds adjacent to them on the eastern slope, and as far south as the Rio Grande. Westward it extends to the coast ranges of Washington, Oregon, and California, and follows the highlands some distance into Mexico."* Throughout the vast region bounded on the east by the Wahsatch Mountains and on the west by the Sierra there are more than a hundred subordinate ranges and mountain groups, trending north and south, range beyond range, with summits rising from eight to twelve thousand feet above the level of the sea, probably all of which, according to my own observations, is, or has been, inhabited by this species.

* Pacific Railroad Survey, Vol. VIII, page 678.

Compared with the argali, which, considering its size and the vast extent of its range, is probably the most important of all the wild sheep, our species is about the same size, but the horns are less twisted and less divergent. The more important characteristics are, however, essentially the same, some of the best naturalists maintaining that the two are only varied forms of one species. In accordance with this view, Cuvier conjectures that since central Asia seems to be the region where the sheep first appeared, and from which it has been distributed, the argali may have been distributed over this continent from Asia by crossing Bering Strait on ice. This conjecture is not so ill founded as at first sight would appear; for the Strait is only about fifty miles wide, is interrupted by three islands, and is jammed with ice nearly every winter. Furthermore the argali is abundant on the mountains adjacent to the Strait at East Cape, where it is well known to the Tschuckchi hunters and where I have seen many of their horns.

On account of the extreme variability of the sheep under culture, it is generally supposed that the innumerable domestic breeds have all been derived from the few wild species; but the whole question is involved in obscurity. According to Darwin, sheep have been domesticated from a very ancient period, the remains of a small breed, differing from any now known, having been found in the famous Swiss lake-dwellings.

Compared with the best-known domestic breeds, we find that our wild species is much larger, and, instead of an all-wool garment, wears a thick overcoat of hair like that of the deer, and an undercovering of fine wool. The hair, though rather coarse, is comfortably soft and spongy, and lies smooth, as if carefully tended with comb and brush. The predominant color during most of the year is brownish-gray, varying to bluish-gray in the autumn; the belly and a large, conspicuous patch on the buttocks are white; and the tail, which is very short, like that of a deer, is black, with a yellowish border. The wool is white, and grows in beautiful spirals down out of sight among the shining hair, like delicate climbing vines among stalks of corn.

The horns of the male are of immense size, measuring in their greater diameter from five to six and a half inches, and from two and a half to three feet in length around the curve. They are yellowish-white in color, and ridged transversely, like those of the domestic ram. Their cross-section near the base is somewhat triangular in outline, and flattened toward the tip. Rising boldly from the top of the head, they curve gently backward and outward, then forward and outward, until about three fourths of a circle is described, and until the flattened, blunt tips are about two feet or two and a half feet apart. Those of the female are flattened throughout their entire length, are less curved than those of the male, and much smaller, measuring less than a foot along the curve.

A ram and ewe that I obtained near the Modoc lava-beds, to the northeast of Mount Shasta, measured as follows:

	RAM.		EWE.	
	FT.	IN.	FT.	IN.
Height at shoulders	3	6	3	0
Girth around shoulders	3	11	3	3¾
Length from nose to root of tail	5	10¼	4	3½
Length of ears	0	4¾	0	5
Length of tail	0	4½	0	4½
Length of horns around curve	2	9	0	11½
Distance across from tip to tip of horns	2	5½		
Circumference of horns at base	1	4	0	6

The measurements of a male obtained in the Rocky Mountains by Audubon vary but little as compared with the above. The weight of his specimen was 344 pounds,* which is, perhaps, about an average for full-grown males. The females are about a third lighter.

Besides these differences in size, color, hair, etc., as noted above, we may observe that the domestic sheep, in a general way, is expressionless, like a dull bundle of something only half alive, while the wild is as elegant and graceful as a deer, every movement manifesting admirable strength and character. The tame is timid; the wild is bold. The tame is always more or less ruffled and dirty; while the wild is as smooth and clean as the flowers of his mountain pastures.

The earliest mention that I have been able to find of the wild sheep in America is by Father Picolo, a Catholic missionary at Monterey, in the year 1797, who, after describing it, oddly enough,

* Audubon and Bachman's "Quadrupeds of North America."

as "a kind of deer with a sheep-like head, and about as large as a calf one or two years old," naturally hurries on to remark: "I have eaten of these beasts; their flesh is very tender and delicious." Mackenzie, in his northern travels, heard the species spoken of by the Indians as "white buffaloes." And Lewis and Clark tell us that, in a time of great scarcity on the head waters of the Missouri, they saw plenty of wild sheep, but they were "too shy to be shot."

A few of the more energetic of the Pah Ute Indians hunt the wild sheep every season among the more accessible sections of the High Sierra, in the neighborhood of passes, where, from having been pursued, they have become extremely wary; but in the rugged wilderness of peaks and cañons, where the foaming tributaries of the San Joaquin and King's rivers take their rise, they fear no hunter save the wolf, and are more guileless and approachable than their tame kindred.

While engaged in the work of exploring high regions where they delight to roam I have been greatly interested in studying their habits. In the months of November and December, and probably during a considerable portion of midwinter, they all flock together, male and female, old and young. I once found a complete band of this kind numbering upward of fifty, which, on being alarmed, went bounding away across a jagged lava-bed at admirable speed, led by a majestic old ram, with the lambs safe in the middle of the flock.

In spring and summer, the full-grown rams form separate bands of from three to twenty, and are usually found feeding along the edges of glacier meadows, or resting among the castle-like crags of the high summits; and whether quietly feeding, or scaling the wild cliffs, their noble forms and the power and beauty of their movements never fail to strike the beholder with lively admiration.

Their resting-places seem to be chosen with reference to sunshine and a wide outlook, and most of all to safety. Their feeding-grounds are among the most beautiful of the wild gardens, bright with daisies and gentians and mats of purple bryanthus, lying hidden away on rocky headlands and cañon sides, where sunshine is abundant, or down in the shady glacier valleys, along the banks of the streams and lakes, where the plushy sod is greenest. Here they feast all summer, the happy wanderers, perhaps relishing the beauty as well as the taste of the lovely flora on which they feed.

When the winter storms set in, loading their highland pastures with snow, then, like the birds, they gather and go to lower climates, usually descending the eastern flank of the range to the rough, volcanic table-lands and treeless ranges of the Great Basin adjacent to the Sierra. They never make haste, however, and seem to have no dread of storms, many of the strongest only going down leisurely to bare, wind-swept ridges, to feed on bushes and dry bunch-grass, and then returning up into the snow. Once I was snow-bound on Mount Shasta for three days, a little below the timber line. It was a dark and stormy time, well calculated to test the skill and endurance of mountaineers. The snow-laden gale drove on night and day in hissing, blinding floods, and when at length it began to abate, I found that a small band of wild sheep had weathered the storm in the lee of a clump of Dwarf Pines a few yards above my storm-nest, where the snow was eight or ten feet deep. I was warm back of a rock, with blankets, bread, and fire. My brave companions lay in the snow, without food, and with only the partial shelter of the short trees, yet they made no sign of suffering or faint-heartedness.

In the months of May and June, the wild sheep bring forth their young in solitary and almost inaccessible crags, far above the nesting-rocks of the eagle. I have frequently come upon the beds of the ewes and lambs at an elevation of from 12,000 to 13,000 feet above sea-level. These beds are simply oval-shaped hollows, pawed out among loose, disintegrating rock-chips and sand, upon some sunny spot commanding a good outlook, and partially sheltered from the winds that sweep those lofty peaks almost without intermission. Such is the cradle of the little mountaineer, aloft in the very sky; rocked in storms, curtained in clouds, sleeping in thin, icy air; but, wrapped in his hairy coat, and nourished by a strong, warm mother, defended from the talons of the eagle and the teeth of the sly coyote, the bonny lamb grows apace. He soon learns to nibble the tufted rock-grasses and leaves of the white spiræa; his horns begin to shoot, and before summer is done he is strong and agile, and goes forth with the flock, watched by the same divine love that tends the more helpless human lamb in its cradle by the fireside.

Nothing is more commonly remarked by noisy, dusty trail-travelers in the Sierra than the want of animal life—no song-birds, no deer, no squirrels, no game of any kind, they say. But if such could only go away quietly into the wilderness, sauntering afoot and alone with natural deliberation, they would soon learn that these mountain mansions are not without inhabitants, many of whom, confiding and gentle, would not try to shun their acquaintance.

In the fall of 1873 I was tracing the South Fork of the San Joaquin up its wild cañon to its farthest glacier fountains. It was the season of alpine Indian summer. The sun beamed lovingly; the squirrels were nutting in the pine-trees, butterflies hovered about the last of the goldenrods, the willow and maple thickets were yellow, the meadows brown, and the whole sunny, mellow landscape glowed like a countenance in the deepest and sweetest repose. On my way over the glacier-polished rocks along the river, I came to an expanded portion of the cañon, about two miles long and half a mile wide, which formed a level park inclosed with picturesque granite walls like those of Yosemite Valley. Down through the middle of it poured the beautiful river shining and spangling in the golden light, yellow groves on its banks, and strips of brown meadow; while the whole park was astir with wild life, some of which even the noisiest and least observing of travelers must have seen had they been with me. Deer, with their supple, well-grown fawns, bounded from thicket to thicket as I advanced; grouse kept rising from the brown grass with a great whirring of wings, and, alighting on the lower branches of the pines and poplars, allowed a near approach, as if curious to see me. Farther on, a broad-shouldered wildcat showed himself, coming out of a grove, and crossing the river on a flood-jamb of logs, halting for a moment to look back. The bird-like tamias frisked about my feet everywhere among the pine-needles and seedy grass-tufts; cranes waded the shallows of the river-bends, the kingfisher rattled from perch to perch, and the blessed ouzel sang amid the spray of every cascade. Where may lonely wanderer find a more interesting family of mountain-dwellers, earth-born companions and fellow-mortals? It was afternoon when I joined them, and the glorious landscape began to fade in the gloaming before I awoke from their enchantment. Then I sought a camp-ground on the river-bank, made a cupful of tea, and lay down to sleep on a smooth place among the yellow leaves of an aspen grove. Next day I discovered yet grander landscapes and grander life. Following the river over huge, swelling rock-bosses through a majestic cañon, and past innumerable cascades, the scenery in general became gradually wilder and more alpine. The Sugar Pine and Silver Firs gave place to the hardier Cedar and Hemlock Spruce. The cañon walls became more rugged and bare, and gentians and arctic daisies became more abundant in the gardens and strips of meadow along the streams. Toward the middle of the afternoon I came to another valley, strikingly wild and original in all its features, and perhaps never before touched by human foot. As regards area of level bottom-land, it is one of the very smallest of the Yosemite type, but its walls are sublime, rising to a height of from 2000 to 4000 feet above the river. At the head of the valley the main cañon forks, as is found to be the case in all yosemites. The formation of this one is due chiefly to the action of two great glaciers, whose fountains lay to the eastward, on the flanks of Mounts Humphrey and Emerson and a cluster of nameless peaks farther south.

The gray, boulder-chafed river was singing loudly through the valley, but above its massy roar I heard the booming of a waterfall, which drew me eagerly on; and just as I emerged from the tangled groves and brier-thickets at the head of the valley, the main fork of the river came in sight, falling fresh from its glacier fountains in a snowy cascade, between granite walls 2000 feet high. The steep incline down which the glad waters thundered seemed to bar all farther progress. It was not long, however, before I discovered a crooked seam in the rock, by which I was enabled to climb to the edge of a terrace that crosses the cañon, and divides the cataract nearly in the middle. Here I sat down to take breath and make some entries in my note-book, taking advantage, at the same time, of my elevated position above the trees to gaze back over the valley into the heart of the noble landscape, little knowing the while what neighbors were near.

After spending a few minutes in this way, I chanced to look across the fall, and there stood three sheep quietly observing me. Never did the sudden appearance of a mountain, or fall, or human friend more forcibly seize and rivet my attention. Anxiety to observe accurately held me perfectly still. Eagerly I marked the flowing undulations of their firm, braided muscles, their strong legs, ears, eyes, heads, their graceful rounded necks, the color of their hair, and the bold, up-sweeping curves of their noble horns. When they moved I watched every gesture, while they, in no wise disconcerted either by my attention or by the tumultuous roar of the water, advanced deliberately alongside the rapids, between the two divisions of the cataract, turning now and then to look at me. Presently they came to a steep, ice-burnished acclivity, which they ascended by a succession of quick, short, stiff-legged leaps, reaching the top without a struggle. This was the most startling feat of mountaineering I had ever witnessed, and, considering only the mechanics of the thing, my astonishment could hardly have been greater had they displayed wings and taken to flight. "Sure-footed" mules on such ground would have fallen and rolled like loosened boulders. Many a time, where the slopes are far lower, I have been compelled to take off my shoes and stockings, tie them to my belt, and creep barefooted, with the utmost caution. No wonder then, that I watched the progress of these animal mountaineers with keen sympathy, and exulted in the boundless sufficiency of wild nature displayed in their invention, construction, and keeping. A few minutes later I caught sight of a dozen more in one band, near the foot of the upper fall. They were standing on the same side of the river with me, only twenty-five or thirty yards away, looking as unworn and perfect as if created on the spot. It appeared by their tracks, which I had seen in the Little Yosemite, and by their present position, that when I came up the cañon they were all feeding together down in the valley, and in their haste to reach high ground, where they could look about them to ascertain the nature of the strange disturbance, they were divided, three ascending on one side of the river, the rest on the other.

The main band, headed by an experienced chief, now began to cross the wild rapids between the two divisions of the cascade. This was another exciting feat; for, among all the varied experiences of mountaineers, the crossing of boisterous, rock-dashed torrents is found to be one of the most trying to the nerves. Yet these fine fellows walked fearlessly to the brink, and jumped from boulder to boulder, holding themselves in easy poise above the whirling, confusing current, as if they were doing nothing extraordinary.

In the immediate foreground of this rare picture there was a fold of ice-burnished granite, traversed by a few bold lines in which rock-ferns and tufts of bryanthus were growing, the gray cañon walls on the sides, nobly sculptured and adorned with brown cedars and pines; lofty peaks in the distance, and in the middle ground the snowy fall, the voice and soul of the landscape; fringing bushes beating time to its thunder-tones, the brave sheep in front of it, their gray forms slightly obscured in the spray, yet standing out in good, heavy relief against the close white water, with their huge horns rising like the upturned roots of dead pine-trees, while the evening sunbeams streaming up the cañon colored all the picture a rosy purple and made it glorious. After crossing the river, the dauntless climbers, led by their chief, at once began to scale the cañon wall, turning now right, now left, in long, single file, keeping well apart out of one another's way, and leaping in regular succession from crag to crag, now ascending slippery dome-curves, now walking leisurely along the edges of precipices, stopping at times to gaze down at me from some flat-topped rock, with heads held aslant, as if curious to learn what I thought about it, or whether I was likely to follow them. After reaching the top of the wall, which, at this place, is somewhere between 1500 and 2000 feet high, they were still visible against the sky as they lingered, looking down in groups of twos or threes.

Throughout the entire ascent they did not make a single awkward step, or an unsuccessful effort of any kind. I have frequently seen tame sheep in mountains jump upon a sloping rock-surface, hold on tremulously a few seconds, and fall back baffled and irresolute. But in the most trying situations, where the slightest want or inaccuracy would have been fatal, these always seemed to move in comfortable reliance on their strength and skill, the limits of which they never appeared to know. Moreover, each one of the flock, while following the guidance of the most experienced, yet climbed with intelligent independence as a perfect individual, capable of separate existence whenever it should wish or be compelled to withdraw from the little clan. The domestic sheep, on the contrary, is only a fraction of an animal, a whole flock being required to form an individual, just as numerous flowerets are required to make one complete sunflower.

Those shepherds who, in summer, drive their flocks to the mountain pastures, and, while watching them night and day, have seen them frightened by bears and storms, and scattered like wind-driven chaff, will, in some measure, be able to appreciate the self-reliance and strength and noble individuality of Nature's sheep.

Like the Alp-climbing ibex of Europe, our mountaineer is said to plunge headlong down the faces of sheer precipices, and alight on his big horns. I know only two hunters who claim to have actually witnessed this feat; I never was so fortunate. They describe the act as a diving head-foremost. The horns are so large at the base that they cover the upper portion of the head down nearly to a level with the eyes, and the skull is exceedingly strong. I struck an old, bleached specimen on Mount Ritter a dozen blows with my ice-ax without breaking it. Such skulls would not fracture very readily by the wildest rock-diving, but other bones could hardly be expected to hold together in such a performance; and the mechanical difficulties in the way of controlling their movements, after striking upon an irregular surface, are, in themselves, sufficient to show this boulder-like method of progression to be impossible, even in the absence of all other evidence on the subject; moreover, the ewes follow wherever the rams may lead, although their horns are mere spikes. I have found many pairs of the horns of the old rams considerably battered, doubtless a result of fighting. I was particularly interested in the question, after witnessing the performances of this San Joaquin band upon the glaciated rocks at the foot of the falls; and as soon as I procured specimens and examined their feet, all the mystery disappeared. The secret, considered in connection with exceptionally strong muscles, is simply this: the wide posterior portion of the bottom of the foot, instead of wearing down and becoming flat and hard, like the feet of tame sheep and horses, bulges out in a soft, rubber-like pad or cushion, which not only grips and holds well on smooth rocks, but fits into small cavities, and down upon or against slight protuberances. Even the hardest portions of the edge of the hoof are comparatively soft and elastic; furthermore, the toes admit of an extraordinary amount of both lateral and vertical movement, allowing the foot to accommodate itself still more perfectly to the irregularities of rock surfaces, while at the same time increasing the gripping power.

At the base of Sheep Rock, one of the winter strongholds of the Shasta flocks, there lives a stock-raiser who has had the advantage of observing the movements of wild sheep every winter; and, in the course of a conversation with him on the subject of their diving habits, he pointed to the front of a lava headland about 150 feet high, which is only eight or ten degrees out of the perpendicular. "There," said he, "I followed a band of them fellows to the back of that rock yonder, and expected to capture them all, for I thought I had a dead thing on them. I got behind them on a narrow bench that runs along the face of the wall near the top and comes to an end where they couldn't get away without falling and being killed; but they jumped off, and landed all right, as if that were the regular thing with them."

"What!" said I, "jumped 150 feet perpendicular! Did you see them do it?"

"No," he replied, "I didn't see them going down, for I was behind them; but I saw them go off over the brink, and then I went below and found their tracks where they struck on the loose rubbish at the bottom. They just *sailed right off* and landed on their feet right side up. That is the kind of animal *they* is—beats anything else that goes on four legs."

On another occasion, a flock that was pursued by hunters retreated to another portion of this same cliff where it is still higher, and, on being followed, they were seen jumping down in perfect order, one behind another, by two men who happened to be chopping where they had a fair view of them and could watch their progress from top to bottom of the precipice. Both ewes and rams made the frightful descent without evincing any extraordinary concern, hugging the rock closely, and controlling the velocity of their half falling, half leaping movements by striking at short intervals and holding back with their cushioned, rubber feet upon small ledges and roughened inclines until near the bottom, when they "sailed off" into the free air and alighted on their feet, but with their bodies so nearly in a vertical position that they appeared to be diving.

It appears, therefore, that the methods of this wild mountaineering become clearly comprehensible as soon as we make ourselves acquainted with the rocks, and the kind of feet and muscles brought to bear upon them.

The Modoc and Pah Ute Indians are, or rather have been, the most successful hunters of the wild sheep in the regions that have come under my own observation. I have seen large numbers of heads and horns in the caves of Mount Shasta and the Modoc lava-beds, where the Indians had been feasting in stormy weather; also in the cañons of the Sierra opposite Owen's Valley; while the heavy obsidian arrow-heads found on some of the highest peaks show that this warfare has long been going on.

In the more accessible ranges that stretch across the desert regions of western Utah and Nevada, considerable numbers of Indians used to hunt in company like packs of wolves, and being perfectly acquainted with the topography of their hunting-grounds, and with the habits and instincts of the game, they were pretty successful. On the tops of nearly every one of the Nevada mountains that I have visited, I found small, nest-like inclosures built of stones, in which, as I afterward learned, one or more Indians would lie in wait while their companions scoured the ridges below, knowing that the alarmed sheep would surely run to the summit, and when they could be made to approach with the wind they were shot at short range.

Still larger bands of Indians used to make extensive hunts upon some dominant mountain much frequented by the sheep, such as Mount Grant on the Wassuck Range to the west of Walker Lake. On some particular spot, favorably situated with reference to the well-known trails of the sheep, they built a high-walled corral, with long guiding wings diverging from the gateway; and into this inclosure they sometimes succeeded in driving the noble game. Great numbers of Indians were of course required, more, indeed, than they could usually muster, counting in squaws, children, and all; they were compelled, therefore, to build rows of dummy hunters out of stones, along the ridge-tops which they wished to prevent the sheep from crossing. And, without discrediting the sagacity of the game, these dummies were found effective; for, with a few live Indians moving about excitedly among them, they could hardly be distinguished at a little distance from men, by any one not in the secret. The whole ridge-top then seemed to be alive with hunters.

The only animal that may fairly be regarded as a companion or rival of the sheep is the so-called Rocky Mountain goat (*Aplocerus montana*, Rich.), which, as its name indicates, is more antelope than goat. He, too, is a brave and hardy climber, fearlessly crossing the wildest summits, and braving the severest storms, but he is shaggy, short-legged, and much less dignified in demeanor than the sheep. His jet-black horns are only about five or six inches in length, and the long, white hair with which he is covered obscures the expression of his limbs. I have never yet seen a single specimen in the Sierra, though possibly a few flocks may have lived on Mount Shasta a comparatively short time ago.

The ranges of these two mountaineers are pretty distinct, and they see but little of each other; the sheep being restricted mostly to the dry, inland mountains; the goat or chamois to the wet, snowy glacier-laden mountains of the northwest coast of the continent in Oregon, Washington, British Columbia, and Alaska. Probably more than 200 dwell on the icy, volcanic cone of Mount Rainier; and while I was exploring the glaciers of Alaska I saw flocks of these admirable mountaineers nearly every day, and often followed their trails through the mazes of bewildering crevasses, in which they are excellent guides.

Three species of deer are found in California,—the black-tailed, white-tailed, and mule deer. The first mentioned (*Cervus columbianus*) is by far the most abundant, and occasionally meets the sheep during the summer on high glacier meadows, and along the edge of the timber line; but being a forest animal, seeking shelter and rearing its young in dense thickets, it seldom visits the wild sheep in its higher homes. The antelope, though not a mountaineer, is occasionally met in winter by the sheep while feeding along the edges of the sage-plains and bare volcanic hills to the east of the Sierra. So also is the mule deer, which is almost restricted in its range to this eastern region. The white-tailed species belongs to the coast ranges.

Perhaps no wild animal in the world is without enemies, but highlanders, as a class, have fewer than lowlanders. The wily panther, slipping and crouching among long grass and bushes, pounces upon the antelope and deer, but seldom crosses the bald, craggy thresholds of the sheep. Neither can the bears be regarded as enemies; for, though they seek to vary their every-day diet of nuts and berries by an occasional meal of mutton, they prefer to hunt tame and helpless flocks. Eagles and coyotes, no doubt, capture an unprotected lamb at times, or some unfortunate beset in deep, soft snow, but these cases are little more than accidents. So, also, a few perish in long-continued snow-storms, though, in all my mountaineering, I have not found more than five or six that seemed to have met their fate in this way. A little band of three were discovered snow-bound in Bloody Cañon a few years ago, and were killed with an ax by mountaineers, who chanced to be crossing the range in winter.

Man is the most dangerous enemy of all, but even from him our brave mountain-dweller has little to fear in the remote solitudes of the High Sierra. The golden plains of the Sacramento and San Joaquin were lately thronged with bands of elk and antelope, but, being fertile and accessible, they were required for human pastures. So, also, are many of the feeding-grounds of the deer—hill, valley, forest, and meadow—but it will be long before man will care to take the highland castles of the sheep. And when we consider here how rapidly entire species of noble animals, such as the elk, moose, and buffalo, are being pushed to the very verge of extinction, all lovers of wildness will rejoice with me in the rocky security of *Ovis montana*, the bravest of all the Sierra mountaineers.

CHAPTER XV

IN THE SIERRA FOOT-HILLS

MURPHY'S CAMP IS a curious old mining-town in Calaveras County, at an elevation of 2400 feet above the sea, situated like a nest in the center of a rough, gravelly region, rich in gold. Granites, slates, lavas, limestone, iron ores, quartz veins, auriferous gravels, remnants of dead fire-rivers and dead water-rivers are developed here side by side within a radius of a few miles, and placed invitingly open before the student like a book, while the people and the region beyond the camp furnish mines of study of never-failing interest and variety.

When I discovered this curious place, I was tracing the channels of the ancient pre-glacial rivers, instructive sections of which have been laid bare here and in the adjacent regions by the miners. Rivers, according to the poets, "go on forever"; but those of the Sierra are young as yet and have scarcely learned the way down to the sea; while at least one generation of them have died and vanished together with most of the basins they drained. All that remains of them to tell their history is a series of interrupted fragments of channels, mostly choked with gravel, and buried beneath broad, thick sheets of lava. These are known as the "Dead Rivers of California," and the gravel deposited in them is comprehensively called the "Blue Lead." In some places the channels of the present rivers trend in the same direction, or nearly so, as those of the ancient rivers; but, in general, there is little correspondence between them, the entire drainage having been changed, or, rather, made new. Many of the hills of the ancient landscapes have become hollows, and the old hollows have become hills. Therefore the fragmentary channels, with their loads of auriferous gravel, occur in all kinds of unthought-of places, trending obliquely, or even at right angles to the present drainage, across the tops of lofty ridges or far beneath them, presenting impressive illustrations of the magnitude of the changes accomplished since those ancient streams were annihilated. The last volcanic period preceding the regeneration of the Sierra landscapes seems to have come on over all the range almost simultaneously, like the glacial period, notwithstanding lavas of different age occur together in many places, indicating numerous periods of activity in the Sierra fire-fountains. The most important of the ancient river-channels in this region is a section that extends from the south side of the town beneath Coyote Creek and the ridge beyond it to the Cañon of the Stanislaus; but on account of its depth below the general surface of the present valleys the rich gold gravels it is known to contain cannot be easily worked on a large scale. Their extraordinary richness may be inferred from the fact that many claims were profitably worked in them by sinking shafts to a depth of 200 feet or more, and hoisting the dirt by a windlass. Should the dip of this ancient channel be such as to make the Stanislaus Cañon available as a dump, then the grand deposit might be worked by the hydraulic method, and although a long, expensive tunnel would be required, the scheme might still prove profitable, for there is "millions in it."

The importance of these ancient gravels as gold fountains is well known to miners. Even the superficial placers of the present streams have derived much of their gold from them. According to all accounts, the Murphy placers have been very rich–"terrific rich," as they say here. The hills have been cut and scalped, and every gorge and gulch and valley torn to pieces and disemboweled, expressing a fierce and desperate energy hard to understand. Still, any kind of effort-making is better than inaction, and there is something sublime in seeing men working in dead earnest at anything, pursuing an object with glacier-like energy and persistence. Many a brave fellow has recorded a most eventful chapter of life on these Calaveras rocks. But most of the pioneer miners are sleeping now, their wild day done, while the few survivors linger languidly in the washed-out gulches or sleepy village like harried bees around the ruins of their hive. "We have no industry left *now*," they told me, "and no men; everybody and everything hereabouts has gone to decay. We are only bummers—out of the game, a thin scatterin' of poor, dilapidated cusses, compared with what we used to be in the grand old gold-days. We were giants then, and you can look around here and see our tracks." But although these lingering pioneers are perhaps more exhausted than the mines, and about as dead as the dead rivers, they are yet a rare and interesting set of men, with much gold mixed with the rough, rocky gravel of their characters; and they manifest a breeding and intelligence little looked for in such surroundings as theirs. As the heavy, long-continued grinding of the glaciers brought out the features of the Sierra, so the intense experiences of the gold period have brought out the features of these old miners, forming a richness and variety of character little known as yet. The sketches of Bret Harte, Hayes, and Miller have not exhausted this field by any means. It is interesting to note the extremes possible in one and the same character: harshness and gentleness, manliness and childishness, apathy and fierce endeavor. Men who, twenty years ago, would not cease their shoveling to save their lives, now play in the streets with children. Their long, Micawber-like waiting after the exhaustion of the placers has brought on an exaggerated form of dotage. I heard a group of brawny pioneers in the street eagerly discussing the quantity of tail required for a boy's kite; and one graybeard undertook the sport of flying it, volunteering the information that he was a boy, "always was a boy, and d—n a man who was not a boy inside, however ancient outside!" Mines, morals, politics, the immortality of the soul, etc., were discussed beneath shade-trees and in saloons, the time for each being governed apparently by the temperature. Contact with Nature, and the habits of observation acquired in gold-seeking, had made them all, to some extent, collectors, and, like wood-rats, they had gathered all kinds of odd specimens into their cabins, and now required me to examine them. They were themselves the oddest and most interesting specimens. One of them offered to show me around the old diggings, giving me fair warning before setting out that I might not like him, "because," said he, "people say I'm eccentric. I notice everything, and gather beetles and snakes and anything that's queer; and so some don't like me, and call me eccentric. I'm always trying to find out things. Now, there's a weed; the Indians eat it for greens. What do you call those long-bodied flies with big heads?" "Dragon-flies," I suggested. "Well, their jaws work sidewise, instead of up and down, and grasshoppers' jaws work the same way, and therefore I think they are the same species. I always notice everything like that, and just because I do, they say I'm eccentric," etc.

Anxious that I should miss none of the wonders of their old gold-field, the good people had much to say about the marvelous beauty of Cave City Cave, and advised me to explore it. This I was very glad to do, and finding a guide who knew the way to the mouth of it, I set out from Murphy the next morning.

The most beautiful and extensive of the mountain caves of

California occur in a belt of metamorphic limestone that is pretty generally developed along the western flank of the Sierra from the McCloud River on the north to the Kaweah on the south, a distance of over 400 miles, at an elevation of from 2000 to 7000 feet above the sea. Besides this regular belt of caves, the California landscapes are diversified by long imposing ranks of sea-caves, rugged and variable in architecture, carved in the coast headlands and precipices by centuries of wave-dashing; and innumerable lava-caves, great and small, originating in the unequal flowing and hardening of the lava sheets in which they occur, fine illustrations of which are presented in the famous Modoc Lava Beds, and around the base of icy Shasta. In this comprehensive glance we may also notice the shallow wind-worn caves in stratified sandstones along the margins of the plains; and the cave-like recesses in the Sierra slates and granites, where bears and other mountaineers find shelter during the fall of sudden storms. In general, however, the grand massive uplift of the Sierra, as far as it has been laid bare to observation, is about as solid and caveless as a boulder.

Fresh beauty opens one's eyes wherever it is really seen, but the very abundance and completeness of the common beauty that besets our steps prevents its being absorbed and appreciated. It is a good thing, therefore, to make short excursions now and then to the bottom of the sea among dulse and coral, or up among the clouds on mountain-tops, or in balloons, or even to creep like worms into dark holes and caverns underground, not only to learn something of what is going on in those out-of-the-way places, but to see better what the sun sees on our return to common every-day beauty.

Our way from Murphy's to the cave lay across a series of picturesque, moory ridges in the chaparral region between the brown foot-hills and the forests, a flowery stretch of rolling hill-waves breaking here and there into a kind of rocky foam on the higher summits, and sinking into delightful bosky hollows embowered with vines. The day was a fine specimen of California summer, pure sunshine, unshaded most of the time by a single cloud. As the sun rose higher, the heated air began to flow in tremulous waves from every southern slope. The sea-breeze that usually comes up the foot-hills at this season, with cooling on its wings, was scarcely perceptible. The birds were assembled beneath leafy shade, or made short, languid flights in search of food, all save the majestic buzzard; with broad wings outspread he sailed the warm air unwearily from ridge to ridge, seeming to enjoy the fervid sunshine like a butterfly. Squirrels, too, whose spicy ardor no heat or cold may abate, were nutting among the pines, and the innumerable hosts of the insect kingdom were throbbing and wavering unwearied as sunbeams.

This brushy, berry-bearing region used to be a deer and bear pasture, but since the disturbances of the gold period these fine animals have almost wholly disappeared. Here, also, once roamed the mastodon and elephant, whose bones are found entombed in the river gravels and beneath thick folds of lava. Toward noon, as we were riding slowly over bank and brae, basking in the unfeverish sun-heat, we witnessed the upheaval of a new mountain-range, a Sierra of clouds abounding in landscapes as truly sublime and beautiful—if only we have a mind to think so and eyes to see—as the more ancient rocky Sierra beneath it, with its forests and waterfalls; reminding us that, as there is a lower world of caves, so, also, there is an upper world of clouds. Huge, bossy cumuli developed with astonishing rapidity from mere buds, swelling with visible motion into colossal mountains, and piling higher, higher, in long massive ranges, peak beyond peak, dome over dome, with many a picturesque valley and shadowy cave between; while the dark firs and pines of the upper benches of the Sierra were projected against their pearl bosses with exquisite clearness of outline. These cloud mountains vanished in the azure as quickly as they were developed, leaving no detritus; but they were not a whit less real or interesting on this account. The more enduring hills over which we rode were vanishing as surely as they, only not so fast, a difference which is great or small according to the standpoint from which it is contemplated.

At the bottom of every dell we found little homesteads embosomed in wild brush and vines wherever the recession of the hills left patches of arable ground. These secluded flats are settled mostly by Italians and Germans, who plant a few vegetables and grape-vines at odd times, while their main business is mining and prospecting. In spite of all the natural beauty of these dell cabins, they can hardly be called homes. They are only a better kind of camp, gladly abandoned whenever the hoped-for gold harvest has been gathered. There is an air of profound unrest and melancholy about the best of them. Their beauty is thrust upon them by exuberant Nature, apart from which they are only a few logs and boards rudely jointed and without either ceiling or floor, a rough fireplace with corresponding cooking utensils, a shelf-bed, and stool. The ground about them is strewn with battered prospecting-pans, picks, sluice-boxes, and quartz specimens from many a ledge, indicating the trend of their owners' hard lives.

The ride from Murphy's to the cave is scarcely two hours long, but we lingered among quartz-ledges and banks of dead river gravel until long after noon. At length emerging from a narrow-throated gorge, a small house came in sight set in a thicket of fig-trees at the base of a limestone hill. "That," said my guide, pointing to the house, "is Cave City, and the cave is in that gray hill." Arriving at the one house of this one-house city, we were boisterously welcomed by three drunken men who had come to town to hold a spree. The mistress of the house tried to keep order, and in reply to our inquiries told us that the cave guide was then in the cave with a party of ladies. "And must we wait until he returns?" we asked. No, that was unnecessary; we might take candles and go into the cave alone, provided we shouted from time to time so as to be found by the guide, and were careful not to fall over the rocks or into the dark pools. Accordingly taking a trail from the house, we were led around the base of the hill to the mouth of the cave, a small inconspicuous archway, mossy around the edges and shaped like the door of a water-ouzel's nest, with no appreciable hint or advertisement of the grandeur of the many crystal chambers within. Lighting our candles, which seemed to have no illuminating power in the thick darkness, we groped our way onward as best we could along narrow lanes and alleys, from chamber to chamber, around rustic columns and heaps of fallen rocks, stopping to rest now and then in particularly beautiful places—fairy alcoves furnished with admirable variety of shelves and tables, and round bossy stools covered with sparkling crystals. Some of the corridors were muddy, and in plodding along these we seemed to be in the streets of some prairie village in spring-time. Then we would come to handsome marble stairways conducting right and left into upper chambers ranged above one another three or four stories high, floors, ceilings, and walls lavishly decorated with innumerable crystalline forms. After thus wandering exploringly, and alone for a mile or so, fairly enchanted, a murmur of voices and a gleam of light betrayed the approach of the guide and his party, from whom, when they came up, we received a most hearty and natural stare, as we stood half concealed in a side recess among stalagmites. I ventured to ask the dripping, crouching company how they had enjoyed their saunter, anxious to learn how the strange sunless scenery of the underworld had impressed them. "Ah, it's nice! It's splendid!" they all replied and echoed. "The Bridal Chamber back here is just glorious! This morning we came down from the Calaveras Big Tree Grove, and the trees are nothing to it." After making this curious comparison they hastened sunward, the guide promising to join us shortly on the bank of a deep pool, where we were to wait for him. This is a charming little lakelet of unknown depth, never yet stirred by a breeze, and its eternal calm excites the imagination even more profoundly than the silvery lakes of the glaciers rimmed with meadows and snow and reflecting sublime mountains.

Our guide, a jolly, rollicking Italian, led us into the heart of the hill, up and down, right and left, from chamber to chamber more and more magnificent, all a-glitter like a glacier cave with icicle-like stalactites and stalagmites combined in forms of indescribable beauty. We were shown one large room that was occasionally used as a dancing-hall; another that was used as a chapel, with natural pulpit and crosses and pews, sermons in every stone, where a priest had said mass. Mass-saying is not so generally developed in connection with natural wonders as dancing. One of the first conceits excited by the giant Sequoias was to cut one of them down and dance on its stump. We have also seen dancing in the spray of Niagara; dancing in the famous Bower Cave above Coulterville; and nowhere have I seen so much dancing as in Yosemite. A dance on the inaccessible South Dome would likely follow the making of an easy way to the top of it.

It was delightful to witness here the infinite deliberation of Nature, and the simplicity of her methods in the production of such mighty results, such perfect repose combined with restless enthusiastic energy. Though cold and bloodless as a landscape of polar ice, building was going on in the dark with incessant activity. The archways and ceilings were everywhere hung with down-growing crystals, like inverted groves of leafless saplings, some of them large, others delicately attenuated, each tipped with a single drop of water, like the terminal bud of a pine-tree. The only appreciable sounds were the dripping and tinkling of water falling into pools or faintly plashing on the crystal floors.

In some places the crystal decorations are arranged in graceful flowing folds deeply plicated like stiff silken drapery. In others straight lines of the ordinary stalactite forms are combined with reference to size and tone in a regularly graduated system like the strings of a harp with musical tones corresponding thereto; and on these stone harps we played by striking the crystal strings with a stick. The delicious liquid tones they gave forth seemed perfectly divine as they sweetly whispered and wavered through the majestic halls and died away in faintest cadence,—the music of fairy-land. Here we lingered and reveled, rejoicing to find so much music in stony silence, so much splendor in darkness, so many mansions in the depths of the mountains, buildings ever in process of construction, yet ever finished, developing from perfection to perfection, profusion without overabundance; every particle visible or invisible in glorious motion, marching to the music of the spheres in a region regarded as the abode of eternal stillness and death.

The outer chambers of mountain caves are frequently selected as homes by wild beasts. In the Sierra, however, they seem to prefer homes and hiding-places in chaparral and beneath shelving precipices, as I have never seen their tracks in any of the caves. This is the more remarkable because notwithstanding the darkness and oozing water there is nothing uncomfortably cellar-like or sepulchral about them.

When we emerged into the bright landscapes of the sun everything looked brighter, and we felt our faith in Nature's beauty strengthened, and saw more clearly that beauty is universal and immortal, above, beneath, on land and sea, mountain and plain, in heat and cold, light and darkness.

CHAPTER XVI

THE BEE-PASTURES

WHEN CALIFORNIA WAS wild, it was one sweet bee-garden throughout its entire length, north and south, and all the way across from the snowy Sierra to the ocean.

Wherever a bee might fly within the bounds of this virgin wilderness—through the redwood forests, along the banks of the rivers, along the bluffs and headlands fronting the sea, over valley and plain, park and grove, and deep, leafy glen, or far up the piny slopes of the mountains—throughout every belt and section of climate up to the timber line, bee-flowers bloomed in lavish abundance. Here they grew more or less apart in special sheets and patches of no great size, there in broad, flowing folds hundreds of miles in length—zones of polleny forests, zones of flowery chaparral, stream-tangles of rubus and wild rose, sheets of golden compositæ, beds of violets, beds of mint, beds of bryanthus and clover, and so on, certain species blooming somewhere all the year round.

But of late years ploughs and sheep have made sad havoc in these glorious pastures, destroying tens of thousands of the flowery acres like a fire, and banishing many species of the best honey-plants to rocky cliffs and fence-corners, while, on the other hand, cultivation thus far has given no adequate compensation, at least in kind; only acres of alfalfa for miles of the richest wild pasture, ornamental roses and honeysuckles around cottage doors for cascades of wild roses in the dells, and small, square orchards and orange-groves for broad mountain-belts of chaparral.

The Great Central Plain of California, during the months of March, April, and May, was one smooth, continuous bed of honey-bloom, so marvelously rich that, in walking from one end of it to the other, a distance of more than 400 miles, your foot would press about a hundred flowers at every step. Mints, gilias, nemophilas, castilleias, and innumerable compositæ were so crowded together that, had ninety-nine per cent. of them been taken away, the plain would still have seemed to any but Californians extravagantly flowery. The radiant, honeyful corollas, touching and overlapping, and rising above one another, glowed in the living light like a sunset sky—one sheet of purple and gold, with the bright Sacramento pouring through the midst of it from the north, the San Joaquin from the south, and their many tributaries sweeping in at right angles from the mountains, dividing the plain into sections fringed with trees.

Along the rivers there is a strip of bottom-land, countersunk beneath the general level, and wider toward the foot-hills, where magnificent oaks, from three to eight feet in diameter, cast grateful masses of shade over the open, prairie-like levels. And close along the water's edge there was a fine jungle of tropical luxuriance, composed of wild-rose and bramble bushes and a great variety of climbing vines, wreathing and interlacing the branches and trunks of willows and alders, and swinging across from summit to summit in heavy festoons. Here the wild bees reveled in fresh bloom long after the flowers of the drier plain had withered and gone to seed. And in midsummer, when the "blackberries" were ripe, the Indians came from the mountains to feast—men, women, and babies in long, noisy trains, often joined by the farmers of the neighborhood, who gathered this wild fruit with commendable appreciation of its superior flavor, while their home orchards were full of ripe peaches, apricots, nectarines, and figs, and their vineyards were laden with grapes. But, though these luxuriant, shaggy river-beds were thus distinct from the smooth, treeless plain, they made no heavy dividing lines in general views. The whole appeared as one continuous sheet of bloom bounded only by the mountains.

When I first saw this central garden, the most extensive and regular of all the bee-pastures of the State, it seemed all one sheet of plant gold, hazy and vanishing in the distance, distinct as a new map along the foot-hills at my feet.

Descending the eastern slopes of the Coast Range through beds of gilias and lupines, and around many a breezy hillock and bush-crowned headland, I at length waded out into the midst of it. All the ground was covered, not with grass and green leaves, but with radiant corollas, about ankle-deep next the foot-hills, knee-deep or more five or six miles out. Here were bahia, madia, madaria, burrielia, chrysopsis, corethrogyne, grindelia, etc., growing in close social congregations of various shades of yellow, blending finely with the purples of clarkia, orthocarpus, and œnothera, whose delicate petals were drinking the vital sunbeams without giving back any sparkling glow.

Because so long a period of extreme drought succeeds the rainy season, most of the vegetation is composed of annuals, which spring up simultaneously, and bloom together at about the same height above the ground, the general surface being but slightly ruffled by the taller phacelias, pentstemons, and groups of *Salvia carduacea*, the king of the mints.

Sauntering in any direction, hundreds of these happy sun-plants brushed against my feet at every step, and closed over them as if I were wading in liquid gold. The air was sweet with fragrance, the larks sang their blessed songs, rising on the wing as I advanced, then sinking out of sight in the polleny sod, while myriads of wild bees stirred the lower air with their monotonous hum—monotonous, yet forever fresh and sweet as every-day sunshine. Hares and spermophiles showed themselves in considerable numbers in shallow places, and small bands of antelopes were almost constantly in sight, gazing curiously from some slight elevation, and then bounding swiftly away with unrivaled grace of motion. Yet I could discover no crushed flowers to mark their track, nor, indeed, any destructive action of any wild foot or tooth whatever.

The great yellow days circled by uncounted, while I drifted toward the north, observing the countless forms of life thronging about me, lying down almost anywhere on the approach of night. And what glorious botanical beds I had! Oftentimes on awaking I would find several new species leaning over me and looking me full in the face, so that my studies would begin before rising.

About the first of May I turned eastward, crossing the San Joaquin River between the mouths of the Tuolumne and Merced, and by the time I had reached the Sierra foot-hills most of the vegetation had gone to seed and become as dry as hay.

All the seasons of the great plain are warm or temperate, and bee-flowers are never wholly wanting; but the grand springtime—the annual resurrection—is governed by the rains, which usually set in about the middle of November or the beginning of December. Then the seeds, that for six months have lain on the ground dry and fresh as if they had been gathered into barns, at once unfold their treasured life. The general brown and purple of the ground, and the dead vegetation of the preceding year, give place to the green of mosses and liverworts and myriads of young leaves. Then one species after another comes into flower, gradually overspreading the green with yellow and purple, which lasts until May.

The "rainy season" is by no means a gloomy, soggy period of constant cloudiness and rain. Perhaps nowhere else in North America, perhaps in the world, are the months of December, January, February, and March so full of bland, plant-building sunshine. Referring to my notes of the winter and spring of 1868–69, every day of which I spent out of doors, on that section of the plain lying between the Tuolumne and Merced rivers, I find that the first rain of the season fell on December 18th. January had only six rainy days—that is, days on which rain fell; February three, March five, April three, and May three, completing the so-called rainy season, which was about an average one. The ordinary rain-storm of this region is seldom very cold or violent. The winds, which in settled weather come from the northwest, veer round into the opposite direction, the sky fills gradually and evenly with one general cloud, from which the rain falls steadily, often for days in succession, at a temperature of about 45° or 50°.

More than seventy-five per cent. of all the rain of this season came from the northwest, down the coast over southeastern Alaska, British Columbia, Washington, and Oregon, though the local winds of these circular storms blow from the southeast. One magnificent local storm from the northwest fell on March 21. A massive, round-browed cloud came swelling and thundering over the flowery plain in most imposing majesty, its bossy front burning white and purple in the full blaze of the sun, while warm rain poured from its ample fountains like a cataract, beating down flowers and bees, and flooding the dry watercourses as suddenly as those of Nevada are flooded by the so-called "cloud-bursts." But in less than half an hour not a trace of the heavy, mountain-like cloud-structure was left in the sky, and the bees were on the wing, as if nothing more gratefully refreshing could have been sent them.

By the end of January four species of plants were in flower, and five or six mosses had already adjusted their hoods and were in the prime of life; but the flowers were not sufficiently numerous as yet to affect greatly the general green of the young leaves. Violets made their appearance in the first week of February, and toward the end of this month the warmer portions of the plain were already golden with myriads of the flowers of rayed compositæ.

This was the full springtime. The sunshine grew warmer and richer, new plants bloomed every day; the air became more tuneful with humming wings, and sweeter with the fragrance of the opening flowers. Ants and ground squirrels were getting ready for their summer work, rubbing their benumbed limbs, and sunning themselves on the husk-piles before their doors, and spiders were busy mending their old webs, or weaving new ones.

In March, the vegetation was more than doubled in depth and color; claytonia, calandrinia, a large white gilia, and two nemophilas were in bloom, together with a host of yellow compositæ, tall enough now to bend in the wind and show wavering ripples of shade.

In April, plant-life, as a whole, reached its greatest height, and the plain, over all its varied surface, was mantled with a close, furred plush of purple and golden corollas. By the end of this month, most of the species had ripened their seeds, but undecayed, still seemed to be in bloom from the numerous corolla-like involucres and whorls of chaffy scales of the compositæ. In May, the bees found in flower only a few deep-set liliaceous plants and eriogonums.

June, July, August, and September is the season of rest and sleep,—a winter of dry heat,—followed in October by a second outburst of bloom at the very driest time of the year. Then, after the shrunken mass of leaves and stalks of the dead vegetation crinkle and turn to dust beneath the foot, as if it had been baked in an oven, *Hemizonia virgata*, a slender, unobtrusive little plant, from six inches to three feet high, suddenly makes its appearance in patches miles in extent, like a resurrection of the bloom of April. I have counted upward of 3000 flowers, five eighths of an inch in diameter, on a single plant. Both its leaves and stems are so slender as to be nearly invisible, at a distance of a few yards, amid so showy a multitude of flowers. The ray and disk flowers are both yellow, the stamens purple, and the texture of the rays is rich and velvety, like the petals of garden pansies. The prevailing wind turns all the heads round to the southeast, so that in facing northwestward we have the flowers looking us in the face. In my estimation, this little plant, the last born of the brilliant host of compositæ that glorify the plain, is the most interesting of all. It remains in flower until November, uniting with two or three species of wiry eriogonums, which continue the floral chain around December to the spring flowers of January. Thus, although the main bloom and honey season is only about three months long, the floral circle, however thin around some of the hot, rainless months, is never completely broken.

How long the various species of wild bees have lived in this honey-garden, nobody knows; probably ever since the main body of the present flora gained possession of the land, toward the close of the glacial period. The first brown honey-bees brought to California are said to have arrived in San Francisco in March, 1853. A bee-keeper by the name of Shelton purchased a lot, consisting of twelve swarms, from some one at Aspinwall, who had brought them from New York. When landed at San Francisco, all the hives contained live bees, but they finally dwindled to one hive, which was taken to San José. The little immigrants flourished and multiplied in the bountiful pastures of the Santa Clara Valley, sending off three swarms the first season. The owner was killed shortly afterward, and in settling up his estate, two of the swarms were sold at auction for $105 and $110 respectively. Other importations were made, from time to time, by way of the Isthmus, and, though great pains were taken to insure success, about one half usually died on the way. Four swarms were brought safely across the plains in 1859, the hives being placed in the rear end of a wagon, which was stopped in the afternoon to allow the bees to fly and feed in the floweriest places that were within reach until dark, when the hives were closed.

In 1855, two years after the time of the first arrivals from New York, a single swarm was brought over from San José, and let fly in the Great Central Plain. Bee-culture, however, has never gained much attention here, notwithstanding the extraordinary abundance of honey-bloom, and the high price of honey during the early years. A few hives are found here and there among settlers who chanced to have learned something about the business before coming to the State. But sheep, cattle, grain, and fruit raising are the chief industries, as they require less skill and care, while the profits thus far have been greater. In 1856 honey sold here at from one and a half to two dollars per pound. Twelve years later the price had fallen to twelve and a half cents. In 1868 I sat down to dinner with a band of ravenous sheep-shearers at a ranch on the San Joaquin, where fifteen or twenty hives were kept, and our host advised us not to spare the large pan of honey he had placed on the table, as it was the cheapest article he had to offer. In all my walks, however, I have never come upon a regular bee-ranch in the Central Valley like those so common and so skillfully managed in the southern counties of the State. The few pounds of honey and wax produced are consumed at home, and are scarcely taken into account among the coarser products of the farm. The swarms that escape from their careless owners have a weary, perplexing time of it in seeking suitable homes. Most of them make their way to the foot-hills of the mountains, or to the trees that line the banks of the rivers, where some hollow log or trunk may be found. A friend of mine, while out hunting on the San Joaquin, came upon an old coon trap, hidden among some tall grass, near the edge of the river, upon which he sat down to rest. Shortly afterward his attention was attracted to a crowd of angry bees that were flying excitedly about his head, when he discovered that he was sitting upon their hive, which was found to contain more than 200 pounds of honey. Out in the broad, swampy delta of the Sacramento and San Joaquin rivers, the little wanderers have been known to build their combs in a bunch of rushes, or stiff, wiry grass, only slightly protected from the weather, and in danger every spring of being carried away by floods. They have the advantage, however, of a vast extent of fresh pasture, accessible only to themselves.

The present condition of the Grand Central Garden is very different from that we have sketched. About twenty years ago, when the gold placers had been pretty thoroughly exhausted, the attention of fortune-seekers—not home-seekers—was, in great part, turned away from the mines to the fertile plains, and many began experiments in a kind of restless, wild agriculture. A load of lumber would be hauled to some spot on the free wilderness, where water could be easily found, and a rude box-cabin built. Then a gang-plough was procured, and a dozen mustang ponies, worth ten or fifteen dollars apiece, and with these hundreds of acres were stirred as easily as if the land had been under cultivation for years, tough, perennial roots being almost wholly absent. Thus a ranch was established, and from these bare wooden huts, as centers of desolation, the wild flora vanished in ever-widening circles. But the arch destroyers are the shepherds, with their flocks of hoofed locusts, sweeping over the ground like a fire, and trampling down every rod that escapes the plough as completely as if the whole plain were a cottage garden-plot without a fence. But notwithstanding these destroyers, a thousand swarms of bees may be pastured here for every one now gathering honey. The greater portion is still covered every season with a repressed growth of bee-flowers, for most of the species are annuals, and many of them are not relished by sheep or cattle, while the rapidity of their growth enables them to develop and mature their seeds before any foot has time to crush them. The ground is, therefore, kept sweet, and the race is perpetuated, though only as a suggestive shadow of the magnificence of its wildness.

The time will undoubtedly come when the entire area of this noble valley will be tilled like a garden, when the fertilizing waters of the mountains, now flowing to the sea, will be distributed to every acre, giving rise to prosperous towns, wealth, arts, etc. Then, I suppose, there will be few left, even among botanists, to deplore the vanished primeval flora. In the mean time, the pure waste going on—the wanton destruction of the innocents—is a sad sight to see, and the sun may well be pitied in being compelled to look on.

The bee-pastures of the Coast Ranges last longer and are more varied than those of the great plain, on account of differences of soil and climate, moisture, and shade, etc. Some of the mountains are upward of 4000 feet in height, and small streams, springs, oozy bogs, etc., occur in great abundance and variety in the wooded regions, while open parks, flooded with sunshine, and hill-girt valleys lying at different elevations, each with its own peculiar climate and exposure, possess the required conditions for the development of species and families of plants widely varied.

Next the plain there is, first, a series of smooth hills, planted with a rich and showy vegetation that differs but little from that of the plain itself—as if the edge of the plain had been lifted and bent into flowing folds, with all its flowers in place, only toned down a little as to their luxuriance, and a few new species introduced, such as the hill lupines, mints, and gilias. The colors show finely when thus held to view on the slopes; patches of red, purple, blue, yellow, and white, blending around the edges, the whole appearing at a little distance like a map colored in sections.

Above this lies the park and chaparral region, with oaks, mostly evergreen, planted wide apart, and blooming shrubs from three to ten feet high; manzanita and ceanothus of several species, mixed with rhamnus, cercis, pickeringia, cherry, amelanchier, and adenostoma, in shaggy, interlocking thickets, and many species of hosackia, clover, monardella, castilleia, etc., in the openings.

The main ranges send out spurs somewhat parallel to their axes, inclosing level valleys, many of them quite extensive, and containing a great profusion of sun-loving bee-flowers in their wild state; but these are, in great part, already lost to the bees by cultivation.

Nearer the coast are the giant forests of the redwoods, extending from near the Oregon line to Santa Cruz. Beneath the cool, deep shade of these majestic trees the ground is occupied by ferns, chiefly woodwardia and aspidiums, with only a few flowering plants—oxalis, trientalis, erythronium, fritillaria, smilax, and other shade-lovers. But all along the redwood belt there are sunny openings on hill-slopes looking to the south, where the giant trees stand back, and give the ground to the small sunflowers and the bees. Around the lofty redwood walls of these little bee-acres there is usually a fringe of Chestnut Oak, Laurel, and Madroño, the last of which is a surpassingly beautiful tree, and a great favorite with the bees. The trunks of the largest specimens are seven or eight feet thick, and about fifty feet high; the bark red and chocolate colored, the leaves plain, large, and glossy, like those of *Magnolia grandiflora*, while the flowers are yellowish-white, and urn-shaped, in well-proportioned panicles, from five to ten inches long. When in full bloom, a single tree seems to be visited at times by a whole hive of bees at once, and the deep hum of such a multitude makes the listener guess that more than the ordinary work of honey-winning must be going on.

How perfectly enchanting and care-obliterating are these withdrawn gardens of the woods—long vistas opening to the sea—sunshine sifting and pouring upon the flowery ground in a tremulous, shifting mosaic, as the light-ways in the leafy wall open and close with the swaying breeze—shining leaves and flowers, birds and bees, mingling together in springtime harmony, and soothing fragrance exhaling from a thousand thousand fountains! In these balmy, dissolving days, when the deep heart-beats of Nature are felt thrilling rocks and trees and everything alike, common business and friends are happily forgotten, and even the natural honey-work of bees, and the care of birds for their young, and mothers for their children, seem slightly out of place.

To the northward, in Humboldt and the adjacent counties, whole hillsides are covered with rhododendron, making a glorious melody of bee-bloom in the spring. And the Western azalea, hardly less flowery, grows in massy thickets three to eight feet high around the edges of groves and woods as far south as San Luis Obispo, usually accompanied by manzanita; while the valleys, with their varying moisture and shade, yield a rich variety of the smaller honey-flowers, such as mentha, lycopus, micromeria, audibertia, trichostema, and other mints; with vaccinium, wild strawberry, geranium, calais, and goldenrod; and in the cool glens along the stream-banks, where the shade of trees is not too deep, spiræa, dogwood, heteromeles, and calycanthus, and many species of rubus form interlacing tangles, some portion of which continues in bloom for months.

Though the coast region was the first to be invaded and settled by white men, it has suffered less from a bee point of view than either of the other main divisions, chiefly, no doubt, because of the unevenness of the surface, and because it is owned and protected instead of lying exposed to the flocks of the wandering "sheep-men." These remarks apply more particularly to the north half of the coast. Farther south there is less moisture, less forest shade, and the honey flora is less varied.

The Sierra region is the largest of the three main divisions of the bee-lands of the State, and the most regularly varied in its subdivisions, owing to their gradual rise from the level of the Central Plain to the alpine summits. The foot-hill region is about as dry and sunful, from the end of May until the setting in of the winter rains, as the plain. There are no shady forests, no damp glens, at all like those lying at the same elevations in the Coast Mountains. The social compositæ of the plain, with a few added species, form the bulk of the herbaceous portion of the vegetation up to a height of 1500 feet or more, shaded lightly here and there with oaks and Sabine Pines, and interrupted by patches of ceanothus and buckeye. Above this, and just below the forest region, there is a dark, heath-like belt of chaparral, composed almost exclusively of *Adenostoma fasciculata*, a bush belonging to the rose family, from five to eight feet high, with small, round leaves in fascicles, and bearing a multitude of small white flowers in panicles on the ends of the upper branches. Where it occurs at all, it usually covers all the ground with a close, impenetrable growth, scarcely broken for miles.

Up through the forest region, to a height of about 9000 feet above sea-level, there are ragged patches of manzanita, and five or six species of ceanothus, called deer-brush or California lilac. These are the most important of all the honey-bearing bushes of the Sierra. *Chamæbatia foliolosa*, a little shrub about a foot high, with flowers like the strawberry, makes handsome carpets beneath the pines, and seems to be a favorite with the bees; while pines themselves furnish unlimited quantities of pollen and honey-dew. The product of a single tree, ripening its pollen at the right time of year, would be sufficient for the wants of a whole hive. Along the streams there is a rich growth of lilies, larkspurs, pedicularis, castilleias, and clover. The alpine region contains the flowery glacier meadows, and countless small gardens in all sorts of places full of potentilla of several species, spraguea, ivesia, epilobium, and goldenrod, with beds of bryanthus and the charming cassiope covered with sweet bells. Even the tops of the mountains are blessed with flowers,—dwarf phlox, polemonium, ribes, hulsea, etc. I have seen wild bees and butterflies feeding at a height of 13,000 feet above the sea. Many, however, that go up these dangerous heights never come down again. Some, undoubtedly, perish in storms, and I have found thousands lying dead or benumbed on the surface of the glaciers, to which they had perhaps been attracted by the white glare, taking them for beds of bloom.

From swarms that escaped their owners in the lowlands, the honey-bee is now generally distributed throughout the whole length of the Sierra, up to an elevation of 8000 feet above sea-level. At this height they flourish without care, though the snow every winter is deep. Even higher than this several bee-trees have been cut which contained over 200 pounds of honey.

The destructive action of sheep has not been so general on the mountain pastures as on those of the great plain, but in many places it has been more complete, owing to the more friable character of the soil, and its sloping position. The slant digging and down-raking action of hoofs on the steeper slopes of moraines has uprooted and buried many of the tender plants from year to year, without allowing them time to mature their seeds. The shrubs, too, are badly bitten, especially the various species of ceanothus. Fortunately, neither sheep nor cattle care to feed on the manzanita, spiræa, or adenostoma; and these fine honey-bushes are too stiff and tall, or grow in places too rough and inaccessible, to be trodden under foot. Also the cañon walls and gorges, which form so considerable a part of the area of the range, while inaccessible to domestic sheep, are well fringed with honey-shrubs, and contain thousands of lovely bee-gardens, lying hid in narrow side-cañons and recesses fenced with avalanche taluses, and on the top of flat, projecting headlands, where only bees would think to look for them.

But, on the other hand, a great portion of the woody plants that escape the feet and teeth of the sheep are destroyed by the shepherds by means of running fires, which are set everywhere during the dry autumn for the purpose of burning off the old fallen trunks and underbrush, with a view to improving the pastures, and making more open ways for the flocks. These destructive sheep-fires sweep through nearly the entire forest belt of the range, from one extremity to the other, consuming not only the underbrush, but the young trees and seedlings on which the permanence of the forests depends; thus setting in motion a long train of evils which will certainly reach far beyond bees and bee-keepers.

The plough has not yet invaded the forest region to any appreciable extent, neither has it accomplished much in the foot-hills. Thousands of bee-ranches might be established along the margin of the plain, and up to a height of 4000 feet, wherever water could be obtained. The climate at this elevation admits of the making of permanent homes, and by moving the hives to higher pastures as the lower pass out of bloom, the annual yield of honey would be nearly doubled. The foot-hill pastures, as we have seen, fail about the end of May, those of the chaparral belt and lower forests are in full bloom in June, those of the upper and alpine region in July, August, and September. In Scotland, after the best of the Lowland bloom is past, the bees are carried in carts to the Highlands, and set free on the heather hills. In France, too, and in Poland, they are carried from pasture to pasture among orchards and fields in the same way, and along the rivers in barges to collect the honey of the delightful vegetation of the banks. In Egypt they are taken far up the Nile, and floated slowly home again, gathering the honey-harvest of the various fields on the way, timing their movements in accord with the seasons. Were similar methods pursued in California the productive season would last nearly all the year.

The average elevation of the north half of the Sierra is, as we have seen, considerably less than that of the south half, and small streams, with the bank and meadow gardens dependent upon them, are less abundant. Around the head waters of the Yuba, Feather, and Pitt rivers, the extensive table-lands of lava are sparsely planted with pines, through which the sunshine reaches the ground with little interruption. Here flourishes a scattered, tufted growth of golden applopappus, linosyris, bahia, wyetheia, arnica, artemisia, and similar plants; with manzanita, cherry, plum, and thorn in ragged patches on the cooler hill-slopes. At the extremities of the Great Central Plain, the Sierra and Coast Ranges curve around and lock together in a labyrinth of mountains and valleys, throughout which their floras are mingled, making at the north, with its temperate climate and copious rainfall, a perfect paradise for bees, though, strange to say, scarcely a single regular bee-ranch has yet been established in it.

Of all the upper flower fields of the Sierra, Shasta is the most honeyful, and may yet surpass in fame the celebrated honey hills of Hybla and hearthy Hymettus. Regarding this noble mountain from a bee point of view, encircled by its many climates, and sweeping aloft from the torrid plain into the frosty azure, we find the first 5000 feet from the summit generally snow-clad, and therefore about as honeyless as the sea. The base of this arctic region is girdled by a belt of crumbling lava measuring about 1000 feet in vertical breadth, and is mostly free from snow in summer. Beautiful lichens enliven the faces of the cliffs with their bright colors, and in some of the warmer nooks there are a few tufts of alpine daisies, wall-flowers and pentstemons; but, notwithstanding these bloom freely in the late summer, the zone as a whole is almost as honey-less as the icy summit, and its lower edge may be taken as the honey-line. Immediately below this comes the forest zone, covered with a rich growth of conifers, chiefly Silver Firs, rich in pollen and honey-dew, and diversified with countless garden openings, many of them less than a hundred yards across. Next, in orderly succession, comes the great bee zone. Its area far surpasses that of the icy summit and both the other zones combined, for it goes sweeping majestically around the entire mountain, with a breadth of six or seven miles and a circumference of nearly a hundred miles.

Shasta, as we have already seen, is a fire-mountain created by a succession of eruptions of ashes and molten lava, which, flowing over the lips of its several craters, grew outward and upward like the trunk of a knotty exogenous tree. Then followed a strange contrast. The glacial winter came on, loading the cooling mountain with ice, which flowed slowly outward in every direction, radiating from the summit in the form of one vast conical glacier—a down-crawling mantle of ice upon a fountain of smoldering fire, crushing and grinding for centuries its brown, flinty lavas with incessant activity, and thus degrading and remodeling the entire mountain. When, at length, the glacial period began to draw near its close, the ice-mantle was gradually melted off around the bottom, and, in receding and breaking into its present fragmentary condition, irregular rings and heaps of moraine matter were stored upon its flanks. The glacial erosion of most of the Shasta lavas produces detritus, composed of rough, sub-angular boulders of moderate size and of porous gravel and sand, which yields freely to the transporting power of running water. Magnificent floods from the ample fountains of ice and snow working with sublime energy upon this prepared glacial detritus, sorted it out and carried down immense quantities from the higher slopes, and re-formed it in smooth, delta-like beds around the base; and it is these flood-beds joined together that now form the main honey-zone of the old volcano.

Thus, by forces seemingly antagonistic and destructive, has Mother Nature accomplished her beneficent designs—now a flood of fire, now a flood of ice, now a flood of water; and at length an outburst of organic life, a milky way of snowy petals and wings, girdling the rugged mountain like a cloud, as if the vivifying sunbeams beating against its sides had broken into a foam of plant-bloom and bees, as sea-waves break and bloom on a rock shore.

In this flowery wilderness the bees rove and revel, rejoicing in the bounty of the sun, clambering eagerly through bramble and hucklebloom, ringing the myriad bells of the manzanita, now humming aloft among polleny willows and firs, now down on the ashy ground among gilias and buttercups, and anon plunging deep into snowy banks of cherry and buckthorn. They consider the lilies and roll into them, and, like lilies, they toil not, for they are impelled by sun-power, as water-wheels by water-power; and when the one has plenty of high-pressure water, the other plenty of sunshine, they hum and quiver alike. Sauntering in the Shasta bee-lands in the sun-days of summer, one may readily infer the time of day from the comparative energy of bee-movements alone—drowsy and moderate in the cool of the morning, increasing in energy with the ascending sun, and, at high noon, thrilling and quivering in wild ecstasy, then gradually declining again to the stillness of night. In my excursions among the glaciers I occasionally meet bees that are hungry, like mountaineers who venture too far and remain too long above the bread-line; then they droop and wither like autumn leaves. The Shasta bees are perhaps better fed than any others in the Sierra. Their field-work is one perpetual feast; but, however exhilarating the sunshine or bountiful the supply of flowers, they are always dainty feeders. Humming-moths and humming-birds seldom set foot upon a flower, but poise on the wing in front of it, and reach forward as if they were sucking through straws. But bees, though as dainty as they, hug their favorite flowers with profound cordiality, and push their blunt, polleny faces against them, like babies on their mother's bosom. And fondly, too, with eternal love, does Mother Nature clasp her small bee-babies, and suckle them, multitudes at once, on her warm Shasta breast.

Besides the common honey-bee there are many other species here—fine mossy, burly fellows, who were nourished on the mountains thousands of sunny seasons before the advent of the domestic species. Among these are the bumblebees, mason-bees, carpenter-bees, and leaf-cutters. Butterflies, too, and moths of every size and pattern; some broad-winged like bats, flapping slowly, and sailing in easy curves; others like small, flying violets, shaking about loosely in short, crooked flights close to the flowers, feasting luxuriously night and day. Great numbers of deer also delight to dwell in the brushy portions of the bee-pastures.

Bears, too, roam the sweet wilderness, their blunt, shaggy forms harmonizing well with the trees and tangled bushes, and with the bees, also, notwithstanding the disparity in size. They are fond of all good things, and enjoy them to the utmost, with but little troublesome discrimination—flowers and leaves as well as berries, and the bees themselves as well as their honey. Though the California bears have as yet had but little experience with honey-bees, they often succeed in reaching their bountiful stores, and it seems doubtful whether bees themselves enjoy honey with so great a relish. By means of their powerful teeth and claws they can gnaw and tear open almost any hive conveniently accessible. Most honey-bees, however, in search of a home are wise enough to make choice of a hollow in a living tree, a considerable distance above the ground, when such places are to be had; then they are pretty secure, for though the smaller black and brown bears climb well, they are unable to break into strong hives while compelled to exert themselves to keep from falling, and at the same time to endure the stings of the fighting bees without having their paws free to rub them off. But woe to the black bumblebees discovered in their mossy nests in the ground! With a few strokes of their huge paws the bears uncover the entire establishment, and, before time is given for a general buzz, bees old and young, larvæ, honey, stings, nest, and all are taken in one ravishing mouthful.

Not the least influential of the agents concerned in the superior sweetness of the Shasta flora are its storms—storms I mean that are strictly local, bred and born on the mountain. The magical rapidity with which they are grown on the mountain-top, and bestow their charity in rain and snow, never fails to astonish the inexperienced lowlander. Often in calm, glowing days, while the bees are still on the wing, a storm-cloud may be seen far above in the pure ether, swelling its pearl bosses, and growing silently, like a plant. Presently a clear, ringing discharge of thunder is heard, followed by a rush of wind that comes sounding over the bending woods like the roar of the ocean, mingling raindrops, snow-flowers, honey-flowers, and bees in wild storm harmony.

Still more impressive are the warm, reviving days of spring in the mountain pastures. The blood of the plants throbbing beneath the life-giving sunshine seems to be heard and felt. Plant growth goes on before our eyes, and every tree in the woods, and every bush and flower is seen as a hive of restless industry. The deeps of the sky are mottled with singing wings of every tone and color; clouds of brilliant chrysididæ dancing and swirling in exquisite rhythm, golden-barred vespidæ, dragon-flies, butterflies, grating cicadas, and jolly, rattling grasshoppers, fairly enameling the light.

On bright, crisp mornings a striking optical effect may frequently be observed from the shadows of the higher mountains while the sunbeams are pouring past overhead. Then every insect, no matter what may be its own proper color, burns white in the light. Gauzy-winged hymenoptera, moths, jet-black beetles, all are transformed alike in pure, spiritual white, like snowflakes.

In Southern California, where bee-culture has had so much skillful attention of late years, the pasturage is not more abundant, or more advantageously varied as to the number of its honey-plants and their distribution over mountain and plain, than that of many other portions of the State where the industrial currents flow in other channels. The famous White Sage (*Audibertia*), belonging to the mint family, flourishes here in all its glory, blooming in May, and yielding great quantities of clear, pale honey, which is greatly prized in every market it has yet reached. This species grows chiefly in the valleys and low hills. The Black Sage on the mountains is part of a dense, thorny chaparral, which is composed chiefly of adenostoma, ceanothus, manzanita, and cherry—not differing greatly from that of the southern portion of the Sierra, but more dense and continuous, and taller, and remaining longer in bloom. Stream-side gardens, so charming a feature of both the Sierra and Coast Mountains, are less numerous in Southern California, but they are exceedingly rich in honey-flowers, wherever found,—melilotus, columbine, collinsia, verbena, zauschneria, wild rose, honeysuckle, philadelphus, and lilies rising from the warm, moist dells in a very storm of exuberance. Wild buckwheat of many species is developed in abundance over the dry, sandy valleys and lower slopes of the mountains, toward the end of summer, and is, at this time, the main dependence of the bees, reinforced here and there by orange groves, alfalfa fields, and small home gardens.

The main honey months, in ordinary seasons, are April, May, June, July, and August; while the other months are usually flowery enough to yield sufficient for the bees.

According to Mr. J. T. Gordon, President of the Los Angeles County Bee-keepers' Association, the first bees introduced into the county were a single hive, which cost $150 in San Francisco, and arrived in September, 1854.* In April, of the following year, this hive sent out two swarms, which were sold for $100 each. From this small beginning the bees gradually multiplied to about 3000 swarms in the year 1873. In 1876 it was estimated that there were between 15,000 and 20,000 hives in the county, producing an annual yield of about 100 pounds to the hive—in some exceptional cases, a much greater yield.

In San Diego County, at the beginning of the season of 1878, there were about 24,000 hives, and the shipments from the one port of San Diego for the same year from July 17 to November 10, were 1071 barrels, 15,544 cases, and nearly 90 tons. The largest bee-ranches

* Fifteen hives of Italian bees were introduced into Los Angeles County in 1855, and in 1876 they had increased to 500. The marked superiority claimed for them over the common species is now attracting considerable attention.

have about a thousand hives, and are carefully and skillfully managed, every scientific appliance of merit being brought into use. There are few bee-keepers, however, who own half as many as this, or who give their undivided attention to the business. Orange culture, at present, is heavily overshadowing every other business.

A good many of the so-called bee-ranches of Los Angeles and San Diego counties are still of the rudest pioneer kind imaginable. A man unsuccessful in everything else hears the interesting story of the profits and comforts of bee-keeping, and concludes to try it, buys a few colonies, or gets them from some overstocked ranch shares, takes them back to the foot of some cañon, where the pasturage is fresh, squats on the land, with, or without, the permission of the owner, sets up his hives, makes a box-cabin for himself, scarcely bigger than a bee-hive, and awaits his fortune.

Bees suffer sadly from famine during the dry years which occasionally occur in the southern and middle portions of the State. If the rainfall amounts only to three or four inches, instead of from twelve to twenty, as in ordinary seasons, then sheep and cattle die in thousands, and so do these small, winged cattle, unless they are carefully fed, or removed to other pastures. The year 1877 will long be remembered as exceptionally rainless and distressing. Scarcely a flower bloomed on the dry valleys away from the stream-sides, and not a single grain-field depending upon rain was reaped. The seed only sprouted, came up a little way, and withered. Horses, cattle, and sheep grew thinner day by day, nibbling at bushes and weeds, along the shallowing edges of streams, many of which were dried up altogether, for the first time since the settlement of the country.

In the course of a trip I made during the summer of that year through Monterey, San Luis Obispo, Santa Barbara, Ventura, and Los Angeles counties, the deplorable effects of the drought were everywhere visible—leafless fields, dead and dying cattle, dead bees, and half-dead people with dusty, doleful faces. Even the birds and squirrels were in distress, though their suffering was less painfully apparent than that of the poor cattle. These were falling one by one in slow, sure starvation along the banks of the hot, sluggish streams, while thousands of buzzards correspondingly fat were sailing above them, or standing gorged on the ground beneath the trees, waiting with easy faith for fresh carcasses. The quails, prudently considering the hard times, abandoned all thought of pairing. They were too poor to marry, and so continued in flocks all through the year without attempting to rear young. The ground-squirrels, though an exceptionally industrious and enterprising race, as every farmer knows, were hard pushed for a living; not a fresh leaf or seed was to be found save in the trees, whose bossy masses of dark green foliage presented a striking contrast to the ashen baldness of the ground beneath them. The squirrels, leaving their accustomed feeding-grounds, betook themselves to the leafy oaks to gnaw out the acorn stores of the provident woodpeckers, but the latter kept up a vigilant watch upon their movements. I noticed four woodpeckers in league against one squirrel, driving the poor fellow out of an oak that they claimed. He dodged round the knotty trunk from side to side, as nimbly as he could in his famished condition, only to find a sharp bill everywhere. But the fate of the bees that year seemed the saddest of all. In different portions of Los Angeles and San Diego counties, from one half to three fourths of them died of sheer starvation. Not less than 18,000 colonies perished in these two counties alone, while in the adjacent counties the death-rate was hardly less.

Even the colonies nearest to the mountains suffered this year, for the smaller vegetation on the foot-hills was affected by the drought almost as severely as that of the valleys and plains, and even the hardy, deep-rooted chaparral, the surest dependence of the bees, bloomed sparingly, while much of it was beyond reach. Every swarm could have been saved, however, by promptly supplying them with food when their own stores began to fail, and before they became enfeebled and discouraged; or by cutting roads back into the mountains, and taking them into the heart of the flowery chaparral. The Santa Lucia, San Rafael, San Gabriel, San Jacinto, and San Bernardino ranges are almost untouched as yet save by the wild bees. Some idea of their resources, and of the advantages and disadvantages they offer to bee-keepers, may be formed from an excursion that I made into the San Gabriel Range about the beginning of August of "the dry year." This range, containing most of the characteristic features of the other ranges just mentioned, overlooks the Los Angeles vineyards and orange groves from the north, and is more rigidly inaccessible in the ordinary meaning of the word than any other that I ever attempted to penetrate. The slopes are exceptionally steep and insecure to the foot, and they are covered with thorny bushes from five to ten feet high. With the exception of little spots not visible in general views, the entire surface is covered with them, massed in close hedge growth, sweeping gracefully down into every gorge and hollow, and swelling over every ridge and summit in shaggy, ungovernable exuberance, offering more honey to the acre for half the year than the most crowded clover-field. But when beheld from the open San Gabriel Valley, beaten with dry sunshine, all that was seen of the range seemed to wear a forbidding aspect. From base to summit all seemed gray, barren, silent, its glorious chaparral appearing like dry moss creeping over its dull, wrinkled ridges and hollows.

Setting out from Pasadena, I reached the foot of the range about sundown; and being weary and heated with my walk across the shadeless valley, concluded to camp for the night. After resting a few moments, I began to look about among the flood-boulders of Eaton Creek for a camp-ground, when I came upon a strange, dark-looking man who had been chopping cord-wood. He seemed surprised at seeing me, so I sat down with him on the live-oak log he had been cutting, and made haste to give a reason for my appearance in his solitude, explaining that I was anxious to find out something about the mountains, and meant to make my way up Eaton Creek next morning. Then he kindly invited me to camp with him, and led me to his little cabin, situated at the foot of the mountains, where a small spring oozes out of a bank overgrown with wild-rose bushes. After supper, when the daylight was gone, he explained that he was out of candles; so we sat in the dark, while he gave me a sketch of his life in a mixture of Spanish and English. He was born in Mexico, his father Irish, his mother Spanish. He had been a miner, rancher, prospector, hunter, etc., rambling always, and wearing his life away in mere waste; but now he was going to settle down. His past life, he said, was of "no account," but the future was promising. He was going to "make money and marry a Spanish woman." People mine here for water as for gold. He had been running a tunnel into a spur of the mountain back of his cabin. "My prospect is good," he said, "and if I chance to strike a good, strong flow, I'll soon be worth $5000 or $10,000. For that flat out there," referring to a small, irregular patch of bouldery detritus, two or three acres in size, that had been deposited by Eaton Creek during some flood season,—"that flat is large enough for a nice orange-grove, and the bank behind the cabin will do for a vineyard, and after watering my own trees and vines

I will have some water left to sell to my neighbors below me, down the valley. And then," he continued, "I can keep bees, and make money that way, too, for the mountains above here are just full of honey in the summer-time, and one of my neighbors down here says that he will let me have a whole lot of hives, on shares, to start with. You see I've a good thing; I'm all right now." All this prospective affluence in the sunken, boulder-choked flood-bed of a mountain-stream! Leaving the bees out of the count, most fortune-seekers would as soon think of settling on the summit of Mount Shasta. Next morning, wishing my hopeful entertainer good luck, I set out on my shaggy excursion.

About half an hour's walk above the cabin, I came to "The Fall," famous throughout the valley settlements as the finest yet discovered in the San Gabriel Mountains. It is a charming little thing, with a low, sweet voice, singing like a bird, as it pours from a notch in a short ledge, some thirty-five or forty feet into a round mirror-pool. The face of the cliff back of it, and on both sides, is smoothly covered and embossed with mosses, against which the white water shines out in showy relief, like a silver instrument in a velvet case. Hither come the San Gabriel lads and lassies, to gather ferns and dabble away their hot holidays in the cool water, glad to escape from their commonplace palm-gardens and orange-groves. The delicate maidenhair grows on fissured rocks within reach of the spray, while broad-leaved maples and sycamores cast soft, mellow shade over a rich profusion of bee-flowers, growing among boulders in front of the pool—the fall, the flowers, the bees, the ferny rocks, and leafy shade forming a charming little poem of wildness, the last of a series extending down the flowery slopes of Mount San Antonio through the rugged, foam-beaten bosses of the main Eaton Cañon.

From the base of the fall I followed the ridge that forms the western rim of the Eaton basin to the summit of one of the principal peaks, which is about 5000 feet above sea-level. Then, turning eastward, I crossed the middle of the basin, forcing a way over its many subordinate ridges and across its eastern rim, having to contend almost everywhere with the floweriest and most impenetrable growth of honey-bushes I had ever encountered since first my mountaineering began. Most of the Shasta chaparral is leafy nearly to the ground; here the main stems are naked for three or four feet, and interspiked with dead twigs, forming a stiff *chevaux de frise* through which even the bears make their way with difficulty. I was compelled to creep for miles on all fours, and in following the bear-trails often found tufts of hair on the bushes where they had forced themselves through.

For 100 feet or so above the fall the ascent was made possible only by tough cushions of club-moss that clung to the rock. Above this the ridge weathers away to a thin knife-blade for a few hundred yards, and thence to the summit of the range it carries a bristly mane of chaparral. Here and there small openings occur on rocky places, commanding fine views across the cultivated valley to the ocean. These I found by the tracks were favorite outlooks and resting-places for the wild animals—bears, wolves, foxes, wildcats, etc.—which abound here, and would have to be taken into account in the establishment of bee-ranches. In the deepest thickets I found wood-rat villages—groups of huts four to six feet high, built of sticks and leaves in rough, tapering piles, like musk-rat cabins. I noticed a good many bees, too, most of them wild. The tame honey-bees seemed languid and wing-weary, as if they had come all the way up from the flowerless valley.

After reaching the summit I had time to make only a hasty survey of the basin, now glowing in the sunset gold, before hastening down into one of the tributary cañons in search of water. Emerging from a particularly tedious breadth of chaparral, I found myself free and erect in a beautiful park-like grove of Mountain Live Oak, where the ground was planted with aspidiums and brier-roses, while the glossy foliage made a close canopy overhead, leaving the gray dividing trunks bare to show the beauty of their interlacing arches. The bottom of the cañon was dry where I first reached it, but a bunch of scarlet mimulus indicated water at no great distance, and I soon discovered about a bucketful in a hollow of the rock. This, however, was full of dead bees, wasps, beetles, and leaves, well steeped and simmered, and would, therefore, require boiling and filtering through fresh charcoal before it could be made available. Tracing the dry channel about a mile farther down to its junction with a larger tributary cañon, I at length discovered a lot of boulder pools, clear as crystal, brimming full, and linked together by glistening streamlets just strong enough to sing audibly. Flowers in full bloom adorned their margins, lilies ten feet high, larkspur, columbines, and luxuriant ferns, leaning and overarching in lavish abundance, while a noble old Live Oak spread its rugged arms over all. Here I camped, making my bed on smooth cobblestones.

Next day, in the channel of a tributary that heads on Mount San Antonio, I passed about fifteen or twenty gardens like the one in which I slept—lilies in every one of them, in the full pomp of bloom. My third camp was made near the middle of the general basin, at the head of a long system of cascades from ten to 200 feet high, one following the other in close succession down a rocky, inaccessible cañon, making a total descent of nearly 1700 feet. Above the cascades the main stream passes through a series of open, sunny levels, the largest of which are about an acre in size, where the wild bees and their companions were feasting on a showy growth of zauschneria, painted cups, and monardella; and gray squirrels were busy harvesting the burs of the Douglas Spruce, the only conifer I met in the basin.

The eastern slopes of the basin are in every way similar to those we have described, and the same may be said of other portions of the range. From the highest summit, far as the eye could reach, the landscape was one vast bee-pasture, a rolling wilderness of honey-bloom, scarcely broken by bits of forest or the rocky outcrops of hill-tops and ridges.

Behind the San Bernardino Range lies the wild "sage-brush country," bounded on the east by the Colorado River, and extending in a general northerly direction to Nevada and along the eastern base of the Sierra beyond Mono Lake.

The greater portion of this immense region, including Owen's Valley, Death Valley, and the Sink of the Mohave, the area of which is nearly one fifth that of the entire State, is usually regarded as a desert, not because of any lack in the soil, but for want of rain, and rivers available for irrigation. Very little of it, however, is desert in the eyes of a bee.

Looking now over all the available pastures of California, it appears that the business of bee-keeping is still in its infancy. Even in the more enterprising of the southern counties, where so vigorous a beginning has been made, less than a tenth of their honey resources have as yet been developed; while in the Great Plain, the Coast Ranges, the Sierra Nevada, and the northern region about Mount Shasta, the business can hardly be said to exist at all. What the limits of its developments in the future may be, with the advantages of cheaper transportation and the invention of better methods in general, it is not easy to guess. Nor, on the other hand, are we able to measure the influence on bee interests likely to follow the destruction of the forests, now rapidly falling before fire and the ax. As to the sheep evil, that can hardly become greater than it is at the present day. In short, notwithstanding the wide-spread deterioration and destruction of every kind already effected, California, with her incomparable climate and flora, is still, as far as I know, the best of all the bee-lands of the world.

THE YOSEMITE

Affectionately dedicated
to my friend,
Robert Underwood Johnson,
faithful lover and defender
of our glorious forests
and originator of
the Yosemite National Park.

I

THE APPROACH TO THE VALLEY

WHEN I SET out on the long excursion that finally led to California I wandered afoot and alone, from Indiana to the Gulf of Mexico, with a plant-press on my back, holding a generally southward course, like the birds when they are going from summer to winter. From the west coast of Florida I crossed the gulf to Cuba, enjoyed the rich tropical flora there for a few months, intending to go thence to the north end of South America, make my way through the woods to the headwaters of the Amazon, and float down that grand river to the ocean. But I was unable to find a ship bound for South America—fortunately perhaps, for I had incredibly little money for so long a trip and had not yet fully recovered from a fever caught in the Florida swamps. Therefore I decided to visit California for a year or two to see its wonderful flora and the famous Yosemite Valley. All the world was before me and every day was a holiday, so it did not seem important to which one of the world's wildernesses I first should wander.

Arriving by the Panama steamer, I stopped one day in San Francisco and then inquired for the nearest way out of town. "But where do you want to go?" asked the man to whom I had applied for this important information. "To any place that is wild," I said. This reply startled him. He seemed to fear I might be crazy and therefore the sooner I was out of town the better, so he directed me to the Oakland ferry.

So on the first of April, 1868, I set out afoot for Yosemite. It was the bloom-time of the year over the lowlands and coast ranges; the landscapes of the Santa Clara Valley were fairly drenched with sunshine, all the air was quivering with the songs of the meadow-larks, and the hills were so covered with flowers that they seemed to be painted. Slow indeed was my progress through these glorious gardens, the first of the California flora I had seen. Cattle and cultivation were making few scars as yet, and I wandered enchanted in long wavering curves, knowing by my pocket map that Yosemite Valley lay to the east and that I should surely find it.

THE SIERRA FROM THE WEST

Looking eastward from the summit of the Pacheco Pass one shining morning, a landscape was displayed that after all my wanderings still appears as the most beautiful I have ever beheld. At my feet lay the Great Central Valley of California, level and flowery, like a lake of pure sunshine, forty or fifty miles wide, five hundred miles long, one rich furred garden of yellow *Compositæ*. And from the eastern boundary of this vast golden flower-bed rose the mighty Sierra, miles in height, and so gloriously colored and so radiant, it seemed not clothed with light, but wholly composed of it, like the wall of some celestial city. Along the top and extending a good way down, was a rich pearl-gray belt of snow; below it a belt of blue and dark purple, marking the extension of the forests; and stretching along the base of the range a broad belt of rose-purple; all these colors, from the blue sky to the yellow valley smoothly blending as they do in a rainbow, making a wall of light ineffably fine. Then it seemed to me that the Sierra should be called, not the Nevada or Snowy Range, but the Range of Light. And after ten years of wandering and wondering in the heart of it, rejoicing in its glorious floods of light, the white beams of the morning streaming through the passes, the noonday radiance on the crystal rocks, the flush of the alpenglow, and the irised spray of countless waterfalls, it still seems above all others the Range of Light.

In general views no mark of man is visible upon it, nor anything to suggest the wonderful depth and grandeur of its sculpture. None of its magnificent forest-crowned ridges seems to rise much above the general level to publish its wealth. No great valley or river is seen, or group of well-marked features of any kind standing out as distinct pictures. Even the summit peaks, marshaled in glorious array so high in the sky, seem comparatively regular in form. Nevertheless the whole range five hundred miles long is furrowed with cañons 2,000 to 5,000 feet deep, in which once flowed majestic glaciers, and in which now flow and sing the bright rejoicing rivers.

CHARACTERISTICS OF THE CAÑONS

Though of such stupendous depth, these cañons are not gloomy gorges, savage and inaccessible. With rough passages here and there they are flowery pathways conducting to the snowy, icy fountains; mountain streets full of life and light, graded and sculptured by the ancient glaciers, and presenting throughout all their courses a rich variety of novel and attractive scenery—the most attractive that has yet been discovered in the mountain-ranges of the world. In many places, especially in the middle region of the western flank, the main cañons widen into spacious valleys or parks diversified like landscape gardens with meadows and groves and thickets of blooming bushes, while the lofty walls, infinitely varied in form, are fringed with ferns, flowering plants, shrubs of many species, and tall evergreens and oaks that find footholds on small benches and tables, all enlivened and made glorious with rejoicing streams that come chanting in chorus over the cliffs and through side cañons in falls of every conceivable form, to join the river that flows in tranquil, shining beauty down the middle of each one of them.

THE INCOMPARABLE YOSEMITE

The most famous and accessible of these cañon valleys, and also the one that presents their most striking and sublime features on the grandest scale, is the Yosemite, situated in the basin of the Merced River at an elevation of 4000 feet above the level of the sea. It is about seven miles long, half a mile to a mile wide, and nearly a mile deep in the solid granite flank of the range. The walls are made up of rocks, mountains in size, partly separated from each other by side cañons, and they are so sheer in front, and so compactly and harmoniously arranged on a level floor, that the Valley, comprehensively seen, looks like an immense hall or temple lighted from above.

But no temple made with hands can compare with Yosemite. Every rock in its walls seems to glow with life. Some lean back in majestic repose; others, absolutely sheer or nearly so for thousands of feet, advance beyond their companions in thoughtful attitudes, giving welcome to storms and calms alike, seemingly aware, yet heedless, of everything going on about them. Awful in stern, immovable majesty, how softly these rocks are adorned, and how fine and reassuring the company they keep: their feet among beautiful groves and meadows, their brows in the sky, a thousand flowers leaning confidingly against their feet, bathed in floods of water, floods of light, while the snow and waterfalls, the winds and avalanches and clouds shine and sing and wreathe about them as the years go by, and myriads of small winged creatures—birds, bees, butterflies—give glad animation and help to make all the air into music. Down through the middle of the Valley flows the crystal Merced, River of Mercy, peacefully quiet, reflecting lilies and trees and the onlooking rocks; things frail and fleeting and types of endurance meeting here and blending in countless forms, as if into this one mountain mansion Nature had gathered her choicest treasures, to draw her lovers into close and confiding communion with her.

THE APPROACH TO THE VALLEY

Sauntering up the foot-hills to Yosemite by any of the old trails or roads in use before the railway was built from the town of Merced up the river to the boundary of Yosemite Park, richer and wilder become the forests and streams. At an elevation of 6000 feet above the level of the sea the silver firs are 200 feet high, with branches whorled around the colossal shafts in regular order, and every branch beautifully primate like a fern frond. The Douglas spruce, the yellow and sugar pines and brown-barked Libocedrus here reach their finest developments of beauty and grandeur. The majestic Sequoia is here, too, the king of conifers, the noblest of all the noble race. These colossal trees are as wonderful in fineness of beauty and proportion as in stature—an assemblage of conifers surpassing all that have ever yet been discovered in the forests of the world. Here indeed is the tree-lover's paradise; the woods, dry and wholesome, letting in the light in shimmering masses of half sunshine, half shade; the night air as well as the day air indescribably spicy and exhilarating; plushy fir-boughs for campers' beds, and cascades to sing us to sleep. On the highest ridges, over which these old Yosemite ways passed, the silver fir (*Abies magnifica*) forms the bulk of the woods, pressing forward in glorious array to the very brink of the Valley walls on both sides, and beyond the Valley to a height of from 8000 to 9000 feet above the level of the sea. Thus it appears that Yosemite, presenting such stupendous faces of bare granite, is nevertheless imbedded in magnificent forests, and the main species of pine, fir, spruce and libocedrus are also found in the Valley itself, but there are no "big trees" (*Sequoia gigantea*) in the Valley or about the rim of it. The nearest are about ten and twenty miles beyond the lower end of the valley on small tributaries of the Merced and Tuolumne Rivers.

THE FIRST VIEW: THE BRIDAL VEIL

From the margin of these glorious forests the first general view of the Valley used to be gained—a revelation in landscape affairs that enriches one's life forever. Entering the Valley, gazing overwhelmed with the multitude of grand objects about us, perhaps the first to fix our attention will be the Bridal Veil, a beautiful waterfall on our right. Its brow, where it first leaps free from the cliff, is about 900 feet above us; and as it sways and sings in the wind, clad in gauzy, sun-sifted spray, half falling, half floating, it seems infinitely gentle and fine; but the hymns it sings tell the solemn fateful power hidden beneath its soft clothing.

The Bridal Veil shoots free from the upper edge of the cliff by the velocity the stream has acquired in descending a long slope above the head of the fall. Looking from the top of the rock-avalanche talus on the west side, about one hundred feet above the foot of the fall, the under surface of the water arch is seen to be finely grooved and striated; and the sky is seen through the arch between rock and water, making a novel and beautiful effect.

Under ordinary weather conditions the fall strikes on flat-topped slabs, forming a kind of ledge about two-thirds of the way down from the top, and as the fall sways back and forth with great variety of motions among these flat-topped pillars, kissing and plashing notes as well as thunder-like detonations are produced, like those of the Yosemite Fall, though on a smaller scale.

The rainbows of the Veil, or rather the spray- and foam-bows, are superb, because the waters are dashed among angular blocks of granite at the foot, producing abundance of spray of the best quality for iris effects, and also for a luxuriant growth of grass and maidenhair on the side of the talus, which lower down is planted with oak, laurel and willows.

GENERAL FEATURES OF THE VALLEY

On the other side of the Valley, almost immediately opposite the Bridal Veil, there is another fine fall, considerably wider than the Veil when the snow is melting fast and more than 1000 feet in height, measured from the brow of the cliff where it first springs out into the air to the head of the rocky talus on which it strikes and is broken up into ragged cascades. It is called the Ribbon Fall or Virgin's Tears. During the spring floods it is a magnificent object, but the suffocating blasts of spray that fill the recess in the wall which it occupies prevent a near approach. In autumn, however, when its feeble current falls in a shower, it may then pass for tears with the sentimental onlooker fresh from a visit to the Bridal Veil.

Just beyond this glorious flood the El Capitan Rock, regarded by many as the most sublime feature of the Valley, is seen through the pine groves, standing forward beyond the general line of the wall in most imposing grandeur, a type of permanence. It is 3300 feet high, a plain, severely simple, glacier-sculptured face of granite, the end of one of the most compact and enduring of the mountain ridges, unrivaled in height and breadth and flawless strength.

Across the Valley from here, next to the Bridal Veil, are the picturesque Cathedral Rocks, nearly 2700 feet high, making a noble display of fine yet massive sculpture. They are closely related to El Capitan, having been eroded from the same mountain ridge by the great Yosemite Glacier when the Valley was in process of formation.

Next to the Cathedral Rocks on the south side towers the Sentinel Rock to a height of more than 3000 feet, a telling monument of the glacial period.

Almost immediately opposite the Sentinel are the Three Brothers, an immense mountain mass with three gables fronting the Valley, one above another, the topmost gable nearly 4000 feet high. They were named for three brothers, sons of old Tenaya, the Yosemite chief, captured here during the Indian War, at the time of the discovery of the Valley in 1852.

Sauntering up the Valley through meadow and grove, in the company of these majestic rocks, which seem to follow us as we advance, gazing, admiring, looking for new wonders ahead where all about us is so wonderful, the thunder of the Yosemite Fall is heard, and when we arrive in front of the Sentinel Rock it is revealed in all its glory from base to summit, half a mile in height, and seeming to spring out into the Valley sunshine direct from the sky. But even this fall, perhaps the most wonderful of its kind in the world, cannot at first hold our attention, for now the wide upper portion of the Valley is displayed to view, with the finely modeled North Dome, the Royal Arches and Washington Column on our left; Glacier Point, with its massive, magnificent sculpture on the right; and in the middle, directly in front, looms Tissiack or Half Dome, the most beautiful and most sublime of all the wonderful Yosemite rocks, rising in serene majesty from flowery groves and meadows to a height of 4750 feet.

THE UPPER CAÑONS

Here the Valley divides into three branches, the Tenaya, Nevada, and Illilouette Cañons, extending back into the fountains of the High Sierra, with scenery every way worthy the relation they bear to Yosemite.

In the south branch, a mile or two from the main Valley, is the Illilouette Fall, 600 feet high, one of the most beautiful of all the Yosemite choir, but to most people inaccessible as yet on account of its rough, steep, boulder-choked cañon. Its principal fountains of ice and snow lie in the beautiful and interesting mountains of the Merced group, while its broad open basin between its fountain mountains and cañon is noted for the beauty of its lakes and forests and magnificent moraines.

Returning to the Valley, and going up the north branch of Tenaya Cañon, we pass between the North Dome and Half Dome, and in less than an hour come to Mirror Lake, the Dome Cascades, and Tenaya Fall. Beyond the Fall, on the north side of the cañon, is the sublime El Capitan-like rock called Mount Watkins; on the south the vast granite wave of Clouds' Rest, a mile in height; and between them the fine Tenaya Cascade with silvery plumes outspread on smooth glacier-polished folds of granite, making a vertical descent in all of about 700 feet.

Just beyond the Dome Cascades, on the shoulder of Mount Watkins, there is an old trail once used by Indians on their way across the range to Mono, but in the cañon above this point there is no trail of any sort. Between Mount Watkins and Clouds' Rest the cañon is accessible only to mountaineers, and it is so dangerous that I hesitate to advise even good climbers, anxious to test their nerve and skill, to attempt to pass through it. Beyond the Cascades no great difficulty will be encountered. A succession of charming lily gardens and meadows occurs in filled-up lake basins among the rock-waves in the bottom of the cañon, and everywhere the surface of the granite has a smooth-wiped appearance, and in many places reflects the sunbeams like glass, a phenomenon due to glacial action, the cañon having been the channel of one of the main tributaries of the ancient Yosemite Glacier.

About ten miles above the Valley we come to the beautiful Tenaya Lake, and here the cañon terminates. A mile or two above the lake stands the grand Sierra Cathedral, a building of one stone, hewn from the living rock, with sides, roof, gable, spire and ornamental pinnacles, fashioned and finished symmetrically like a work of art, and set on a well-graded plateau about 9000 feet high, as if Nature in making so fine a building had also been careful that it should be finely seen. From every direction its peculiar form and graceful, majestic beauty of expression never fail to charm. Its height from its base to the ridge of the roof is about 2500 feet, and among the pinnacles that adorn the front grand views may be gained of the upper basins of the Merced and Tuolumne Rivers.

Passing the Cathedral we descend into the delightful, spacious Tuolumne Valley, from which excursions may be made to Mounts Dana, Lyell, Ritter, Conness, and Mono Lake, and to the many curious peaks that rise above the meadows on the south, and to the Big Tuolumne Cañon, with its glorious abundance of rocks and falling, gliding, tossing water. For all these the beautiful meadows near the Soda Springs form a delightful center.

NATURAL FEATURES NEAR THE VALLEY

Returning now to Yosemite and ascending the middle or Nevada branch of the Valley, occupied by the main Merced River, we come within a few miles to the Vernal and Nevada Falls, 400 and 600 feet high, pouring their white, rejoicing waters in the midst of the most novel and sublime rock scenery to be found in all the world. Tracing the river beyond the head of the Nevada Fall we are led into the Little Yosemite, a valley like the great Yosemite in form, sculpture and vegetation. It is about three miles long, with walls 1500 to 2000 feet high, cascades coming over them, and the river flowing through the meadows and groves of the level bottom in tranquil, richly-embowered reaches.

Beyond this Little Yosemite in the main cañon, there are three other little yosemites, the highest situated a few miles below the base of Mount Lyell, at an elevation of about 7800 feet above the sea. To describe these, with all their wealth of Yosemite furniture, and the wilderness of lofty peaks above them, the home of the avalanche and treasury of the fountain snow, would take us far beyond the bounds of a single book. Nor can we here consider the formation of these mountain landscapes—how the crystal rocks were brought to light by glaciers made up of crystal snow, making beauty whose influence is so mysterious on every one who sees it.

Of the small glacier lakes so characteristic of these upper regions, there are no fewer than sixty-seven in the basin of the main middle branch, besides countless smaller pools. In the basin of the Illilouette there are sixteen, in the Tenaya basin and its branches thirteen, in the Yosemite Creek basin fourteen, and in the Pohono or Bridal Veil one, making a grand total of one hundred and eleven lakes whose waters come to sing at Yosemite. So glorious is the background of the great Valley, so harmonious its relations to its widespreading fountains.

The same harmony prevails in all the other features of the adjacent landscapes. Climbing out of the Valley by the subordinate cañons, we find the ground rising from the brink of the walls: on the south side to the fountains of the Bridal Veil Creek, the basin of which is noted for the beauty of its meadows and its superb forests of silver fir; on the north side through the basin of the Yosemite Creek to the dividing ridge along the Tuolumne Cañon and the fountains of the Hoffmann Range.

DOWN THE YOSEMITE CREEK

In general views the Yosemite Creek basin seems to be paved with domes and smooth, whaleback masses of granite in every stage of development—some showing only their crowns; others rising high and free above the girdling forests, singly or in groups. Others are developed only on one side, forming bold outstanding bosses usually well fringed with shrubs and trees, and presenting the polished surfaces given them by the glacier that brought them into relief. On the upper portion of the basin broad moraine beds have been deposited and on these fine, thrifty forests are growing. Lakes and meadows and small spongy bogs may be found hiding here and there in the woods or back in the fountain recesses of Mount Hoffmann, while a thousand gardens are planted along the banks of the streams.

All the wide, fan-shaped upper portion of the basin is covered with a network of small rills that go cheerily on their way to their grand fall in the Valley, now flowing on smooth pavements in sheets thin as glass, now diving under willows and laving their red roots, oozing through green, plushy bogs, plashing over small falls and dancing down slanting cascades, calming again, gliding through patches of smooth glacier meadows with sod of alpine agrostis mixed with blue and white violets, and daisies, breaking, tossing among rough boulders and fallen trees, resting in calm pools, flowing together until, all united, they go to their fate with stately, tranquil gestures like a full-grown river. At the crossing of the Mono Trail, about two miles above the head of the Yosemite Fall, the stream is nearly forty feet wide, and when the snow is melting rapidly in the spring it is about four feet deep, with a current of two and a half miles an hour. This is about the volume of water that forms the Fall in May and June when there had been much snow the preceding winter; but it varies greatly from month to month. The snow rapidly vanishes from the open portion of the basin, which faces southward, and only a few of the tributaries reach back to perennial snow and ice fountains in the shadowy amphitheaters on the precipitous northern slopes of Mount Hoffmann. The total descent made by the stream from its highest sources to its confluence with the Merced in the Valley is about 6000 feet, while the distance is only about ten miles, an average fall of 600 feet per mile. The last mile of its course lies between the sides of sunken domes and swelling folds of the granite that are clustered and pressed together like a mass of bossy cumulus clouds. Through this shining way Yosemite Creek goes to its fate, swaying and swirling with easy, graceful gestures and singing the last of its mountain songs before it reaches the dizzy edge of Yosemite to fall 2600 feet into another world, where climate, vegetation, inhabitants, all are different. Emerging from this last cañon the stream glides, in flat, lace-like folds, down a smooth incline into a small pool where it seems to rest and compose itself before taking the grand plunge. Then calmly, as if leaving a lake, it slips over the polished lip of the pool down another incline and out over the brow of the precipice in a magnificent curve thick-sown with rainbow spray.

THE YOSEMITE FALL

Long ago before I had traced this fine stream to its head back of Mount Hoffmann, I was eager to reach the extreme verge to see how it behaved in flying so far through the air; but after enjoying this view and getting safely away I have never advised any one to follow my steps. The last incline down which the stream journeys so gracefully is so steep and smooth one must slip cautiously forward on hands and feet alongside the rushing water, which so near one's head is very exciting. But to gain a perfect view one must go yet farther, over a curving brow to a slight shelf on the extreme brink. This shelf, formed by the flaking off of a fold of granite, is about three inches wide, just wide enough for a safe rest for one's heels. To me it seemed nerve-trying to slip to this narrow foothold and poise on the edge of such a precipice so close to the confusing whirl of the waters; and after casting longing glances over the shining brow of the fall and listening to its sublime psalm, I concluded not to attempt to go nearer, but, nevertheless, against reasonable judgment, I did. Noticing some tufts of artemisia in a cleft of rock, I filled my mouth with the leaves, hoping their bitter taste might help to keep caution keen and prevent giddiness. In spite of myself I reached the little ledge, got my heels well set, and worked sidewise twenty or thirty feet to a point close to the out-plunging current. Here the view is perfectly free down into the heart of the bright irised throng of comet-like streamers into which the whole ponderous volume of the fall separates, two or three hundred feet below the brow. So glorious a display of pure wildness, acting at close range while cut off from all the world beside, is terribly impressive. A less nerve-trying view may be obtained from a fissured portion of the edge of the cliff about forty yards to the eastward of the fall. Seen from this point towards noon, in the spring, the rainbow on its brow seems to be broken up and mingled with the rushing comets until all the fall is stained with iris colors, leaving no white water visible. This is the best of the safe views from above, the huge steadfast rocks, the flying waters, and the rainbow light forming one of the most glorious pictures conceivable.

The Yosemite Fall is separated into an upper and a lower fall with a series of falls and cascades between them, but when viewed in front from the bottom of the Valley they all appear as one.

So grandly does this magnificent fall display itself from the floor of the Valley, few visitors take the trouble to climb the walls to gain nearer views, unable to realize how vastly more impressive it is near by than at a distance of one or two miles.

A WONDERFUL ASCENT

The views developed in a walk up the zigzags of the trail leading to the foot of the Upper Fall are about as varied and impressive as those displayed along the favorite Glacier Point Trail. One rises as if on wings. The groves, meadows, fern-flats and reaches of the river gain new interest, as if never seen before; all the views changing in a most striking manner as we go higher from point to point. The foreground also changes every few rods in the most surprising manner, although the earthquake talus and the level bench on the face of the wall over which the trail passes seem monotonous and commonplace as seen from the bottom of the Valley. Up we climb with glad exhilaration, through shaggy fringes of laurel, ceanothus, glossy-leaved manzanita and live-oak, from shadow to shadow across bars and patches of sunshine, the leafy openings making charming frames for the Valley pictures beheld through them, and for the glimpses of the high peaks that appear in the distance. The higher we go the farther we seem to be from the summit of the vast granite wall. Here we pass a projecting buttress whose grooved and rounded surface tells a plain story of the time when the Valley, now filled with sunshine, was filled with ice, when the grand old Yosemite Glacier, flowing river-like from its distant fountains, swept through it, crushing, grinding, wearing its way ever deeper, developing and fashioning these sublime rocks. Again we cross a white, battered gully, the pathway of rock avalanches or snow avalanches. Farther on we come to a gentle stream slipping down the face of the cliff in lace-like strips, and dropping from ledge to ledge too small to be called a fall—trickling, dripping, oozing, a pathless wanderer from one of the upland meadows lying a little way back of the Valley rim, seeking a way century after century to the depths of the Valley without any appreciable channel. Every morning after a cool night, evaporation being checked, it gathers strength and sings like a bird, but as the day advances and the sun strikes its thin currents outspread on the heated precipices, most of its waters vanish ere the bottom of the Valley is reached. Many a fine, hanging-garden aloft on breezy inaccessible heights owes to it its freshness and fullness of beauty; ferneries in shady nooks, filled with Adiantum, Woodwardia, Woodsia, Aspidium, Pellaea, and Cheilanthes, rosetted and tufted and ranged in lines, daintily overlapping, thatching the stupendous cliffs with softest beauty, some of the delicate fronds seeming to float on the warm moist air, without any connection with rock or stream. Nor is there any lack of colored plants wherever they can find a place to cling to; lilies and mints, the showy cardinal mimulus, and glowing cushions of the golden bahia, enlivened with butterflies and bees and all the other small, happy humming creatures that belong to them.

After the highest point on the lower division of the trail is gained it leads up into the deep recess occupied by the great fall, the noblest display of falling water to be found in the Valley, or perhaps in the world. When it first comes in sight it seems almost within reach of one's hand, so great in the spring is its volume and velocity, yet it is still nearly a third of a mile away and appears to recede as we advance. The sculpture of the walls about it is on a scale of grandeur, according nobly with the fall—plain and massive, though elaborately finished, like all the other cliffs about the Valley.

In the afternoon an immense shadow is cast athwart the plateau in front of the fall, and over the chaparral bushes that clothe the slopes and benches of the walls to the eastward, creeping upward until the fall is wholly overcast, the contrast between the shaded and illumined sections being very striking in these near views.

Under this shadow, during the cool centuries immediately following the breaking-up of the Glacial Period, dwelt a small residual glacier, one of the few that lingered on this sun-beaten side of the Valley after the main trunk glacier had vanished. It sent down a long winding current through the narrow cañon on the west side of the fall, and must have formed a striking feature of the ancient scenery of the Valley; the lofty fall of ice and fall of water side by side, yet separate and distinct.

The coolness of the afternoon shadow and the abundant dewy spray make a fine climate for the plateau ferns and grasses, and for the beautiful azalea bushes that grow here in profusion and bloom in September, long after the warmer thickets down on the floor of the Valley have withered and gone to seed. Even close to the fall, and behind it at the base of the cliff, a few venturesome plants may be found undisturbed by the rock-shaking torrent.

The basin at the foot of the fall into which the current directly pours, when it is not swayed by the wind, is about ten feet deep and fifteen to twenty feet in diameter. That it is not much deeper is surprising, when the great height and force of the fall is considered. But the rock where the water strikes probably suffers less erosion than it would were the descent less than half as great, since the current is outspread, and much of its force is spent ere it reaches the bottom—being received on the air as upon an elastic cushion, and borne outward and dissipated over a surface more than fifty yards wide.

This surface, easily examined when the water is low, is intensely clean and fresh looking. It is the raw, quick flesh of the mountain wholly untouched by the weather. In summer droughts, when the snowfall of the preceding winter has been light, the fall is reduced to a mere shower of separate drops without any obscuring spray. Then we may safely go back of it and view the crystal shower from beneath, each drop wavering and pulsing as it makes its way through the air, and flashing off jets of colored light of ravishing beauty. But all this is invisible from the bottom of the Valley, like a thousand other interesting things. One must labor for beauty as for bread, here as elsewhere.

THE GRANDEUR OF THE YOSEMITE FALL

During the time of the spring floods the best near view of the fall is obtained from Fern Ledge on the east side above the blinding spray at a height of about 400 feet above the base of the fall. A climb of about 1400 feet from the Valley has to be made, and there is no trail, but to any one fond of climbing this will make the ascent all the more delightful. A narrow part of the ledge extends to the side of the fall and back of it, enabling us to approach it as closely as we wish. When the afternoon sunshine is streaming through the throng of comets, ever wasting, ever renewed, the marvelous fineness, firmness and variety of their forms are beautifully revealed. At the top of the fall they seem to burst forth in irregular spurts from some grand, throbbing mountain heart. Now and then one mighty throb sends forth a mass of solid water into the free air far beyond the others, which rushes alone to the bottom of the fall with long streaming tail, like combed silk, while the others, descending in clusters, gradually mingle and lose their identity. But they all rush past us with amazing velocity and display of power, though apparently drowsy and deliberate in their movements when observed from a distance of a mile or two. The heads of these comet-like masses are composed of nearly solid water, and are dense white in color like pressed snow, from the friction they suffer in rushing through the air, the portion worn off forming the tail, between the white lustrous threads and films of which faint, grayish pencilings appear, while the outer, finer sprays of water-dust, whirling in sunny eddies, are pearly gray throughout. At the bottom of the fall there is but little distinction of form visible. It is mostly a hissing, clashing, seething, upwhirling mass of scud and spray, through which the light sifts in gray and purple tones, while at times when the sun strikes at the required angle, the whole wild and apparently lawless, stormy, striving mass is changed to brilliant rainbow hues, manifesting finest harmony. The middle portion of the fall is the most openly beautiful; lower, the various forms into which the waters are wrought are more closely and voluminously veiled, while higher, towards the head, the current is comparatively simple and undivided. But even at the bottom, in the boiling clouds of spray, there is no confusion, while the rainbow light makes all divine, adding glorious beauty and peace to glorious power. This noble fall has far the richest, as well as the most powerful, voice of all the falls of the Valley, its tones varying from the sharp hiss and rustle of the wind in the glossy leaves of the live-oaks and the soft, sifting, hushing tones of the pines, to the loudest rush and roar of storm winds and thunder among the crags of the summit peaks. The low bass, booming, reverberating tones, heard under favorable circumstances five or six miles away, are formed by the dashing and exploding of heavy masses mixed with air upon two projecting ledges on the face of the cliff, the one on which we are standing and another about 200 feet above it. The torrent of massive comets is continuous at time of high water, while the explosive, booming notes are wildly intermittent, because, unless influenced by the wind, most of the heavier masses shoot out from the face of the precipice, and pass the ledges upon which at other times they are exploded. Occasionally the whole fall is swayed away from the front of the cliff, then suddenly dashed flat against it, or vibrated from side to side like a pendulum, giving rise to endless variety of forms and sounds.

THE NEVADA FALL

The Nevada Fall is 600 feet high and is usually ranked next to the Yosemite in general interest among the five main falls of the Valley. Coming through the Little Yosemite in tranquil reaches, the river is first broken into rapids on a moraine boulder-bar that crosses the lower end of the Valley. Thence it pursues its way to the head of the fall in a rough, solid rock channel, dashing on side angles, heaving in heavy surging masses against elbow knobs, and swirling and swashing in pot-holes without a moment's rest. Thus, already chafed and dashed to foam, overfolded and twisted, it plunges over the brink of the precipice as if glad to escape into the open air. But before it reaches the bottom it is pulverized yet finer by impinging upon a sloping portion of the cliff about halfway down, thus making it the whitest of all the falls of the Valley, and altogether one of the most wonderful in the world.

On the north side, close to its head, a slab of granite projects over the brink, forming a fine point for a view, over its throng of streamers and wild plunging, into its intensely white bosom, and, through the broad drifts of spray, to the river far below, gathering its spent waters and rushing on again down the cañon in glad exultation into Emerald Pool, where at length it grows calm and gets rest for what still lies before it. All the features of the view correspond with the waters in grandeur and wildness. The glacier-sculptured walls of the cañon on either hand, with the sublime mass of the Glacier Point Ridge in front, form a huge triangular pit-like basin, which, filled with the roaring of the falling river, seems as if it might be the hopper of one of the mills of the gods in which the mountains were being ground.

THE VERNAL FALL

The Vernal, about a mile below the Nevada, is 400 feet high, a staid, orderly, graceful, easy-going fall, proper and exact in every movement and gesture, with scarce a hint of the passionate enthusiasm of the Yosemite or of the impetuous Nevada, whose chafed and twisted waters hurrying over the cliff seem glad to escape into the open air, while its deep, booming, thunder-tones reverberate over the listening landscape. Nevertheless it is a favorite with most visitors, doubtless because it is more accessible than any other, more closely approached and better seen and heard. A good stairway ascends the cliff beside it and the level plateau at the head enables one to saunter safely along the edge of the river as it comes from Emerald Pool and to watch its waters, calmly bending over the brow of the precipice, in a sheet eighty feet wide, changing in color from green to purplish gray and white until dashed on a boulder talus. Thence issuing from beneath its fine broad spray-clouds we see the tremendously adventurous river still unspent, beating its way down the wildest and deepest of all its cañons in gray roaring rapids, dear to the ouzel, and below the confluence of the Illilouette, sweeping around the shoulder of the Half Dome on its approach to the head of the tranquil levels of the Valley.

THE ILLILOUETTE FALL

The Illilouette in general appearance most resembles the Nevada. The volume of water is less than half as great, but it is about the same height (600 feet) and its waters received the same kind of preliminary tossing in a rocky, irregular channel. Therefore it is a very white and fine-grained fall. When it is in full springtime bloom it is partly divided by rocks that roughen the lip of the precipice, but this division amounts only to a kind of fluting and grooving of the column, which has a beautiful effect. It is not nearly so grand a fall as the upper Yosemite, or so symmetrical as the Vernal, or so airily graceful and simple as the Bridal Veil, nor does it ever display so tremendous an outgush of snowy magnificence as the Nevada; but in the exquisite fineness and richness of texture of its flowing folds it surpasses them all.

One of the finest effects of sunlight on falling water I ever saw in Yosemite or elsewhere I found on the brow of this beautiful fall. It was in the Indian summer, when the leaf colors were ripe and the great cliffs and domes were transfigured in the hazy golden air. I had scrambled up its rugged talus-dammed cañon, oftentimes stopping to take breath and look back to admire the wonderful views to be had there of the great Half Dome, and to enjoy the extreme purity of the water, which in the motionless pools on this stream is almost perfectly invisible; the colored foliage of the maples, dogwoods, *Rubus* tangles, etc., and the late goldenrods and asters. The voice of the fall was now low, and the grand spring and summer floods had waned to sifting, drifting gauze and thin-broidered folds of linked and arrowy lace-work. When I reached the foot of the fall sunbeams were glinting across its head, leaving all the rest of it in shadow; and on its illumined brow a group of yellow spangles of singular form and beauty were playing, flashing up and dancing in large flame-shaped masses, wavering at times, then steadying, rising and falling in accord with the shifting forms of the water. But the color of the dancing spangles changed not at all. Nothing in clouds or flowers, on bird-wings or the lips of shells, could rival it in fineness. It was the most divinely beautiful mass of rejoicing yellow light I ever beheld—one of Nature's precious gifts that perchance may come to us but once in a lifetime.

THE MINOR FALLS

There are many other comparatively small falls and cascades in the Valley. The most notable are the Yosemite Gorge Fall and Cascades, Tenaya Fall and Cascades, Royal Arch Falls, the two Sentinel Cascades and the falls of Cascade and Tamarack Creeks, a mile or two below the lower end of the Valley. These last are often visited. The others are seldom noticed or mentioned; although in almost any other country they would be visited and described as wonders.

The six intermediate falls in the gorge between the head of the Lower and the base of the Upper Yosemite Falls, separated by a few deep pools and strips of rapids, and three slender, tributary cascades on the west side form a series more strikingly varied and combined than any other in the Valley, yet very few of all the Valley visitors ever see them or hear of them. No available standpoint commands a view of them all. The best general view is obtained from the mouth of the gorge near the head of the Lower Fall. The two lowest of the series, together with one of the three tributary cascades, are visible from this standpoint, but in reaching it the last twenty or thirty feet of the descent is rather dangerous in time of high water, the shelving rocks being then slippery on account of spray, but if one should chance to slip when the water is low, only a bump or two and a harmless plash would be the penalty. No part of the gorge, however, is safe to any but cautious climbers.

Though the dark gorge hall of these rejoicing waters is never flushed by the purple light of morning or evening, it is warmed and cheered by the white light of noonday, which, falling into so much foam and spray of varying degrees of fineness, makes marvelous displays of rainbow colors. So filled, indeed, is it with this precious light, at favorable times it seems to take the place of common air. Laurel bushes shed fragrance into it from above and live-oaks, those fearless mountaineers, hold fast to angular seams and lean out over it with their fringing sprays and bright mirror leaves.

One bird, the ouzel, loves this gorge and flies through it merrily, or cheerily, rather, stopping to sing on foam-washed bosses where other birds could find no rest for their feet. I have even seen a gray squirrel down in the heart of it beside the wild rejoicing water.

One of my favorite night walks was along the rim of this wild gorge in times of high water when the moon was full, to see the lunar bows in the spray.

For about a mile above Mirror Lake the Tenaya Cañon is level, and richly planted with fir, Douglas spruce and libocedrus, forming a remarkably fine grove, at the head of which is the Tenaya Fall. Though seldom seen or described, this is, I think, the most picturesque of all the small falls. A considerable distance above it, Tenaya Creek comes hurrying down, white and foamy, over a flat pavement inclined at an angle of about eighteen degrees. In time of high water this sheet of rapids is nearly seventy feet wide, and is varied in a very striking way by three parallel furrows that extend in the direction of its flow. These furrows, worn by the action of the stream upon cleavage joints, vary in width, are slightly sinuous, and have large boulders firmly wedged in them here and there in narrow places, giving rise, of course, to a complicated series of wild dashes, doublings, and upleaping arches in the swift torrent. Just before it reaches the head of the fall the current is divided, the left division making a vertical drop of about eighty feet in a romantic, leafy, flowery, mossy nook, while the other forms a rugged cascade.

The Royal Arch Fall in time of high water is a magnificent object, forming a broad ornamental sheet in front of the arches. The two Sentinel Cascades, 3000 feet high, are also grand spectacles when the snow is melting fast in the spring, but by the middle of summer they have diminished to mere streaks scarce noticeable amid their sublime surroundings.

THE BEAUTY OF THE RAINBOWS

The Bridal Veil and Vernal Falls are famous for their rainbows; and special visits to them are often made when the sun shines into the spray at the most favorable angle. But amid the spray and foam and fine-ground mist ever rising from the various falls and cataracts there is an affluence and variety of iris bows scarcely known to visitors who stay only a day or two. Both day and night, winter and summer, this divine light may be seen wherever water is falling, dancing, singing; telling the heart-peace of Nature amid the wildest displays of her power. In the bright spring mornings the black-walled recess at the foot of the Lower Yosemite Fall is lavishly filled with irised spray; and not simply does this span the dashing foam, but the foam itself, the whole mass of it, beheld at a certain distance, seems to be colored, and drifts and wavers from color to color, mingling with the foliage of the adjacent trees, without suggesting any relationship to the ordinary rainbow. This is perhaps the largest and most reservoir-like fountain of iris colors to be found in the Valley.

Lunar rainbows or spray-bows also abound in the glorious affluence of dashing, rejoicing, hurrahing, enthusiastic spring floods, their colors as distinct as those of the sun and regularly and obviously banded, though less vivid. Fine specimens may be found any night at the foot of the Upper Yosemite Fall, glowing gloriously amid the gloomy shadows and thundering waters, whenever there is plenty of moonlight and spray. Even the secondary bow is at times distinctly visible.

The best point from which to observe them is on Fern Ledge. For some time after moonrise, at time of high water, the arc has a span of about five hundred feet, and is set upright; one end planted in the boiling spray at the bottom, the other in the edge of the fall, creeping lower, of course, and becoming less upright as the moon rises higher. This grand arc of color, glowing in mild, shapely beauty in so weird and huge a chamber of night shadows, and amid the rush and roar and tumultuous dashing of this thunder-voiced fall, is one of the most impressive and most cheering of all the blessed mountain evangels.

Smaller bows may be seen in the gorge on the plateau between the Upper and Lower Falls. Once toward midnight, after spending a few hours with the wild beauty of the Upper Fall, I sauntered along the edge of the gorge, looking in here and there, wherever the footing felt safe, to see what I could learn of the night aspects of the smaller falls that dwell there. And down in an exceedingly black, pit-like portion of the gorge, at the foot of the highest of the intermediate falls, into which the moonbeams were pouring through a narrow opening, I saw a well-defined spray-bow, beautifully distinct in colors, spanning the pit from side to side, while pure white foam-waves beneath the beautiful bow were constantly springing up out of the dark into the moonlight like dancing ghosts.

AN UNEXPECTED ADVENTURE

A wild scene, but not a safe one, is made by the moon as it appears through the edge of the Yosemite Fall when one is behind it. Once, after enjoying the night-song of the waters and watching the formation of the colored bow as the moon came round the domes and sent her beams into the wild uproar, I ventured out on the narrow bench that extends back of the fall from Fern Ledge and began to admire the dim-veiled grandeur of the view. I could see the fine gauzy threads of the fall's filmy border by having the light in front; and wishing to look at the moon through the meshes of some of the denser portions of the fall, I ventured to creep farther behind it while it was gently wind-swayed, without taking sufficient thought about the consequences of its swaying back to its natural position after the wind-pressure should be removed. The effect was enchanting: fine, savage music sounding above, beneath, around me; while the moon, apparently in the very midst of the rushing waters, seemed to be struggling to keep her place, on account of the ever-varying form and density of the water masses through which she was seen, now darkly veiled or eclipsed by a rush of thick-headed comets, now flashing out through openings between their tails. I was in fairyland between the dark wall and the wild throng of illumined waters, but suffered sudden disenchantment; for, like the witch-scene in Alloway Kirk, "in an instant all was dark." Down came a dash of spent comets, thin and harmless-looking in the distance, but they felt desperately solid and stony when they struck my shoulders, like a mixture of choking spray and gravel and big hailstones. Instinctively dropping on my knees, I gripped an angle of the rock, curled up like a young fern frond with my face pressed against my breast, and in this attitude submitted as best I could to my thundering bath. The heavier masses seemed to strike like cobblestones, and there was a confused noise of many waters about my ears—hissing, gurgling, clashing sounds that were not heard as music. The situation was quickly realized. How fast one's thoughts burn in such times of stress! I was weighing chances of escape. Would the column be swayed a few inches away from the wall, or would it come yet closer? The fall was in flood and not so lightly would its ponderous mass be swayed. My fate seemed to depend on a breath of the "idle wind." It was moved gently forward, the pounding ceased, and I was once more visited by glimpses of the moon. But fearing I might be caught at a disadvantage in making too hasty a retreat, I moved only a few feet along the bench to where a block of ice lay. I wedged myself between the ice and the wall, and lay face downwards, until the steadiness of the light gave encouragement to rise and get away. Somewhat nerve-shaken, drenched, and benumbed, I made out to build a fire, warmed myself, ran home, reached my cabin before day-light, got an hour or two of sleep, and awoke sound and comfortable, better, not worse, for my hard midnight bath.

CLIMATE AND WEATHER

Owing to the westerly trend of the Valley and its vast depth there is a great difference between the climates of the north and south sides—greater than between many countries far apart; for the south wall is in shadow during the winter months, while the north is bathed in sunshine every clear day. Thus there is mild spring weather on one side of the Valley while winter rules the other. Far up the north-side cliffs many a nook may be found closely embraced by sun-beaten rock-bosses in which flowers bloom every month of the year. Even butterflies may be seen in these high winter gardens except when snow-storms are falling and a few days after they have ceased. Near the head of the lower Yosemite Fall in January I found the ant lions lying in wait in their warm sand-cups, rock ferns being unrolled, club mosses covered with fresh-growing points, the flowers of the laurel nearly open, and the honeysuckle rosetted with bright young leaves; every plant seemed to be thinking about summer. Even on the shadow-side of the Valley the frost is never very sharp. The lowest temperature I ever observed during four winters was 7° Fahrenheit. The first twenty-four days of January had an average temperature at 9 A.M. of 32°, minimum 22°; at 3 P.M. the average was 40° 30′, the minimum 32°. Along the top of the walls, 7000 and 8000 feet high, the temperature was, of course, much lower. But the difference in temperature between the north and south sides is due not so much to the winter sunshine as to the heat of the preceding summer, stored up in the rocks, which rapidly melts the snow in contact with them. For though summer sun-heat is stored in the rocks of the south side also, the amount is much less because the rays fall obliquely on the south wall even in summer and almost vertically on the north.

The upper branches of the Yosemite streams are buried every winter beneath a heavy mantle of snow, and set free in the spring in magnificent floods. Then, all the fountains, full and overflowing, every living thing breaks forth into singing, and the glad exulting streams, shining and falling in the warm sunny weather, shake everything into music, making all the mountain-world a song.

The great annual spring thaw usually begins in May in the forest region, and in June and July on the high Sierra, varying somewhat both in time and fullness with the weather and the depth of the snow. Toward the end of summer the streams are at their lowest ebb, few even of the strongest singing much above a whisper as they slip and ripple through gravel and boulder-beds from pool to pool in the hollows of their channels, and drop in pattering showers like rain, and slip down precipices and fall in sheets of embroidery, fold over fold. But, however low their singing, it is always ineffably fine in tone, in harmony with the restful time of the year.

The first snow of the season that comes to the help of the streams usually falls in September or October, sometimes even in the latter part of August, in the midst of yellow Indian summer, when the goldenrods and gentians of the glacier meadows are in their prime. This Indian-summer snow, however, soon melts, the chilled flowers spread their petals to the sun, and the gardens as well as the streams are refreshed as if only a warm shower had fallen. The snow-storms that load the mountains to form the main fountain supply for the year seldom set in before the middle or end of November.

WINTER BEAUTY OF THE VALLEY

When the first heavy storms stopped work on the high mountains, I made haste down to my Yosemite den, not to "hole up" and sleep the white months away; I was out every day, and often all night, sleeping but little, studying the so-called wonders and common things ever on show, wading, climbing, sauntering among the blessed storms and calms, rejoicing in almost everything alike that I could see or hear: the glorious brightness of frosty mornings; the sunbeams pouring over the white domes and crags into the groves and waterfalls, kindling marvelous iris fires in the hoarfrost and spray; the great forests and mountains in their deep noon sleep; the good-night alpenglow; the stars; the solemn gazing moon, drawing the huge domes and headlands one by one glowing white out of the shadows hushed and breathless like an audience in awful enthusiasm, while the meadows at their feet sparkle with frost-stars like the sky; the sublime darkness of storm-nights, when all the lights are out; the clouds in whose depths the frail snow-flowers grow; the behavior and many voices of the different kinds of storms, trees, birds, waterfalls, and snow-avalanches in the ever-changing weather.

Every clear, frosty morning loud sounds are heard booming and reverberating from side to side of the Valley at intervals of a few minutes, beginning soon after sunrise and continuing an hour or two like a thunder-storm. In my first winter in the Valley I could not make out the source of this noise. I thought of falling boulders, rock-blasting, etc. Not till I saw what looked like hoarfrost dropping from the side of the Fall was the problem explained. The strange thunder is made by the fall of sections of ice formed of spray that is frozen on the face of the cliff along the sides of the Upper Yosemite Fall—a sort of crystal plaster, a foot or two thick, cracked off by the sun-beams, awakening all the Valley like cock-crowing, announcing the finest weather, shouting aloud Nature's infinite industry and love of hard work in creating beauty.

EXPLORING AN ICE CONE

This frozen spray gives rise to one of the most interesting winter features of the Valley—a cone of ice at the foot of the fall, four or five hundred feet high. From the Fern Ledge standpoint its crater-like throat is seen, down which the fall plunges with deep, gasping explosions of compressed air, and, after being well churned in the stormy interior, the water bursts forth through arched openings at its base, apparently scourged and weary and glad to escape, while belching spray, spouted up out of the throat past the descending current, is wafted away in irised drifts to the adjacent rocks and groves. It is built during the night and early hours of the morning; only in spells of exceptionally cold and cloudy weather is the work continued through the day. The greater part of the spray material falls in crystalline showers direct to its place, something like a small local snow-storm; but a considerable portion is first frozen on the face of the cliff along the sides of the fall and stays there until expanded and cracked off in irregular masses, some of them tons in weight, to be built into the walls of the cone; while in windy, frosty weather, when the fall is swayed from side to side, the cone is well drenched and the loose ice masses and spray-dust are all firmly welded and frozen together. Thus the finest of the downy wafts and curls of spray-dust, which in mild nights fall about as silently as dew, are held back until sunrise to make a store of heavy ice to reinforce the waterfall's thunder-tones.

While the cone is in process of formation, growing higher and wider in the frosty weather, it looks like a beautiful smooth, pure-white hill; but when it is wasting and breaking up in the spring its surface is strewn with leaves, pine branches, stones, sand, etc., that have been brought over the fall, making it look like a heap of avalanche detritus.

Anxious to learn what I could about the structure of this curious hill I often approached it in calm weather and tried to climb it, carrying an ax to cut steps. Once I nearly succeeded in gaining the summit. At the base I was met by a current of spray and wind that made seeing and breathing difficult. I pushed on backward, however, and soon gained the slope of the hill, where by creeping close to the surface most of the choking blast passed over me and I managed to crawl up with but little difficulty. Thus I made my way nearly to the summit, halting at times to peer up through the wild whirls of spray at the veiled grandeur of the fall, or to listen to the thunder beneath me; the whole hill was sounding as if it were a huge, bellowing drum. I hoped that by waiting until the fall was blown aslant I should be able to climb to the lip of the crater and get a view of the interior; but a suffocating blast, half air, half water, followed by the fall of an enormous mass of frozen spray from a spot high up on the wall, quickly discouraged me. The whole cone was jarred by the blow and some fragments of the mass sped past me dangerously near; so I beat a hasty retreat, chilled and drenched, and lay down on a sunny rock to dry.

Once during a wind-storm when I saw that the fall was frequently blown westward, leaving the cone dry, I ran up to Fern Ledge hoping to gain a clear view of the interior. I set out at noon. All the way up the storm notes were so loud about me that the voice of the fall was almost drowned by them. Notwithstanding the rocks and bushes everywhere were drenched by the wind-driven spray, I approached the brink of the precipice overlooking the mouth of the ice cone, but I was almost suffocated by the drenching, gusty spray, and was compelled to seek shelter. I searched for some hiding-place in the wall from whence I might run out at some opportune moment when the fall with its whirling spray and torn shreds of comet tails and trailing, tattered skirts was borne westward, as I had seen it carried several times before, leaving the cliffs on the east side and the ice hill bare in the sunlight. I had not long to wait, for, as if ordered so for my special accommodation, the mighty downrush of comets with their whirling drapery swung westward and remained aslant for nearly half an hour. The cone was admirably lighted and deserted by the water, which fell most of the time on the rocky western slopes mostly outside of the cone. The mouth into which the fall pours was, as near as I could guess, about one hundred feet in diameter north and south and about two hundred feet east and west, which is about the shape and size of the fall at its best in its normal condition at this season.

The crater-like opening was not a true oval, but more like a huge coarse mouth. I could see down the throat about one hundred feet or perhaps farther.

The fall precipice overhangs from a height of 400 feet above the base; therefore the water strikes some distance from the base of the cliff, allowing space for the accumulation of a considerable mass of ice between the fall and the wall.

II

WINTER STORMS AND SPRING FLOODS

THE BRIDAL VEIL and the Upper Yosemite Falls, on account of their height and exposure, are greatly influenced by winds. The common summer winds that come up the river cañon from the plains are seldom very strong; but the north winds do some very wild work, worrying the falls and the forests, and hanging snow-banners on the comet-peaks. One wild winter morning I was awakened by a storm-wind that was playing with the falls as if they were mere wisps of mist and making the great pines bow and sing with glorious enthusiasm. The Valley had been visited a short time before by a series of fine snow-storms, and the floor and the cliffs and all the region round about were lavishly adorned with its best winter jewelry, the air was full of fine snow-dust, and pine branches, tassels and empty cones were flying in an almost continuous flock.

Soon after sunrise, when I was seeking a place safe from flying branches, I saw the Lower Yosemite Fall thrashed and pulverized from top to bottom into one glorious mass of rainbow dust; while a thousand feet above it the main Upper Fall was suspended on the face of the cliff in the form of an inverted bow, all silvery white and fringed with short wavering strips. Then, suddenly assailed by a tremendous blast, the whole mass of the fall was blown into threads and ribbons, and driven back over the brow of the cliff whence it came, as if denied admission to the Valley. This kind of storm-work was continued about ten or fifteen minutes; then another change in the play of the huge exulting swirls and billows and up-heaving domes of the gale allowed the baffled fall to gather and arrange its tattered waters, and sink down again in its place. As the day advanced, the gale gave no sign of dying, excepting brief lulls, the Valley was filled with its weariless roar, and the cloudless sky grew garish-white from myriads of minute, sparkling snow-spicules. In the afternoon, while I watched the Upper Fall from the shelter of a big pine tree, it was suddenly arrested in its descent at a point about halfway down, and was neither blown upward nor driven aside, but simply held stationary in mid-air, as if gravitation below that point in the path of its descent had ceased to act. The ponderous flood, weighing hundreds of tons, was sustained, hovering, hesitating, like a bunch of thistledown, while I counted one hundred and ninety. All this time the ordinary amount of water was coming over the cliff and accumulating in the air, swedging and widening and forming an irregular cone about seven hundred feet high, tapering to the top of the wall, the whole standing still, resting on the invisible arm of the North Wind. At length, as if commanded to go on again, scores of arrowy comets shot forth from the bottom of the suspended mass as if escaping from separate outlets.

The brow of El Capitan was decked with long snow-streamers like hair, Clouds' Rest was fairly enveloped in drifting gossamer films, and the Half Dome loomed up in the garish light like a majestic, living creature clad in the same gauzy, wind-woven drapery, while upward currents meeting at times overhead made it smoke like a volcano.

AN EXTRAORDINARY STORM AND FLOOD

Glorious as are these rocks and waters arrayed in storm robes, or chanting rejoicing in every-day dress, they are still more glorious when rare weather conditions meet to make them sing with floods. Only once during all the years I have lived in the Valley have I seen it in full flood bloom. In 1871 the early winter weather was delightful; the days all sunshine, the nights all starry and calm, calling forth fine crops of frost-crystals on the pines and withered ferns and grasses for the morning sunbeams to sift through. In the afternoon of December 16, when I was sauntering on the meadows, I noticed a massive crimson cloud growing in solitary grandeur above the Cathedral Rocks, its form scarcely less striking than its color. It had a picturesque, bulging base like an old sequoia, a smooth, tapering stem, and a bossy, down-curling crown like a mushroom; all its parts were colored alike, making one mass of translucent crimson. Wondering what the meaning of that strange, lonely red cloud might be, I was up betimes next morning looking at the weather, but all seemed tranquil as yet. Towards noon gray clouds, with a close, curly grain like bird's-eye maple began to grow, and late at night rain fell, which soon changed to snow. Next morning the snow on the meadows was about ten inches deep, and it was still falling in a fine, cordial storm. During the night of the 18th heavy rain fell on the snow, but as the temperature was 34°, the snow-line was only a few hundred feet above the bottom of the Valley, and one had only to climb a little higher than the tops of the pines to get out of the rain-storm into the snow-storm. The streams, instead of being increased in volume by the storm, were diminished, because the snow sponged up part of their waters and choked the smaller tributaries. But about midnight the temperature suddenly rose to 42°, carrying the snow-line far beyond the Valley walls, and next morning Yosemite was rejoicing in a glorious flood. The comparatively warm rain falling on the snow was at first absorbed and held back, and so also was that portion of the snow that the rain melted, and all that was melted by the warm wind, until the whole mass of snow was saturated and became sludgy, and at length slipped and rushed simultaneously from a thousand slopes in wildest extravagance, heaping and swelling flood over flood, and plunging into the Valley in stupendous avalanches.

Awakened by the roar, I looked out and at once recognized the extraordinary character of the storm. The rain was still pouring in torrent abundance and the wind at gale speed was doing all it could with the flood-making rain.

The section of the north wall visible from my cabin was fairly streaked with new falls—wild roaring singers that seemed strangely out of place. Eager to get into the midst of the show, I snatched a piece of bread for breakfast and ran out. The mountain waters, suddenly liberated, seemed to be holding a grand jubilee. The two Sentinel Cascades rivaled the great falls at ordinary stages, and across the Valley by the Three Brothers I caught glimpses of more falls than I could readily count; while the whole Valley throbbed and trembled, and was filled with an awful, massive, solemn, sea-like roar. After gazing a while enchanted with the network of new falls that were adorning and transfiguring every rock in sight, I tried to reach the upper meadows, where the Valley is widest, that I might be able to see the walls on both sides, and thus gain general views. But the river was over its banks and the meadows were flooded, forming an almost continuous lake dotted with blue sludgy islands, while innumerable streams roared like lions across my path and were sweeping forward rocks and logs with tremendous energy over ground where tiny gilias had been growing but a short time before. Climbing into the talus slopes, where these savage torrents were broken among earthquake boulders, I managed to cross them, and force my way up the Valley to Hutchings' Bridge, where I crossed the river and waded to the middle of the upper meadow. Here most of the new falls were in sight, probably the most glorious assemblage of waterfalls ever displayed from any one standpoint. On that portion of the south wall between Hutchings' and the Sentinel there were ten falls plunging and booming from a height of nearly three thousand feet, the smallest of which might have been heard miles away. In the neighborhood of Glacier Point there were six; between the Three Brothers and Yosemite Fall, nine; between Yosemite and Royal Arch Falls, ten; from Washington Column to Mount Watkins, ten; on the slopes of Half Dome and Clouds' Rest, facing Mirror Lake and Tenaya Cañon, eight; on the shoulder of Half Dome, facing the Valley, three: fifty-six new falls occupying the upper end of the Valley, besides a countless host of silvery threads gleaming everywhere. In all the Valley there must have been upwards of a hundred. As if celebrating some great event, falls and cascades in Yosemite costume were coming down everywhere from fountain basins, far and near; and, though newcomers, they behaved and sang as if they had lived here always.

All summer-visitors will remember the comet forms of the Yosemite Fall and the laces of the Bridal Veil and Nevada. In the falls of this winter jubilee the lace forms predominated, but there was no lack of thunder-toned comets. The lower portion of one of the Sentinel Cascades was composed of two main white torrents with the space between them filled in with chained and beaded gauze of intricate pattern, through the singing threads of which the purplish-gray rock could be dimly seen. The series above Glacier Point was still more complicated in structure, displaying every form that one could imagine water might be dashed and combed and woven into. Those on the north wall between Washington Column and the Royal Arch Fall were so nearly related they formed an almost continuous sheet, and these again were but slightly separated from those about Indian Cañon. The group about the Three Brothers and El Capitan, owing to the topography and cleavage of the cliffs back of them, was more broken and irregular. The Tissiack Cascades were comparatively small, yet sufficient to give that noblest of mountain rocks a glorious voice. In the midst of all this extravagant rejoicing the great Yosemite Fall was scarce heard until about three o'clock in the afternoon. Then I was startled by a sudden thundering crash as if a rock avalanche had come to the help of the roaring waters. This was the flood-wave of Yosemite Creek, which had just arrived, delayed by the distance it had to travel, and by the choking snows of its widespread fountains. Now, with volume tenfold increased beyond its springtime fullness, it took its place as leader of the glorious choir.

And the winds, too, were singing in wild accord, playing on every tree and rock, surging against the huge brows and domes and outstanding battlements, deflected hither and thither and broken into a thousand cascading, roaring currents in the cañons, and low bass, drumming swirls in the hollows. And these again, reacting on the clouds, eroded immense cavernous spaces in their gray depths and swept forward the resulting detritus in ragged trains like the moraines of glaciers. These cloud movements in turn published the work of the winds, giving them a visible body, and enabling us to trace them. As if endowed with independent motion, a detached cloud would rise hastily to the very top of the wall as if on some important errand, examining the faces of the cliffs, and then perhaps as suddenly descend to sweep imposingly along the meadows, trailing its draggled fringes through the pines, fondling its waving spires with infinite gentleness, or, gliding behind a grove or a single tree, bringing it into striking relief, as it bowed and waved in solemn rhythm. Sometimes, as the busy clouds drooped and condensed or dissolved to misty gauze, half of the Valley would be suddenly veiled, leaving here and there some lofty headland cut off from all visible connection with the walls, looming alone, dim, spectral, as if belonging to the sky—visitors, like the new falls, come to take part in the glorious festival. Thus for two days and nights in measureless extravagance the storm went on, and mostly without spectators, at least of a terrestrial kind. I saw nobody out—bird, bear, squirrel, or man. Tourists had vanished months before, and the hotel people and laborers were out of sight, careful about getting cold, and satisfied with views from windows. The bears, I suppose, were in their cañon-boulder dens, the squirrels in their knot-hole nests, the grouse in close fir groves, and the small singers in the Indian Cañon chaparral, trying to keep warm and dry. Strange to say, I did not see even the water-ouzels, though they must have greatly enjoyed the storm.

This was the most sublime waterfall flood I ever saw—clouds, winds, rocks, waters, throbbing together as one. And then to contemplate what was going on simultaneously with all this in other mountain temples; the Big Tuolumne Cañon—how the white waters and the winds were singing there! And in Hetch Hetchy Valley and the great King's River yosemite, and in all the other Sierra cañons and valleys from Shasta to the southernmost fountains of the Kern, thousands of rejoicing flood waterfalls chanting together in jubilee dress.

III
SNOW-STORMS

AS HAS BEEN already stated, the first of the great snow-storms that replenish the Yosemite fountains seldom sets in before the end of November. Then, warned by the sky, wide-awake mountaineers, together with the deer and most of the birds, make haste to the lowlands or foot-hills; and burrowing marmots, mountain beavers, wood-rats, and other small mountain people, go into winter quarters, some of them not again to see the light of day until the general awakening and resurrection of the spring in June or July. The fertile clouds, drooping and condensing in brooding silence, seem to be thoughtfully examining the forests and streams with reference to the work that lies before them. At length, all their plans perfected, tufted flakes and single starry crystals come in sight, solemnly swirling and glinting to their blessed appointed places; and soon the busy throng fills the sky and makes darkness like night. The first heavy fall is usually from about two to four feet in depth; then with intervals of days or weeks of bright weather storm succeeds storm, heaping snow on snow, until thirty to fifty feet has fallen. But on account of its settling and compacting, and waste from melting and evaporation, the average depth actually found at any time seldom exceeds ten feet in the forest regions, or fifteen feet along the slopes of the summit peaks. After snow-storms come avalanches, varying greatly in form, size, behavior and in the songs they sing; some on the smooth slopes of the mountains are short and broad; others long and river-like in the side cañons of yosemites and in the main cañons, flowing in regular channels and booming like waterfalls, while countless smaller ones fall everywhere from laden trees and rocks and lofty cañon walls. Most delightful it is to stand in the middle of Yosemite on still clear mornings after snow-storms and watch the throng of avalanches as they come down, rejoicing, to their places, whispering, thrilling like birds, or booming and roaring like thunder. The noble yellow pines stand hushed and motionless as if under a spell until the morning sunshine begins to sift through their laden spires; then the dense masses on the ends of the leafy branches begin to shift and fall, those from the upper branches striking the lower ones in succession, enveloping each tree in a hollow conical avalanche of fairy fineness; while the relieved branches spring up and wave with startling effect in the general stillness, as if each tree was moving of its own volition. Hundreds of broad cloud-shaped masses may also be seen, leaping over the brows of the cliffs from great heights, descending at first with regular avalanche speed until, worn into dust by friction, they float in front of the precipices like irised clouds. Those which descend from the brow of El Capitan are particularly fine; but most of the great Yosemite avalanches flow in regular channels like cascades and waterfalls. When the snow first gives way on the upper slopes of their basins, a dull rushing, rumbling sound is heard which rapidly increases and seems to draw nearer with appalling intensity of tone. Presently the white flood comes bounding into sight over bosses and sheer places, leaping from bench to bench, spreading and narrowing and throwing off clouds of whirling dust like the spray of foaming cataracts. Compared with waterfalls and cascades, avalanches are short-lived, few of them lasting more than a minute or two, and the sharp, clashing sounds so common in falling water are mostly wanting; but in their low massy thunder-tones and purple-tinged whiteness, and in their dress, gait, gestures and general behavior, they are much alike.

AVALANCHES

Besides these common after-storm avalanches that are to be found not only in the Yosemite but in all the deep, sheer-walled cañons of the Range there are two other important kinds, which may be called annual and century avalanches, which still further enrich the scenery. The only place about the Valley where one may be sure to see the annual kind is on the north slope of Clouds' Rest. They are composed of heavy, compacted snow, which has been subjected to frequent alternations of freezing and thawing. They are developed on cañon and mountainsides at an elevation of from nine to ten thousand feet, where the slopes are inclined at an angle too low to shed off the dry winter snow, and which accumulates until the spring thaws sap their foundations and make them slippery; then away in grand style go the ponderous icy masses without any fine snow-dust.

Those of Clouds' Rest descend like thunderbolts for more than a mile.

The great century avalanches and the kind that mow wide swaths through the upper forests occur on mountainsides about ten or twelve thousand feet high, where under ordinary weather conditions the snow accumulated from winter to winter lies at rest for many years, allowing trees, fifty to a hundred feet high, to grow undisturbed on the slopes beneath them. On their way down through the woods they seldom fail to make a perfectly clean sweep, stripping off the soil as well as the tree, clearing paths two or three hundred yards wide from the timber line to the glacier meadows or lakes, and piling their uprooted trees, head downward, in rows along the sides of the gaps like lateral moraines. Scars and broken branches of the trees standing on the sides of the gaps record the depth of the overwhelming flood; and when we come to count the annual wood-rings on the uprooted trees we learn that some of these immense avalanches occur only once in a century or even at still wider intervals.

A RIDE ON AN AVALANCHE

Few Yosemite visitors ever see snow avalanches and fewer still know the exhilaration of riding on them. In all my mountaineering I have enjoyed only one avalanche ride, and the start was so sudden and the end came so soon I had but little time to think of the danger that attends this sort of travel, though at such times one thinks fast. One fine Yosemite morning after a heavy snowfall, being eager to see as many avalanches as possible and wide views of the forest and summit peaks in their new white robes before the sunshine had time to change them, I set out early to climb by a side cañon to the top of a commanding ridge a little over three thousand feet above the Valley. On account of the looseness of the snow that blocked the cañon I knew the climb would require a long time, some three or four hours as I estimated; but it proved far more difficult than I had anticipated. Most of the way I sank waist deep, almost out of sight in some places. After spending the whole day to within half an hour or so of sundown, I was still several hundred feet below the summit. Then my hopes were reduced to getting up in time to see the sunset. But I was not to get summit views of any sort that day, for deep trampling near the cañon head, where the snow was strained, started an avalanche, and I was swished down to the foot of the cañon as if by enchantment. The wallowing ascent had taken nearly all day, the descent only about a minute. When the avalanche started I threw myself on my back and spread my arms to try to keep from sinking. Fortunately, though the grade of the cañon is very steep, it is not interrupted by precipices large enough to cause outbounding or free plunging. On no part of the rush was I buried. I was only moderately imbedded on the surface or at times a little below it, and covered with a veil of back-streaming dust particles; and as the whole mass beneath and about me joined in the flight there was no friction, though I was tossed here and there and lurched from side to side. When the avalanche swedged and came to rest I found myself on top of the crumpled pile without a bruise or scar. This was a fine experience. Hawthorne says somewhere that steam has spiritualized travel; though unspiritual smells, smoke, etc., still attend steam travel. This flight in what might be called a milky way of snow-stars was the most spiritual and exhilarating of all the modes of motion I have ever experienced. Elijah's flight in a chariot of fire could hardly have been more gloriously exciting.

THE STREAMS IN OTHER SEASONS

In the spring, after all the avalanches are down and the snow is melting fast, then all the Yosemite streams, from their fountains to their falls, sing their grandest songs. Countless rills make haste to the rivers, running and singing soon after sunrise, louder and louder with increasing volume until sundown; then they gradually fail through the frosty hours of the night. In this way the volume of the upper branches of the river is nearly doubled during the day, rising and falling as regularly as the tides of the sea. Then the Merced overflows its banks, flooding the meadows, sometimes almost from wall to wall in some places, beginning to rise towards sundown just when the streams on the fountains are beginning to diminish, the difference in time of the daily rise and fall being caused by the distance the upper flood streams have to travel before reaching the Valley. In the warmest weather they seem fairly to shout for joy and clash their upleaping waters together like clapping of hands; racing down the cañons with white manes flying in glorious exuberance of strength, compelling huge, sleeping boulders to wake up and join in their dance and song, to swell their exulting chorus.

In early summer, after the flood season, the Yosemite streams are in their prime, running crystal clear, deep and full but not overflowing their banks—about as deep through the night as the day, the difference in volume so marked in spring being now too slight to be noticed. Nearly all the weather is cloudless and everything is at its brightest—lake, river, garden and forest with all their life. Most of the plants are in full flower. The blessed ouzels have built their mossy huts and are now singing their best songs with the streams.

In tranquil, mellow autumn, when the year's work is about done and the fruits are ripe, birds and seeds out of their nests, and all the landscape is glowing like a benevolent countenance, then the streams are at their lowest ebb, with scarce a memory left of their wild spring floods. The small tributaries that do not reach back to the lasting snow fountains of the summit peaks shrink to whispering, tinkling currents. After the snow is gone from the basins, excepting occasional thunder-showers, they are now fed only by small springs whose waters are mostly evaporated in passing over miles of warm pavements, and in feeling their way slowly from pool to pool through the midst of boulders and sand. Even the main rivers are so low they may easily be forded, and their grand falls and cascades, now gentle and approachable, have waned to sheets of embroidery.

IV

SNOW-BANNERS

BUT IT IS on the mountain-tops, when they are laden with loose, dry snow and swept by a gale from the north, that the most magnificent storm scenery is displayed. The peaks along the axis of the Range are then decorated with resplendent banners, some of them more than a mile long, shining, streaming, waving with solemn exuberant enthusiasm as if celebrating some surpassingly glorious event.

The snow of which these banners are made falls on the high Sierra in most extravagant abundance, sometimes to a depth of fifteen or twenty feet, coming from the fertile clouds not in large tangled flakes such as one oftentimes sees in Yosemite, seldom even in complete crystals, for many of the starry blossoms fall before they are ripe, while most of those that attain perfect development as six-petaled flowers are more or less broken by glinting and chafing against one another on the way down to their work. This dry frosty snow is prepared for the grand banner-waving celebrations by the action of the wind. Instead of at once finding rest like that which falls into the tranquil depths of the forest, it is shoved and rolled and beaten against boulders and out-jutting rocks, swirled in pits and hollows like sand in river pot-holes, and ground into sparkling dust. And when storm-winds find this snow-dust in a loose condition on the slopes above the timber line they toss it back into the sky and sweep it onward from peak to peak in the form of smooth regular banners, or in cloudy drifts, according to the velocity and direction of the wind, and the conformation of the slopes over which it is driven. While thus flying through the air a small portion escapes from the mountains to the sky as vapor; but far the greater part is at length locked fast in bossy overcurling cornices along the ridges, or in stratified sheets in the glacier cirques, some of it to replenish the small residual glaciers and remain silent and rigid for centuries before it is finally melted and sent singing down home to the sea.

But, though snow-dust and storm-winds abound on the mountains, regular shapely banners are, for causes we shall presently see, seldom produced. During the five winters that I spent in Yosemite

I made many excursions to high points above the walls in all kinds of weather to see what was going on outside; from all my lofty outlooks I saw only one banner-storm that seemed in every way perfect. This was in the winter of 1873, when the snow-laden peaks were swept by a powerful norther. I was awakened early in the morning by a wild storm-wind and of course I had to make haste to the middle of the Valley to enjoy it. Rugged torrents and avalanches from the main wind-flood overhead were roaring down the side cañons and over the cliffs, arousing the rocks and the trees and the streams alike into glorious hurrahing enthusiasm, shaking the whole Valley into one huge song. Yet inconceivable as it must seem even to those who love all Nature's wildness, the storm was telling its story on the mountains in still grander characters.

A WONDERFUL WINTER SCENE

I had long been anxious to study some points in the structure of the ice-hill at the foot of the Upper Yosemite Fall, but, as I have already explained, blinding spray had hitherto prevented me from getting sufficiently near it. This morning the entire body of the Fall was oftentimes torn into gauzy strips and blown horizontally along the face of the cliff, leaving the ice-hill dry; and while making my way to the top of Fern Ledge to seize so favorable an opportunity to look down its throat, the peaks of the Merced group came in sight over the boulder of the South Dome, each waving a white glowing banner against the dark blue sky, as regular in form and firm and fine in texture as if it were made of silk. So rare and splendid a picture, of course, smothered everything else and I at once began to scramble and wallow up the snow-choked Indian Cañon to a ridge about 8000 feet high, commanding a general view of the main summits along the axis of the Range, feeling assured I should find them bannered still more gloriously; nor was I in the least disappointed. I reached the top of the ridge in four or five hours, and through an opening in the woods the most imposing wind-storm effect I ever beheld came full in sight; unnumbered mountains rising sharply into the cloudless sky, their bases solid white, their sides plashed with snow, like ocean rocks with foam, and on every summit a magnificent silvery banner, from two thousand to six thousand feet in height, slender at the point of attachment, and widening gradually until about a thousand or fifteen hundred feet in breadth, and as shapely and as substantial looking in texture as the banners of the finest silk, all streaming and waving free and clear in the sun-glow with nothing to blur the sublime picture they made.

Fancy yourself standing beside me on this Yosemite Ridge. There is a strange garish glitter in the air and the gale drives wildly overhead, but you feel nothing of its violence, for you are looking out through a sheltered opening in the woods, as through a meadow. In the immediate foreground there is a forest of silver firs, their foliage warm yellow-green, and the snow beneath them is strewn with their plumes, plucked off by the storm; and beyond a broad, ridgy, cañon-furrowed, dome-dotted middle ground, darkened here and there with belts of pines, you behold the lofty snow-laden mountains in glorious array, waving their banners with jubilant enthusiasm as if shouting aloud for joy. They are twenty miles away, but you would not wish them nearer, for every feature is distinct, and the whole wonderful show is seen in its right proportions, like a painting on the sky.

And now after this general view, mark how sharply the ribs and buttresses and summits of the mountains are defined, excepting the portions veiled by the banners; how gracefully and nobly the banners are waving in accord with the throbbing of the wind-flood; how trimly each is attached to the very summit of its peak like a streamer at a mast-head; how bright and glowing white they are, and how finely their fading fringes are penciled on the sky! See how solid white and opaque they are at the point of attachment and how filmy and translucent toward the end, so that the parts of the peaks past which they are streaming look dim as if seen through a veil of ground glass. And see how some of the longest of the banners on the highest peaks are streaming perfectly free from peak to peak across intervening notches or passes, while others overlap and partly hide one another.

As to their formation, we find that the main causes of the wondrous beauty and perfection of those we are looking at are the favorable direction and force of the wind, the abundance of snow-dust, and the form of the north sides of the peaks. In general, the north sides are concave in both their horizontal and vertical sections, having been sculptured into this shape by the residual glaciers that lingered in the protecting northern shadows, while the sun-beaten south sides, having never been subjected to this kind of glaciation, are convex or irregular. It is essential, therefore, not only that the wind should move with great velocity and steadiness to supply a sufficiently copious and continuous stream of snow-dust, but that it should come from the north. No perfect banner is ever hung on the Sierra peaks by the south wind. Had the gale to-day blown from the south, leaving the other conditions unchanged, only swirling, interfering, cloudy drifts would have been produced; for the snow, instead of being spouted straight up and over the tops of the peaks in condensed currents to be drawn out as streamers, would have been driven over the convex southern slopes from peak to peak like white pearly fog.

It appears, therefore, that shadows in great part determine not only the forms of lofty ice mountains, but also those of the snow-banners that the wild winds hang up on them.

EARTHQUAKE STORMS

The avalanche taluses, leaning against the walls at intervals of a mile or two, are among the most striking and interesting of the secondary features of the Valley. They are from about three to five hundred feet high, made up of huge, angular, well-preserved, unshifting boulders, and instead of being slowly weathered from the cliffs like ordinary taluses, they were all formed suddenly and simultaneously by a great earthquake that occurred at least three centuries ago. And though thus hurled into existence in a few seconds or minutes, they are the least changeable of all the Sierra soil-beds. Excepting those which were launched directly into the channels of swift rivers, scarcely one of their wedged and interlacing boulders has moved since the day of their creation; and though mostly made up of huge blocks of granite, many of them from ten to fifty feet cube, weighing thousands of tons with only a few small chips, trees and shrubs make out to live and thrive on them and even delicate herbaceous plants—draperia, collomia, zauschneria, etc., soothing and coloring their wild rugged slopes with gardens and groves.

I was long in doubt on some points concerning the origin of these taluses. Plainly enough they were derived from the cliffs above them, because they are of the size of scars on the wall, the rough angular surface of which contrasts with the rounded, glaciated, unfractured parts. It was plain, too, that instead of being made up of material slowly and gradually weathered from the cliffs like ordinary taluses, almost every one of them had been formed suddenly in a single avalanche, and had not been increased in size during the last three or four centuries, for trees three or four hundred years old are growing on them, some standing at the top close to the wall without a bruise or broken branch, showing that scarcely a single boulder had ever fallen among them. Furthermore, all these taluses throughout the Range seemed by the trees and lichens growing on them to be of the same age. All the phenomena thus pointed straight to a grand ancient earthquake. But for years I left the question open, and went on from cañon to cañon, observing again and again; measuring the heights of taluses throughout the Range on both flanks, and the variations in the angles of their surface slopes; studying the way their boulders had been assorted and related and brought to rest, and their correspondence in size with the cleavage joints of the cliffs from whence they were derived, cautious about making up my mind. But at last all doubt as to their formation vanished.

At half-past two o'clock of a moonlit morning in March, I was awakened by a tremendous earthquake, and though I had never before enjoyed a storm of this sort, the strange thrilling motion could not be mistaken, and I ran out of my cabin, both glad and frightened, shouting, "A noble earthquake! A noble earthquake!" feeling sure I was going to learn something. The shocks were so violent and varied, and succeeded one another so closely, that I had to balance myself carefully in walking as if on the deck of a ship among waves, and it seemed impossible that the high cliffs of the Valley could escape being shattered. In particular, I feared that the sheer-fronted Sentinel Rock, towering above my cabin, would be shaken down, and I took shelter back of a large yellow pine, hoping that it might protect me from at least the smaller outbounding boulders. For a minute or two the shocks became more and more violent—flashing horizontal thrusts mixed with a few twists and battering, explosive, upheaving jolts,—as if Nature were wrecking her Yosemite temple, and getting ready to build a still better one.

I was now convinced before a single boulder had fallen that earthquakes were the talus-makers and positive proof soon came. It was a calm moonlight night, and no sound was heard for the first minute or so, save low, muffled, underground, bubbling rumblings, and the whispering and rustling of the agitated trees, as if Nature were holding her breath. Then, suddenly, out of the strange silence and strange motion there came a tremendous roar. The Eagle Rock on the south wall, about a half a mile up the Valley, gave way and I saw it falling in thousands of the great boulders I had so long been studying, pouring to the Valley floor in a free curve luminous from friction, making a terribly sublime spectacle—an arc of glowing, passionate fire, fifteen hundred feet span, as true in form and as serene in beauty as a rainbow in the midst of the stupendous, roaring rock-storm. The sound was so tremendously deep and broad and earnest, the whole earth like a living creature seemed to have at last found a voice and to be calling to her sister planets. In trying to tell something of the size of this awful sound it seems to me that if all the thunder of all the storms I had ever heard were condensed into one roar it would not equal this rock-roar at the birth of a mountain talus. Think, then, of the roar that arose to heaven at the simultaneous birth of all the thousands of ancient cañon-taluses throughout the length and breadth of the Range!

The first severe shocks were soon over, and eager to examine the new-born talus I ran up the Valley in the moonlight and climbed upon it before the huge blocks, after their fiery flight, had come to complete rest. They were slowly settling into their places, chafing, grating against one another, groaning, and whispering; but no motion was visible except in a stream of small fragments pattering down the face of the cliff. A cloud of dust particles, lighted by the moon, floated out across the whole breadth of the Valley, forming a ceiling that lasted until after sunrise, and the air was filled with the odor of crushed Douglas spruces from a grove that had been mowed down and mashed like weeds.

After the ground began to calm I ran across the meadow to the river to see in what direction it was flowing and was glad to find that *down* the Valley was still down. Its waters were muddy from portions of its banks having given way, but it was flowing around its curves and over its ripples and shallows with ordinary tones and gestures. The mud would soon be cleared away and the raw slips on the banks would be the only visible record of the shaking it suffered.

The Upper Yosemite Fall, glowing white in the moonlight,

seemed to know nothing of the earthquake, manifesting no change in form or voice, as far as I could see or hear.

After a second startling shock, about half-past three o'clock, the ground continued to tremble gently, and smooth, hollow rumbling sounds, not always distinguishable from the rounded, bumping, explosive tones of the falls, came from deep in the mountains in a northern direction.

The few Indians fled from their huts to the middle of the Valley, fearing that angry spirits were trying to kill them; and, as I afterward learned, most of the Yosemite tribe, who were spending the winter at their village on Bull Creek forty miles away, were so terrified that they ran into the river and washed themselves,—getting themselves clean enough to say their prayers, I suppose, or to die. I asked Dick, one of the Indians with whom I was acquainted, "What made the ground shake and jump so much?" He only shook his head and said, "No good. No good," and looked appealingly to me to give him hope that his life was to be spared.

In the morning I found the few white settlers assembled in front of the old Hutchings Hotel comparing notes and meditating flight to the lowlands, seemingly as sorely frightened as the Indians. Shortly after sunrise a low, blunt, muffled rumbling, like distant thunder, was followed by another series of shocks, which, though not nearly so severe as the first, made the cliffs and domes tremble like jelly, and the big pines and oaks thrill and swish and wave their branches with startling effect. Then the talkers were suddenly hushed, and the solemnity on their faces was sublime. One in particular of these winter neighbors, a somewhat speculative thinker with whom I had often conversed, was a firm believer in the cataclysmic origin of the Valley; and I now jokingly remarked that his wild tumble-down-and-engulfment hypothesis might soon be proved, since these underground rumblings and shakings might be the forerunners of another Yosemite-making cataclysm, which would perhaps double the depth of the Valley by swallowing the floor, leaving the ends of the roads and trails dangling three or four thousand feet in the air. Just then came the third series of shocks, and it was fine to see how awfully silent and solemn he became. His belief in the existence of a mysterious abyss, into which the suspended floor of the Valley and all the domes and battlements of the walls might at any moment go roaring down, mightily troubled him. To diminish his fears and laugh him into something like reasonable faith, I said, "Come, cheer up; smile a little and clap your hands, now that kind Mother Earth is trotting us on her knee to amuse us and make us good." But the well-meant joke seemed irreverent and utterly failed, as if only prayerful terror could rightly belong to the wild beauty-making business. Even after all the heavier shocks were over I could do nothing to reassure him. On the contrary, he handed me the keys of his little store to keep, saying that with a companion of like mind he was going to the lowlands to stay until the fate of poor, trembling Yosemite was settled. In vain I rallied them on their fears, calling attention to the strength of the granite walls of our Valley home, the very best and solidest masonry in the world, and less likely to collapse and sink than the sedimentary lowlands to which they were looking for safety; and saying that in any case they sometime would have to die, and so grand a burial was not to be slighted. But they were too seriously panic-stricken to get comfort from anything I could say.

During the third severe shock the trees were so violently shaken that the birds flew out with frightened cries. In particular, I noticed two robins flying in terror from a leafless oak, the branches of which swished and quivered as if struck by a heavy battering-ram. Exceedingly interesting were the flashing and quivering of the elastic needles of the pines in the sunlight and the waving up and down of the branches while the trunks stood rigid. There was no swaying, waving or swiveling as in wind-storms, but quick, quivering jerks, and at times the heavy tasseled branches moved as if they had all been pressed down against the trunk and suddenly let go, to spring up and vibrate until they came to rest again. Only the owls seemed to be undisturbed. Before the rumbling echoes had died away a hollow-voiced owl began to hoot in philosophical tranquillity from near the edge of the new talus as if nothing extraordinary had occurred, although, perhaps, he was curious to know what all the noise was about. His "hoot-too-hoot-too-whoo" might have meant, "what's a' the steer, kimmer?"

It was long before the Valley found perfect rest. The rocks trembled more or less every day for over two months, and I kept a bucket of water on my table to learn what I could of the movements. The blunt thunder in the depths of the mountains was usually followed by sudden jarring, horizontal thrusts from the northward, often succeeded by twisting, upjolting movements. More than a month after the first great shock, when I was standing on a fallen tree up the Valley near Lamon's winter cabin, I heard a distinct bubbling thunder from the direction of Tenaya Cañon. Carlo, a large intelligent St. Bernard dog standing beside me seemed greatly astonished, and looked intently in that direction with mouth open and uttered a low *Wouf!* as if saying, "What's that?" He must have known that it was not thunder, though like it. The air was perfectly still, not the faintest breath of wind perceptible, and a fine, mellow, sunny hush pervaded everything, in the midst of which came that subterranean thunder. Then, while we gazed and listened, came the corresponding shocks, distinct as if some mighty hand had shaken the ground. After the sharp horizontal jars died away, they were followed by a gentle rocking and undulating of the ground so distinct that Carlo looked at the log on which he was standing to see who was shaking it. It was the season of flooded meadows and the pools about me, calm as sheets of glass, were suddenly thrown into low ruffling waves.

Judging by its effects, this Yosemite, or Inyo earthquake, as it is sometimes called, was gentle as compared with the one that gave rise to the grand talus system of the Range and did so much for the cañon scenery. Nature, usually so deliberate in her operations, then created, as we have seen, a new set of features, simply by giving the mountains a shake—changing not only the high peaks and cliffs, but the streams. As soon as these rock avalanches fell, the streams began to sing new songs; for in many places thousands of boulders were hurled into their channels, roughening and half-damming them, compelling the waters to surge and roar in rapids where before they glided smoothly. Some of the streams were completely dammed; driftwood, leaves, etc., gradually filling the interstices between the boulders, thus giving rise to lakes and level reaches; and these again, after being gradually filled in, were changed to meadows, through which the streams are now silently meandering; while at the same time some of the taluses took the places of old meadows and groves. Thus rough places were made smooth, and smooth places rough. But, on the whole, by what at first sight seemed pure confounded confusion and ruin, the landscapes were enriched; for gradually every talus was covered with groves and gardens, and made a finely proportioned and ornamental base for the cliffs. In this work of beauty, every boulder is prepared and measured and put in its place more thoughtfully than are the stones of temples. If for a moment you are inclined to regard these taluses as mere draggled, chaotic dumps, climb to the top of one of them, and run down without any haggling, puttering hesitation, boldly jumping from boulder to boulder with even speed. You will then find your feet playing a tune, and quickly discover the music and poetry of these magnificent rock piles—a fine lesson; and all Nature's wildness tells the same story—the shocks and outbursts of earthquakes, volcanoes, geysers, roaring, thundering waves and floods, the silent uprush of sap in plants, storms of every sort—each and all are the orderly beauty-making love-beats of Nature's heart.

V

THE TREES OF THE VALLEY

THE MOST INFLUENTIAL of the Valley trees is the yellow pine (*Pinus ponderosa*). It attains its noblest dimensions on beds of water-washed, coarsely-stratified moraine material, between the talus slopes and meadows, dry on the surface, well-watered below and where not too closely assembled in groves the branches reach nearly to the ground, forming grand spires 200 to 220 feet in height. The largest that I have measured is standing alone almost opposite the Sentinel Rock, or a little to the westward of it. It is a little over eight feet in diameter and about 220 feet high. Climbing these grand trees, especially when they are waving and singing in worship in wind-storms, is a glorious experience. Ascending from the lowest branch to the topmost is like stepping up stairs through a blaze of white light, every needle thrilling and shining as if with religious ecstasy.

Unfortunately there are but few sugar pines in the Valley, though in the King's yosemite they are in glorious abundance. The incense cedar (*Libocedrus decurrens*) with cinnamon-colored bark and yellow-green foliage is one of the most interesting of the Yosemite trees. Some of them are 150 feet high, from six to ten feet in diameter, and they are never out of sight as you saunter among the yellow pines. Their bright brown shafts and towers of flat, frond-like branches make a striking feature of the landscapes throughout all the seasons. In midwinter, when most of the other trees are asleep, this cedar puts forth its flowers in millions,—the pistillate pale green and inconspicuous, but the staminate bright yellow, tingeing all the branches and making the trees as they stand in the snow look like gigantic goldenrods. The branches, outspread in flat plumes and beautifully fronded, sweep gracefully downward and outward, except those near the top, which aspire; the lowest, especially in youth and middle age, droop to the ground, overlapping one another, shedding off rain and snow like shingles, and making fine tents for birds and campers. This tree frequently lives more than a thousand years and is well worthy its place beside the great pines and the Douglas spruce.

The two largest specimens I know of the Douglas spruce, about eight feet in diameter, are growing at the foot of the Liberty Cap near the Nevada Fall, and on the terminal moraine of the small residual glacier that lingered in the shady Illilouette Cañon.

After the conifers, the most important of the Yosemite trees are the oaks, two species; the California live-oak (*Quercus agrifolia*), with black trunks, reaching a thickness of from four to nearly seven feet, wide spreading branches and bright deeply-scalloped leaves. It occupies the greater part of the broad sandy flats of the upper end of the Valley, and is the species that yields the acorns so highly prized by the Indians and woodpeckers.

The other species is the mountain live-oak, or gold-cup oak (*Quercus chrysolepis*), a sturdy mountaineer of a tree, growing mostly on the earthquake taluses and benches of the sunny north wall of the Valley. In tough, unwedgeable, knotty strength, it is the oak of oaks, a magnificent tree.

The largest and most picturesque specimen in the Valley is near the foot of the Tenaya Fall, a romantic spot seldom seen on account of the rough trouble of getting to it. It is planted on three huge boulders and yet manages to draw sufficient moisture and food from this craggy soil to maintain itself in good health. It is twenty feet in circumference, measured above a large branch between three and four feet in diameter that has been broken off. The main knotty trunk seems to be made up of craggy granite boulders like those on which it stands, being about the same color as the mossy, lichened boulders and about as rough. Two moss-lined caves near the ground open back into the trunk, one on the north side, the other on the west, forming picturesque, romantic seats. The largest of the main branches is eighteen feet and nine inches in circumference, and some of the long pendulous branchlets droop over the stream at the foot of the fall where it is gray with spray. The leaves are glossy yellow-green, ever in motion from the wind from the fall. It is a fine place to dream in, with falls, cascades, cool rocks lined with hypnum three inches thick; shaded with maple, dogwood, alder, willow; grand clumps of chain-ferns where no hand may touch them; light filtering through translucent leaves; oaks fifty feet high; lilies eight feet high in a filled lake basin near by, and the finest libocedrus groves and tallest ferns and goldenrods.

In the main river cañon below the Vernal Fall and on the shady south side of the Valley there are a few groves of the silver fir (*Abies concolor*), and superb forests of the magnificent species around the rim of the Valley.

On the tops of the domes is found the sturdy, storm-enduring red cedar (*Juniperus occidentalis*). It never makes anything like a forest here, but stands out separate and independent in the wind, clinging by slight joints to the rock, with scarce a handful of soil in sight of it, seeming to depend chiefly on snow and air for nourishment, and yet it has maintained tough health on this diet for two thousand years or more. The largest hereabouts are from five to six feet in diameter and fifty feet in height.

The principal river-side trees are poplar, alder, willow, broad-leaved maple, and Nuttall's flowering dogwood. The poplar (*Populus trichocarpa*), often called balm-of-Gilead from the gum on its buds, is a tall tree, towering above its companions and gracefully embowering the banks of the river. Its abundant foliage turns bright yellow in the fall, and the Indian-summer sunshine sifts through it in delightful tones over the slow-gliding waters when they are at their lowest ebb.

Some of the involucres of the flowering dogwood measure six to eight inches in diameter, and the whole tree when in flower looks as if covered with snow. In the spring when the streams are in flood it is the whitest of trees. In Indian summer the leaves become bright crimson, making a still grander show than the flowers.

The broad-leaved maple and mountain maple are found mostly in the cool cañons at the head of the Valley, spreading their branches in beautiful arches over the foaming streams.

Scattered here and there are a few other trees, mostly small—the mountain mahogany, cherry, chestnut-oak, and laurel. The California nutmeg (*Torreya californica*), a handsome evergreen, belonging to the yew family, forms small groves near the cascades a mile or two below the foot of the Valley.

VI

THE FOREST TREES IN GENERAL

FOR THE USE of the ever-increasing number of Yosemite visitors who make extensive excursions into the mountains beyond the Valley, a sketch of the forest trees in general will probably be found useful. The different species are arranged in zones and sections, which brings the forest as a whole within the comprehension of every observer. These species are always found as controlled by the climates of different elevations, by soil and by the comparative strength of each species in taking and holding possession of the ground; and so appreciable are these relations the traveler need never be at a loss in determining within a few hundred feet his elevation above sea-level by the trees alone; for, notwithstanding some of the species range upward for several thousand feet and all pass one another more or less, yet even those species possessing the greatest vertical range are available in measuring the elevation; inasmuch as they take on new forms corresponding with variations and altitude. Entering the lower fringe of the forest composed of Douglas oaks and Sabine pines, the trees grow so far apart that not one-twentieth of the surface of the ground is in shade at noon. After advancing fifteen or twenty miles towards Yosemite and making an ascent of from two to three thousand feet you reach the lower margin of the main pine belt, composed of great sugar pine, yellow pine, incense cedar and sequoia. Next you come to the magnificent silver-fir belt and lastly to the upper pine belt, which sweeps up to the feet of the summit peaks in a dwarfed fringe, to a height of from ten to twelve thousand feet. That this general order of distribution depends on climate as affected by height above the sea, is seen at once, but there are other harmonies that become manifest only after observation and study. One of the most interesting of these is the arrangement of the forest in long curving bands, braided together into lace-like patterns in some places and outspread in charming variety. The key to these striking arrangements is the system of ancient glaciers; where they flowed the trees followed, tracing their courses along the sides of cañons, over ridges, and high plateaus. The cedar of Lebanon, said Sir Joseph Hooker, occurs upon one of the moraines of an ancient glacier. All the forests of the Sierra are growing upon moraines, but moraines vanish like the glaciers that make them. Every storm that falls upon them wastes them, carrying away their decaying, disintegrating material into new formations, until they are no longer recognizable without tracing their transitional forms down from the range still in process of formation,—in some places through those that are more and more ancient and more obscured by vegetation and all kinds of post-glacial weathering. It appears, therefore, that the Sierra forests indicate the extent and positions of ancient moraines as well as they do belts of climate.

One will have no difficulty in knowing the Nut Pine (*Pinus sabiniana*), for it is the first conifer met in ascending the Range from the west, springing up here and there among Douglas oaks and thickets of ceanothus and manzanita; its extreme upper limit being about 4000 feet above the sea, its lower about from 500 to 800 feet. It is remarkable for its loose, airy, wide-branching habit, and thin gray foliage. Full-grown specimens are from forty to fifty feet in height and from two to three feet in diameter. The trunk usually divides into three or four main branches about fifteen or twenty feet from the ground that, after bearing away from one another, shoot straight up and form separate summits. Their slender, grayish needles are from eight to twelve inches long, and inclined to droop, contrasting with the rigid, dark-colored trunk and branches. No other tree of my acquaintance so substantial in its body has foliage so thin and pervious to the light. The cones are from five to eight inches long and about as large in thickness; rich chocolate-brown in color and protected by strong, down-curving hooks which terminate the scales. Nevertheless the little Douglas squirrel can open them. Indians climb the trees like bears and beat off the cones or recklessly cut off the more fruitful branches with hatchets, while the squaws gather and roast them until the scales open sufficiently to allow the hard-shell seeds to be beaten out. The curious little *Pinus attenuata* is found at an elevation of from 1500 to 3000 feet, growing in close groves and belts. It is exceedingly slender and graceful, although trees that chance to stand alone send out very long, curved branches, making a striking contrast to the ordinary grove form. The foliage is of the same peculiar gray-green color as that of the nut pine, and is worn about as loosely, so that the body of the tree is scarcely obscured by it. At the age of seven or eight years it begins to bear cones in whorls on the main axis, and as they never fall off the trunk is soon picturesquely dotted with them. Branches also soon become fruitful. The average size of the tree is about thirty or forty feet in height and twelve to fourteen inches in diameter. The cones are about four inches long, and covered with a sort of varnish and gum, rendering them impervious to moisture.

No observer can fail to notice the admirable adaptation of this curious pine to the fire-swept regions where alone it is found. After a running fire has scorched and killed it the cones open and the ground beneath it is as one sown broadcast, with all the seeds ripened during its whole life. Then up spring a crowd of bright, hopeful seedlings, giving beauty for ashes in lavish abundance.

THE SUGAR PINE, KING OF PINE TREES

Of all the world's eighty or ninety species of pine trees, the Sugar Pine (*Pinus lambertiana*) is king, surpassing all others, not merely in size but in lordly beauty and majesty. In the Yosemite region it grows at an elevation of from 3000 to 7000 feet above the sea and attains most perfect development at a height of about 5000 feet. The largest specimens are commonly about 220 feet high and from six to eight feet in diameter four feet from the ground, though some grand old patriarch may be met here and there that has enjoyed six or eight centuries of storms and attained a thickness of ten or even twelve feet, still sweet and fresh in every fiber. The trunk is a remarkably smooth, round, delicately-tapered shaft, straight and regular as if turned in a lathe, mostly without limbs, purplish brown in color and usually enlivened with tufts of a yellow lichen. Toward the head of this magnificent column long branches sweep gracefully outward and downward, sometimes forming a palm-like crown, but far more impressive than any palm crown I ever beheld. The needles are about three inches long in fascicles of five, and arranged in rather close tassels at the ends of slender branchlets that clothe the long outsweeping limbs. How well they sing in the wind, and how strikingly harmonious an effect is made by the long cylindrical cones, depending loosely from the ends of the long branches! The cones are about fifteen to eighteen inches long, and three in diameter; green, shaded with dark purple on their sunward sides. They are ripe in

September and October of the second year from the flower. Then the flat, thin scales open and the seeds take wing, but the empty cones become still more beautiful and effective as decorations, for their diameter is nearly doubled by the spreading of the scales, and their color changes to yellowish brown while they remain, swinging on the tree all the following winter and summer, and continue effectively beautiful even on the ground many years after they fall. The wood is deliciously fragrant, fine in grain and texture and creamy yellow, as if formed of condensed sunbeams. The sugar from which the common name is derived is, I think, the best of sweets. It exudes from the heart-wood where wounds have been made by forest fires or the ax, and forms irregular, crisp, candy-like kernels of considerable size, something like clusters of resin beads. When fresh it is white, but because most of the wounds on which it is found have been made by fire the sap is stained and a hardened sugar becomes brown. Indians are fond of it, but on account of its laxative properties only small quantities may be eaten. No tree lover will ever forget his first meeting with the sugar pine. In most pine trees there is the sameness of expression which to most people is apt to become monotonous, for the typical spiral form of conifers, however beautiful, affords little scope for appreciable individual character. The sugar pine is as free from conventionalities as the most picturesque oaks. No two are alike, and though they toss out their immense arms in what might seem extravagant gestures they never lose their expression of serene majesty. They are the priests of pines and seem ever to be addressing the surrounding forest. The yellow pine is found growing with them on warm hillsides, and the silver fir on cool northern slopes; but, noble as these are, the sugar pine is easily king, and spreads his arms above them in blessing while they rock and wave in sign of recognition. The main branches are sometimes forty feet long, yet persistently simple, seldom dividing at all, excepting near the end; but anything like a bare cable appearance is prevented by the small, tasseled branchlets that extend all around them; and when these superb limbs sweep out symmetrically on all sides, a crown sixty or seventy feet wide is formed, which, gracefully poised on the summit of the noble shaft, is a glorious object. Commonly, however, there is a preponderance of limbs toward the east, away from the direction of the prevailing winds.

Although so unconventional when full-grown, the sugar pine is a remarkably proper tree in youth—a strict follower of coniferous fashions—slim, erect, with leafy branches kept exactly in place, each tapering in outline and terminating in a spirey point. The successive forms between the cautious neatness of youth and the bold freedom of maturity offer a delightful study. At the age of fifty or sixty years, the shy, fashionable form begins to be broken up. Specialized branches push out and bend with the great cones, giving individual character, that becomes more marked from year to year. Its most constant companion is the yellow pine. The Douglas spruce, libocedrus, sequoia, and the silver fir are also more or less associated with it; but on many deep-soiled mountainsides, at an elevation of about 5000 feet above the sea, it forms the bulk of the forest, filling every swell and hollow and down-plunging ravine. The majestic crowns, approaching each other in bold curves, make a glorious canopy through which the tempered sunbeams pour, silvering the needles, and gilding the massive boles and the flowery, park-like ground into a scene of enchantment.

On the most sunny slopes the white-flowered, fragrant chamaebatia is spread like a carpet, brightened during early summer with the crimson sarcodes, the wild rose, and innumerable violets and gilias. Not even in the shadiest nooks will you find any rank, untidy weeds or unwholesome darkness. In the north sides of ridges the boles are more slender, and the ground is mostly occupied by an underbrush of hazel, ceanothus, and flowering dogwood, but not so densely as to prevent the traveler from sauntering where he will; while the crowning branches are never impenetrable to the rays of the sun, and never so interblended as to lose their individuality.

THE YELLOW OR SILVER PINE

The Silver Pine (*Pinus ponderosa*), or Yellow Pine, as it is commonly called, ranks second among the pines of the Sierra as a lumber tree, and almost rivals the sugar pine in stature and nobleness of port. Because of its superior powers of enduring variations of climate and soil, it has a more extensive range than any other conifer growing on the Sierra. On the western slope it is first met at an elevation of about 2000 feet, and extends nearly to the upper limit of the timber line. Thence, crossing the range by the lowest passes, it descends to the eastern base, and pushes out for a considerable distance into the hot, volcanic plains, growing bravely upon well-watered moraines, gravelly lake basins, climbing old volcanoes and dropping ripe cones among ashes and cinders.

The average size of full-grown trees on the western slope, where it is associated with the sugar pine, is a little less than 200 feet in height and from five to six feet in diameter, though specimens considerably larger may easily be found. Where there is plenty of free sunshine and other conditions are favorable, it presents a striking contrast in form to the sugar pine, being a symmetrical spire, formed of a straight round trunk, clad with innumerable branches that are divided over and over again. Unlike the Yosemite form about one-half of the trunk is commonly branchless, but where it grows at all close three-fourths or more is naked, presenting then a more slender and elegant shaft than any other tree in the woods. The bark is mostly arranged in massive plates, some of them measuring four or five feet in length by eighteen inches in width, with a thickness of three or four inches, forming a quite marked and distinguishing feature. The needles are of a fine, warm, yellow-green color, six to eight inches long, firm and elastic, and crowded in handsome, radiant tassels on the upturning ends of the branches. The cones are about three or four inches long, and two and a half wide, growing in close, sessile clusters among the leaves.

The species attains its noblest form in filled-up lake basins, especially in those of the older yosemites, and as we have seen, so prominent a part does it form of their groves that it may well be called the Yosemite Pine.

The Jeffrey variety attains its finest development in the northern portion of the Range, in the wide basins of the McCloud and Pitt Rivers, where it forms magnificent forests scarcely invaded by any other tree. It differs from the ordinary form in size, being only about half as tall, in its redder and more closely-furrowed bark, grayish-green foliage, less divided branches, and much larger cones; but intermediate forms come in which make a clear separation impossible, although some botanists regard it as a distinct species. It is this variety of ponderosa that climbs storm-swept ridges above, and wanders out among the volcanoes of the Great Basin. Whether exposed to extremes of heat or cold, it is dwarfed like many other trees, and becomes all knots and angles, wholly unlike the majestic forms we have been sketching. Old specimens, bearing cones about as big as pineapples, may sometimes be found clinging to rifted rocks at an elevation of 7000 or 8000 feet, whose highest branches scarce reach above one's shoulders.

I have often feasted on the beauty of these noble trees when they were towering in all their winter grandeur, laden with snow—one mass of bloom; in summer, too, when the brown, staminate clusters hang thick among the shimmering needles, and the big purple burrs are ripening in the mellow light; but it is during cloudless windstorms that these colossal pines are most impressively beautiful. Then they bow like willows, their leaves streaming forward all in one direction, and, when the sun shines upon them at the required angle, entire groves glow as if every leaf were burnished silver. The fall of tropic light on the crown of a palm is a truly glorious spectacle, the fervid sun-flood breaking upon the glossy leaves in long lance-rays, at the foot of an enthusiastic cataract, like mountain water among boulders. But to me there is something more impressive in the fall of light upon these noble, silver pine pillars: it is beaten to the finest dust and shed off in myriads of minute sparkles that seem to radiate from the very heart of the tree, as if like rain, falling upon fertile soil, it had been absorbed to reappear in flowers of light. This species also gives forth the finest wind music. After listening to it in all kinds of winds, night and day, season after season, I think I could approximate to my position on the mountain by this pine music alone. If you would catch the tone of separate needles climb a tree in breezy weather. Every needle is carefully tempered and gives forth no uncertain sound, each standing out with no interference excepting during heavy gales; then you may detect the click of one needle from another, readily distinguishable from the free wind-like hum.

When a sugar pine and one of this species equal in size are observed together, the latter is seen to be more simple in manners, more lively and graceful, and its beauty is of a kind more easily appreciated; on the other hand it is less dignified and original in demeanor. The yellow pine seems ever eager to shoot aloft, higher and higher. Even while it is drowsing in autumn sun-gold you may still detect a skyward aspiration, but the sugar pine seems too unconsciously noble and too complete in every way to leave room for even a heavenward care.

THE DOUGLAS SPRUCE

The Douglas Spruce (*Pseudotsuga douglasii*) is one of the largest and longest-lived of the giants that flourish throughout the main pine belt, often attaining a height of nearly 200 feet, and a diameter of six or seven feet. Where the growth is not too close, the stout, spreading branches, covering more than half of the trunk, are hung with innumerable slender, drooping sprays, handsomely feathered with the short leaves which radiate at right angles all around them. This vigorous tree is ever beautiful, welcoming the mountain winds and the snow as well as the mellow summer light; and it maintains its youthful freshness undiminished from century to century through a thousand storms. It makes its finest appearance during the months of June and July, when the brown buds at the ends of the sprays swell and open, revealing the young leaves, which at first are bright yellow, making the tree appear as if covered with gay blossoms; while the pendulous bracted cones, three or four inches long, with their shell-like scales, are a constant adornment.

The young trees usually are assembled in family groups, each sapling exquisitely symmetrical. The primary branches are whorled regularly around the axis, generally in fives, while each is draped with long, feathery sprays that descend in lines as free and as finely drawn as those of falling water.

In Oregon and Washington it forms immense forests, growing tall and mast-like to a height of 300 feet, and is greatly prized as a lumber tree. Here it is scattered among other trees, or forms small groves, seldom ascending higher than 5500 feet, and never making what would be called a forest. It is not particular in its choice of soil: wet or dry, smooth or rocky, it makes out to live well on them all. Two of the largest specimens, as we have seen, are in Yosemite; one of these, more than eight feet in diameter, is growing on a moraine; the other, nearly as large, on angular blocks of granite. No other tree in the Sierra seems so much at home on earthquake taluses and many of these huge boulder-slopes are almost exclusively occupied by it.

THE INCENSE CEDAR

Incense Cedar, (*Libocedrus decurrens*), already noticed among the

Yosemite trees, is quite generally distributed throughout the pine belt without exclusively occupying any considerable area, or even making extensive groves. On the warmer mountain slopes it ascends to about 5000 feet, and reaches the climate most congenial to it at a height of about 4000 feet, growing vigorously at this elevation in all kinds of soil and, in particular, it is capable of enduring more moisture about its roots than any of its companions excepting only the sequoia.

Casting your eye over the general forest from some ridge-top you can identify it by the color alone of its spirey summits, a warm yellow-green. In its youth up to the age of seventy or eighty years, none of its companions forms so strictly tapered a cone from top to bottom. As it becomes older it oftentimes grows strikingly irregular and picturesque. Large branches push out at right angles to the trunk, forming stubborn elbows and shoot up parallel with the axis. Very old trees are usually dead at the top. The flat fragrant plumes are exceedingly beautiful: no waving fern-frond is finer in form and texture. In its prime the whole tree is thatched with them, but if you would see the libocedrus in all its glory you must go to the woods in midwinter when it is laden with myriads of yellow flowers about the size of wheat grains, forming a noble illustration of Nature's immortal virility and vigor. The mature cones, about three-fourths of an inch long, borne on the ends of the plumy branchlets, serve to enrich still more the surpassing beauty of this winter-blooming tree-goldenrod.

THE SILVER FIRS

We come now to the most regularly planted and most clearly defined of the main forest belts, composed almost exclusively of two Silver Firs—*Abies concolor* and *Abies magnifica*—extending with but little interruption 450 miles at an elevation of from 5000 to 9000 feet above the sea. In its youth the Silver Fir is a charmingly symmetrical tree with its flat plumy branches arranged in regular whorls around the whitish-gray axis which terminates in a stout, hopeful shoot, pointing straight to the zenith, like an admonishing finger. The leaves are arranged in two horizontal rows along branchlets that commonly are less than eight years old, forming handsome plumes, pinnated like the fronds of ferns. The cones are a grayish-green when ripe, cylindrical, from three to four inches long, and one and a half to two inches wide, and stand upright on the upper horizontal branches. Full-grown trees in favorable situations are usually about 200 feet high and five or six feet in diameter. As old age creeps on, the rough bark becomes rougher and grayer, the branches loose their exact regularity of form. Many that are snow-bent are broken off and the axis often becomes double or otherwise irregular from accidents to the terminal bud or shoot. Nevertheless, throughout all the vicissitudes of its three or four centuries of life, come what may, the noble grandeur of these species, however obscured, is never lost.

The magnificent Silver Fir, or California Red Fir (*Abies magnifica*) is the most symmetrical of all the Sierra giants, far surpassing its companion species in this respect and easily distinguished from it by the purplish-red bark, which is also more closely furrowed than that of the white, and by its larger cones, its more regularly whorled and fronded branches, and its shorter leaves, which grow all around the branches and point upward instead of being arranged in two horizontal rows. The branches are mostly whorled in fives, and stand out from the straight, red-purple bole in level, or in old trees in drooping, collars, every branch regularly pinnated like fern-fronds, making broad plumes, singularly rich and sumptuous-looking. The flowers are in their prime about the middle of June; the male red, growing on the underside of the branches in crowded profusion, giving a very rich color to all the trees; the female greenish-yellow, tinged with pink, standing erect on the upper side of the topmost branches, while the tufts of young leaves, about as brightly colored as those of the Douglas spruce, make another grand show, with cones mature in a single season from the flowers. When mature they are about six to eight inches long, three or four in diameter, covered with a fine gray down and streaked and beaded with transparent balsam, very rich and precious-looking, and stand erect like casts on the topmost branches. The inside of the cone is, if possible, still more beautiful. The scales and bracts are tinged with red and the seedlings are purple with bright iridescence. Both of the silver firs live between two and three centuries when the conditions about them are at all favorable. Some venerable patriarch may be seen heavily storm-marked, towering in severe majesty above the rising generation, with a protecting grove of hopeful saplings pressing close around his feet, each dressed with such loving care that not a leaf seems wanting. Other groups are made up of trees near the prime of life, nicely arranged as if Nature had carved them with discrimination from all the rest of the woods. It is from this tree, called Red Fir by the lumbermen, that mountaineers cut boughs to sleep on when they are so fortunate as to be within its limit. Two or three rows of the sumptuous plushy-fronded branches, overlapping along the middle, and a crescent of smaller plumes mixed to one's taste with ferns and flowers for a pillow, form the very best bed imaginable. The essence of the pressed leaves seems to fill every pore of one's body. Falling water makes a soothing hush, while the spaces between the grand spires afford noble openings through which to gaze dreamily into the starry sky. The fir woods are fine sauntering-grounds at almost any time of the year, but finest in autumn when the noble trees are hushed in the hazy light and drip with balsam; and the flying, whirling seeds, escaping from the ripe cones, mottle the air like flocks of butterflies. Even in the richest part of these unrivaled forests where so many noble trees challenge admiration we linger fondly among the colossal firs and extol their beauty again and again, as if no other tree in the world could henceforth claim our love. It is in these woods the great granite domes arise that are so striking and characteristic a feature of the Sierra. Here, too, we find the best of the garden-meadows full of lilies. A dry spot a little way back from the margin of a silver fir lily-garden makes a glorious campground, especially where the slope is toward the east with a view of the distant peaks along the summit of the Range. The tall lilies are brought forward most impressively like visitors by the light of your camp-fire and the nearest of the trees with their whorled branches tower above you like larger lilies and the sky seen through the garden-opening seems one vast meadow of white lily stars.

THE TWO-LEAVED PINE

The Two-Leaved Pine (*Pinus contorta*, var. *murrayana*), above the Silver Fir zone, forms the bulk of the alpine forests up to a height of from 8000 to 9500 feet above the sea, growing in beautiful order on moraines scarcely changed as yet by post-glacial weathering.

Compared with the giants of the lower regions this is a small tree, seldom exceeding a height of eighty or ninety feet. The largest I ever measured was ninety feet high and a little over six feet in diameter. The average height of mature trees throughout the entire belt is probably not far from fifty or sixty feet with a diameter of two feet. It is a well-proportioned, rather handsome tree with grayish-brown bark and crooked, much-divided branches which cover the greater part of the trunk, but not so densely as to prevent it being seen. The lower limbs, like those of most other conifers that grow in snowy regions, curve downward, gradually take a horizontal position about halfway up the trunk, then aspire more and more toward the summit. The short, rigid needles in fascicles of two are arranged in comparatively long cylindrical tassels at the ends of the tough upcurving branches. The cones are about two inches long, growing in clusters among the needles without any striking effect except while very young, when the flowers are of a vivid crimson color and the whole tree appears to be dotted with brilliant flowers. The staminate flowers are still more showy on account of their great abundance, often giving a reddish-yellow tinge to the whole mass of foliage and filling the air with pollen. No other pine on the Range is so regularly planted as this one, covering moraines that extend along the sides of the high rocky valleys for miles without interruption. The thin bark is streaked and sprinkled with resin as though it had been showered upon the forest like rain.

Therefore this tree more than any other is subject to destruction by fire. During strong winds extensive forests are destroyed, the flames leaping from tree to tree in continuous belts that go surging and racing onward above the bending wood like prairie-grass fires. During the calm season and Indian summer the fire creeps quietly along the ground, feeding on the needles and cones; arriving at the foot of a tree, the resin bark is ignited and the heated air ascends in a swift current, increasing in velocity and dragging the flames upward. Then the leaves catch, forming an immense column of fire, beautifully spired on the edges and tinted a rose-purple hue. It rushes aloft thirty or forty feet above the top of the tree, forming a grand spectacle, especially at night. It lasts, however, only a few seconds, vanishing with magical rapidity, to be succeeded by others along the fire-line at irregular intervals, tree after tree, up-flashing and darting, leaving the trunks and branches scarcely scarred. The heat, however, is sufficient to kill the tree and in a few years the bark shrivels and falls off. Forests miles in extent are thus killed and left standing, with the branches on, but peeled and rigid, appearing gray in the distance like misty clouds. Later the branches drop off, leaving a forest of bleached spires. At length the roots decay and the forlorn gray trunks are blown down during some storm and piled one upon another, encumbering the ground until, dry and seasoned, they are consumed by another fire and leave the ground ready for a fresh crop.

In sheltered lake-hollows, on beds of alluvium, this pine varies so far from the common form that frequently it could be taken for a distinct species, growing in damp sods like grasses from forty to eighty feet high, bending all together to the breeze and whirling in eddying gusts more lively than any other tree in the woods. I frequently found species fifty feet high less than five inches in diameter. Being so slender and at the same time clad with leafy boughs, it is often bent and weighed down to the ground when laden with soft snow; thus forming fine ornamental arches, many of them to last until the melting of the snow in the spring.

THE MOUNTAIN PINE

The Mountain Pine (*Pinus monticola*) is the noblest tree of the alpine zone—hardy and long-lived, towering grandly above its companions and becoming stronger and more imposing just where other species begin to crouch and disappear. At its best it is usually about ninety feet high and five or six feet in diameter, though you may find specimens here and there considerably larger than this. It is as massive and as suggestive of enduring strength as an oak. About two-thirds of the trunk is commonly free of limbs, but close, fringy tufts of spray occur nearly all the way down to the ground. On trees that occupy exposed situations near its upper limit the bark is deep reddish-brown and rather deeply furrowed, the main furrows running nearly parallel to each other and connected on the old trees by conspicuous cross-furrows. The cones are from four to eight inches long, smooth, slender, cylindrical and somewhat curved. They grow in clusters of from three to six or seven and become pendulous as they increase in weight. This species is nearly related to the sugar pine and, though not half so tall, it suggests its noble relative in the way that it extends its long branches in general habit. It is first met on the upper margin of the silver fir zone, singly, in what appears as chance situations without making much impression on the general forest. Continuing up through the forests of the two-leaved pine it begins to show its distinguishing characteristic in the most marked way at an elevation of about 10,000 feet, extending its tough, rather slender arms across the air, welcoming the storms and feeding on them and reaching sometimes to the grand old age of 1000 years.

THE WESTERN JUNIPER

The Juniper or Red Cedar (*Juniperus occidentalis*) is preëminently a rock tree, occupying the baldest domes and pavements in the upper silver fir and alpine zones, at a height of from 7000 to 9500 feet. In such situations, rooted in narrow cracks or fissures, where there is scarcely a handful of soil, it is frequently over eight feet in diameter and not much more in height. The tops of old trees are almost always dead, and large stubborn-looking limbs push out horizontally, most of them broken and dead at the end, but densely covered, and imbedded here and there with tufts or mounds of gray-green scale-like foliage. Some trees are mere storm-beaten stumps about as broad as long, decorated with a few leafy sprays, reminding one of the crumbling towers of old castles scantily draped with ivy. Its homes on bare, barren dome and ridge-top seem to have been chosen for safety against fire, for, on isolated mounds of sand and gravel free from grass and bushes on which fire could feed, it is often found growing tall and unscathed to a height of forty to sixty feet, with scarce a trace of the rocky angularity and broken limbs so characteristic a feature throughout the greater part of its range. It never makes anything like a forest; seldom even a grove. Usually it stands out separate and independent, clinging by slight joints to the rocks, living chiefly on snow and thin air and maintaining sound health on this diet for 2000 years or more. Every feature or every gesture it makes expresses steadfast, dogged endurance. The bark is of a bright cinnamon color and is handsomely braided and reticulated on thrifty trees, flaking off in thin, shining ribbons that are sometimes used by the Indians for tent matting. Its fine color and picturesqueness are appreciated by artists, but to me the juniper seems a singularly strange and taciturn tree. I have spent many a day and night in its company and always have found it silent and rigid. It seems to be a survivor of some ancient race, wholly unacquainted with its neighbors. Its broad stumpiness, of course, makes wind-waving or even shaking out of the question, but it is not this rocky rigidity that constitutes its silence. In calm, sun-days the sugar pine preaches like an enthusiastic apostle without moving a leaf. On level rocks the juniper dies standing and wastes insensibly out of existence like granite, the wind exerting about as little control over it, alive or dead, as it does over a glacier boulder.

I have spent a good deal of time trying to determine the age of these wonderful trees, but as all of the very old ones are honeycombed with dry rot I never was able to get a complete count of the largest. Some are undoubtedly more than 2000 years old, for though on deep moraine soil they grow about as fast as some of the pines, on bare pavements and smoothly glaciated, overswept ridges in the dome region they grow very slowly. One on the Starr King Ridge only two feet eleven inches in diameter was 1140 years old forty years ago. Another on the same ridge, only one foot seven and a half inches in diameter, had reached the age of 834 years. The first fifteen inches from the bark of the medium-size tree six feet in diameter, on the north Tenaya pavement, had 859 layers of wood. Beyond this the count was stopped by dry rot and scars. The largest I examined was thirty-three feet in girth, or nearly ten feet in diameter and, although I have failed to get anything like a complete count, I learned enough from this and many other specimens to convince me that most of the trees eight or ten feet thick, standing on pavements, are more than twenty centuries old rather than less. Barring accidents, for all I can see they would live forever; even when overthrown by avalanches, they refuse to lie at rest, lean stubbornly on their big branches as if anxious to rise, and while a single root holds to the rock, put forth fresh leaves with a grim, never-say-die expression.

THE MOUNTAIN HEMLOCK

As the juniper is the most stubborn and unshakeable of trees in the Yosemite region, the Mountain Hemlock (*Tsuga mertensiana*) is the most graceful and pliant and sensitive. Until it reaches a height of fifty or sixty feet it is sumptuously clothed down to the ground with drooping branches, which are divided again and again into delicate waving sprays, grouped and arranged in ways that are indescribably beautiful, and profusely adorned with small brown cones. The flowers also are peculiarly beautiful and effective; the female dark rich purple, the male blue, of so fine and pure a tone that the best azure of the mountain sky seems to be condensed in them. Though apparently the most delicate and feminine of all the mountain trees, it grows best where the snow lies deepest, at a height of from 9000 to 9500 feet, in hollows on the northern slopes of mountains and ridges. But under all circumstances, sheltered from heavy winds or in bleak exposure to them, well fed or starved, even at its highest limit, 10,500 feet above the sea, on exposed ridge-tops where it has to crouch and huddle close in low thickets, it still contrives to put forth its sprays and branches in forms of invincible beauty, while on moist, well-drained moraines it displays a perfectly tropical luxuriance of foliage, flowers and fruit. The snow of the first winter storm is frequently soft, and lodges in the dense leafy branches, weighing them down against the trunk, and the slender, drooping axis, bending lower and lower as the load increases, at length reaches the ground, forming an ornamental arch. Then, as storm succeeds storm and snow is heaped on snow, the whole tree is at last buried, not again to see the light of day or move leaf or limb until set free by the spring thaws in June or July. Not only the young saplings are thus carefully covered and put to sleep in the whitest of white beds for five or six months of the year, but trees thirty feet high or more. From April to May, when the snow by repeated thawing and freezing is firmly compacted, you may ride over the prostrate groves without seeing a single branch or leaf of them. No other of our alpine conifers so finely veils its strength; poised in thin, white sunshine, clad with branches from head to foot, it towers in unassuming majesty, drooping as if unaffected with the aspiring tendencies of its race, loving the ground, conscious of heaven and joyously receptive of its blessings, reaching out its branches like sensitive tentacles, feeling the light and reveling in it. The largest specimen I ever found was nineteen feet seven inches in circumference. It was growing on the edge of Lake Hollow, north of Mount Hoffmann, at an elevation of 9250 feet above the level of the sea, and was probably about a hundred feet in height. Fine groves of mature trees, ninety to a hundred feet in height, are growing near the base of Mount Conness. It is widely distributed from near the south extremity of the high Sierra northward along the Cascade Mountains of Oregon and Washington and the coast ranges of British Columbia to Alaska, where it was first discovered in 1827. Its northernmost limit, so far as I have observed, is in the icy fiords of Prince William Sound in latitude 61°, where it forms pure forests at the level of the sea, growing tall and majestic on the banks of glaciers. There, as in the Yosemite region, it is ineffably beautiful, the very loveliest of all the American conifers.

THE WHITE-BARK PINE

The Dwarf Pine, or White-Bark Pine (*Pinus albicaulis*), forms the extreme edge of the timber line throughout nearly the whole extent of the Range on both flanks. It is first met growing with the two-leaved pine on the upper margin of the alpine belt, as an erect tree from fifteen to thirty feet high and from one to two feet in diameter; thence it goes straggling up the flanks of the summit peaks, upon moraines or crumbling ledges, wherever it can get a foothold, to an elevation of from 10,000 to 12,000 feet, where it dwarfs to a mass of crumpled branches, covered with slender shoots, each tipped with a short, close-packed, leaf tassel. The bark is smooth and purplish, in some places almost white. The flowers are bright scarlet and rose-purple, giving a very flowery appearance little looked for in such a tree. The cones are about three inches long, an inch and a half in diameter, grow in rigid clusters, and are dark chocolate in color while young, and bear beautiful pearly-white seeds about the size of peas, most of which are eaten by chipmunks and the Clarke's crows. Pines are commonly regarded as sky-living trees that must necessarily aspire or die. This species forms a marked exception, crouching and creeping in compliance with the most rigorous demands of climate; yet enduring bravely to a more advanced age than many of its lofty relatives in the sun-lands far below it. Seen from a distance it would never be taken for a tree of any kind. For example, on Cathedral Peak there is a scattered growth of this pine, creeping like mosses over the roof, nowhere giving hint of an ascending axis. While, approached quite near, it still appears matty and healthy, and one experiences no difficulty in walking over the top of it, yet it is seldom absolutely prostrate, usually attaining a height of three or four feet with a main trunk, and with branches outspread above it, as if in ascending they had been checked by a ceiling against which they had been compelled to spread horizontally. The winter snow *is* a sort of ceiling, lasting half the year; while the pressed surface is made yet smoother by violent winds armed with cutting sand-grains that bear down any shoot which offers to rise much above the general level, and that carve the dead trunks and branches in beautiful patterns.

During stormy nights I have often camped snugly beneath the interlacing arches of this little pine. The needles, which have accumulated for centuries, make fine beds, a fact well known to other mountaineers, such as deer and wild sheep, who paw out oval hollows and lie beneath the larger trees in safe and comfortable concealment. This lowly dwarf reaches a far greater age than would be guessed. A specimen that I examined, growing at an elevation of 10,700 feet, yet looked as though it might be plucked up by the roots, for it was only three and a half inches in diameter and its topmost tassel reached hardly three feet above the ground. Cutting it half through and counting the annual rings with the aid of a lens, I found its age to be no less than 255 years. Another specimen about the same height, with a trunk six inches in diameter, I found to be 426 years old, forty years ago; and one of its supple branchlets hardly an eighth of an inch in diameter inside the bark, was seventy-five years old, and so filled with oily balsam and seasoned by storms that I tied it in knots like a whip-cord.

THE NUT PINE

In going across the Range from the Tuolumne River Soda Springs to Mono Lake one makes the acquaintance of the curious little Nut Pine, (*Pinus monophylla*). It dots the eastern flank of the Sierra to which it is mostly restricted in grayish bush-like patches, from the margin of the sage-plains to an elevation of from 7000 to 8000 feet. A more contented, fruitful and unaspiring conifer could not be conceived. All the species we have been sketching make departures more or less distant from the typical spire form, but none goes so far as this. Without any apparent cause it keeps near the ground, throwing out crooked, divergent branches like an orchard apple-tree, and seldom pushes a single shoot higher than fifteen or twenty feet above the ground.

The average thickness of the trunk is, perhaps, about ten or twelve inches. The leaves are mostly undivided, like round awls, instead of being separated, like those of other pines, into twos and threes and fives. The cones are green while growing, and are usually found over all the tree, forming quite a marked feature as seen against the bluish-gray foliage. They are quite small, only about two inches in length, and seem to have but little space for seeds; but when we come to open them, we find that about half the entire bulk of the cone is made up of sweet, nutritious nuts, nearly as large as hazelnuts. This is undoubtedly the most important food-tree on the Sierra, and furnishes the Mona, Carson, and Walker River Indians with more and better nuts than all the other species taken together. It is the Indian's own tree, and many a white man have they killed for cutting it down. Being so low, the cones are readily beaten off with poles, and the nuts procured by roasting them until the scales open. In bountiful seasons a single Indian may gather thirty or forty bushels.

VII

THE BIG TREES

BETWEEN THE HEAVY pine and silver fir zones towers the Big Tree (*Sequoia gigantea*), the king of all the conifers in the world, "the noblest of the noble race." The groves nearest Yosemite Valley are about twenty miles to the westward and southward and are called the Tuolumne, Merced and Mariposa groves. It extends, a widely interrupted belt, from a very small grove on the middle fork of the American River to the head of Deer Creek, a distance of about 260 miles, its northern limit being near the thirty-ninth parallel, the southern a little below the thirty-sixth. The elevation of the belt above the sea varies from about 5000 to 8000 feet. From the American River to Kings River the species occurs only in small isolated groups so sparsely distributed along the belt that three of the gaps in it are from forty to sixty miles wide. But from Kings River southward the sequoia is not restricted to mere groves but extends across the wide rugged basins of the Kaweah and Tule Rivers in noble forests, a distance of nearly seventy miles, the continuity of this part of the belt being broken only by the main cañons. The Fresno, the largest of the northern grove, has an area of three or four square miles, a short distance to the southward of the famous Mariposa grove. Along the south rim of the cañon of the south fork of Kings River there is a majestic sequoia forest about six miles long by two wide. This is the northernmost group that may fairly be called a forest. Descending the divide between the Kings and Kaweah Rivers you come to the grand forests that form the main continuous portion of the belt. Southward the giants become more and more irrepressibly jubilant, heaving their massive crowns into the sky from every ridge and slope, waving onward in graceful compliance with the complicated topography of the region. The finest of the Kaweah section of the belt is on the broad ridge between Marble Creek and the middle fork, and is called the Giant Forest. It extends from the granite headlands, overlooking the hot San Joaquin plains, to within a few miles of the cool glacial fountains of the summit peaks. The extreme upper limit of the belt is reached between the middle and south forks of the Kaweah at a height of 8400 feet, but the finest block of big tree forests in the entire belt is on the north fork of Tule River, and is included in the Sequoia National Park.

In the northern groves there are comparatively few young trees or saplings. But here for every old storm-beaten giant there are many in their prime and for each of these a crowd of hopeful young trees and saplings, growing vigorously on moraines, rocky ledges, along water courses and meadows. But though the area occupied by the big tree increases so greatly from north to south, there is no marked increase in the size of the trees. The height of 275 feet or thereabouts and a diameter of about twenty feet, four feet from the ground is, perhaps, about the average size of what may be called full-grown trees, where they are favorably located. The specimens twenty-five feet in diameter are not very rare and a few are nearly three hundred feet high. In the Calaveras grove there are four trees over 300 feet in height, the tallest of which as measured by the Geological Survey is 325 feet. The very largest that I have yet met in the course of my explorations is a majestic old fire-scarred monument in the Kings River forest. It is thirty-five feet and eight inches in diameter inside the bark, four feet above the ground. It is burned half through, and I spent a day in clearing away the charred surface with a sharp ax and counting the annual wood-rings with the aid of a pocket lens. I succeeded in laying bare a section all the way from the outside to the heart and counted a little over four thousand rings, showing that this tree was in its prime about twenty-seven feet in diameter at the beginning of the Christian era. No other tree in the world, as far as I know, has looked down on so many centuries as the sequoia or opens so many impressive and suggestive views into history. Under the most favorable conditions these giants probably live 5000 years or more, though few of even the larger trees are half as old. The age of one that was felled in Calaveras grove, for the sake of having its stump for a dancing-floor, was about 1300 years, and its diameter measured across the stump twenty-four feet inside the bark. Another that was felled in the Kings River forest was about the same size but nearly a thousand years older (2200 years), though not a very old-looking tree.

So harmonious and finely balanced are even the mightiest of these monarchs in all their proportions that there is never anything overgrown or monstrous about them. Seeing them for the first time you are more impressed with their beauty than their size, their grandeur being a great part invisible; but sooner or later it becomes manifest to the loving eye, stealing slowly on the senses like the grandeur of Niagara or of the Yosemite Dome. When you approach them and walk around them you begin to wonder at their colossal size and try to measure them. They bulge considerably at the base, but not more than is required for beauty and safety and the only reason that this bulging seems in some cases excessive is that only a comparatively small section is seen in near views. One that I measured in the Kings River forest was twenty-five feet in diameter at the ground and ten feet in diameter 200 feet above the ground, showing the fineness of the taper of the trunk as a whole. No description can give anything like an adequate idea of their singular majesty, much less of their beauty. Except the sugar pine, most of their neighbors with pointed tops seem ever trying to go higher, while the big tree, soaring above them all, seems satisfied. Its grand domed head seems to be poised about as lightly as a cloud, giving no impression of seeking to rise higher. Only when it is young does it show like other conifers a heavenward yearning, sharply aspiring with a long quick-growing top. Indeed, the whole tree for the first century or two, or until it is a hundred or one hundred and fifty feet high, is arrow-head in form, and, compared with the solemn rigidity of age, seems as sensitive to the wind as a squirrel's tail. As it grows older, the lower branches are gradually dropped and the upper ones thinned out until comparatively few are left. These, however, are developed to a great size, divide again and again and terminate in bossy, rounded masses of leafy branchlets, while the head becomes dome-shaped, and is the first to feel the touch of the rosy beams of the morning, the last to bid the sun good night. Perfect specimens, unhurt by running fires or lightning, are singularly regular and symmetrical in general form, though not in the least conventionalized, for they show extraordinary variety in the unity and harmony of their general outline. The immensely strong, stately shafts are free of limbs for one hundred and fifty feet or so. The large limbs reach out with equal boldness in every direction, showing no weather side, and no other tree has foliage so densely massed, so finely molded in outline and so perfectly subordinate to an ideal type. A particularly knotty, angular, ungovernable-looking branch, from five to seven or eight feet in diameter and perhaps a thousand years old, may occasionally be seen pushing out from the trunk as if determined to break across the bounds of the regular curve, but like all the others it dissolves in bosses of branchlets and sprays as soon as the general outline is approached. Except in picturesque old age, after being struck by lightning or broken by thousands of snow-storms, the regularity of forms is one of their most distinguishing characteristics. Another is the simple beauty of the trunk and its great thickness as compared with its height and the width of the branches, which makes them look more like finely modeled and sculptured architectural columns than the stems of trees, while the great limbs look like rafters, supporting the magnificent dome-head. But though so consummately beautiful, the big tree always seems unfamiliar, with peculiar physiognomy, awfully solemn and earnest; yet with all its strangeness it impresses us as being more at home than any of its neighbors, holding the best right to the ground as the oldest, strongest inhabitant. One soon becomes acquainted with new species of pine and fir and spruce as with friendly people, shaking their outstretched branches like shaking hands and fondling their little ones, while the venerable aboriginal sequoia, ancient of other days, keeps you at a distance, looking as strange in aspect and behavior among its neighbor trees as would the mastodon among the homely bears and deers. Only the Sierra juniper is at all like it, standing rigid and unconquerable on glacier pavements for thousands of years, grim and silent, with an air of antiquity about as pronounced as that of the sequoia.

The bark of the largest tree is from one to two feet thick, rich cinnamon brown, purplish on young trees, forming magnificent masses of color with the underbrush. Toward the end of winter the trees are in bloom, while the snow is still eight or ten feet deep. The female flowers are about three-eighths of an inch long, pale green, and grow in countless thousands on the ends of sprays. The male are still more abundant, pale yellow, a fourth of an inch long and when the pollen is ripe they color the whole tree and dust the air and the ground. The cones are bright grass-green in color, about two and a half inches long, one and a half wide, made up of thirty or forty strong, closely-packed, rhomboidal scales, with four to eight seeds at the base of each. The seeds are wonderfully small and light, being only from an eighth to a fourth of an inch long and wide, including a filmy surrounding wing, which causes them to glint and waver in falling and enables the wind to carry them considerable distances.

Unless harvested by the squirrels, the cones discharge their seed and remain on the tree for many years. In fruitful seasons the trees are fairly laden. On two small branches one and a half and two inches in diameter I counted 480 cones. No other California conifer produces nearly so many seeds, except, perhaps, the other sequoia, the Redwood of the Coast Mountains. Millions are ripened annually by a single tree, and in a fruitful year the product of one of the northern groves would be enough to plant all the mountain ranges in the world.

As soon as any accident happens to the crown, such as being smashed off by lightning, the branches beneath the wound, no matter how situated, seem to be excited, like a colony of bees that have lost their queen, and become anxious to repair the damage. Limbs that have grown outward for centuries at right angles to the trunk begin to turn upward to assist in making a new crown, each speedily assuming the special form of true summits. Even in the case of mere stumps, burned half through, some mere ornamental tuft will try to go aloft and do its best as a leader in forming a new head. Groups of two or three are often found standing close together, the seeds from which they sprang having probably grown on ground cleared for their reception by the fall of a large tree of a former generation. They are called "loving couples," "three graces," etc. When these trees are young they are seen to stand twenty or thirty feet apart, by the time they are full-grown their trunks will touch and crowd against each other and in some cases even appear as one.

It is generally believed that the sequoia was once far more widely distributed over the Sierra; but after long and careful study I have come to the conclusion that it never was, at least since the close of the glacial period, because a diligent search along the margins of the groves, and in the gaps between fails to reveal a single trace of its previous existence beyond its present bounds. Notwithstanding, I feel confident that if every sequoia in the Range were to die today, numerous monuments of their existence would remain, of so imperishable a nature as to be available for the student more than ten thousand years hence.

In the first place, no species of coniferous tree in the Range keeps its members so well together as the sequoia; a mile is, perhaps, the greatest distance of any straggler from the main body, and all of those stragglers that have come under my observation are young, instead of old monumental trees, relics of a more extended growth.

Again, the great trunks of the sequoia last for centuries after they fall. I have a specimen block of sequoia wood, cut from a fallen tree, which is hardly distinguishable from a similar section cut from a living tree, although the one cut from the fallen trunk has certainly lain on the damp forest floor more than 380 years, probably thrice as long. The time-measure in the case is simply this: When the ponderous trunk to which the old vestige belonged fell, it sunk itself into the ground, thus making a long, straight ditch, and in the middle of this ditch a silver fir four feet in diameter and 380 years old was growing, as I determined by cutting it half through and counting the rings, thus demonstrating that the remnant of the trunk that made the ditch has lain on the ground *more* than 380 years. For it is evident that, to find the whole time, we must add to the 380 years the time that the vanished portion of the trunk lay in the ditch before being burned out of the way, plus the time that passed before the seed from which the monumental fir sprang fell into the prepared soil and took root. Now, because sequoia trunks are never wholly consumed in one forest fire, and those fires recur only at considerable intervals, and because sequoia ditches after being cleared are often left unplanted for centuries, it becomes evident that the trunk-remnant in question may probably have lain a thousand years or more. And this instance is by no means a rare one.

Again, admitting that upon those areas supposed to have been once covered with sequoia forests, every tree may have fallen, and every trunk may have been burned or buried, leaving not a remnant, many of the ditches made by the fall of the ponderous trunks, and the bowls made by their upturning roots, would remain patent for thousands of years after the last vestige of the trunks that made them had vanished. Much of this ditch-writing would no doubt be quickly effaced by the flood-action of overflowing streams and rain-washing; but no inconsiderable portion would remain enduringly engraved on ridge-tops beyond such destructive action; for, where all the conditions are favorable, it is almost imperishable. Now these historic ditches and root-bowls occur in all the present sequoia groves and forests, but, as far as I have observed, not the faintest vestige of one presents itself outside of them.

We therefore conclude that the area covered by sequoia has not been diminished during the last eight or ten thousand years, and probably not at all in post-glacial time. Nevertheless, the questions may be asked: Is the species verging toward extinction? What are its relations to climate, soil, and associated trees?

All the phenomena bearing on these questions also throw light, as we shall endeavor to show, upon the peculiar distribution of the species, and sustain the conclusion already arrived at as to the question of former extension. In the northern groups, as we have seen, there are few young trees or saplings growing up around the old ones to perpetuate the race, and inasmuch as those aged sequoias, so nearly childless, are the only ones commonly known, the species, to most observers, seems doomed to speedy extinction, as being nothing more than an expiring remnant, vanquished in the so-called struggle for life by pines and firs that have driven it into its last strongholds in moist glens where the climate is supposed to be exceptionally favorable. But the story told by the majestic continuous forests of the south creates a very different impression. No tree in the forest is more enduringly established in concordance with both climate and soil. It grows heartily everywhere—on moraines, rocky ledges, along watercourses, and in the deep, moist alluvium of meadows with, as we have seen, a multitude of seedlings and saplings crowding up around the aged, abundantly able to maintain the forest in prime vigor. So that if all the trees of any section of the main sequoia forest were ranged together according to age, a very promising curve would be presented, all the way up from last year's seedlings to giants, and with the young and middle-aged portion of the curve many times longer than the old portion. Even as far north as the Fresno, I counted 536 saplings and seedlings, growing promisingly upon a landslip not exceeding two acres in area. This soil-bed was about seven years old, and had been seeded almost simultaneously by pines, firs, libocedrus, and sequoia, presenting a simple and instructive illustration of the struggle for life among the rival species; and it was interesting to note that the conditions thus far affecting them have enabled the young sequoias to gain a marked advantage. Toward the south where the sequoia becomes most exuberant and numerous, the rival trees become less so; and where they mix with sequoias they grow up beneath them like slender grasses among stalks of Indian corn. Upon a bed of sandy flood-soil I counted ninety-four sequoias, from one to twelve feet high, on a patch of ground once occupied by four large sugar pines which lay crumbling beneath them—an instance of conditions which have enabled sequoias to crowd out the pines. I also noted eighty-six vigorous saplings upon a piece of fresh ground prepared for their reception by fire. Thus fire, the great destroyer of the sequoia, also furnishes the bare ground required for its growth from the seed. Fresh ground is, however, furnished in sufficient quantities for the renewal of the forests without the aid of fire by the fall of old trees. The soil is thus upturned and mellowed, and many trees are planted for every one that falls.

It is constantly asserted in a vague way that the Sierra was vastly wetter than now, and that the increasing drought will of itself extinguish the sequoia, leaving its ground to other trees supposed capable of flourishing in a drier climate. But that the sequoia can and does grow on as dry ground as any of its present rivals is manifest in a thousand places. "Why, then," it will be asked, "are sequoias always found only in well-watered places?" Simply because a growth of sequoias creates those streams. The thirsty mountaineer knows well that in every sequoia grove he will find running water, but it is a mistake to suppose that the water is the cause of the grove being there; on the contrary, the grove is the cause of the water being there. Drain off the water and the trees will remain, but cut off the trees, and the streams will vanish. Never was cause more completely mistaken for effect than in the case of these related phenomena of sequoia woods and perennial streams.

When attention is called to the method of sequoia stream-making, it will be apprehended at once. The roots of this immense tree fill the ground, forming a thick sponge that absorbs and holds back the rain and melting snow, only allowing it to ooze and flow gently. Indeed, every fallen leaf and rootlet, as well as long clasping root, and prostrate trunk, may be regarded as a dam hoarding the bounty of storm-clouds, and dispensing it as blessings all through the summer, instead of allowing it to go headlong in short-lived floods.

Since, then, it is a fact that thousands of sequoias are growing thriftily on what is termed dry ground, and even clinging like mountain pines to rifts in granite precipices, and since it has also been shown that the extra moisture found in connection with the denser growths is an effect of their presence, instead of a cause of their presence, then the notions as to the former extension of the species and its near approach to extinction, based upon its supposed dependence on greater moisture, are seen to be erroneous.

The decrease in the rain and snowfall since the close of the glacial period in the Sierra is much less than is commonly guessed. The highest post-glacial water-marks are well preserved in all the upper river channels, and they are not greatly higher than the spring flood-marks of the present; showing conclusively that no extraordinary decrease has taken place in the volume of the upper tributaries of post-glacial Sierra streams since they came into existence. But, in the meantime, eliminating all this complicated question of climatic change, the plain fact remains that the present rain and snowfall is abundantly sufficient for the luxuriant growth of sequoia forests. Indeed, all my observations tend to show that in a prolonged drought the sugar pines and firs would perish before the sequoia, not alone because of the greater longevity of individual trees, but because the species can endure more drought, and make the most of whatever moisture falls.

Again, if the restriction and irregular distribution of the species be interpreted as a result of the desiccation of the Range, then instead of increasing as it does in individuals toward the south where the rainfall is less, it should diminish. If, then, the peculiar distribution of sequoia has not been governed by superior conditions of soil as to fertility or moisture, by what has it been governed?

In the course of my studies I observed that the northern groves, the only ones I was at first acquainted with, were located on just those portions of the general forest soil-belt that were first laid bare toward the close of the glacial period when the ice-sheet began to break up into individual glaciers. And while searching the wide basin of the San Joaquin, and trying to account for the absence of sequoia where every condition seemed favorable for its growth, it occurred to me that this remarkable gap in the sequoia belt fifty miles wide is located exactly in the basin of the vast, ancient *mer de glace* of the San Joaquin and Kings River basins which poured its frozen floods to the plain through this gap as its channel. I then perceived that the next great gap in the belt to the northward, forty miles wide, extending between the Calaveras and Tuolumne groves, occurs in the basin of the great ancient *mer de glace* of the Tuolumne and Stanislaus basins; and that the smaller gap between the Merced and Mariposa groves occurs in the basin of the smaller glacier of the

Merced. The wider the ancient glacier, the wider the corresponding gap in the sequoia belt.

Finally, pursuing my investigations across the basins of the Kaweah and Tule, I discovered that the sequoia belt attained its greatest development just where, owing to the topographical peculiarities of the region, the ground had been best protected from the main ice-rivers that continued to pour past from the summit fountains long after the smaller local glaciers had been melted.

Taking now a general view of the belt, beginning at the south, we see that the majestic ancient glaciers were shed off right and left down the valleys of Kern and Kings Rivers by the lofty protective spurs outspread embracingly above the warm sequoia-filled basins of the Kaweah and Tule. Then, next northward, occurs the wide sequoia-less channel, or basin of the ancient San Joaquin and Kings River *mer de glace*; then the warm, protected spots of Fresno and Mariposa groves; then the sequoia-less channel of the ancient Merced glacier; next the warm, sheltered ground of the Merced and Tuolumne groves; then the sequoia-less channel of the grand ancient *mer de glace* of the Tuolumne and Stanislaus; then the warm old ground of the Calaveras and Stanislaus groves. It appears, therefore, that just where, at a certain period in the history of the Sierra, the glaciers were not, there the sequoia is, and just where the glaciers were, there the sequoia is not.

But although all the observed phenomena bearing on the post-glacial history of this colossal tree point to the conclusion that it never was more widely distributed on the Sierra since the close of the glacial epoch; that its present forests are scarcely past prime, if, indeed, they have reached prime; that the post-glacial day of the species is probably not half done; yet, when from a wider outlook the vast antiquity of the genus is considered, and its ancient richness in species and individuals,—comparing our Sierra Giant and *Sequoia sempervirens* of the Coast Range, the only other living species of sequoia, with the twelve fossil species already discovered and described by Heer and Lesquereux, some of which flourished over vast areas in the Arctic regions and in Europe and our own territories, during tertiary and cretaceous times—then, indeed, it becomes plain that our two surviving species, restricted to narrow belts within the limits of California, are mere remnants of the genus, both as to species and individuals, and that they may be verging to extinction. But the verge of a period beginning in cretaceous times may have a breadth of tens of thousands of years, not to mention the possible existence of conditions calculated to multiply and re-extend both species and individuals.

There is no absolute limit to the existence of any tree. Death is due to accidents, not, as that of animals, to the wearing out of organs. Only the leaves die of old age. Their fall is foretold in their structure; but the leaves are renewed every year, and so also are the essential organs—wood, roots, bark, buds. Most of the Sierra trees die of disease, insects, fungi, etc., but nothing hurts the big tree. I never saw one that was sick or showed the slightest sign of decay. Barring accidents, it seems to be immortal. It is a curious fact that all the very old sequoias had lost their heads by lightning strokes. "All things come to him who waits." But of all living things, sequoia is perhaps the only one able to wait long enough to make sure of being struck by lightning.

So far as I am able to see at present only fire and the ax threaten the existence of these noblest of God's trees. In Nature's keeping they are safe, but through the agency of man destruction is making rapid progress, while in the work of protection only a good beginning has been made. The Fresno grove, the Tuolumne, Merced and Mariposa groves are under the protection of the Federal Government in the Yosemite National Park. So are the General Grant and Sequoia National Parks; the latter, established twenty-one years ago, has an area of 240 square miles and is efficiently guarded by a troop of cavalry under the direction of the Secretary of the Interior; so also are the small General Grant National Park, established at the same time with an area of four square miles, and the Mariposa grove, about the same size and the small Merced and Tuolumne group. Perhaps more than half of all the big trees have been thoughtlessly sold and are now in the hands of speculators and mill men. It appears, therefore, that far the largest and important section of protected big trees is in the great Sequoia National Park, now easily accessible by rail to Lemon Cove and thence by a good stage road into the giant forest of the Kaweah and thence by trail to other parts of the park; but large as it is it should be made much larger. Its natural eastern boundary is the High Sierra and the northern and southern boundaries are the Kings and Kern Rivers. Thus could be included the sublime scenery on the headwaters of these rivers and perhaps nine-tenths of all the big trees in existence. All private claims within these bounds should be gradually extinguished by purchase by the Government. The big tree, leaving all its higher uses out of the count, is a tree of life to the dwellers of the plain dependent on irrigation, a never-failing spring, sending living waters to the lowland. For every grove cut down a stream is dried up. Therefore all California is crying, "Save the trees of the fountains." Nor, judging by the signs of the times, is it likely that the cry will cease until the salvation of all that is left of *Sequoia gigantea* is made sure.

VIII

THE FLOWERS

YOSEMITE WAS ALL one glorious flower garden before ploughs and scythes and trampling, biting horses came to make its wide open spaces look like farmers' pasture fields. Nevertheless, countless flowers still bloom every year in glorious profusion on the grand talus slopes, wall benches and tablets, and in all the fine, cool side-cañons up to the rim of the Valley, and beyond, higher and higher, to the summits of the peaks. Even on the open floor and in easily-reached side-nooks many common flowering plants have survived and still make a brave show in the spring and early summer. Among these we may mention tall œnotheras, *Pentstemon lutea*, and *P. douglasii* with fine blue and red flowers; Spraguea, scarlet zauschneria, with its curious radiant rosettes characteristic of the sandy flats; mimulus, eunanus, blue and white violets, geranium, columbine, erythraea, larkspur, collomia, draperia, gilias, heleniums, bahia, goldenrods, daisies, honeysuckle; heuchera, bolandra, saxifrages, gentians; in cool cañon nooks and on Clouds' Rest and the base of Starr King Dome you may find *Primula suffrutescens*, the only wild primrose discovered in California, and the only known shrubby species in the genus. And there are several fine orchids, habenaria, and cypripedium, the latter very rare, once common in the Valley near the foot of Glacier Point, and in a bog on the rim of the Valley near a place called Gentry's Station, now abandoned. It is a very beautiful species, the large oval lip white, delicately veined with purple; the other petals and the sepals purple, strap-shaped, and elegantly curled and twisted.

Of the lily family, fritillaria, smilacina, chlorogalum and several fine species of brodiæa, Ithuriel's spear, and others less prized are common, and the favorite calochortus, or Mariposa lily, a unique genus of many species, something like the tulips of Europe but far finer. Most of them grow on the warm foot-hills below the Valley, but two charming species, *C. cœruleus* and *C. nudus*, dwell in springy places on the Wawona road a few miles beyond the brink of the walls.

The snow plant (*Sarcodes sanguinea*) is more admired by tourists than any other in California. It is red, fleshy and watery and looks like a gigantic asparagus shoot. Soon after the snow is off the ground it rises through the dead needles and humus in the pine and fir woods like a bright glowing pillar of fire. In a week or so it grows to a height of eight or twelve inches with a diameter of an inch and a half or two inches; then its long fringed bracts curl aside, allowing the twenty- or thirty-five-lobed, bell-shaped flowers to open and look straight out from the axis. It is said to grow up through the snow; on the contrary, it always waits until the ground is warm, though with other early flowers it is occasionally buried or half-buried for a day or two by spring storms. The entire plant—flowers, bracts, stem, scales, and roots—is fiery red. Its color should appeal to one's blood. Nevertheless, it is a singularly cold and unsympathetic plant. Everybody admires it as a wonderful curiosity, but nobody loves it as lilies, violets, roses, daisies are loved. Without fragrance, it stands beneath the pines and firs lonely and silent, as if unacquainted with any other plant in the world; never moving in the wildest storms; rigid as if lifeless, though covered with beautiful rosy flowers.

Far the most delightful and fragrant of the Valley flowers is the Washington lily, white, moderate in size, with from three- to ten-flowered racemes. I found one specimen in the lower end of the Valley at the foot of the Wawona grade that was eight feet high, the raceme two feet long, with fifty-two flowers, fifteen of them open; the others had faded or were still in the bud. This famous lily is distributed over the sunny portions of the sugar-pine woods, never in large meadow-garden companies like the large and the small tiger lilies (*pardalinum* and *parvum*), but widely scattered, standing up to the waist in dense ceanothus and manzanita chaparral, waving its lovely flowers above the blooming wilderness of brush, and giving their fragrance to the breeze. It is now becoming scarce in the most accessible parts of its range on account of the high price paid for its bulbs by gardeners through whom it has been distributed far and wide over the flower-loving world. For, on account of its pure color and delicate, delightful fragrance, all lily lovers at once adopted it as a favorite.

The principal shrubs are manzanita and ceanothus, several species of each, azalea, *Rubus nutkanus*, brier rose, choke-cherry, philadelphus, calycanthus, garrya, rhamnus, etc.

The manzanita never fails to attract particular attention. The species common in the Valley is usually about six or seven feet high, round-headed with innumerable branches, red or chocolate-color bark, pale green leaves set on edge, and a rich profusion of small, pink, narrow-throated, urn-shaped flowers, like those of arbutus. The knotty, crooked, angular branches are about as rigid as bones, and the red bark is so thin and smooth on both trunk and branches, they look as if they had been peeled and polished and painted. In the spring large areas on the mountain up to a height of eight or nine thousand feet are brightened with the rosy flowers, and in autumn with their red fruit. The pleasantly acid berries, about the size of peas, look like little apples, and a hungry mountaineer is glad to eat them, though half their bulk is made up of hard seeds. Indians, bears, coyotes, foxes, birds and other mountain people live on them for weeks and months. The different species of ceanothus usually associated with manzanita are flowery fragrant and altogether delightful shrubs, growing in glorious abundance, not only in the Valley, but high up in the forest on sunny or half-shaded ground. In the sugar-pine woods the most beautiful species is *C. integerrimus*, often called Californian lilac, or deer brush. It is five or six feet high with slender branches, glossy foliage, and abundance of blue flowers in close, showy panicles. Two species, *C. prostratus* and *C. procumbens*, spread smooth, blue-flowered mats and rugs beneath the pines, and offer fine beds to tired mountaineers. The commonest species, *C. cordulatus*, is most common in the silver-fir woods. It is white-flowered and thorny, and makes dense thickets of tangled chaparral, difficult to wade through or to walk over. But it is pressed flat every winter by ten or fifteen feet of snow. The western azalea makes glorious beds of bloom along the river-bank and meadows. In the Valley it is from two to five feet high, has fine green leaves, mostly hidden beneath its rich profusion of large, fragrant white and yellow flowers, which are in their prime in June, July and August, according to the elevation, ranging from 3000 to 6000 feet. Near the azalea-bordered streams the small wild rose, resembling *R. blanda*, makes large thickets deliciously fragrant, especially on a dewy morning and after showers. Not far from these azalea and rose gardens, *Rubus nutkanus* covers the ground with broad, soft, velvety leaves, and pure-white flowers as large as those of its neighbor and relative, the rose, and much finer in texture, followed at the end of summer by soft red berries good for everybody. This is the commonest and the most beautiful of the whole blessed, flowery, fruity Rubus genus.

There are a great many interesting ferns in the Valley and about it. Naturally enough the greater number are rock ferns—pellæa, cheilanthes, polypodium, adiantum, woodsia, cryptogramma, etc., with small tufted fronds, lining cool glens and fringing the seams of the cliffs. The most important of the larger species are woodwardia, aspidium, asplenium, and, above all, the common pteris. *Woodwardia radicans* is a superb, broad-shouldered fern five to eight feet high, growing in vase-shaped clumps where the ground is nearly level and on some of the benches of the north wall of the Valley where it is watered by a broad trickling stream. It thatches the sloping rocks, frond overlapping frond like roof shingles. The broad-fronded, hardy *Pteris aquilina*, the commonest of ferns, covers large areas on the floor of the Valley. No other fern does so much for the color glory of autumn, with its browns and reds and yellows, even after lying dead beneath the snow all winter. It spreads a rich brown mantle over the desolate ground in the spring before the grass has sprouted, and at the first touch of sun-heat its young fronds come rearing up full of faith and hope through the midst of the last year's ruins.

Of the five species of pellæa, *P. breweri* is the hardiest as to enduring high altitudes and stormy weather and at the same time it is the most fragile of the genus. It grows in dense tufts in the clefts of storm-beaten rocks, high up on the mountainside on the very edge of the fern line. It is a handsome little fern about four or five inches high, has pale-green pinnate fronds, and shining bronze-colored stalks about as brittle as glass. Its companions on the lower part of its range are *Cryptogramma acrostichoides* and *Phegopteris alpestris*, the latter with soft, delicate fronds, not in the least like those of Rock fern, though it grows on the rocks where the snow lies longest. *Pellæa bridgesii*, with blue-green, narrow, simply-pinnate fronds, is about the same size as Breweri and ranks next to it as a mountaineer, growing in fissures, wet or dry, and around the edges of boulders that are resting on glacier pavements with no fissures whatever. About a thousand feet lower we find the smaller, more abundant *P. densa* on ledges and boulder-strewn, fissured pavements, watered until late in summer from oozing currents, derived from lingering snowbanks. It is, or rather was, extremely abundant between the foot of the Nevada and the head of the Vernal Fall, but visitors with great industry have dug out almost every root, so that now one has to scramble in out-of-the-way places to find it. The three species of Cheilanthes in the Valley—*C. californica, C. gracillima*, and *myriophylla*, with beautiful two-to-four-pinnate fronds, an inch to five inches long, adorn the stupendous walls however dry and sheer. The exceedingly delicate californica is so rare that I have found it only once. The others are abundant and are sometimes accompanied by the little gold fern, *Gymnogramme triangularis*, and rarely by the curious little *Botrychium simplex*, some of them less than an inch high. The finest of all the rock ferns is *Adiantum pedatum*, lover of waterfalls and the finest spray-dust. The homes it loves best are over-leaning, cave-like hollows, beside the larger falls, where it can wet its fingers with their dewy spray. Many of these moss-lined chambers contain thousands of these delightful ferns, clinging to mossy walls by the slightest hold, reaching out their delicate finger-fronds on dark, shining stalks, sensitive and tremulous, throbbing in unison with every movement and tone of the falling water, moving each division of the frond separately at times, as if fingering the music.

May and June are the main bloom-months of the year. Both the flowers and falls are then at their best. By the first of August the midsummer glories of the Valley are past their prime. The young birds are then out of their nests. Most of the plants have gone to seed; berries are ripe; autumn tints begin to kindle and burn over meadow and grove, and a soft mellow haze in the morning sunbeams heralds the approach of Indian summer. The shallow river is now at rest, its flood-work done. It is now but little more than a series of pools united by trickling, whispering currents that steal softly over brown pebbles and sand with scarce an audible murmur. Each pool has a character of its own and, though they are nearly currentless, the night air and tree shadows keep them cool. Their shores curve in and out in bay and promontory, giving the appearance of miniature lakes, their banks in most places embossed with brier and azalea, sedge and grass and fern; and above these in their glory of autumn colors a mingled growth of alder, willow, dogwood and balm-of-Gilead; mellow sunshine overhead, cool shadows beneath; light filtered and strained in passing through the ripe leaves like that which passes through colored windows. The surface of the water is stirred, perhaps, by whirling water-beetles, or some startled trout, seeking shelter beneath fallen logs or roots. The falls, too, are quiet;

no wind stirs, and the whole Valley floor is a mosaic of greens and purples, yellows and reds. Even the rocks seem strangely soft and mellow, as if they, too, had ripened.

IX

THE BIRDS

THE SONGS OF the Yosemite winds and waterfalls are delightfully enriched with bird song, especially in the nesting time of spring and early summer. The most familiar and best known of all is the common robin, who may be seen every day, hopping about briskly on the meadows and uttering his cheery, enlivening call. The black-headed grosbeak, too, is here, with the Bullock oriole, and western tanager, brown song-sparrow, hermit thrush, the purple finch,—a fine singer, with head and throat of a rosy-red hue,—several species of warblers and vireos, kinglets, flycatchers, etc.

But the most wonderful singer of all the birds is the water-ouzel that dives into foaming rapids and feeds at the bottom, holding on in a wonderful way, living a charmed life.

Several species of humming-birds are always to be seen, darting and buzzing among the showy flowers. The little red-bellied nut-hatches, the chickadees, and little brown creepers, threading the furrows of the bark of the pines, searching for food in the crevices. The large Steller's jay makes merry in the pine-tops; flocks of beautiful green swallows skim over the streams, and the noisy Clarke's crow may oftentimes be seen on the highest points around the Valley; and in the deep woods beyond the walls you may frequently hear and see the dusky grouse and the pileated woodpecker, or woodcock almost as large as a pigeon. The junco or snow-bird builds its nest on the floor of the Valley among the ferns; several species of sparrow are common and the beautiful lazuli bunting, a common bird in the underbrush, flitting about among the azalea and ceanothus bushes and enlivening the groves with his brilliant color; and on gravelly bars the spotted sandpiper is sometimes seen. Many woodpeckers dwell in the Valley; the familiar flicker, the Harris woodpecker and the species which so busily stores up acorns in the thick bark of the yellow pines.

The short, cold days of winter are also sweetened with the music and hopeful chatter of a considerable number of birds. No cheerier choir ever sang in snow. First and best of all is the water-ouzel, a dainty, dusky little bird about the size of a robin, that sings a sweet fluty song all winter and all summer, in storms and calms, sunshine and shadow, haunting the rapids and waterfalls with marvelous constancy, building his nest in the cleft of a rock bathed in spray. He is not web-footed, yet he dives fearlessly into foaming rapids, seeming to take the greater delight the more boisterous the stream, always as cheerful and calm as any linnet in a grove. All his gestures as he flits about amid the loud uproar of the falls bespeak the utmost simplicity and confidence—bird and stream one and inseparable. What a pair! yet they are well related. A finer bloom than the foam bell in an eddying pool is this little bird. We may miss the meaning of the loud-resounding torrent, but the flute-like voice of the bird—only love is in it.

A few robins, belated on their way down from the upper meadows, linger in the Valley and make out to spend the winter in comparative comfort, feeding on the mistletoe berries that grow on the oaks. In the depths of the great forests, on the high meadows, in the severest altitudes, they seem as much at home as in the fields and orchards about the busy habitations of man, ascending the Sierra as the snow melts, following the green footsteps of Spring, until in July or August the highest glacier meadows are reached on the summit of the Range. Then, after the short summer is over, and their work in cheering and sweetening these lofty wilds is done, they gradually make their way down again in accord with the weather, keeping below the snow-storms, lingering here and there to feed on huckleberries and frost-nipped wild cherries growing on the upper slopes. Thence down to the vineyards and orchards of the lowlands to spend the winter; entering the gardens of the great towns as well as parks and fields, where the blessed wanderers are too often slaughtered for food—surely a bad use to put so fine a musician to; better make stove wood of pianos to feed the kitchen fire.

The kingfisher winters in the Valley, and the flicker and, of course, the carpenter woodpecker, that lays up large stores of acorns in the bark of trees; wrens also, with a few brown and gray linnets, and flocks of the arctic bluebird, making lively pictures among the snow-laden mistletoe bushes. Flocks of pigeons are often seen, and about six species of ducks, as the river is never wholly frozen over. Among these are the mallard and the beautiful wood-duck, now less common on account of being so often shot at. Flocks of wandering geese used to visit the Valley in March and April, and perhaps do so still, driven down by hunger or stress of weather while on their way across the Range. When pursued by the hunters I have frequently seen them try to fly over the walls of the Valley until tired out and compelled to re-alight. Yosemite magnitudes seem to be as deceptive to geese as to men, for after circling to a considerable height and forming regular harrow-shaped ranks they would suddenly find themselves in danger of being dashed against the face of the cliff, much nearer the bottom than the top. Then turning in confusion with loud screams they would try again and again until exhausted and compelled to descend. I have occasionally observed large flocks on their travels crossing the summits of the Range at a height of 12,000 to 13,000 feet above the level of the sea, and even in so rare an atmosphere as this they seemed to be sustaining themselves without extra effort. Strong, however, as they are of wind and wing, they cannot fly over Yosemite walls, starting from the bottom.

A pair of golden eagles have lived in the Valley ever since I first visited it, hunting all winter along the northern cliffs and down the river cañon. Their nest is on a ledge of the cliff over which pours the Nevada Fall. Perched on the top of a dead spar, they were always interested observers of the geese when they were being shot at. I once noticed one of the geese compelled to leave the flock on account of being sorely wounded, although it still seemed to fly pretty well. Immediately the eagles pursued it and no doubt struck it down, although I did not see the result of the hunt. Anyhow, it flew past me up the Valley, closely pursued.

One wild, stormy winter morning after five feet of snow had fallen on the floor of the Valley and the flying flakes driven by a strong wind still thickened the air, making darkness like the approach of night, I sallied forth to see what I might learn and enjoy. It was impossible to go very far without the aid of snow-shoes, but I found no great difficulty in making my way to a part of the river where one of my ouzels lived. I found him at home busy about his breakfast, apparently unaware of anything uncomfortable in the weather. Presently he flew out to a stone against which the icy current was beating, and turning his back to the wind, sang as delightfully as a lark in springtime.

After spending an hour or two with my favorite, I made my way across the Valley, boring and wallowing through the loose snow, to learn as much as possible about the way the other birds were spending their time. In winter one can always find them because they are then restricted to the north side of the Valley, especially the Indian Cañon groves, which from their peculiar exposure are the warmest.

I found most of the robins cowering on the lee side of the larger branches of the trees, where the snow could not fall on them, while two or three of the more venturesome were making desperate efforts to get at the mistletoe berries by clinging to the underside of the snow-crowned masses, back downward, something like woodpeckers. Every now and then some of the loose snow was dislodged and sifted down on the hungry birds, sending them screaming back to their companions in the grove, shivering and muttering like cold, hungry children.

Some of the sparrows were busy scratching and pecking at the feet of the larger trees where the snow had been shed off, gleaning seeds and benumbed insects, joined now and then by a robin weary of his unsuccessful efforts to get at the snow-covered mistletoe berries. The brave woodpeckers were clinging to the snowless sides of the larger boles and overarching branches of the camp trees, making short flights from side to side of the grove, pecking now and then at the acorns they had stored in the bark, and chattering aimlessly as if unable to keep still, evidently putting in the time in a very dull way. The hardy nut-hatches were threading the open furrows of the barks in their usual industrious manner and uttering their quaint notes, giving no evidence of distress. The Steller's jays were, of course, making more noise and stir than all the other birds combined; ever coming and going with loud bluster, screaming as if each had a lump of melting sludge in his throat, and taking good care to improve every opportunity afforded by the darkness and confusion of the storm to steal from the acorn stores of the woodpeckers. One of the golden eagles made an impressive picture as he stood bolt upright on the top of a tall pine-stump, braving the storm, with his back to the wind and a tuft of snow piled on his broad shoulders, a monument of passive endurance. Thus every stormbound bird seemed more or less uncomfortable, if not in distress. The storm was reflected in every gesture, and not one cheerful note, not to say song, came from a single bill. Their cowering, joyless endurance offered striking contrasts to the spontaneous, irrepressible gladness of the ouzel, who could no more help giving out sweet song than a rose sweet fragrance. He must sing, though the heavens fall.

X

THE SOUTH DOME

WITH THE EXCEPTION of a few spires and pinnacles, the South Dome is the only rock about the Valley that is strictly inaccessible without artificial means, and its inaccessibility is expressed in severe terms. Nevertheless many a mountaineer, gazing admiringly, tried hard to invent a way to the top of its noble crown—all in vain, until in the year 1875, George Anderson, an indomitable Scotchman, undertook the adventure. The side facing Tenaya Cañon is an absolutely vertical precipice from the summit to a depth of about 1600 feet, and on the opposite side it is nearly vertical for about as great a depth. The southwest side presents a very steep and finely drawn curve from the top down a thousand feet or more, while on the northeast, where it is united with the Clouds' Rest Ridge, one may easily reach a point called the Saddle, about seven hundred feet below the summit. From the Saddle the Dome rises in a graceful curve a few degrees too steep for unaided climbing, besides being defended by overleaning ends of the concentric dome layers of the granite.

A year or two before Anderson gained the summit, John Conway, the master trail-builder of the Valley, and his little sons, who climbed smooth rocks like lizards, made a bold effort to reach the top by climbing barefooted up the grand curve with a rope which they fastened at irregular intervals by means of eye-bolts driven into joints of the rock. But finding that the upper part would require laborious drilling, they abandoned the attempt, glad to escape from the dangerous position they had reached, some 300 feet above the Saddle. Anderson began with Conway's old rope, which had been left in place, and resolutely drilled his way to the top, inserting eye-bolts five to six feet apart, and making his rope fast to each in succession, resting his feet on the last bolt while he drilled a hole for the next above. Occasionally some irregularity in the curve, or slight foothold, would enable him to climb a few feet without a rope, which he would pass and begin drilling again, and thus the whole work was accomplished in a few days. From this slender beginning he proposed to construct a substantial stairway which he hoped to complete in time for the next year's travel, but while busy getting out timber for his stairway and dreaming of the wealth he hoped to gain from tolls, he was taken sick and died all alone in his little cabin.

On the 10th of November, after returning from a visit to Mount Shasta, a month or two after Anderson had gained the summit, I made haste to the Dome, not only for the pleasure of climbing, but to see what I might learn. The first winter storm-clouds had blossomed and the mountains and all the high points about the Valley were mantled in fresh snow. I was, therefore, a little apprehensive of danger from the slipperiness of the rope and the rock. Anderson himself tried to prevent me from making the attempt, refusing to believe that any one could climb his rope in the snow-muffled condition in which it then was. Moreover, the sky was overcast and solemn snow-clouds began to curl around the summit, and my late experiences on icy Shasta came to mind. But reflecting that I had matches in my pocket, and that a little fire-wood might be found, I concluded that in case of a storm the night could be spent on the Dome without suffering anything worth minding, no matter what the clouds might bring forth. I therefore pushed on and gained the top.

It was one of those brooding, changeful days that come between the Indian summer and winter, when the leaf colors have grown dim and the clouds come and go among the cliffs like living creatures looking for work: now hovering aloft, now caressing rugged rock-brows with great gentleness, or, wandering afar over the tops of the forests, touching the spires of fir and pine with their soft silken fringes as if trying to tell the glad news of the coming of snow.

The first view was perfectly glorious. A massive cloud of pure pearl luster, apparently as fixed and calm as the meadows and groves in the shadow beneath it, was arched across the Valley from wall to wall, one end resting on the grand abutment of El Capitan, the other on Cathedral Rock. A little later, as I stood on the tremendous verge overlooking Mirror Lake, a flock of smaller clouds, white as snow, came from the north, trailing their downy skirts over the dark forests, and entered the Valley with solemn god-like gestures through Indian Cañon and over the North Dome and Royal Arches, moving swiftly, yet with majestic deliberation. On they came, nearer and nearer, gathering and massing beneath my feet and filling the Tenaya Cañon. Then the sun shone free, lighting the pearly gray surface of the cloud-like sea and making it glow. Gazing, admiring, I was startled to see for the first time the rare optical phenomenon of the "Specter of the Brocken." My shadow, clearly outlined, about half a mile long, lay upon this glorious white surface with startling effect. I walked back and forth, waved my arms and struck all sorts of attitudes, to see every slightest movement enormously exaggerated. Considering that I have looked down so many times from mountain-tops on seas of all sorts of clouds, it seems strange that I should have seen the "Brocken Specter" only this once. A grander surface and a grander standpoint, however, could hardly have been found in all the Sierra.

After this grand show the cloud-sea rose higher, wreathing the Dome, and for a short time submerging it, making darkness like night, and I began to think of looking for a campground in a cluster of dwarf pines. But soon the sun shone free again, the clouds, sinking lower and lower, gradually vanished, leaving the Valley with its Indian-summer colors apparently refreshed, while to the eastward the summit-peaks, clad in new snow, towered along the horizon in glorious array.

Though apparently it is perfectly bald, there are four clumps of pines growing on the summit, representing three species, *Pinus albicaulis*, *P. contorta* and *P. ponderosa*, var. *jeffreyi*—all three, of course, repressed and storm-beaten. The alpine spiræa grows here also and blossoms profusely with potentilla, erigeron, eriogonum, pentstemon, solidago, and an interesting species of onion, and four or five species of grasses and sedges. None of these differs in any respect from those of other summits of the same height, excepting the curious little narrow-leaved, waxen-bulbed onion, which I had not seen elsewhere.

Notwithstanding the enthusiastic eagerness of tourists to reach the crown of the Dome the views of the Valley from this lofty standpoint are less striking than from many other points comparatively low, chiefly on account of the foreshortening effect produced by looking down from so great a height. The North Dome is dwarfed almost beyond recognition, the grand sculpture of the Royal Arches is scarcely noticeable, and the whole range of walls on both sides seem comparatively low, especially when the Valley is flooded with noon sunshine; while the Dome itself, the most sublime feature of all the Yosemite views, is out of sight beneath one's feet. The view of Little Yosemite Valley is very fine, though inferior to one obtained from the base of the Starr King Cone, but the summit landscapes towards Mounts Ritter, Lyell, Dana, Conness, and the Merced Group, are very effective and complete.

No one has attempted to carry out Anderson's plan of making the Dome accessible. For my part I should prefer leaving it in pure wildness, though, after all, no great damage could be done by tramping over it. The surface would be strewn with tin cans and bottles, but the winter gales would blow the rubbish away. Avalanches might strip off any sort of stairway or ladder that might be built. Blue jays and Clarke's crows have trodden the Dome for many a day, and so have beetles and chipmunks, and Tissiack would hardly be more "conquered" or spoiled should man be added to her list of visitors. His louder scream and heavier scrambling would not stir a line of her countenance.

When the sublime ice-floods of the glacial period poured down the flank of the Range over what is now Yosemite Valley, they were compelled to break through a dam of domes extending across from Mount Starr King to North Dome; and as the period began to draw near a close the shallowing ice-currents were divided and the South Dome was, perhaps, the first to emerge, burnished and shining like a mirror above the surface of the icy sea; and though it has sustained the wear and tear of the elements tens of thousands of years, it yet remains a telling monument of the action of the great glaciers that brought it to light. Its entire surface is still covered with glacial hieroglyphics whose interpretation is the reward of all who devoutly study them.

XI

THE ANCIENT YOSEMITE GLACIERS: HOW THE VALLEY WAS FORMED

ALL CALIFORNIA HAS been glaciated, the low plains and valleys as well as the mountains. Traces of an ice-sheet, thousands of feet in thickness, beneath whose heavy folds the present landscapes have been molded, may be found everywhere, though glaciers now exist only among the peaks of the High Sierra. No other mountain chain on this or any other of the continents that I have seen is so rich as the Sierra in bold, striking, well-preserved glacial monuments. Indeed, every feature is more or less tellingly glacial. Not a peak, ridge, dome, cañon, yosemite, lake-basin, stream or forest will you see that does not in some way explain the past existence and modes of action of flowing, grinding, sculpturing, soil-making, scenery-making ice. For, notwithstanding the post-glacial agents—the air, rain, snow, frost, river, avalanche, etc.—have been at work upon the greater portion of the Range for tens of thousands of stormy years, each engraving its own characters more and more deeply over those of the ice, the latter are so enduring and so heavily emphasized, they still rise in sublime relief, clear and legible, through every after-inscription. The landscapes of North Greenland, Antarctica, and some of those of our own Alaska, are still being fashioned beneath a slow-crawling mantle of ice, from a quarter of a mile to probably more than a mile in thickness, presenting noble illustrations of the ancient condition of California, when its sublime scenery lay hidden in process of formation. On the Himalaya, the mountains of Norway and Switzerland, the Caucasus, and on most of those of Alaska, their ice-mantle has been melted down into separate glaciers that flow river-like through the valleys, illustrating a similar past condition in the Sierra, when every cañon and valley was the channel of an ice-stream, all of which may be easily traced back to their fountains, where some sixty-five or seventy of their topmost residual branches still linger beneath protecting mountain shadows.

The change from one to another of those glacial conditions was slow as we count time. When the great cycle of snow years, called the Glacial Period, was nearly complete in California, the ice-mantle, wasting from season to season faster than it was renewed, began to withdraw from the lowlands and gradually became shallower everywhere. Then the highest of the Sierra domes and dividing ridges, containing distinct glaciers between them, began to appear above the icy sea. These first river-like glaciers remained united in one continuous sheet toward the summit of the Range for many centuries. But as the snowfall diminished, and the climate became milder, this upper part of the ice-sheet was also in turn separated into smaller distinct glaciers, and these again into still smaller ones, while at the same time all were growing shorter and shallower, though fluctuations of the climate now and then occurred that brought their receding ends to a standstill, or even enabled them to advance for a few tens or hundreds of years.

Meanwhile, hardy, home-seeking plants and animals, after long waiting, flocked to their appointed places, pushing bravely on higher and higher, along every sun-warmed slope, closely following the retreating ice, which, like shreds of summer clouds, at length vanished from the new-born mountains, leaving them in all their main, telling features nearly as we find them now.

Tracing the ways of glaciers, learning how Nature sculptures mountain-waves in making scenery-beauty that so mysteriously influences every human being, is glorious work.

The most striking and attractive of the glacial phenomena in the upper Yosemite region are the polished glacier pavements, because they are so beautiful, and their beauty is of so rare a kind, so unlike any portion of the loose, deeply weathered lowlands where people make homes and earn their bread. They are simply flat or gently undulating areas of hard resisting granite, which present the unchanged surface upon which with enormous pressure the ancient glaciers flowed. They are found in most perfect condition in the subalpine region, at an elevation of from eight thousand to nine thousand feet. Some are miles in extent, only slightly interrupted by spots that have given way to the weather, while the best preserved portions reflect the sunbeams like calm water or glass, and shine as if polished afresh every day, notwithstanding they have been exposed to corroding rains, dew, frost, and snow measureless thousands of years.

The attention of wandering hunters and prospectors, who see so many mountain wonders, is seldom commanded by other glacial phenomena, moraines however regular and artificial-looking, cañons however deep or strangely modeled, rocks however high; but when they come to these shining pavements they stop and stare in wondering admiration, kneel again and again to examine the brightest spots, and try hard to account for their mysterious shining smoothness. They may have seen the winter avalanches of snow descending in awful majesty through the woods, scouring the rocks and sweeping away like weeds the trees that stood in their way, but conclude that this cannot be the work of avalanches, because the scratches and fine polished strim show that the agent, whatever it was, moved along the sides of high rocks and ridges and up over the tops of them as well as down their slopes. Neither can they see how water may possibly have been the agent, for they find the same strange polish upon ridges and domes thousands of feet above the reach of any conceivable flood. Of all the agents of whose work they know anything, only the wind seems capable of moving across the face of the country in the directions indicated by the scratches and grooves. The Indian name of Lake Tenaya is "Pyweak"—the lake of shining rocks. One of the Yosemite tribe, Indian Tom, came to me and asked if I could tell him what had made the Tenaya rocks so smooth. Even dogs and horses, when first led up the mountains, study geology to this extent that they gaze wonderingly at the strange brightness of the ground and smell it, and place their feet cautiously upon it as if afraid of falling or sinking.

In the production of this admirable hard finish, the glaciers in many places flowed with a pressure of more than a thousand tons to the square yard, planing down granite, slate, and quartz alike, and bringing out the veins and crystals of the rocks with beautiful distinctness. Over large areas below the sources of the Tuolumne and Merced the granite is porphyritic; feldspar crystals an inch or two in length in many places form the greater part of the rock, and these, when planed off level with the general surface, give rise to a beautiful mosaic on which the happy sunbeams plash and glow in passionate enthusiasm. Here lie the brightest of all the Sierra landscapes. The Range both to the north and south of this region was, perhaps, glaciated about as heavily, but because the rocks are less resisting, their polished surfaces have mostly given way to the weather, leaving only small imperfect patches. The lowest remnants of the old glacial surface occur at an elevation of from 3000 to 5000 feet above the sea-level, and twenty to thirty miles below the axis of the Range. The short, steeply inclined cañons of the eastern flank also contain enduring, brilliantly striated and polished rocks, but these are less magnificent than those of the broad western flank.

One of the best general views of the brightest and best of the Yosemite park landscapes that every Yosemite tourist should see, is to be had from the top of Fairview Dome, a lofty conoidal rock near Cathedral Peak that long ago I named the Tuolumne Glacier Monument, one of the most striking and best preserved of the domes. Its burnished crown is about 1500 feet above the Tuolumne Meadows and 10,000 above the sea. At first sight it seems inaccessible, though a good climber will find it may be scaled on the south side. About halfway up you will find it so steep that there is danger of slipping, but feldspar crystals, two or three inches long, of which the rock is full, having offered greater resistance to atmospheric erosion than the mass of the rock in which they are imbedded, have been brought into slight relief in some places, roughening the surface here and there, and affording helping footholds.

The summit is burnished and scored like the sides and base, the scratches and striæ indicating that the mighty Tuolumne Glacier swept over it as if it were only a mere boulder in the bottom of its channel. The pressure it withstood must have been enormous. Had it been less solidly built it would have been carried away, ground into moraine fragments, like the adjacent rock in which it lay imbedded; for, great as it is, it is only a hard residual knot like the Yosemite domes, brought into relief by the removal of less resisting rock about it; an illustration of the survival of the strongest and most favorably situated.

Hardly less wonderful is the resistance it has offered to the trying mountain weather since first its crown rose above the icy sea. The whole quantity of post-glacial wear and tear it has suffered has not degraded it a hundredth of an inch, as may readily be shown by the polished portions of the surface. A few erratic boulders, nicely poised on its crown, tell an interesting story. They came from the summit-peaks twelve miles away, drifting like chips on the frozen sea, and were stranded here when the top of the monument emerged from the ice, while their companions, whose positions chanced to be above the slopes of the sides where they could not find rest, were carried farther on by falling back on the shallowing ice current.

The general view from the summit consists of a sublime assemblage of ice-born rocks and mountains, long wavering ridges, meadows, lakes, and forest-covered moraines, hundreds of square miles of them. The lofty summit-peaks rise grandly along the sky to the east, the gray pillared slopes of the Hoffmann Range toward the west, and a billowy sea of shining rocks like the Monument, some of them almost as high and which from their peculiar sculpture seem to be rolling westward in the middle ground, something like breaking waves. Immediately beneath you are the Big Tuolumne Meadows, smooth lawns with large breadths of woods on either side, and watered by the young Tuolumne River, rushing cool and clear from its many snow- and ice-fountains. Nearly all the upper part of the basin of the Tuolumne Glacier is in sight, one of the greatest and most influential of all the Sierra ice-rivers. Lavishly flooded by many a noble affluent from the ice-laden flanks of Mounts Dana, Lyell, McClure, Gibbs, Conness, it poured its majestic outflowing current full against the end of the Hoffmann Range, which divided and deflected it to right and left, just as a river of water is divided against an island in the middle of its channel. Two distinct glaciers were thus formed, one of which flowed through the great Tuolumne Cañon and Hetch Hetchy Valley, while the other swept upward in a deep current two miles wide across the divide, five hundred feet high between the basins of the Tuolumne and Merced, into the Tenaya Basin, and thence down through the Tenaya Cañon and Yosemite.

The map-like distinctness and freshness of this glacial landscape cannot fail to excite the attention of every beholder, no matter how little of its scientific significance may be recognized. These bald, westward-leaning rocks, with their rounded backs and shoulders toward the glacier fountains of the summit-mountains, and their split, angular fronts looking in the opposite direction, explain the tremendous grinding force with which the ice-flood passed over them, and also the direction of its flow. And the mountain peaks around the sides of the upper general Tuolumne Basin, with their sharp unglaciated summits and polished rounded sides, indicate the height to which the glaciers rose; while the numerous moraines, curving and swaying in beautiful lines, mark the boundaries of the main trunk and its tributaries as they existed toward the close of the glacial winter. None of the commercial highways of the land or sea, marked with buoys and lamps, fences, and guide-boards, is so unmistakably indicated as are these broad, shining trails of the vanished Tuolumne Glacier and its far-reaching tributaries.

I should like now to offer some nearer views of a few characteristic specimens of these wonderful old ice-streams, though it is not easy to make a selection from so vast a system intimately interblended. The main branches of the Merced Glacier are, perhaps, best suited to our purpose, because their basins, full of telling inscriptions, are the ones most attractive and accessible to the Yosemite visitors who like to look beyond the valley walls. They number five, and may well be called Yosemite glaciers, since they were the agents Nature used in developing and fashioning the grand Valley. The names I have given them are, beginning with the northernmost, Yosemite Creek, Hoffmann, Tenaya, South Lyell, and Illilouette Glaciers. These all converged in admirable poise around from northeast to southeast, welded themselves together into the main Yosemite Glacier, which, grinding gradually deeper, swept down through the Valley, receiving small tributaries on its way from the Indian, Sentinel, and Pohono Cañons; and at length flowed out of the Valley, and on down the Range in a general westerly direction. At the time that the tributaries mentioned above were well defined as to their boundaries, the upper portion of the valley walls, and the highest rocks about them, such as the Domes, the uppermost of the Three Brothers and the Sentinel, rose above the surface of the ice. But during the Valley's earlier history, all its rocks, however lofty, were buried beneath a continuous sheet, which swept on above and about them like the wind, the upper portion of the current flowing steadily, while the lower portion went mazing and swedging down in the crooked and dome-blocked cañons toward the head of the Valley.

Every glacier of the Sierra fluctuated in width and depth and length, and consequently in degree of individuality, down to the latest glacial days. It must, therefore, be borne in mind that the following description of the Yosemite glaciers applies only to their separate condition, and to that phase of their separate condition that they presented toward the close of the glacial period after most of their work was finished, and all the more telling features of the Valley and the adjacent region were brought into relief.

The comparatively level, many-fountained Yosemite Creek Glacier was about fourteen miles in length by four or five in width, and from five hundred to a thousand feet deep. Its principal tributaries, drawing their sources from the northern spurs of the Hoffmann Range, at first pursued a westerly course; then, uniting with each other, and a series of short affluents from the western rim of the basin, the trunk thus formed swept around to the southward in a magnificent curve, and poured its ice over the north wall of Yosemite in cascades about two miles wide. This broad and comparatively shallow glacier formed a sort of crawling, wrinkled ice-cloud, that gradually became more regular in shape and river-like as it grew older. Encircling peaks began to overshadow its highest fountains, rock islets rose here and there amid its ebbing currents, and its picturesque banks, adorned with domes and round-backed ridges, extended in massive grandeur down to the brink of the Yosemite walls.

In the meantime the chief Hoffmann tributaries, slowly receding to the shelter of the shadows covering their fountains, continued to live and work independently, spreading soil, deepening lake-basins and giving finishing touches to the sculpture in general. At length these also vanished, and the whole basin is now full of light. Forests flourish luxuriantly upon its ample moraines, lakes and meadows shine and bloom amid its polished domes, and a thousand gardens adorn the banks of its streams.

It is to the great width and even slope of the Yosemite Creek Glacier that we owe the unrivaled height and sheerness of the Yosemite Falls. For had the positions of the ice-fountains and the structure of the rocks been such as to cause down-thrusting concentration of the Glacier as it approached the Valley, then, instead of a high vertical fall we should have had a long slanting cascade, which after all would perhaps have been as beautiful and interesting, if we only had a mind to see it so.

The short, comparatively swift-flowing Hoffmann Glacier, whose fountains extend along the south slopes of the Hoffmann Range, offered a striking contrast to the one just described. The erosive energy of the latter was diffused over a wide field of sunken, boulder-like domes and ridges. The Hoffmann Glacier, on the contrary, moved right ahead on a comparatively even surface, making a descent of nearly five thousand feet in five miles, steadily contracting and deepening its current, and finally united with the Tenaya Glacier as one of its most influential tributaries in the development and sculpture of the great Half Dome, North Dome and the rocks adjacent to them about the head of the Valley.

The story of its death is not unlike that of its companion already described, though the declivity of its channel, and its uniform exposure to sun-heat prevented any considerable portion of its current from becoming torpid, lingering only well up on the mountain slopes to finish their sculpture and encircle them with a zone of moraine soil for forests and gardens. Nowhere in all this wonderful region will you find more beautiful trees and shrubs and flowers covering the traces of ice.

The rugged Tenaya Glacier wildly crevassed here and there above the ridges it had to cross, instead of drawing its sources direct from the summit of the Range, formed, as we have seen, one of the outlets of the great Tuolumne Glacier, issuing from this noble fountain like a river from a lake, two miles wide, about fourteen miles long, and from 1500 to 2000 feet deep.

In leaving the Tuolumne region it crossed over the divide, as mentioned above, between the Tuolumne and Tenaya basins, making an ascent of five hundred feet. Hence, after contracting its wide current and receiving a strong affluent from the fountains about Cathedral Peak, it poured its massive flood over the northeastern rim of its basin in splendid cascades. Then, crushing heavily against the Clouds' Rest Ridge, it bore down upon the Yosemite domes with concentrated energy.

Toward the end of the ice period, while its Hoffmann companion continued to grind rock-meal for coming plants, the main trunk became torpid, and vanished, exposing wide areas of rolling rock-waves and glistening pavements, on whose channelless surface water ran wild and free. And because the trunk vanished almost simultaneously throughout its whole extent, terminal moraines are found in its cañon channel; nor, since its walls are, in most places, too steeply inclined to admit of the deposition of moraine matter, do we find much of the two main laterals. The lowest of its residual glaciers lingered beneath the shadow of the Yosemite Half Dome; others along the base of Coliseum Peak above Lake Tenaya and along the precipitous wall extending from the lake to the Big Tuolumne Meadows. The latter, on account of the uniformity and continuity of their protecting shadows, formed moraines of considerable length and regularity that are liable to be mistaken for portions of the left lateral of the Tuolumne tributary glacier.

Spend all the time you can spare or steal on the tracks of this grand old glacier, charmed and enchanted by its magnificent cañon, lakes and cascades and resplendent glacier pavements.

The Nevada Glacier was longer and more symmetrical than the last, and the only one of the Merced system whose sources extended directly back to the main summits on the axis of the Range. Its numerous fountains were ranged side by side in three series, at an elevation of from 10,000 to 12,000 feet above the sea. The first, on the right side of the basin, extended from the Matterhorn to Cathedral Peak; that on the left through the Merced group, and these two parallel series were united by a third that extended around the head of the basin in a direction at right angles to the others.

The three ranges of high peaks and ridges that supplied the snow for these fountains, together with the Clouds' Rest Ridge, nearly inclose a rectangular basin, that was filled with a massive sea of ice, leaving an outlet toward the west through which flowed the main trunk glacier, three-fourths of a mile to a mile and a half wide, fifteen miles long, and from 1000 to 1500 feet deep, and entered Yosemite between the Half Dome and Mount Starr King.

Could we have visited Yosemite Valley at this period of its history, we should have found its ice cascades vastly more glorious than their tiny water representatives of the present day. One of the grandest of these was formed by that portion of the Nevada Glacier that poured over the shoulder of the Half Dome.

This glacier, as a whole, resembled an oak, with a gnarled swelling base and wide-spreading branches. Picturesque rocks of every conceivable form adorned its banks, among which glided the numerous tributaries, mottled with black and red and gray boulders, from the fountain peaks, while ever and anon, as the deliberate centuries passed away, dome after dome raised its burnished crown above the ice-flood to enrich the slowly opening landscapes.

The principal moraines occur in short irregular sections along the sides of the cañons, their fragmentary condition being due to interruptions caused by portions of the sides of the cañon walls being too steep for moraine matter to lie on, and to down-sweeping torrents and avalanches. The left lateral of the trunk may be traced about five miles from the mouth of the first main tributary to the Illilouette Cañon. The corresponding section of the right lateral, extending from Cathedral tributary to the Half Dome, is more complete because of the more favorable character of the north side of the cañon. A short side-glacier came in against it from the slopes of Clouds' Rest; but being fully exposed to the sun, it was melted long before the main trunk, allowing the latter to deposit this portion of its moraine undisturbed. Some conception of the size and appearance of this fine moraine may be gained by following the Clouds' Rest trail from Yosemite, which crosses it obliquely and conducts past several sections made by streams. Slate boulders may be seen that must have come from the Lyell group, twelve miles distant. But the bulk of the moraine is composed of porphyritic granite derived from Feldspar and Cathedral Valleys.

On the sides of the moraines we find a series of terraces, indicating fluctuations in the level of the glacier, caused by variations of snowfall, temperature, etc., showing that the climate of the glacial period was diversified by cycles of milder or stormier seasons similar to those of post-glacial time.

After the depth of the main trunk diminished to about five hundred feet, the greater portion became torpid, as is shown by the moraines, and lay dying in its crooked channel like a wounded snake, maintaining for a time a feeble squirming motion in places of exceptional depth, or where the bottom of the cañon was more deeply inclined. The numerous fountain-wombs, however, continued fruitful long after the trunk had vanished, giving rise to an imposing array of short residual glaciers, extending around the rim of the general basin a distance of nearly twenty-four miles. Most of these have but recently succumbed to the new climate, dying in turn as determined by elevation, size, and exposure, leaving only a few feeble survivors beneath the coolest shadows, which are now slowly completing the sculpture of one of the noblest of the Yosemite basins.

The comparatively shallow glacier that at this time filled the Illilouette Basin, though once far from shallow, more resembled a lake than a river of ice, being nearly half as wide as it was long. Its greatest length was about ten miles, and its depth perhaps nowhere much exceeded 1000 feet. Its chief fountains, ranged along the west side of the Merced group, at an elevation of about 10,000 feet, gave birth to fine tributaries that flowed in a westerly direction, and united in the center of the basin. The broad trunk at first flowed northwestward, then curved to the northward, deflected by the lofty wall forming its western bank, and finally united with the grand Yosemite trunk, opposite Glacier Point.

All the phenomena relating to glacial action in this basin are remarkably simple and orderly, on account of the sheltered positions occupied by its ice-fountains, with reference to the disturbing effects of larger glaciers from the axis of the main Range earlier in the period. From the eastern base of the Starr King cone you may obtain a fine view of the principal moraines sweeping grandly out into the middle of the basin from the shoulders of the peaks, between which the ice-fountains lay. The right lateral of the tributary, which took its rise between Red and Merced Mountains, measures two hundred and fifty feet in height at its upper extremity, and displays three well-defined terraces, similar to those of the South Lyell Glacier. The comparative smoothness of the uppermost terrace shows that it is considerably more ancient than the others, many of the boulders of which it is composed having crumbled. A few miles to the westward, this moraine has an average slope of twenty-seven degrees, and an elevation above the bottom of the channel of six hundred and sixty feet. Near the middle of the main basin, just where the regularly formed medial and lateral moraines flatten out and disappear, there is a remarkably smooth field of gravel, planted with arctostaphylos, that looks at the distance of a mile like a delightful meadow. Stream sections show the gravel deposit to be composed of the same material as the moraines, but finer, and more water-worn from the action of converging torrents issuing from the tributary glaciers after the trunk was melted. The southern boundary of the basin is a strikingly perfect wall, gray on the top, and white down the sides and at the base with snow, in which many a crystal brook takes rise. The northern boundary is made up of smooth undulating masses of gray granite, that lift here and there into beautiful domes of which the Starr King cluster is the finest, while on the east tower the majestic fountain-peaks with wide cañons and neve amphitheaters between them, whose variegated rocks show out gloriously against the sky.

The ice-ploughs of this charming basin, ranged side by side in orderly gangs, furrowed the rocks with admirable uniformity, producing irrigating channels for a brood of wild streams, and abundance of rich soil adapted to every requirement of garden and grove.

No other section of the Yosemite uplands is in so perfect a state of glacial cultivation. Its domes, and peaks, and swelling rock-waves, however majestic in themselves, are yet submissively subordinate to the garden center. The other basins we have been describing are combinations of sculptured rocks, embellished with gardens and groves; the Illilouette is one grand garden and forest, embellished with rocks, each of the five beautiful in its own way, and all as harmoniously related as are the five petals of a flower. After uniting in the Yosemite Valley, and expending the down-thrusting energy derived from their combined weight and the declivity of their channels, the grand trunk flowed on through and out of the Valley. In effecting its exit a considerable ascent was made, traces of which may still be seen on the abraded rocks at the lower end of the Valley, while the direction pursued after leaving the Valley is surely indicated by the immense lateral moraines extending from the ends of the walls at an elevation of from 1500 to 1800 feet. The right lateral moraine was disturbed by a large tributary glacier that occupied the basin of Cascade Creek, causing considerable complication in its structure. The left is simple in form for several miles of its length, or to the point where a tributary came in from the southeast. But both are greatly obscured by the forests and underbrush growing upon them, and by the denuding action of rains and melting snows, etc. It is, therefore, the less to be wondered at that these moraines, made up of material derived from the distant fountain-mountains, and from the Valley itself, were not sooner recognized.

The ancient glacier systems of the Tuolumne, San Joaquin, Kern, and Kings River Basins were developed on a still grander scale and are so replete with interest that the most sketchy outline descriptions of each, with the works they have accomplished, would fill many a volume. Therefore I can do but little more than invite everybody who is free to go and see for himself.

The action of flowing ice, whether in the form of river-like glaciers or broad mantles, especially the part it played in sculpturing the earth, is as yet but little understood. Water rivers work openly where people dwell, and so does the rain, and the sea, thundering on all the shores of the world; and the universal ocean of air, though invisible, speaks aloud in a thousand voices, and explains its modes of working and its power. But glaciers, back in their white solitudes, work apart from men, exerting their tremendous energies in silence and darkness. Outspread, spirit-like, they brood above the predestined landscapes, work on unwearied through immeasurable ages, until, in the fullness of time, the mountains and valleys are brought forth, channels furrowed for rivers, basins made for lakes and meadows, and arms of the sea, soils spread for forests and fields; then they shrink and vanish like summer clouds.

XII

HOW BEST TO SPEND ONE'S YOSEMITE TIME

ONE-DAY EXCURSIONS

NO. I.

IF I WERE so time-poor as to have only one day to spend in Yosemite I should start at daybreak, say at three o'clock in midsummer, with a pocketful of any sort of dry breakfast stuff, for Glacier Point, Sentinel Dome, the head of Illilouette Fall, Nevada Fall, the top of Liberty Cap, Vernal Fall and the wild boulder-choked River Cañon. The trail leaves the Valley at the base of the Sentinel Rock, and as you slowly saunter from point to point along its many accommodating zigzags nearly all the Valley rocks and falls are seen in striking, ever-changing combinations. At an elevation of about five hundred feet a particularly fine, wide-sweeping view down the Valley is obtained, past the sheer face of the Sentinel and between the Cathedral Rocks and El Capitan. At a height of about 1500 feet the great Half Dome comes full in sight, overshadowing every other feature of the Valley to the eastward. From Glacier Point you look down 3000 feet over the edge of its sheer face to the meadows and groves and innumerable yellow pine spires, with the meandering river sparkling and spangling through the midst of them. Across the Valley a great telling view is presented of the Royal Arches, North Dome, Indian Cañon, Three Brothers and El Capitan, with the dome-paved basin of Yosemite Creek and Mount Hoffmann in the background. To the eastward, the Half Dome close beside you looking higher and more wonderful than ever; southeastward the Starr King, girdled with silver firs, and the spacious garden-like basin of the Illilouette and its deeply sculptured fountain-peaks, called "The Merced Group"; and beyond all, marshaled along the eastern horizon, the icy summits on the axis of the Range and broad swaths of forests growing on ancient moraines, while the Nevada, Vernal and Yosemite Falls are not only full in sight but are distinctly heard as if one were standing beside them in their spray.

The views from the summit of Sentinel Dome are still more

extensive and telling. Eastward the crowds of peaks at the head of the Merced, Tuolumne and San Joaquin Rivers are presented in bewildering array; westward, the vast forests, yellow foot-hills and the broad San Joaquin plains and the Coast Ranges, hazy and dim in the distance.

From Glacier Point go down the trail into the lower end of the Illilouette basin, cross Illilouette Creek and follow it to the Fall where from an out-jutting rock at its head you will get a fine view of its rejoicing waters and wild cañon and the Half Dome. Thence returning to the trail, follow it to the head of the Nevada Fall. Linger here an hour or two, for not only have you glorious views of the wonderful fall, but of its wild, leaping, exulting rapids and, greater than all, the stupendous scenery into the heart of which the white passionate river goes wildly thundering, surpassing everything of its kind in the world. After an unmeasured hour or so of this glory, all your body aglow, nerve currents flashing through you never before felt, go to the top of the Liberty Cap, only a glad saunter now that your legs as well as head and heart are awake and rejoicing with everything. The Liberty Cap, a companion of the Half Dome, is sheer and inaccessible on three of its sides but on the east a gentle, ice-burnished, juniper-dotted slope extends to the summit where other wonderful views are displayed where all are wonderful: the south side and shoulders of Half Dome and Clouds' Rest, the beautiful Little Yosemite Valley and its many domes, the Starr King cluster of domes, Sentinel Dome, Glacier Point, and, perhaps the most tremendously impressive of all, the views of the hopper-shaped cañon of the river from the head of the Nevada Fall to the head of the Valley.

Returning to the trail you descend between the Nevada Fall and the Liberty Cap with fine side views of both the fall and the rock, pass on through clouds of spray and along the rapids to the head of the Vernal Fall, about a mile below the Nevada. Linger here if night is still distant, for views of this favorite fall and the stupendous rock scenery about it. Then descend a stairway by its side, follow a dim trail through its spray, and a plain one along the border of the boulder-dashed rapids and so back to the wide, tranquil Valley.

ONE-DAY EXCURSIONS

NO. 2.

Another grand one-day excursion is to the Upper Yosemite Fall, the top of the highest of the Three Brothers, called Eagle Peak on the Geological Survey maps; the brow of El Capitan; the head of the Ribbon Fall; across the beautiful Ribbon Creek Basin; and back to the Valley by the Big Oak Flat wagon-road.

The trail leaves the Valley on the east side of the largest of the earthquake taluses immediately opposite the Sentinel Rock and as it passes within a few rods of the foot of the great fall, magnificent views are obtained as you approach it and pass through its spray, though when the snow is melting fast you will be well drenched. From the foot of the Fall the trail zigzags up a narrow cañon between the fall and a plain mural cliff that is burnished here and there by glacial action.

You should stop a while on a flat iron-fenced rock a little below the head of the fall beside the enthusiastic throng of starry comet-like waters to learn something of their strength, their marvelous variety of forms, and above all, their glorious music, gathered and composed from the snow-storms, hail-, rain- and wind-storms that have fallen on their glacier-sculptured, domey, ridgy basin. Refreshed and exhilarated, you follow your trail-way through silver fir and pine woods to Eagle Peak, where the most comprehensive of all the views to be had on the north-wall heights are displayed. After an hour or two of gazing, dreaming, studying the tremendous topography, etc., trace the rim of the Valley to the grand El Capitan ridge and go down to its brow, where you will gain everlasting impressions of Nature's steadfastness and power combined with ineffable fineness of beauty.

Dragging yourself away, go to the head of the Ribbon Fall, thence across the beautiful Ribbon Creek Basin to the Big Oak Flat stage-road, and down its fine grades to the Valley, enjoying glorious Yosemite scenery all the way to the foot of El Capitan and your camp.

TWO-DAY EXCURSIONS

NO. I.

For a two-day trip I would go straight to Mount Hoffmann, spend the night on the summit, next morning go down by May Lake to Tenaya Lake and return to the Valley by Clouds' Rest and the Nevada and Vernal Falls. As on the foregoing excursion, you leave the Valley by the Yosemite Falls trail and follow it to the Tioga wagon-road, a short distance east of Porcupine Flat. From that point push straight up to the summit. Mount Hoffmann is a mass of gray granite that rises almost in the center of the Yosemite Park, about eight or ten miles in a straight line from the Valley. Its southern slopes are low and easily climbed, and adorned here and there with castle-like crumbling piles and long jagged crests that look like artificial masonry; but on the north side it is abruptly precipitous and banked with lasting snow. Most of the broad summit is comparatively level and thick sown with crystals, quartz, mica, hornblende, feldspar, granite, zircon, tourmaline, etc., weathered out and strewn closely and loosely as if they had been sown broadcast. Their radiance is fairly dazzling in sunlight, almost hiding the multitude of small flowers that grow among them. At first sight only these radiant crystals are likely to be noticed, but looking closely you discover a multitude of very small gilias, phloxes, mimulus, etc., many of them with more petals than leaves. On the borders of little streams larger plants flourish—lupines, daisies, asters, goldenrods, hairbell, mountain columbine, potentilla, astragalus and a few gentians; with charming heathworts—bryanthus, cassiope, kalmia, vaccinium in boulder-fringing rings or bank covers. You saunter among the crystals and flowers as if you were walking among stars. From the summit nearly all the Yosemite Park is displayed like a map: forests, lakes, meadows, and snowy peaks. Northward lies Yosemite's wide basin with its domes and small lakes, shining like larger crystals; eastward the rocky, meadowy Tuolumne region, bounded by its snowy peaks in glorious array; southward Yosemite and westward the vast forest. On no other Yosemite Park mountain are you more likely to linger. You will find it a magnificent sky camp. Clumps of dwarf pine and mountain hemlock will furnish resin roots and branches for fuel and light, and the rills, sparkling water. Thousands of the little plant people will gaze at your camp-fire with the crystals and stars, companions and guardians as you lie at rest in the heart of the vast serene night.

The most telling of all the wide Hoffmann views is the basin of the Tuolumne with its meadows, forests and hundreds of smooth rock-waves that appear to be coming rolling on towards you like high heaving waves ready to break, and beyond these the great mountains. But best of all are the dawn and the sunrise. No mountain-top could be better placed for this most glorious of mountain views—to watch and see the deepening colors of the dawn and the sunbeams streaming through the snowy High Sierra passes, awakening the lakes and crystals, the chilled plant people and winged people, and making everything shine and sing in pure glory.

With your heart aglow, spangling Lake Tenaya and Lake May will beckon you away for walks on their ice-burnished shores. Leave Tenaya at the west and cross to the south side of the outlet, and gradually work your way up in an almost straight south direction to the summit of the divide between Tenaya Creek and the main upper Merced River or Nevada Creek and follow the divide to Clouds' Rest. After a glorious view from the crest of this lofty granite wave you will find a trail on its western end that will lead you down past Nevada and Vernal Falls to the Valley in good time, provided you left your Hoffmann sky camp early.

TWO-DAY EXCURSIONS

NO. 2.

Another grand two-day excursion is the same as the first of the one-day trips, as far as the head of Illilouette Fall. From there trace the beautiful stream up through the heart of its magnificent forests and gardens to the cañons between the Red and Merced Peaks, and pass the night where I camped forty-one years ago. Early next morning visit the small glacier on the north side of Merced Peak, the first of the sixty-five that I discovered in the Sierra.

Glacial phenomena in the Illilouette Basin are on the grandest scale, and in the course of my explorations I found that the cañon and moraines between the Merced and Red Mountains were the most interesting of them all. The path of the vanished glacier shone in many places as if washed with silver, and pushing up the cañon on this bright road I passed lake after lake in solid basins of granite and many a meadow along the cañon stream that links them together. The main lateral moraines that bound the view below the cañon are from a hundred to nearly two hundred feet high and wonderfully regular, like artificial embankments, covered with a magnificent growth of silver fir and pine. But this garden and forest luxuriance is speedily left behind, and patches of bryanthus, cassiope and arctic willows begin to appear. The small lakes which a few miles down the Valley are so richly bordered with flowery meadows have at an elevation of 10,000 feet only small brown mats of carex, leaving bare rocks around more than half their shores. Yet, strange to say, amid all this arctic repression the mountain pine on ledges and buttresses of Red Mountain seems to find the climate best suited to it. Some specimens that I measured were over a hundred feet high and twenty-four feet in circumference, showing hardly a trace of severe storms, looking as fresh and vigorous as the giants of the lower zones. Evening came on just as I got fairly into the main cañon. It is about a mile wide and a little less than two miles long. The crumbling spurs of Red Mountain bound it on the north, the somber cliffs of Merced Mountain on the south and a deeply-serrated, splintered ridge curving around from mountain to mountain shuts it in on the east. My camp was on the brink of one of the lakes in a thicket of mountain hemlock, partly sheltered from the wind. Early next morning I set out to trace the ancient glacier to its head. Passing around the north shore of my camp lake I followed the main stream from one lakelet to another. The dwarf pines and hemlocks disappeared and the stream was bordered with icicles. The main lateral moraines that extend from the mouth of the cañon are continued in straggling masses along the walls. Tracing the streams back to the highest of its little lakes, I noticed a deposit of fine gray mud, something like the mud worn from a grindstone. This suggested its glacial origin, for the stream that was carrying it issued from a raw-looking moraine that seemed to be in process of formation. It is from sixty to over a hundred feet high in front, with a slope of about thirty-eight degrees. Climbing to the top of it, I discovered a very small but well-characterized glacier swooping down from the shadowy cliffs of the mountain to its terminal moraine. The ice appeared on all the lower portion of the glacier; farther up it was covered with snow. The uppermost crevasse or "bergeschrund" was from twelve to fourteen feet wide. The melting snow and ice formed a network of rills that ran gracefully down the surface of the glacier, merrily singing in their shining channels. After this discovery I made excursions over all the High Sierra and discovered that what at first sight looked like snow-fields were in great part glaciers which were completing the sculpture of the summit peaks.

Rising early,—which will be easy, as your bed will be rather cold and you will not be able to sleep much anyhow,—after visiting the glacier, climb the Red Mountain and enjoy the magnificent views from the summit. I counted forty lakes from one standpoint on this mountain, and the views to the westward over the Illilouette Basin, the most superbly forested of all the basins whose waters drain into Yosemite, and those of the Yosemite rocks, especially the Half Dome and the upper part of the north wall, are very fine. But, of course, far the most imposing view is the vast array of snowy peaks along the axis of the Range. Then from the top of this peak, light and free and exhilarated with mountain air and mountain beauty, you should run lightly down the northern slope of the mountain, descend the cañon between Red and Gray Mountains, thence northward along the bases of Gray Mountain and Mount Clark and go down into the head of Little Yosemite, and thence down past the Nevada and Vernal Falls to the Valley, a truly glorious two-day trip!

A THREE-DAY EXCURSION

The best three-day excursion, as far as I can see, is the same as the first of the two-day trips until you reach Lake Tenaya. There instead of returning to the Valley, follow the Tioga road around the northwest side of the lake, over to the Tuolumne Meadows and up to the west base of Mount Dana. Leave the road there and make straight for the highest point on the timber line between Mounts Dana and Gibbs and camp there.

On the morning of the third day go to the top of Mount Dana in time for the glory of the dawn and the sunrise over the gray Mono Desert and the sublime forest of High Sierra peaks. When you leave the mountain go far enough down the north side for a view of the Dana Glacier, then make your way back to the Tioga road, follow it along the Tuolumne Meadows to the crossing of Budd Creek where you will find the Sunrise trail branching off up the mountainside through the forest in a southwesterly direction past the west side of Cathedral Peak, which will lead you down to the Valley by the Vernal and Nevada Falls. If you are a good walker you can leave the trail where it begins to descend a steep slope in the silver fir woods, and bear off to the right and make straight for the top of Clouds' Rest. The walking is good and almost level and from the west end of Clouds' Rest take the Clouds' Rest trail which will lead direct to the Valley by the Nevada and Vernal Falls. To any one not desperately time-poor this trip should have four days instead of three; camping the second night at the Soda Springs; thence to Mount Dana and return to the Soda Springs, camping the third night there; thence by the Sunrise trail to Cathedral Peak, visiting the beautiful Cathedral lake which lies about a mile to the west of Cathedral Peak, eating your luncheon, and thence to Clouds' Rest and the Valley as above. This is one of the most interesting of all the comparatively short trips that can be made in the whole Yosemite region. Not only do you see all the grandest of the Yosemite rocks and waterfalls and the High Sierra with their glaciers, glacier lakes and glacier meadows, etc., but sections of the magnificent silver fir, two-leaved pine, and dwarf pine zones; with the principal alpine flowers and shrubs, especially sods of dwarf vaccinium covered with flowers and fruit though less than an inch high, broad mats of dwarf willow scarce an inch high with catkins that rise straight from the ground, and glorious beds of blue gentians,—grandeur enough and beauty enough for a lifetime.

THE UPPER TUOLUMNE EXCURSION

We come now to the grandest of all the Yosemite excursions, one that requires at least two or three weeks. The best time to make it is from about the middle of July. The visitor entering the Yosemite in July has the advantage of seeing the falls not, perhaps, in their very flood prime but next thing to it; while the glacier-meadows will be in their glory and the snow on the mountains will be firm enough to make climbing safe. Long ago I made these Sierra trips, carrying only a sackful of bread with a little tea and sugar and was thus independent and free, but now that trails or carriage roads lead out of the Valley in almost every direction it is easy to take a pack animal, so that the luxury of a blanket and a supply of food can easily be had.

The best way to leave the Valley will be by the Yosemite Fall trail, camping the first night on the Tioga road opposite the east end of the Hoffmann Range. Next morning climb Mount Hoffmann; thence push on past Tenaya Lake into the Tuolumne Meadows and establish a central camp near the Soda Springs, from which glorious excursions can be made at your leisure. For here in this upper Tuolumne Valley is the widest, smoothest, most serenely spacious, and in every way the most delightful summer pleasure-park in all the High Sierra. And since it is connected with Yosemite by two good trails, and a fairly good carriage road that passes between Yosemite and Mount Hoffmann, it is also the most accessible. It is in the heart of the High Sierra east of Yosemite, 8500 to 9000 feet above the level of the sea. The gray, picturesque Cathedral Range bounds it on the south; a similar range or spur, the highest peak of which is Mount Conness, on the north; the noble Mounts Dana, Gibbs, Mammoth, Lyell, McClure and others on the axis of the Range on the east; a heaving, billowy crowd of glacier-polished rocks and Mount Hoffmann on the west. Down through the open sunny meadow-levels of the Valley flows the Tuolumne River, fresh and cool from its many glacial fountains, the highest of which are the glaciers that lie on the north sides of Mount Lyell and Mount McClure.

Along the river a series of beautiful glacier-meadows extend with but little interruption, from the lower end of the Valley to its head, a distance of about twelve miles, forming charming sauntering-grounds from which the glorious mountains may be enjoyed as they look down in divine serenity over the dark forests that clothe their bases. Narrow strips of pine woods cross the meadow-carpet from side to side, and it is somewhat roughened here and there by moraine boulders and dead trees brought down from the heights by snow avalanches; but for miles and miles it is so smooth and level that a hundred horsemen may ride abreast over it.

The main lower portion of the meadows is about four miles long and from a quarter to half a mile wide; but the width of the Valley is, on an average, about eight miles. Tracing the river, we find that it forks a mile above the Soda Springs, the main fork turning southward to Mount Lyell, the other eastward to Mount Dana and Mount Gibbs. Along both forks strips of meadow extend almost to their heads. The most beautiful portions of the meadows are spread over lake-basins, which have been filled up by deposits from the river. A few of these river-lakes still exist, but they are now shallow and are rapidly approaching extinction. The sod in most places is exceedingly fine and silky and free from weeds and bushes; while charming flowers abound, especially gentians, dwarf daisies, potentillas, and the pink bells of dwarf vaccinium. On the banks of the river and its tributaries cassiope and bryanthus may be found, where the sod curls over stream banks and around boulders. The principal grass of these meadows is a delicate calamagrostis with very slender filiform leaves, and when it is in flower the ground seems to be covered with a faint purple mist, the stems of the panicles being so fine that they are almost invisible, and offer no appreciable resistance in walking through them. Along the edges of the meadows beneath the pines and throughout the greater part of the Valley tall ribbon-leaved grasses grow in abundance, chiefly bromus, triticum and agrostis.

In October the nights are frosty, and then the meadows at sunrise, when every leaf is laden with crystals, are a fine sight. The days are still warm and calm, and bees and butterflies continue to waver and hum about the late-blooming flowers until the coming of the snow, usually in November. Storm then follows storm in quick succession, burying the meadows to a depth of from ten to twenty feet, while magnificent avalanches descend through the forests from the laden heights, depositing huge piles of snow mixed with uprooted trees and boulders. In the open sunshine the snow usually lasts until the end of June but the new season's vegetation is not generally in bloom until late in July. Perhaps the best all round excursion-time after winters of average snowfall is from the middle of July to the middle or end of August. The snow is then melted from the woods and southern slopes of the mountains and the meadows and gardens are in their glory, while the weather is mostly all-reviving, exhilarating sunshine. The few clouds that rise now and then and the showers they yield are only enough to keep everything fresh and fragrant.

The groves about the Soda Springs are favorite camping-grounds on account of the cold, pleasant-tasting water charged with carbonic acid, and because of the views of the mountains across the meadow—the Glacier Monument, Cathedral Peak, Cathedral Spires, Unicorn Peak and a series of ornamental nameless companions, rising in striking forms and nearness above a dense forest growing on the left lateral moraine of the ancient Tuolumne Glacier, which, broad, deep, and far-reaching, exerted vast influence on the scenery of this portion of the Sierra. But there are fine camping-grounds all along the meadows, and one may move from grove to grove every day all summer, enjoying new homes and new beauty to satisfy every roving desire for change.

There are five main capital excursions to be made from here—to the summits of Mounts Dana, Lyell and Conness, and through the Bloody Cañon Pass to Mono Lake and the volcanoes, and down the Tuolumne Cañon, at least as far as the foot of the wonderful series of river cataracts. All of these excursions are sure to be made memorable with joyful health-giving experiences; but perhaps none of them will be remembered with keener delight than the days spent in sauntering on the broad velvet lawns by the river, sharing the sky with the mountains and trees, gaining something of their strength and peace.

The excursion to the top of Mount Dana is a very easy one; for though the mountain is 13,000 feet high, the ascent from the west side is so gentle and smooth that one may ride a mule to the very summit. Across many a busy stream, from meadow to meadow, lies your flowery way; mountains all about you, few of them hidden by irregular foregrounds. Gradually ascending, other mountains come in sight, peak rising above peak with their snow and ice in endless variety of grouping and sculpture. Now your attention is turned to the moraines, sweeping in beautiful curves from the hollows and cañons, now to the granite waves and pavements rising here and there above the heathy sod, polished a thousand years ago and still shining. Towards the base of the mountain you note the dwarfing of the trees, until at a height of about 11,000 feet you find patches of the tough, white-barked pine, pressed so flat by the ten or twenty feet of snow piled upon them every winter for centuries that you may walk over them as if walking on a shaggy rug. And, if curious about such things, you may discover specimens of this hardy tree-mountaineer not more than four feet high and about as many inches in diameter at the ground, that are from two hundred to four hundred years old, still holding bravely to life, making the most of their slender summers, shaking their tasseled needles in the breeze right cheerily, drinking the thin sunshine and maturing their fine purple cones as if they meant to live forever. The general view from the summit is one of the most extensive and sublime to be found in all the Range. To the eastward you gaze far out over the desert plains and mountains of the "Great Basin," range beyond range extending with soft outlines, blue and purple in the distance. More than six thousand feet below you lies Lake Mono, ten miles in diameter from north to south, and fourteen from west to east, lying bare in the treeless desert like a disk of burnished metal, though at times it is swept by mountain storm-winds and streaked with foam. To the southward there is a well-defined range of pale-gray extinct volcanoes, and though the highest of them rises nearly two thousand feet above the lake, you can look down from here into their circular, cup-like craters, from which a comparatively short time ago ashes and cinders were showered over the surrounding sage plains and glacier-laden mountains.

To the westward the landscape is made up of exceedingly strong, gray, glaciated domes and ridge waves, most of them comparatively low, but the largest high enough to be called mountains; separated by cañons and darkened with lines and fields of forest, Cathedral Peak and Mount Hoffmann in the distance; small lakes and innumerable meadows in the foreground. Northward and southward the great snowy mountains, marshaled along the axis of the Range, are seen in all their glory, crowded together in some places like trees in groves, making landscapes of wild, extravagant, bewildering magnificence, yet calm and silent as the sky.

Some eight glaciers are in sight. One of these is the Dana Glacier on the north side of the mountain, lying at the foot of a precipice about a thousand feet high, with a lovely pale-green lake a little below it. This is one of the many, small, shrunken remnants of the vast glacial system of the Sierra that once filled the hollows and valleys of the mountains and covered all the lower ridges below the immediate summit-fountains, flowing to right and left away from the axis of the Range, lavishly fed by the snows of the glacial period.

In the excursion to Mount Lyell the immediate base of the mountain is easily reached on meadow walks along the river. Turning to the southward above the forks of the river, you enter the narrow Lyell branch of the Valley, narrow enough and deep enough to be called a cañon. It is about eight miles long and from 2000 to 3000 feet deep. The flat meadow bottom is from about three hundred to two hundred yards wide, with gently curved margins about fifty yards wide from which rise the simple massive walls of gray granite at an angle of about thirty-three degrees, mostly timbered with a light growth of pine and streaked in many places with avalanche channels. Towards the upper end of the cañon the Sierra crown comes in sight, forming a finely balanced picture framed by the massive cañon walls. In the foreground, when the grass is in flower, you have the purple meadow willow-thickets on the river-banks; in the middle distance huge swelling bosses of granite that form the base of the general mass of the mountain, with fringing lines of dark woods marking the lower curves, smoothly snow-clad except in the autumn.

If you wish to spend two days on the Lyell trip you will find a good campground on the east side of the river, about a mile above a fine cascade that comes down over the cañon wall in telling style and makes good camp music. From here to the top of the mountains is usually an easy day's work. At one place near the summit careful climbing is necessary, but it is not so dangerous or difficult as to deter any one of ordinary skill, while the views are glorious. To the northward are Mammoth Mountain, Mounts Gibbs, Dana, Warren, Conness and others, unnumbered and unnamed; to the southeast the indescribably wild and jagged range of Mount Ritter and the Minarets; southwestward stretches the dividing ridge between the north fork of the San Joaquin and the Merced, uniting with the Obelisk or Merced group of peaks that form the main fountains of the Illilouette branch of the Merced; and to the northwestward extends the Cathedral spur. These spurs like distinct ranges meet at your feet; therefore you look at them mostly in the direction of their extension, and their peaks seem to be massed and crowded against one another, while immense amphitheaters, cañons and subordinate ridges with their wealth of lakes, glaciers, and snow-fields, maze and cluster between them. In making the ascent in June or October the glacier is easily crossed, for then its snow mantle is smooth or mostly melted off. But in midsummer the climbing is exceedingly tedious because the snow is then weathered into curious and beautiful blades, sharp and slender, and set on edge in a leaning position. They lean towards the head of the glacier and extend across from side to side in regular order in a direction at right angles to the direction of greatest declivity, the distance between the crests being about two or three feet, and the depth of the troughs between them about three feet. A more interesting problem than a walk over a glacier thus sculptured and adorned is seldom presented to the mountaineer.

The Lyell Glacier is about a mile wide and less than a mile long, but presents, nevertheless, all the essential characters of large, river-like glaciers—moraines, earth-bands, blue veins, crevasses, etc., while the streams that issue from it are, of course, turbid with rock-mud, showing its grinding action on its bed. And it is all the more interesting since it is the highest and most enduring remnant of the great Tuolumne Glacier, whose traces are still distinct fifty miles away, and whose influence on the landscape was so profound. The McClure Glacier, once a tributary of the Lyell, is smaller. Thirty-eight years ago I set a series of stakes in it to determine its rate of motion. Towards the end of summer in the middle of the glacier it was only a little over an inch in twenty-four hours.

The trip to Mono from the Soda Springs can be made in a day, but many days may profitably be spent near the shores of the lake, out on its islands and about the volcanoes.

In making the trip down the Big Tuolumne Cañon, animals may be led as far as a small, grassy, forested lake-basin that lies below the crossing of the Virginia Creek trail. And from this point any one accustomed to walking on earthquake boulders, carpeted with cañon chaparral, can easily go down as far as the big cascades and return to camp in one day. Many, however, are not able to do this, and it is better to go leisurely, prepared to camp anywhere, and enjoy the marvelous grandeur of the place.

The cañon begins near the lower end of the meadows and extends to the Hetch Hetchy Valley, a distance of about eighteen miles, though it will seem much longer to any one who scrambles through it. It is from twelve hundred to about five thousand feet deep, and is comparatively narrow, but there are several roomy, park-like openings in it, and throughout its whole extent Yosemite features are displayed on a grand scale—domes, El Capitan rocks, gables, Sentinels, Royal Arches, Glacier Points, Cathedral Spires, etc. There is even a Half Dome among its wealth of rock forms, though far less sublime than the Yosemite Half Dome. Its falls and cascades are innumerable. The sheer falls, except when the snow is melting in early spring, are quite small in volume as compared with those of Yosemite and Hetch Hetchy; though in any other country many of them would be regarded as wonders. But it is the cascades or sloping falls on the main river that are the crowning glory of the cañon, and these in volume, extent and variety surpass those of any other cañon in the Sierra. The most showy and interesting of them are mostly in the upper part of the cañon, above the point of entrance of Cathedral Creek and Hoffmann Creek. For miles the river is one wild, exulting, on-rushing mass of snowy purple bloom, spreading over glacial waves of granite without any definite channel, gliding in magnificent silver plumes, dashing and foaming through huge boulder-dams, leaping high into the air in wheel-like whirls, displaying glorious enthusiasm, tossing from side to side, doubling, glinting, singing in exuberance of mountain energy.

Every one who is anything of a mountaineer should go on through the entire length of the cañon, coming out by Hetch Hetchy. There is not a dull step all the way. With wide variations, it is a Yosemite Valley from end to end.

Besides these main, far-reaching, much-seeing excursions from the main central camp, there are numberless, lovely little saunters and scrambles and a dozen or so not so very little. Among the best of these are to Lambert and Fair View Domes; to the topmost spires of Cathedral Peak, and to those of the North Church, around the base of which you pass on your way to Mount Conness; to one of the very loveliest of the glacier meadows imbedded in the pine woods about three miles north of the Soda Springs, where forty-two years ago I spent six weeks. It trends east and west, and you can find it easily by going past the base of Lambert's Dome to Dog Lake and thence up northward through the woods about a mile or so, to the shining rock-waves full of ice-burnished, feldspar crystals at the foot of the meadows; to Lake Tenaya; and, last but not least, a rather long and very hearty scramble down by the end of the meadow along the Tioga road toward Lake Tenaya to the crossing of Cathedral Creek, where you turn off and trace the creek down to its confluence with the Tuolumne. This is a genuine scramble much of the way but one of the most wonderfully telling in its glacial rock-forms and inscriptions.

If you stop and fish at every tempting lake and stream you come to, a whole month, or even two months, will not be too long for this grand High Sierra excursion. My own Sierra trip was ten years long.

OTHER TRIPS FROM THE VALLEY

Short carriage trips are usually made in the early morning to Mirror Lake to see its wonderful reflections of the Half Dome and Mount Watkins; and in the afternoon many ride down the Valley to see the Bridal Veil rainbows or up the river cañon to see those of the Vernal Fall; where, standing in the spray, not minding getting drenched, you may see what are called round rainbows, when the two ends of the ordinary bow are lengthened and meet at your feet, forming a complete circle which is broken and united again and again as determined by the varying wafts of spray. A few ambitious scramblers climb to the top of the Sentinel Rock, others walk or ride down the Valley and up to the once-famous Inspiration Point for a last grand view; while a good many appreciative tourists, who have only a day or two, do no climbing or riding but spend their time sauntering on the meadows by the river, watching the falls, and the play of light and shade among the rocks from morning to night, perhaps gaining more than those who make haste up the trails in large noisy parties. Those who have unlimited time find something worth while all the year round on every accessible part of the vast, deeply sculptured walls. At least so I have found it after making the Valley my home for years.

Here are a few specimens of a few of my own short trips which walkers may find useful.

One, up the river cañon, across the bridge between the Vernal and Nevada Falls, through chaparral beds and boulders to the shoulder of Half Dome, along the top of the shoulder to the dome itself, down by a crumbling slot gully and close along the base of the tremendous split front (the most awfully impressive, sheer, precipice view I ever found in all my cañon wanderings), thence up the east shoulder and along the ridge to Clouds' Rest—a glorious sunset—then a grand starry run back home to my cabin; down through the junipers, down through the firs, now in black shadows, now in white light, past roaring Nevada and Vernal, glowering ghost-like beneath their huge frowning cliffs; down the dark, gloomy cañon, through the pines of the Valley, dreamily murmuring in their calm, breezy sleep—a fine wild little excursion for good legs and good eyes—so much sun-, moon- and star-shine in it, and sublime, up-and-down rhythmical, glacial topography.

Another, to the head of Yosemite Fall by Indian Cañon; thence up the Yosemite Creek, tracing it all the way to its highest sources back of Mount Hoffmann, then a wide sweep around the head of its dome-paved basin, passing its many little lakes and bogs, gardens and groves, trilling, warbling rills, and back by the Fall Cañon. This was one of my Sabbath walk, run-and-slide excursions long ago before any trail had been made on the north side of the Valley.

Another fine trip was up, bright and early, by Avalanche Cañon to Glacier Point along the rugged south wall, tracing all its far outs and ins to the head of the Bridal Veil Fall, thence back home, bright and late, by a brushy, bouldery slope between Cathedral rocks and Cathedral spires and along the level Valley floor. This was one of my long, bright-day and bright-night walks thirty or forty years ago when, like river and ocean currents, time flowed undivided, uncounted—a fine free, sauntery, scrambly, botanical, beauty-filled ramble. The walk up the Valley was made glorious by the marvelous brightness of the morning star. So great was her light, she made every tree cast a well-defined shadow on the smooth sandy ground.

Everybody who visits Yosemite wants to see the famous Big Trees. Before the railroad was constructed, all three of the stage-roads that entered the Valley passed through a grove of these trees by the way; namely, the Tuolumne, Merced and Mariposa groves. The Tuolumne grove was passed on the Big Oak Flat road, the Merced grove by the Coulterville road and the Mariposa grove by the Raymond and Wawona road. Now, to see any one of these groves, a special trip has to be made. Most visitors go to the Mariposa grove, the largest of the three. On this Sequoia trip you see not only the giant Big Trees but magnificent forests of silver fir, sugar pine, yellow pine, libocedrus and Douglas spruce. The trip need not require more than two days, spending a night in a good hotel at Wawona, a beautiful place on the south fork of the Merced River, and returning to the Valley or to El Portal, the terminus of the railroad. This extra trip by stage costs fifteen dollars. All the High Sierra excursions that I have sketched cost from a dollar a week to anything you like. None of mine when I was exploring the Sierra cost over a dollar a week, most of them less.

EARLY HISTORY OF THE VALLEY

In the wild gold years of 1849 and '50, the Indian tribes along the western Sierra foot-hills became alarmed at the sudden invasion of their acorn orchard and game fields by miners, and soon began to make war upon them, in their usual murdering, plundering style. This continued until the United States Indian Commissioners succeeded in gathering them into reservations, some peacefully, others by burning their villages and stores of food. The Yosemite or Grizzly Bear tribe, fancying themselves secure in their deep mountain stronghold, were the most troublesome and defiant of all, and it was while the Mariposa battalion, under command of Major Savage, was trying to capture this warlike tribe and conduct them to the Fresno reservation that this deep mountain home, the Yosemite Valley, was discovered. From a camp on the south fork of the Merced, Major Savage sent Indian runners to the bands who were supposed to be hiding in the mountains, instructing them to tell the Indians that if they would come in and make treaty with the Commissioners they would be furnished with food and clothing and be protected, but if they did not come in he would make war upon them and kill them all. None of the Yosemite Indians responded to this general message, but when a special messenger was sent to the chief he appeared the next day. He came entirely alone and stood in dignified silence before one of the guards until invited to enter the camp. He was recognized by one of the friendly Indians as Tenaya, the old chief of the Grizzlies, and, after he had been supplied with food, Major Savage, with the aid of Indian interpreters, informed him of the wishes of the Commissioners. But the old chief was very suspicious of Savage and feared that he was taking this method of getting the tribe into his power for the purpose of revenging his personal wrong. Savage told him if he would go to the Commissioners and make peace with them as the other tribes had done there would be no more war. Tenaya inquired what was the object of taking all the Indians to the San Joaquin plain. "My people," said he, "do not want anything from the Great Father you tell me about. The Great Spirit is our father and he has always supplied us with all we need. We do not want anything from white men. Our women are able to do our work. Go, then. Let us remain in the mountains where we were born, where the ashes of our fathers have been given to the wind. I have said enough."

To this the Major answered abruptly in Indian style: "If you and your people have all you desire, why do you steal our horses and mules? Why do you rob the miners' camps? Why do you murder the white men and plunder and burn their houses?"

Tenaya was silent for some time. He evidently understood what the Major had said, for he replied, "My young men have sometimes taken horses and mules from the whites. This was wrong. It is not wrong to take the property of enemies who have wronged my people. My young men believed that the gold diggers were our enemies. We now know they are not and we shall be glad to live in peace with them. We will stay here and be friends. My people do not want to go to the plains. Some of the tribes who have gone there are very bad. We cannot live with them. Here we can defend ourselves."

To this Major Savage firmly said, "Your people must go to the Commissioners. If they do not your young men will again steal horses and kill and plunder the whites. It was your people who robbed my stores, burned my houses and murdered my men. If they do not make a treaty, your whole tribe will be destroyed. Not one of them will be left alive."

To this the old chief replied, "It is useless to talk to you about who destroyed your property and killed your people. I am old and you can kill me if you will, but it is useless to lie to you who know more than all the Indians. Therefore I will not lie to you but if you will let me return to my people I will bring them in." He was allowed to go. The next day he came back and said his people were on the way to our camp to go with the men sent by the Great Father, who was so good and rich.

Another day passed but no Indians from the deep Valley appeared. The old chief said that the snow was so deep and his village was so far down that it took a long time to climb out of it. After waiting still another day the expedition started for the Valley. When Tenaya was questioned as to the route and distance he said that the snow was so deep that the horses could not go through it. Old Tenaya was taken along as guide. When the party had gone about halfway to the Valley they met the Yosemites on their way to the camp on the south fork. There were only seventy-two of them and when the old chief was asked what had become of the rest of his band, he replied, "This is all of my people that are willing to go with me to the plains. All the rest have gone with their wives and children over the mountains to the Mono and Tuolumne tribes." Savage told Tenaya that he was not telling the truth, for Indians could not cross the mountains in the deep snow, and that he knew they must still be at his village or hiding somewhere near it. The tribe had been estimated to number over two hundred. Major Savage then said to him, "You may return to camp with your people and I will take one of your young men with me to your village to see your people who will not come. They will come if I find them." "You will not find any of my people there," said Tenaya; "I do not know where they are. My tribe is small. Many of the people of my tribe have come from other tribes and if they go to the plains and are seen they will be killed by the friends of those with whom they have quarreled. I was told that I was growing old and it was well that I should go, but that young and strong men can find plenty in the mountains: therefore, why should they go to the hot plains to be penned up like horses and cattle? My heart has been sore since that talk but I am now willing to go, for it is best for my people."

Pushing ahead, taking turns in breaking a way through the snow, they arrived in sight of the great Valley early in the afternoon and, guided by one of Tenaya's Indians, descended by the same route as that followed by the Mariposa trail, and the weary party went into camp on the river-bank opposite El Capitan. After supper, seated around a big fire, the wonderful Valley became the topic of conversation and Dr. Bunell suggested giving it a name. Many were proposed, but after a vote had been taken the name Yosemite, proposed by Dr. Bunell, was adopted almost unanimously to perpetuate the name of the tribe who so long had made their home there. The Indian name of the Valley, however, is Ahwahnee. The Indians had names for all the different rocks and streams of the Valley, but very few of them are now in use by the whites, Pohono, the Bridal Veil, being the principal one. The expedition remained only one day and two nights in the Valley, hurrying out on the approach of a storm and reached the south-fork headquarters on the evening of the third day after starting out. Thus, in three days the round trip had been made to the Valley, most of it had been explored in a general way and some of its principal features had been named. But the Indians had fled up the Tenaya Cañon trail and none of them were seen, except an old woman unable to follow the fugitives.

A second expedition was made in the same year under command of Major Boling. When the Valley was entered no Indians were seen, but the many wigwams with smoldering fires showed that they had been hurriedly abandoned that very day. Later, five young Indians who had been left to watch the movements of the expedition were captured at the foot of the Three Brothers after a lively chase. Three of the five were sons of the old chief and the rock was named for them. All of these captives made good their escape within a few days, except the youngest son of Tenaya, who was shot by his guard while trying to escape. That same day the old chief was captured on the cliff on the east side of Indian Cañon by some of Boling's scouts. As Tenaya walked toward the camp his eye fell upon the dead body of his favorite son. Captain Boling, through an interpreter, expressed his regret at the occurrence, but not a word did Tenaya utter in reply. Later, he made an attempt to escape but was caught as he was about to swim across the river. Tenaya expected to be shot for this attempt and when brought into the presence of Captain Boling he said in great emotion, "Kill me, Sir Captain, yes, kill me as you killed my son, as you would kill my people if they were to come to you. You would kill all my tribe if you had the power. Yes, Sir America, you can now tell your warriors to kill the old chief. You have made my life dark with sorrow. You killed the child of my heart. Why not kill the father? But wait a little and when I am dead I will call my people to come and they shall hear me in their sleep and come to avenge the death of their chief and his son. Yes, Sir America, my spirit will make trouble for you and your people, as you have made trouble to me and my people. With the wizards I will follow the white people and make them fear me. You may kill me, Sir Captain, but you shall not live in peace. I will follow in your footsteps. I will not leave my home, but be with the spirits among the rocks, the waterfalls, in the rivers and in the winds; wherever you go I will be with you. You will not see me but you will fear the spirit of the old chief and grow cold. The Great Spirit has spoken. I am done."

This expedition finally captured the remnants of the tribes at the head of Lake Tenaya and took them to the Fresno reservation, together with their chief, Tenaya. But after a short stay they were allowed to return to the Valley under restrictions. Tenaya promised faithfully to conform to everything required, joyfully left the hot and dry reservation, and with his family returned to his Yosemite home.

The following year a party of miners was attacked by the Indians in the Valley and two of them were killed. This led to another Yosemite expedition. A detachment of regular soldiers from Fort Miller under Lieutenant Moore, U.S.A., was at once dispatched to capture or punish the murderers. Lieutenant Moore entered the Valley in the night and surprised and captured a party of five Indians, but an alarm was given and Tenaya and his people fled from their huts and escaped to the Monos on the east side of the Range. On examination of the five prisoners in the morning it was discovered that each of them had some article of clothing that belonged to the murdered men. The bodies of the two miners were found and buried on the edge of the Bridal Veil meadow. When the captives were accused of the murder of the two white men they admitted that they had killed them to prevent white men from coming to their Valley, declaring that it was their home and that white men had no right to come there without their consent. Lieutenant Moore told them through his interpreter that they had sold their lands to the Government, that it belonged to the white men now, and that they had agreed to live on the reservation provided for them. To this they replied that Tenaya had never consented to the sale of their Valley and had never received pay for it. The other chief, they said, had no right to sell their territory. The lieutenant being fully satisfied that he had captured the real murderers, promptly pronounced judgment and had them placed in line and shot. Lieutenant Moore pursued the fugitives to Mono but was not successful in finding any of them. After being hospitably entertained and protected by the Mono and Paute tribes, they stole a number of stolen horses from their entertainers and made their way by a long, obscure route by the head of the north fork of the San Joaquin, reached their Yosemite home once more, but early one morning, after a feast of horse-flesh, a band of Monos surprised them in their huts, killing Tenaya and nearly all his tribe. Only a small remnant escaped down the river cañon. The Tenaya Cañon and Lake were named for the famous old chief.

Very few visits were made to the Valley before the summer of 1855, when Mr. J. M. Hutchings, having heard of its wonderful scenery, collected a party and made the first regular tourist's visit to the Yosemite and in his California magazine described it in articles illustrated by a good artist, who was taken into the Valley by him for that purpose. This first party was followed by another from Mariposa the same year, consisting of sixteen or eighteen persons. The next year the regular pleasure travel began and a trail on the Mariposa side of the Valley was opened by Mann Brothers. This trail was afterwards purchased by the citizens of the county and made free to the public. The first house built in the Yosemite Valley was erected in the autumn of 1856 and was kept as a hotel the next year by G. A. Hite and later by J. H. Neal and S. M. Cunningham. It was situated directly opposite the Yosemite Fall. A little over half a mile farther up the Valley a canvas house was put up in 1858 by G. A. Hite. Next year a frame house was built and kept as a hotel by Mr. Peck, afterward by Mr. Longhurst and since 1864 by Mr. Hutchings. All these hotels have vanished except the frame house built in 1859, which has been changed beyond recognition. A large hotel built on the brink of the river in front of the old one is now the only hotel in the Valley. A large hotel built by the State and located farther up the Valley was burned. To provide for the overflow of visitors there are three camps with board floors, wood frame, and covered with canvas, well furnished, some of them with electric light. A large first-class hotel is very much needed.

Travel of late years has been rapidly increasing, especially after the establishment, by Act of Congress in 1890, of the Yosemite National Park and the recession in 1905 of the original reservation to the Federal Government by the State. The greatest increase, of course, was caused by the construction of the Yosemite Valley railroad from Merced to the border of the Park, eight miles below the Valley.

It is eighty miles long, and the entire distance, except the first twenty-four miles from the town of Merced, is built through the precipitous Merced River Cañon. The roadbed was virtually blasted out of the solid rock for the entire distance in the cañon. Work was begun in September, 1905, and the first train entered El Portal, the terminus, April 15, 1907. Many miles of the road cost as much as $100,000 per mile. Its business has increased from 4000 tourists in the first year it was operated to 15,000 in 1910.

XIII

LAMON

THE GOOD OLD pioneer, Lamon, was the first of all the early Yosemite settlers who cordially and unreservedly adopted the Valley as his home.

He was born in the Shenandoah Valley, Virginia, May 10, 1817, emigrated to Illinois with his father, John Lamon, at the age of nineteen; afterwards went to Texas and settled on the Brazos, where he raised melons and hunted alligators for a living. "Right interestin' business," he said; "especially the alligator part of it." From the Brazos he went to the Comanche Indian country between Gonzales and Austin, twenty miles from his nearest neighbor. During the first summer, the only bread he had was the breast meat of wild turkeys. When the formidable Comanche Indians were on the war-path he left his cabin after dark and slept in the woods. From Texas he crossed the plains to California and worked in the Calaveras and Mariposa gold-fields.

He first heard Yosemite spoken of as a very beautiful mountain valley and after making two excursions in the summers of 1857 and 1858 to see the wonderful place, he made up his mind to quit roving and make a permanent home in it. In April, 1859, he moved into it, located a garden opposite Mount Half Dome, set out a lot of apple, pear and peach trees, planted potatoes, etc., that he had packed in on a "contrary old mule," and worked for his board in building a hotel which was afterwards purchased by Mr. Hutchings. His neighbors thought he was very foolish in attempting to raise crops in so high and cold a valley, and warned him that he could raise nothing and sell nothing, and would surely starve.

For the first year or two lack of provisions compelled him to move out on the approach of winter, but in 1862 after he had succeeded in raising some fruit and vegetables he began to winter in the Valley.

The first winter he had no companions, not even a dog or cat, and one evening was greatly surprised to see two men coming up the Valley. They were very glad to see him, for they had come from Mariposa in search of him, a report having been spread that he had been killed by Indians. He assured his visitors that he felt safer in his Yosemite home, lying snug and squirrel-like in his 10 x 12 cabin than in Mariposa. When the avalanches began to slip, he wondered where all the wild roaring and booming came from, the flying snow preventing them from being seen. But, upon the whole, he wondered most at the brightness, gentleness, and sunniness of the weather, and hopefully employed the calm days in clearing ground for an orchard and vegetable garden.

In the second winter he built a winter cabin under the Royal Arches, where he enjoyed more sunshine. But no matter how he praised the weather he could not induce any one to winter with him until 1864.

He liked to describe the great flood of 1867, the year before I reached California, when all the walls were striped with thundering waterfalls.

He was a fine, erect, whole-souled man, between six and seven feet high, with a broad, open face, bland and guileless as his pet oxen. No stranger to hunger and weariness, he knew well how to appreciate suffering of a like kind in others, and many there be, myself among the number, who can testify to his simple, unostentatious kindness that found expression in a thousand small deeds.

After gaining sufficient means to enjoy a long afternoon of life in comparative affluence and ease, he died in the autumn of 1876. He sleeps in a beautiful spot near Galen Clark and a monument hewn from a block of Yosemite granite marks his grave.

XIV

GALEN CLARK

GALEN CLARK WAS the best mountaineer I ever met, and one of the kindest and most amiable of all my mountain friends. I first met him at his Wawona ranch forty-three years ago on my first visit to Yosemite. I had entered the Valley with one companion by way of Coulterville, and returned by what was then known as the Mariposa trail. Both trails were buried in deep snow where the elevation was from 5000 to 7000 feet above sea-level in the sugar pine and silver fir regions. We had no great difficulty, however, in finding our way by the trends of the main features of the topography. Botanizing by the way, we made slow, plodding progress, and were again about out of provisions when we reached Clark's hospitable cabin at Wawona. He kindly furnished us with flour and a little sugar and tea, and my companion, who complained of the benumbing poverty of a strictly vegetarian diet, gladly accepted Mr. Clark's offer of a piece of a bear that had just been killed. After a short talk about bears and the forests and the way to the Big Trees, we pushed on up through the Wawona firs and sugar pines, and camped in the now-famous Mariposa grove.

Later, after making my home in the Yosemite Valley, I became well acquainted with Mr. Clark, while he was guardian. He was elected again and again to this important office by different Boards of Commissioners on account of his efficiency and his real love of the Valley.

Although nearly all my mountaineering has been done without companions, I had the pleasure of having Galen Clark with me on three excursions. About thirty-five years ago I invited him to accompany me on a trip through the Big Tuolumne Cañon from Hetch Hetchy Valley. The cañon up to that time had not been explored, and knowing that the difference in the elevation of the river at the head of the cañon and in Hetch Hetchy was about 5000 feet, we expected to find some magnificent cataracts or falls; nor were we disappointed. When we were leaving Yosemite an ambitious young man begged leave to join us. I strongly advised him not to attempt such a long, hard trip, for it would undoubtedly prove very trying to an inexperienced climber. He assured us, however, that he was equal to anything, would gladly meet every difficulty as it came, and cause us no hindrance or trouble of any sort. So at last, after repeating our advice that he give up the trip, we consented to his joining us. We entered the cañon by way of Hetch Hetchy Valley, each carrying his own provisions, and making his own tea, porridge, bed, etc.

In the morning of the second day out from Hetch Hetchy we came to what is now known as "Muir Gorge," and Mr. Clark without hesitation prepared to force a way through it, wading and jumping from one submerged boulder to another through the torrent, bracing and steadying himself with a long pole. Though the river was then rather low, the savage, roaring, surging song it was singing was rather nerve-trying, especially to our inexperienced companion. With careful assistance, however, I managed to get him through, but this hard trial, naturally enough, proved too much and he informed us, pale and trembling, that he could go no farther. I gathered some wood at the upper throat of the gorge, made a fire for him and advised him to feel at home and make himself comfortable, hoped he would enjoy the grand scenery and the songs of the water-ouzels which haunted the gorge, and assured him that we would return some time in the night, though it might be late, as we wished to go on through the entire cañon if possible. We pushed our way through the dense chaparral and over the earthquake taluses with such speed that we reached the foot of the upper cataract while we had still an hour or so of daylight for the return trip. It was long after dark when we reached our adventurous, but nerve-shaken companion who, of course, was anxious and lonely, not being accustomed to solitude, however kindly and flowery and full of sweet bird-song and stream-song. Being tired we simply lay down in restful comfort on the river-bank beside a good fire, instead of trying to go down the gorge in the dark or climb over its high shoulder to our blankets and provisions, which we had left in the morning in a tree at the foot of the gorge. I remember Mr. Clark remarking that if he had his choice that night between provisions and blankets he would choose his blankets.

The next morning in about an hour we had crossed over the ridge through which the gorge is cut, reached our provisions, made tea, and had a good breakfast. As soon as we had returned to Yosemite I obtained fresh provisions, pushed off alone up to the head of

Yosemite Creek basin, entered the cañon by a side cañon, and completed the exploration up to the Tuolumne Meadows.

It was on this first trip from Hetch Hetchy to the upper cataracts that I had convincing proofs of Mr. Clark's daring and skill as a mountaineer, particularly in fording torrents, and in forcing his way through thick chaparral. I found it somewhat difficult to keep up with him in dense, tangled brush, though in jumping on boulder taluses and slippery cobble-beds I had no difficulty in leaving him behind.

After I had discovered the glaciers on Mount Lyell and Mount McClure, Mr. Clark kindly made a second excursion with me to assist in establishing a line of stakes across the McClure glacier to measure its rate of flow. On this trip we also climbed Mount Lyell together, when the snow which covered the glacier was melted into upleaning, icy glades which were extremely difficult to cross, not being strong enough to support our weight, nor wide enough apart to enable us to stride across each blade as it was met. Here again I, being lighter, had no difficulty in keeping ahead of him. While resting after wearisome staggering and falling he stared at the marvelous ranks of leaning blades, and said, "I think I have traveled all sorts of trails and cañons, through all kinds of brush and snow, but this gets me."

Mr. Clark at my urgent request joined my small party on a trip to the Kings River yosemite by way of the high mountains, most of the way without a trail. He joined us at the Mariposa Big Tree grove and intended to go all the way, but finding that, on account of the difficulties encountered, the time required was much greater than he expected, he turned back near the head of the north fork of the Kings River.

In cooking his mess of oatmeal porridge and making tea, his pot was always the first to boil, and I used to wonder why, with all his skill in scrambling through brush in the easiest way, and preparing his meals, he was so utterly careless about his beds. He would lie down anywhere on any ground, rough or smooth, without taking pains even to remove cobbles or sharp-angled rocks protruding through the grass or gravel, saying that his own bones were as hard as any stones and could do him no harm.

His kindness to all Yosemite visitors and mountaineers was marvelously constant and uniform. He was not a good business-man, and in building an extensive hotel and barns at Wawona, before the travel to Yosemite had been greatly developed, he borrowed money, mortgaged his property and lost it all.

Though not the first to see the Mariposa Big Tree grove, he was the first to explore it, after he had heard from a prospector, who had passed through the grove and who gave him the indefinite information, that there were some wonderful big trees up there on the top of the Wawona hill and that he believed they must be of the same kind that had become so famous and well-known in the Calaveras grove farther north. On this information, Galen Clark told me, he went up and thoroughly explored the grove, counting the trees and measuring the largest, and becoming familiar with it. He stated also that he had explored the forest to the southward and had discovered the much larger Fresno grove of about two square miles, six or seven miles distant from the Mariposa grove. Unfortunately most of the Fresno grove has been cut and flumed down to the railroad near Madera.

Mr. Clark was truly and literally a gentle-man. I never heard him utter a hasty, angry, fault-finding word. His voice was uniformly pitched at a rather low tone, perfectly even, although glances of his eyes and slight intonations of his voice often indicated that something funny or mildly sarcastic was coming, but upon the whole he was serious and industrious, and, however deep and fun-provoking a story might be, he never indulged in boisterous laughter.

He was very fond of scenery and once told me after I became acquainted with him that he liked "nothing in the world better than climbing to the top of a high ridge or mountain and looking off." He preferred the mountain ridges and domes in the Yosemite regions on account of the wealth and beauty of the forests. Oftentimes he would take his rifle, a few pounds of bacon, a few pounds of flour, and a single blanket and go off hunting, for no other reason than to explore and get acquainted with the most beautiful points of view within a journey of a week or two from his Wawona home. On these trips he was always alone and could indulge in tranquil enjoyment of Nature to his heart's content. He said that on those trips, when he was a sufficient distance from home in a neighborhood where he wished to linger, he always shot a deer, sometimes a grouse, and occasionally a bear. After diminishing the weight of a deer or bear by eating part of it, he carried as much as possible of the best of the meat to

Wawona, and from his hospitable well-supplied cabin no weary wanderer ever went away hungry or unrested.

The value of the mountain air in prolonging life is well exemplified in Mr. Clark's case. While working in the mines he contracted a severe cold that settled on his lungs and finally caused severe inflammation and bleeding, and none of his friends thought he would ever recover. The physicians told him he had but a short time to live. It was then that he repaired to the beautiful sugar pine woods at Wawona and took up a claim, including the fine meadows there, and building his cabin, began his life of wandering and exploring in the glorious mountains about him, usually going bareheaded. In a remarkably short time his lungs were healed.

He was one of the most sincere tree-lovers I ever knew. About twenty years before his death he made choice of a plot in the Yosemite cemetery on the north side of the Valley, not far from the Yosemite Fall, and selecting a dozen or so of seedling sequoias in the Mariposa grove he brought them to the Valley and planted them around the spot he had chosen for his last rest. The ground there is gravelly and dry; by careful watering he finally nursed most of the seedlings into good, thrifty trees, and doubtless they will long shade the grave of their blessed lover and friend.

XV

HETCH HETCHY VALLEY

YOSEMITE IS SO wonderful that we are apt to regard it as an exceptional creation, the only valley of its kind in the world; but Nature is not so poor as to have only one of anything. Several other yosemites have been discovered in the Sierra that occupy the same relative positions on the Range and were formed by the same forces in the same kind of granite. One of these, the Hetch Hetchy Valley, is in the Yosemite National Park about twenty miles from Yosemite and is easily accessible to all sorts of travelers by a road and trail that leaves the Big Oak Flat road at Bronson Meadows a few miles below Crane Flat, and to mountaineers by way of Yosemite Creek basin and the head of the middle fork of the Tuolumne.

It is said to have been discovered by Joseph Screech, a hunter, in 1850, a year before the discovery of the great Yosemite. After my first visit to it in the autumn of 1871, I have always called it the "Tuolumne Yosemite," for it is a wonderfully exact counterpart of the Merced Yosemite, not only in its sublime rocks and waterfalls but in the gardens, groves and meadows of its flowery park-like floor. The floor of Yosemite is about 4000 feet above the sea; the Hetch Hetchy floor about 3700 feet. And as the Merced River flows through Yosemite, so does the Tuolumne through Hetch Hetchy. The walls of both are of gray granite, rise abruptly from the floor, are sculptured in the same style and in both every rock is a glacier monument.

Standing boldly out from the south wall is a strikingly picturesque rock called by the Indians, Kolana, the outermost of a group 2300 feet high, corresponding with the Cathedral Rocks of Yosemite both in relative position and form. On the opposite side of the Valley, facing Kolana, there is a counterpart of the El Capitan that rises sheer and plain to a height of 1800 feet, and over its massive brow flows a stream which makes the most graceful fall I have ever seen. From the edge of the cliff to the top of an earthquake talus it is perfectly free in the air for a thousand feet before it is broken into cascades among talus boulders. It is in all its glory in June, when the snow is melting fast, but fades and vanishes toward the end of summer. The only fall I know with which it may fairly be compared is the Yosemite Bridal Veil; but it excels even that favorite fall both in height and airy-fairy beauty and behavior. Lowlanders are apt to suppose that mountain streams in their wild career over cliffs lose control of themselves and tumble in a noisy chaos of mist and spray. On the contrary, on no part of their travels are they more harmonious and self-controlled. Imagine yourself in Hetch Hetchy on a sunny day in June, standing waist-deep in grass and flowers (as I have often stood), while the great pines sway dreamily with scarcely perceptible motion. Looking northward across the Valley you see a plain, gray granite cliff rising abruptly out of the gardens and groves to a height of 1800 feet, and in front of it Tueeulala's silvery scarf burning with irised sun-fire. In the first white outburst at the head there is abundance of visible energy, but it is speedily hushed and concealed in divine repose, and its tranquil progress to the base of the cliff is like that of a downy feather in a still room. Now observe the fineness and marvelous distinctness of the various sun-illumined fabrics into which the water is woven; they sift and float from form to form down the face of that grand gray rock in so leisurely and unconfused a manner that you can examine their texture, and patterns and tones of color as you would a piece of embroidery held in the hand. Toward the top of the fall you see groups of booming, comet-like masses, their solid, white heads separate, their tails like combed silk interlacing among delicate gray and purple shadows, ever forming and dissolving, worn out by friction in their rush through the air. Most of these vanish a few hundred feet below the summit, changing to varied forms of cloudlike drapery. Near the bottom the width of the fall has increased from about twenty-five feet to a hundred feet. Here it is composed of yet finer tissues, and is still without a trace of disorder—air, water and sunlight woven into stuff that spirits might wear.

So fine a fall might well seem sufficient to glorify any valley; but here, as in Yosemite, Nature seems in nowise moderate, for a short distance to the eastward of Tueeulala booms and thunders the great Hetch Hetchy Fall, Wapama, so near that you have both of them in full view from the same standpoint. It is the counterpart of the Yosemite Fall, but has a much greater volume of water, is about 1700 feet in height, and appears to be nearly vertical, though considerably inclined, and is dashed into huge outbounding bosses of foam on projecting shelves and knobs. No two falls could be more unlike

—Tueeulala out in the open sunshine descending like thistledown; Wapama in a jagged, shadowy gorge roaring and thundering, pounding its way like an earthquake avalanche.

Besides this glorious pair there is a broad, massive fall on the main river a short distance above the head of the Valley. Its position is something like that of the Vernal in Yosemite, and its roar as it plunges into a surging trout-pool may be heard a long way, though it is only about twenty feet high. On Rancheria Creek, a large stream, corresponding in position with the Yosemite Tenaya Creek, there is a chain of cascades joined here and there with swift flashing plumes like the one between the Vernal and Nevada Falls, making magnificent shows as they go their glacier-sculptured way, sliding, leaping, hurrahing, covered with crisp clashing spray made glorious with sifting sunshine. And besides all these a few small streams come over the walls at wide intervals, leaping from ledge to ledge with birdlike song and watering many a hidden cliff-garden and fernery, but they are too unshowy to be noticed in so grand a place.

The correspondence between the Hetch Hetchy walls in their trends, sculpture, physical structure, and general arrangement of the main rock-masses and those of the Yosemite Valley has excited the wondering admiration of every observer. We have seen that the El Capitan and Cathedral rocks occupy the same relative positions in both valleys; so also do their Yosemite points and North Domes. Again, that part of the Yosemite north wall immediately to the east of the Yosemite Fall has two horizontal benches, about 500 and 1500 feet above the floor, timbered with golden-cup oak. Two benches similarly situated and timbered occur on the same relative portion of the Hetch Hetchy north wall, to the east of Wapama Fall, and on no other. The Yosemite is bounded at the head by the great Half Dome. Hetch Hetchy is bounded in the same way, though its head rock is incomparably less wonderful and sublime in form.

The floor of the Valley is about three and a half miles long, and from a fourth to half a mile wide. The lower portion is mostly a level meadow about a mile long, with the trees restricted to the sides and the river-banks, and partially separated from the main, upper, forested portion by a low bar of glacier-polished granite across which the river breaks in rapids.

The principal trees are the yellow and sugar pines, digger pine, incense cedar, Douglas spruce, silver fir, the California and golden-cup oaks, balsam cottonwood, Nuttall's flowering dogwood, alder, maple, laurel, tumion, etc. The most abundant and influential are the great yellow or silver pines like those of Yosemite, the tallest over two hundred feet in height, and the oaks assembled in magnificent groves with massive rugged trunks four to six feet in diameter, and broad, shady, wide-spreading heads. The shrubs forming conspicuous flowery clumps and tangles are manzanita, azalea, spiræa, brier-rose, several species of ceanothus, calycanthus, philadelphus, wild cherry, etc.; with abundance of showy and fragrant herbaceous plants growing about them or out in the open in beds by themselves—lilies, Mariposa tulips, brodiaeas, orchids, iris, spraguea, draperia, collomia, collinsia, castilleja, nemophila, larkspur, columbine, goldenrods, sunflowers, mints of many species, honeysuckle, etc. Many fine ferns dwell here also, especially the beautiful and interesting rock-ferns—pellaea, and cheilanthes of several species—fringing and rosetting dry rock-piles and ledges; woodwardia and asplenium on damp spots with fronds six or seven feet high; the delicate maidenhair in mossy nooks by the falls, and the sturdy, broad-shouldered pteris covering nearly all the dry ground beneath the oaks and pines.

It appears, therefore, that Hetch Hetchy Valley, far from being a plain, common, rock-bound meadow, as many who have not seen it seem to suppose, is a grand landscape garden, one of Nature's rarest and most precious mountain temples. As in Yosemite, the sublime rocks of its walls seem to glow with life, whether leaning back in repose or standing erect in thoughtful attitudes, giving welcome to storms and calms alike, their brows in the sky, their feet set in the groves and gay flowery meadows, while birds, bees, and butterflies help the river and waterfalls to stir all the air into music—things frail and fleeting and types of permanence meeting here and blending, just as they do in Yosemite, to draw her lovers into close and confiding communion with her.

Sad to say, this most precious and sublime feature of the Yosemite National Park, one of the greatest of all our natural resources for the uplifting joy and peace and health of the people, is in danger of being dammed and made into a reservoir to help supply San Francisco with water and light, thus flooding it from wall to wall and burying its gardens and groves one or two hundred feet deep. This grossly destructive commercial scheme has long been planned and urged (though water as pure and abundant can be got from sources outside of the people's park, in a dozen different places), because of the comparative cheapness of the dam and of the territory which it is sought to divert from the great uses to which it was dedicated in the Act of 1890 establishing the Yosemite National Park.

The making of gardens and parks goes on with civilization all over the world, and they increase both in size and number as their value is recognized. Everybody needs beauty as well as bread, places to play in and pray in, where Nature may heal and cheer and give strength to body and soul alike. This natural beauty-hunger is made manifest in the little window-sill gardens of the poor, though perhaps only a geranium slip in a broken cup, as well as in the carefully tended rose and lily gardens of the rich, the thousands of spacious city parks and botanical gardens, and in our magnificent National parks—the Yellowstone, Yosemite, Sequoia, etc.—Nature's sublime wonderlands, the admiration and joy of the world. Nevertheless, like anything else worth while, from the very beginning, however well guarded, they have always been subject to attack by despoiling gain-seekers and mischief-makers of every degree from Satan to Senators, eagerly trying to make everything immediately and selfishly commercial, with schemes disguised in smug-smiling philanthropy, industriously, shampiously crying, "Conservation, conservation, panutilization," that man and beast may be fed and the dear Nation made great. Thus long ago a few enterprising merchants utilized the Jerusalem temple as a place of business instead of a place of prayer, changing money, buying and selling cattle and sheep and doves; and earlier still, the first forest reservation, including only one tree, was likewise despoiled. Ever since the establishment of the Yosemite National Park, strife has been going on around its borders and I suppose this will go on as part of the universal battle between right and wrong, however much its boundaries may be shorn, or its wild beauty destroyed.

The first application to the Government by the San Francisco Supervisors for the commercial use of Lake Eleanor and the Hetch Hetchy Valley was made in 1903, and on December 22nd of that year it was denied by the Secretary of the Interior, Mr. Hitchcock, who truthfully said:

> Presumably the Yosemite National Park was created such by

> law because of the natural objects of varying degrees of scenic importance located within its boundaries, inclusive alike of its beautiful small lakes, like Eleanor, and its majestic wonders, like Hetch Hetchy and Yosemite Valley. It is the aggregation of such natural scenic features that makes the Yosemite Park a wonderland which the Congress of the United States sought by law to reserve for all coming time as nearly as practicable in the condition fashioned by the hand of the Creator—a worthy object of National pride and a source of healthful pleasure and rest for the thousands of people who may annually sojourn there during the heated months.

In 1907 when Mr. Garfield became Secretary of the Interior the application was renewed and granted; but under his successor, Mr. Fisher, the matter has been referred to a Commission, which as this volume goes to press still has it under consideration.

The most delightful and wonderful camp grounds in the Park are its three great valleys—Yosemite, Hetch Hetchy, and Upper Tuolumne; and they are also the most important places with reference to their positions relative to the other great features—the Merced and Tuolumne Cañons, and the High Sierra peaks and glaciers, etc., at the head of the rivers. The main part of the Tuolumne Valley is a spacious flowery lawn four or five miles long, surrounded by magnificent snowy mountains, slightly separated from other beautiful meadows, which together make a series about twelve miles in length, the highest reaching to the feet of Mount Dana, Mount Gibbs, Mount Lyell and Mount McClure. It is about 8500 feet above the sea, and forms the grand central High Sierra campground from which excursions are made to the noble mountains, domes, glaciers, etc.; across the Range to the Mono Lake and volcanoes and down the Tuolumne Cañon to Hetch Hetchy. Should Hetch Hetchy be submerged for a reservoir, as proposed, not only would it be utterly destroyed, but the sublime cañon way to the heart of the High Sierra would be hopelessly blocked and the great camping ground, as the watershed of a city drinking system, virtually would be closed to the public. So far as I have learned, few of all the thousands who have seen the park and seek rest and peace in it are in favor of this outrageous scheme.

One of my later visits to the Valley was made in the autumn of

1907 with the late William Keith, the artist. The leaf-colors were then ripe, and the great godlike rocks in repose seemed to glow with life. The artist, under their spell, wandered day after day along the river and through the groves and gardens, studying the wonderful scenery; and, after making about forty sketches, declared with enthusiasm that although its walls were less sublime in height, in picturesque beauty and charm Hetch Hetchy surpassed even Yosemite.

That any one would try to destroy such a place seems incredible; but sad experience shows that there are people good enough and bad enough for anything. The proponents of the dam scheme bring forward a lot of bad arguments to prove that the only righteous thing to do with the people's parks is to destroy them bit by bit as they are able. Their arguments are curiously like those of the devil, devised for the destruction of the first garden—so much of the very best Eden fruit going to waste; so much of the best Tuolumne water and Tuolumne scenery going to waste. Few of their statements are even partly true, and all are misleading.

Thus, Hetch Hetchy, they say, is a "low-lying meadow." On the contrary, it is a high-lying natural landscape garden, as the photographic illustrations show.

"It is a common minor feature, like thousands of others." On the contrary it is a very uncommon feature; after Yosemite, the rarest and in many ways the most important in the National Park.

"Damming and submerging it 175 feet deep would enhance its beauty by forming a crystal-clear lake." Landscape gardens, places of recreation and worship, are never made beautiful by destroying and burying them. The beautiful sham lake, forsooth, would be only an eyesore, a dismal blot on the landscape, like many others to be seen in the Sierra. For, instead of keeping it at the same level all the year, allowing Nature centuries of time to make new shores, it would, of course, be full only a month or two in the spring, when the snow is melting fast; then it would be gradually drained, exposing the slimy sides of the basin and shallower parts of the bottom, with the gathered drift and waste, death and decay of the upper basins, caught here instead of being swept on to decent natural burial along the banks of the river or in the sea. Thus the Hetch Hetchy dam-lake would be only a rough imitation of a natural lake for a few of the spring months, an open sepulcher for the others.

"Hetch Hetchy water is the purest of all to be found in the Sierra, unpolluted, and forever unpollutable." On the contrary, excepting that of the Merced below Yosemite, it is less pure than that of most of the other Sierra streams, because of the sewerage of campgrounds draining into it, especially of the Big Tuolumne Meadows campground, occupied by hundreds of tourists and mountaineers, with their animals, for months every summer, soon to be followed by thousands from all the world.

These temple destroyers, devotees of ravaging commercialism, seem to have a perfect contempt for Nature, and, instead of lifting their eyes to the God of the mountains, lift them to the Almighty Dollar.

Dam Hetch Hetchy! As well dam for water-tanks the people's cathedrals and churches, for no holier temple has ever been consecrated by the heart of man.

APPENDIX A

LEGISLATION ABOUT THE YOSEMITE

In the year 1864, Congress passed the following act:—

ACT OF JUNE 30, 1864 (13 STAT., 325).

AN ACT Authorizing a grant to the State of California of the "Yosemite Valley," and of the land embracing the "Mariposa Big Tree Grove."

"*Be it enacted by the Senate and House of Representatives of the United States of America, in Congress assembled*, That there shall be, and is hereby, granted to the State of California, the 'Cleft' or 'Gorge' in the Granite Peak of the Sierra Nevada Mountains, situated in the county of Mariposa, in the State aforesaid, and the headwaters of the Merced River, and known as the Yosemite Valley, with its branches and spurs, in estimated length fifteen miles, and in average width one mile back from the main edge of the precipice, on each side of the Valley, with the stipulation, nevertheless, that the said State shall accept this grant upon the express conditions that the premises shall be held for public use, resort, and recreation; shall be inalienable for all time; but leases not exceeding ten years may be granted for portions of said premises. All incomes derived from leases of privileges to be expended in the preservation and improvement of the property, or the roads leading thereto; the boundaries to be established at the cost of said State by the United States Surveyor-General of California, whose official plat, when affirmed by the Commissioner of the General Land Office, shall constitute the evidence of the locus, extent, and limits of the said Cleft or Gorge; the premises to be managed by the Governor of the State, with eight other Commissioners, to be appointed by the Executive of California, and who shall receive no compensation for their services.

"SEC. 2. *And be it further enacted*, That there shall likewise be, and there is hereby, granted to the said State of California, the tracts embracing what is known as the 'Mariposa Big Tree Grove,' not to exceed the area of four sections, and to be taken in legal subdivisions of one-quarter section each, with the like stipulations as expressed in the first section of this Act as to the State's acceptance, with like conditions as in the first section of this Act as to inalienability, yet with the same lease privileges; the income to be expended in the preservation, improvement, and protection of the property, the premises to be managed by Commissioners, as stipulated in the first section of this Act, and to be taken in legal subdivisions as aforesaid; and the official plat of the United States Surveyor-General, when affirmed by the Commissioner of the General Land Office, to be the evidence of the locus of the said Mariposa Big Tree Grove."

This important act was approved by the President, June 30, 1864, and shortly after the Governor of California, F. F. Low, issued a proclamation taking possession of the Yosemite Valley and Mariposa grove of Big Trees, in the name and on behalf of the State, appointing commissioners to manage them, and warning all persons against trespassing or settling there without authority, and especially forbidding the cutting of timber and other injurious acts.

The first Board of Commissioners were F. Law Olmsted, J. D. Whitney, William Ashburner, I. W. Raymond, E. S. Holden, Alexander Deering, George W. Coulter, and Galen Clark.

ACT OF OCTOBER 1, 1890 (26 STAT., 650).*

AN ACT To set apart certain tracts of land in the State of California as forest reservations.

"*Be it enacted by the Senate and House of Representatives of the United States of America in Congress assembled*, That the tracts of land in the State of California known as described as follows: Commencing at the northwest corner of township two north, range nineteen east Mount Diablo meridian, thence eastwardly on the line between townships two and three north, ranges twenty-four and twenty-five east; thence southwardly on the line between ranges twenty-four

* Sections 1 and 2 of this act pertain to the Yosemite National Park, while section 3 sets apart General Grant National Park, and also a portion of Sequoia National Park.

and twenty-five east to the Mount Diablo base line; thence eastwardly on said base line to the corner to township one south, ranges twenty-five and twenty-six east; thence southwardly on the line between ranges twenty-five and twenty-six east to the southeast corner of township two south, range twenty-five east; thence eastwardly on the line between townships two and three south, range twenty-six east to the corner to townships two and three south, ranges twenty-six and twenty-seven east; thence southwardly on the line between ranges twenty-six and twenty-seven east to the first standard parallel south; thence westwardly on the first standard parallel south to the southwest corner of township four south, range nineteen east; thence northwardly on the line between ranges eighteen and nineteen east to the northwest corner of township two south, range nineteen east; thence westwardly on the line between townships one and two south to the southwest corner of township one south, range nineteen east; thence northwardly on the line between ranges eighteen and nineteen east to the northwest corner of township two north, range nineteen east, the place of beginning, are hereby reserved and withdrawn from settlement, occupancy, or sale under the laws of the United States, and set apart as reserved forest lands; and all persons who shall locate or settle upon, or occupy the same or any part thereof, except as hereinafter provided, shall be considered trespassers and removed therefrom: *Provided, however,* That nothing in this act shall be construed as in anywise affecting the grant of lands made to the State of California by virtue of the act entitled, 'An act authorizing a grant to the State of California of the Yosemite Valley, and of the land embracing the Mariposa Big-Tree Grove,' appeared June thirtieth, eighteen hundred and sixty-four; or as affecting any bona-fide entry of land made within the limits above described under any law of the United States prior to the approval of this act.

"SEC. 2. That said reservation shall be under the exclusive control of the Secretary of the Interior, whose duty it shall be, as soon as practicable, to make and publish such rules and regulations as he may deem necessary or proper for the care and management of the same. Such regulations shall provide for the preservation from injury of all timber, mineral deposits, natural curiosities, or wonders within said reservation, and their retention in their natural condition. The Secretary may, in his discretion, grant leases for building purposes for terms not exceeding ten years of small parcels of ground not exceeding five acres; at such places in said reservation as shall require the erection of buildings for the accommodation of visitors; all of the proceeds of said leases and other revenues that may be derived from any source connected with said reservation to be expended under his direction in the management of the same and the construction of roads and paths therein. He shall provide against the wanton destruction of the fish, and game found within said reservation, and against their capture or destruction, for the purposes of merchandise or profit. He shall also cause all persons trespassing upon the same after the passage of this act to be removed therefrom, and, generally, shall be authorized to take all such measures as shall be necessary or proper to fully carry out the objects and purposes of this act.

"SEC. 3. There shall also be and is hereby reserved and withdrawn from settlement, occupancy, or sale under the laws of the United States, and shall be set apart as reserved forest lands, as hereinbefore provided, and subject to all the limitations and provisions herein contained, the following additional lands, to wit: Township seventeen south, range thirty east of the Mount Diablo meridian, excepting sections thirty-one, thirty-two, thirty-three, and thirty-four of said township, included in a previous bill. And there is also reserved and withdrawn from settlement, occupancy, or sale under the laws of the United States, and set apart as forest lands, subject to like limitations, conditions, and provisions, all of townships fifteen and sixteen south, of ranges twenty-nine and thirty east of the Mount Diablo meridian. And there is also hereby reserved and withdrawn from settlement, occupancy, or sale under the laws of the United States, and set apart as reserved forest lands under like limitations, restrictions, and provisions, sections five and six in township fourteen south, range twenty-eight east of Mount Diablo meridian, and also sections thirty-one and thirty-two of township thirteen south, range twenty-eight east of the same meridian. Nothing in this act shall authorize rules or contracts touching the protection and improvement of said reservations, beyond the sums that may be received by the Secretary of the Interior under the foregoing provisions, or authorize any charge against the Treasury of the United States.

ACT OF THE LEGISLATURE OF THE STATE OF CALIFORNIA, APPROVED MARCH 3, 1905

"SEC. 1. The State of California does hereby recede and regrant unto the United States of America the 'cleft' or 'gorge' in the granite peak of the Sierra Nevada Mountains, situated in the county of Mariposa, State of California, and the headwaters of the Merced River, and known as the Yosemite Valley, with its branches and spurs, granted unto the State of California in trust for public use, resort, and recreation by the act of Congress entitled, 'An act authorizing a grant to the State of California of the Yosemite Valley and of the land embracing the Mariposa Big Tree Grove,' approved June thirtieth, eighteen hundred and sixty-four; and the State of California does hereby relinquish unto the United States of America and resign the trusts created and granted by the said act of Congress.

"SEC. 2. The State of California does hereby recede and regrant unto the United States of America the tracts embracing what is known as the 'Mariposa Big Tree Grove,' planted unto the State of California in trust for public use, resort, and recreation by the act of Congress referred to in section one of this act, and the State of California does hereby relinquish unto the United States of America and resign the trusts created and granted by the said act of Congress.

"SEC. 3. This act shall take effect from and after acceptance by the United States of America of the recessions and regrants herein made, thereby forever releasing the State of California from further cost of maintaining the said premises, the same to be held for all time by the United States of America for public use, resort, and recreation, and imposing on the United States of America the cost of maintaining the same as a national park: *Provided, however,* That the recession and regrant hereby made shall not affect vested rights and interests of third persons."

APPENDIX B

TABLE OF DISTANCES

From the Guardian's office, in the village, the distances to various points are in miles as follows:

	Miles.
Bridal Veil Fall	4.04
Cascade Falls	7.67
Clouds' Rest, Summit	11.81
Columbia Rock, on Eagle Peak Trail	1.98
Dana, Mt., Summit	40.34
Eagle Peak	6.59
El Capitan Bridge	3.63
Glacier Point, direct trail	4.45
Glacier Point, by Nevada Falls	16.98
Lyell, Mt., Summit	38.20
Merced Bridge	2.03
Mirror Lake, by Hunt's avenue	2.91
Nevada Fall (Hotel)	4.63
Nevada Fall, Bridge above	5.45
Pohono Bridge	5.29
Register Rock	3.24
Ribbon Fall	3.99
Rocky Point (base of Three Brothers)	1.45
Tenayah Creek Bridge	2.26
Tenayah Lake	16.00
Yosemite Falls, foot	0.90
Yosemite Falls, foot Upper Fall	2.67
Yosemite Falls, top	4.33
Soda Springs (Eagle Peak Trail)	24.50
Sentinel Dome	5.57
Union Point, on Glacier Point Trail	3.13
Vernal Fall	3.50

APPENDIX C

MAXIMUM RATES FOR TRANSPORTATION

The following rates for transportation in and about the Valley have been established by the Board of Commissioners:

SADDLE-HORSES

From	*Route to*	*Amount.*
Valley	Glacier Point and Sentinel Dome, and return, direct, same day	$3 00
Valley	Glacier Point, Sentinel Dome, and Fissures, and return, direct, same day	3 75
Valley	Glacier Point, Sentinel Dome, and Fissures, passing night at Glacier Point	3 00
Valley	Glacier Point, Sentinel Dome, Nevada Fall, and Casa Nevada, passing night at Casa Nevada	3 00
Valley	Glacier Point, Sentinel Dome, Nevada Fall, Vernal Fall, and thence to Valley same day	4 00
Glacier Point	Valley direct	2 00
Glacier Point	Sentinel Dome, Nevada Fall, and Casa Nevada, passing night at Casa Nevada	2 00
Glacier Point	Sentinel Dome, Nevada Fall, Vernal Fall, and thence to Valley same day	3 00
Valley	Summits, Vernal and Nevada Falls, direct, and return to Valley same day	3 00
Valley	Glacier Point by Casa Nevada, passing night at Glacier Point	3 00
Valley	Summits, Vernal and Nevada Falls, Sentinel Dome, Glacier Point, and thence to Valley same day	4 00
Valley	Clouds' Rest and return to Casa Nevada	3 00
Valley	Clouds' Rest and return to Valley same day	5 00

Casa Nevada	Clouds' Rest and return to Casa Nevada or Valley same day	3 00
Casa Nevada	Valley direct	2 00
Casa Nevada	Nevada Fall, Sentinel Dome, and Glacier Point, passing night at Glacier Point	2 00
Valley	Nevada Fall, Sentinel Dome, Glacier Point, and Valley same day	3 00
	Upper Yosemite Fall, Eagle Peak, and return	3 00
	Charge for guide (including horse), when furnished	3 00
	Saddle-horses, on level of Valley, per day	2 50

1. The above charges do not include feed for horses when passing night at Casa Nevada or Glacier Point.

2. Where Valley is specified as starting-point, the above rates prevail from any hotel in Valley, or from the foot of any trail.

3. Any shortening of above trips, without proportionate reduction of rates, shall be at the option of those hiring horses.

4. Trips other than those above specified shall be subject to special arrangement between letter and hirer.

CARRIAGES

From	*Route to*	*Amount.*
Hotels	Mirror Lake and return, direct	$1 00
Hotels	Mirror Lake and return by Tissiack Avenue	1 25
Hotels	Mirror Lake and return to foot of Trail, to Vernal and Nevada Falls	1 00
Hotels	Bridal Veil Fall and return, direct	1 00
Hotels	Pohono Bridge, down either side of Valley, and return on opposite side, stopping at Yosemite and Bridal Veil Falls	1 50

MAXIMUM RATES FOR TRANSPORTATION —*Continued.*

Hotels	Cascade Falls, down either side of Valley, and return on opposite side, stopping at Yosemite and Bridal Veil Falls	2 25
Hotels	Artist Point and return, direct, stopping at Bridal Veil Falls	2 00
Hotels	New Inspiration Point and return, direct, stopping at Bridal Veil Falls.............	2 00
....	Grand Round Drive, including Yosemite and Bridal Veil Falls, excluding Lake and Cascades	2 50
....	Grand Round Drive, including Yosemite and Bridal Veil Falls, Lake, and Cascades..................................	3 50

1. When the value of the seats hired in any vehicle shall exceed $15 for a two-horse team, or $25 for a four-horse team, *for any trip* in the above schedule, the persons hiring the seats shall have the privilege of paying no more than the aggregate sums of $15 and $25 *per trip* for a two-horse and four-horse team, respectively.

2. If saddle-horses should be substituted for any of the above carriage trips, carriage rates will apply to each horse. In no case shall the *per diem* charge of $2.50 for each saddle-horse, on level of Valley, be exceeded.

Any excess of the above rates, as well as any extortion, incivility, misrepresentation, or the riding of unsafe animals, should be promptly reported at the Guardian's office.

TRAVELS IN ALASKA

PART ONE

THE TRIP OF 1879

CHAPTER I

PUGET SOUND AND BRITISH COLUMBIA

AFTER ELEVEN YEARS of study and exploration in the Sierra Nevada of California and the mountain-ranges of the Great Basin, studying in particular their glaciers, forests, and wild life, above all their ancient glaciers and the influence they exerted in sculpturing the rocks over which they passed with tremendous pressure, making new landscapes, scenery, and beauty which so mysteriously influence every human being, and to some extent all life, I was anxious to gain some knowledge of the regions to the northward, about Puget Sound and Alaska. With this grand object in view I left San Francisco in May, 1879, on the steamer Dakota, without any definite plan, as with the exception of a few of the Oregon peaks and their forests all the wild north was new to me.

To the mountaineer a sea voyage is a grand, inspiring, restful change. For forests and plains with their flowers and fruits we have new scenery, new life of every sort; water hills and dales in eternal visible motion for rock waves, types of permanence.

It was curious to note how suddenly the eager countenances of the passengers were darkened as soon as the good ship passed through the Golden Gate and began to heave on the waves of the open ocean. The crowded deck was speedily deserted on account of seasickness. It seemed strange that nearly every one afflicted should be more or less ashamed.

Next morning a strong wind was blowing, and the sea was gray and white, with long breaking waves, across which the Dakota was racing half-buried in spray. Very few of the passengers were on deck to enjoy the wild scenery. Every wave seemed to be making enthusiastic, eager haste to the shore, with long, irised tresses streaming from its tops, some of its outer fringes borne away in scud to refresh the wind, all the rolling, pitching, flying water exulting in the beauty of rainbow light. Gulls and albatrosses, strong, glad life in the midst of the stormy beauty, skimmed the waves against the wind, seemingly without effort, oftentimes flying nearly a mile without a single wing-beat, gracefully swaying from side to side and tracing the curves of the briny water hills with the finest precision, now and then just grazing the highest.

And yonder, glistening amid the irised spray, is a still more striking revelation of warm life in the so-called howling waste,—a half-dozen whales, their broad backs like glaciated bosses of granite heaving aloft in near view, spouting lustily, drawing a long breath, and plunging down home in colossal health and comfort. A merry school of porpoises, a square mile of them, suddenly appear, tossing themselves into the air in abounding strength and hilarity, adding foam to the waves and making all the wilderness wilder. One cannot but feel sympathy with and be proud of these brave neighbors, fellow citizens in the commonwealth of the world, making a living like the rest of us. Our good ship also seemed like a thing of life, its great iron heart beating on through calm and storm, a truly noble spectacle. But think of the hearts of these whales, beating warm against the sea, day and night, through dark and light, on and on for centuries; how the red blood must rush and gurgle in and out, bucketfuls, barrelfuls at a beat!

The cloud colors of one of the four sunsets enjoyed on the voyage were remarkably pure and rich in tone. There was a well-defined range of cumuli a few degrees above the horizon, and a massive, dark-gray rain-cloud above it, from which depended long, bent fringes overlapping the lower cumuli and partially veiling them; and from time to time sunbeams poured through narrow openings and painted the exposed bosses and fringes in ripe yellow tones, which, with the reflections on the water, made magnificent pictures. The scenery of the ocean, however sublime in vast expanse, seems far less beautiful to us dry-shod animals than that of the land seen only in comparatively small patches; but when we contemplate the whole globe as one great dewdrop, striped and dotted with continents and islands, flying through space with other stars all singing and shining together as one, the whole universe appears as an infinite storm of beauty.

The California coast-hills and cliffs look bare and uninviting as seen from the ship, the magnificent forests keeping well back out of sight beyond the reach of the sea winds; those of Oregon and Washington are in some places clad with conifers nearly down to the shore; even the little detached islets, so marked a feature to the northward, are mostly tree-crowned. Up through the Straits of Juan de Fuca the forests, sheltered from the ocean gales and favored with abundant rains, flourish in marvelous luxuriance on the glacier-sculptured mountains of the Olympic Range.

We arrived in Esquimault Harbor, three miles from Victoria, on the evening of the fourth day, and drove to the town through a magnificent forest of Douglas spruce,—with an undergrowth in open spots of oak, madrone, hazel, dogwood, alder, spiræa, willow, and wild rose,—and around many an upswelling *moutonné* rock, freshly glaciated and furred with yellow mosses and lichens.

Victoria, the capital of British Columbia, was in 1879 a small old-fashioned English town on the south end of Vancouver Island. It was said to contain about six thousand inhabitants. The government buildings and some of the business blocks were noticeable, but the attention of the traveler was more worthily attracted to the neat cottage homes found here, embowered in the freshest and floweriest climbing roses and honeysuckles conceivable. Californians may well be proud of their home roses loading sunny verandas, climbing to the tops of the roofs and falling over the gables in white and red cascades. But here, with so much bland fog and dew and gentle laving rain, a still finer development of some of the commonest garden plants is reached. English honeysuckle seems to have found here a most congenial home. Still more beautiful were the wild roses, blooming in wonderful luxuriance along the woodland paths, with corollas two and three inches wide. This rose and three species of spiræa fairly filled the air with fragrance after showers; and how brightly then did the red dogwood berries shine amid the green leaves beneath trees two hundred and fifty feet high.

Strange to say, all of this exuberant forest and flower vegetation was growing upon fresh moraine material scarcely at all moved or in any way modified by post-glacial agents. In the town gardens and orchards, peaches and apples fell upon glacial-polished rocks, and the streets were graded in moraine gravel; and I observed scratched and grooved rock bosses as unweathered and telling as those of the High Sierra of California eight thousand feet or more above sea-level. The Victoria Harbor is plainly glacial in origin, eroded from the solid; and the rock islets that rise here and there in it are unchanged to any appreciable extent by all the waves that have broken over them since first they came to light toward the close of the glacial period. The shores also of the harbor are strikingly grooved and scratched and in every way as glacial in all their characteristics as those of new-born glacial lakes. That the domain of the sea is being slowly extended over the land by incessant wave-action is well known; but in this freshly glaciated region the shores have been so short a time exposed to wave-action that they are scarcely at all wasted. The extension of the sea affected by its own action in post-glacial times is probably less than the millionth part of that affected by glacial action during the last glacier period. The direction of the flow of the ice-sheet to which all the main features of this wonderful region are due was in general southward.

From this quiet little English town I made many short excursions—up the coast to Nanaimo, to Burrard Inlet, now the terminus of the Canadian Pacific Railroad, to Puget Sound, up Fraser River to New Westminster and Yale at the head of navigation, charmed everywhere with the wild, new-born scenery. The most interesting of these and the most difficult to leave was the Puget Sound region, famous the world over for the wonderful forests of gigantic trees about its shores. It is an arm and many-fingered hand of the sea, reaching southward from the Straits of Juan de Fuca about a hundred miles into the heart of one of the noblest coniferous forests on the face of the globe. All its scenery is wonderful—broad river-like reaches sweeping in beautiful curves around bays and capes and jutting promontories, opening here and there into smooth, blue, lake-like expanses dotted with islands and feathered with tall, spirey evergreens, their beauty doubled on the bright mirror-water.

Sailing from Victoria, the Olympic Mountains are seen right ahead, rising in bold relief against the sky, with jagged crests and peaks from six to eight thousand feet high,—small residual glaciers and ragged snow-fields beneath them in wide amphitheaters opening down through the forest-filled valleys. These valleys mark the courses of the Olympic glaciers at the period of their greatest extension, when they poured their tribute into that portion of the great northern ice-sheet that overswept Vancouver Island and filled the strait between it and the mainland.

On the way up to Olympia, then a hopeful little town situated at the end of one of the longest fingers of the Sound, one is often reminded of Lake Tahoe, the scenery of the widest expanses is so lake-like in the clearness and stillness of the water and the luxuriance of the surrounding forests. Doubling cape after cape, passing uncounted islands, new combinations break on the view in endless variety, sufficient to satisfy the lover of wild beauty through a whole life. When the clouds come down, blotting out everything, one feels as if at sea; again lifting a little, some islet may be seen standing alone with the tops of its trees dipping out of sight in gray misty fringes; then the ranks of spruce and cedar bounding the water's edge come to view; and when at length the whole sky is clear the colossal cone of Mt. Rainier may be seen in spotless white, looking down over the dark woods from a distance of fifty or sixty miles, but so high and massive and so sharply outlined, it seems to be just back of a strip of woods only a few miles wide.

Mt. Rainier, or Tahoma (the Indian name), is the noblest of the volcanic cones extending from Lassen Butte and Mt. Shasta along the Cascade Range to Mt. Baker. One of the most telling views of it hereabouts is obtained near Tacoma. From a bluff back of the town it was revealed in all its glory, laden with glaciers and snow down to the forested foot-hills around its finely curved base. Up to this time (1879) it had been ascended but once. From observations made on the summit with a single aneroid barometer, it was estimated to be about 14,500 feet high. Mt. Baker, to the northward, is about 10,700 feet high, a noble mountain. So also are Mt. Adams, Mt. St. Helens, and Mt. Hood. The latter, overlooking the town of Portland, is perhaps the best known. Rainier, about the same height as Shasta, surpasses them all in massive icy grandeur,—the most majestic solitary mountain I had ever yet beheld. How eagerly I gazed and longed to climb it and study its history only the mountaineer may know, but I was compelled to turn away and bide my time.

The species forming the bulk of the woods here is the Douglas spruce (*Pseudotsuga douglasii*), one of the greatest of the western giants. A specimen that I measured near Olympia was about three hundred feet in height and twelve feet in diameter four feet above the ground. It is a widely distributed tree, extending northward through British Columbia, southward through Oregon and California, and eastward to the Rocky Mountains. The timber is used for ship-building, spars, piles, and the framework of houses, bridges, etc. In the California lumber markets it is known as "Oregon pine." In Utah, where it is common on the Wahsatch Mountains, it is called "red pine." In California, on the western slope of the Sierra

Nevada, it forms, in company with the yellow pine, sugar pine, and incense cedar, a pretty well-defined belt at a height of from three to six thousand feet above the sea; but it is only in Oregon and Washington, especially in this Puget Sound region, that it reaches its very grandest development,—tall, straight, and strong, growing down close to tide-water.

All the towns of the Sound had a hopeful, thrifty aspect. Port Townsend, picturesquely located on a grassy bluff, was the port of clearance for vessels sailing to foreign parts. Seattle was famed for its coal-mines, and claimed to be the coming town of the North Pacific Coast. So also did its rival, Tacoma, which had been selected as the terminus of the much-talked-of Northern Pacific Railway. Several coal-veins of astonishing thickness were discovered the winter before on the Carbon River, to the east of Tacoma, one of them said to be no less than twenty-one feet, another twenty feet, another fourteen, with many smaller ones, the aggregate thickness of all the veins being upwards of a hundred feet. Large deposits of magnetic iron ore and brown hematite, together with limestone, had been discovered in advantageous proximity to the coal, making a bright outlook for the Sound region in general in connection with its railroad hopes, its unrivaled timber resources, and its far-reaching geographical relations.

After spending a few weeks in the Puget Sound region with a friend from San Francisco, we engaged passage on the little mail steamer California, at Portland, Oregon, for Alaska. The sail down the broad lower reaches of the Columbia and across its foamy bar, around Cape Flattery, and up the Juan de Fuca Strait, was delightful; and after calling again at Victoria and Port Townsend we got fairly off for icy Alaska.

CHAPTER II

ALEXANDER ARCHIPELAGO AND THE HOME I FOUND IN ALASKA

TO THE LOVER of pure wildness Alaska is one of the most wonderful countries in the world. No excursion that I know of may be made into any other American wilderness where so marvelous an abundance of noble, new-born scenery is so charmingly brought to view as on the trip through the Alexander Archipelago to Fort Wrangell and Sitka. Gazing from the deck of the steamer, one is borne smoothly over calm blue waters, through the midst of countless forest-clad islands. The ordinary discomforts of a sea voyage are not felt, for nearly all the whole long way is on inland waters that are about as waveless as rivers and lakes. So numerous are the islands that they seem to have been sown broadcast; long tapering vistas between the largest of them open in every direction.

Day after day in the fine weather we enjoyed, we seemed to float in true fairyland, each succeeding view seeming more and more beautiful, the one we chanced to have before us the most surprisingly beautiful of all. Never before this had I been embosomed in scenery so hopelessly beyond description. To sketch picturesque bits, definitely bounded, is comparatively easy—a lake in the woods, a glacier meadow, or a cascade in its dell; or even a grand master view of mountains beheld from some commanding outlook after climbing from height to height above the forests. These may be attempted, and more or less telling pictures made of them; but in these coast landscapes there is such indefinite, on-leading expansiveness, such a multitude of features without apparent redundance, their lines graduating delicately into one another in endless succession, while the whole is so fine, so tender, so ethereal, that all penwork seems hopelessly unavailing. Tracing shining ways through fiord and sound, past forests and waterfalls, islands and mountains and far azure headlands, it seems as if surely we must at length reach the very paradise of the poets, the abode of the blessed.

Some idea of the wealth of this scenery may be gained from the fact that the coast-line of Alaska is about twenty-six thousand miles long, more than twice as long as all the rest of the United States. The islands of the Alexander Archipelago, with the straits, channels, canals, sounds, passages, and fiords, form an intricate web of land and water embroidery sixty or seventy miles wide, fringing the lofty icy chain of coast mountains from Puget Sound to Cook Inlet; and, with infinite variety, the general pattern is harmonious throughout its whole extent of nearly a thousand miles. Here you glide into a narrow channel hemmed in by mountain walls, forested down to the water's edge, where there is no distant view, and your attention is concentrated on the objects close about you—the crowded spires of the spruces and hemlocks rising higher and higher on the steep green slopes; stripes of paler green where winter avalanches have cleared away the trees, allowing grasses and willows to spring up; zigzags of cascades appearing and disappearing among the bushes and trees; short, steep glens with brawling streams hidden beneath alder and dogwood, seen only where they emerge on the brown algæ of the shore; and retreating hollows, with lingering snow-banks marking the fountains of ancient glaciers. The steamer is often so near the shore that you may distinctly see the cones clustered on the tops of the trees, and the ferns and bushes at their feet.

But new scenes are brought to view with magical rapidity. Rounding some bossy cape, the eye is called away into far-reaching vistas, bounded on either hand by headlands in charming array, one dipping gracefully beyond another and growing fainter and more ethereal in the distance. The tranquil channel stretching river-like between, may be stirred here and there by the silvery plashing of upspringing salmon, or by flocks of white gulls floating like water-lilies among the sun spangles; while mellow, tempered sunshine is streaming over all, blending sky, land, and water in pale, misty blue. Then, while you are dreamily gazing into the depths of this leafy ocean lane, the little steamer, seeming hardly larger than a duck, turning into some passage not visible until the moment of entering it, glides into a wide expanse—a sound filled with islands, sprinkled and clustered in forms and compositions such as nature alone can invent; some of them so small the trees growing on them seem like single handfuls culled from the neighboring woods and set in the water to keep them fresh, while here and there at wide intervals you may notice bare rocks just above the water, mere dots punctuating grand, outswelling sentences of islands.

The variety we find, both as to the contours and the collocation of the islands, is due chiefly to differences in the structure and composition of their rocks, and the unequal glacial denudation different portions of the coast were subjected to. This influence must have been especially heavy toward the end of the glacial period, when the main ice-sheet began to break up into separate glaciers. Moreover, the mountains of the larger islands nourished local glaciers, some of them of considerable size, which sculptured their summits and sides, forming in some cases wide cirques with cañons or valleys leading down from them into the channels and sounds. These causes have produced much of the bewildering variety of which nature is so fond, but none the less will the studious observer see the underlying harmony—the general trend of the islands in the direction of the flow of the main ice-mantle from the mountains of the Coast Range, more or less varied by subordinate foot-hill ridges and mountains. Furthermore, all the islands, great and small, as well as the headlands and promontories of the mainland, are seen to have a rounded, overrubbed appearance produced by the over-sweeping ice-flood during the period of greatest glacial abundance.

The canals, channels, straits, passages, sounds, etc., are subordinate to the same glacial conditions in their forms, trends, and extent as those which determined the forms, trends, and distribution of the land-masses, their basins being the parts of the pre-glacial margin of the continent, eroded to varying depths below sea-level, and into which, of course, the ocean waters flowed as the ice was melted out of them. Had the general glacial denudation been much less, these ocean ways over which we are sailing would have been valleys and cañons and lakes; and the islands rounded hills and ridges, landscapes with undulating features like those found above sea-level wherever the rocks and glacial conditions are similar. In general, the island-bound channels are like rivers, not only in separate reaches as seen from the deck of a vessel, but continuously so for hundreds of miles in the case of the longest of them. The tide-currents, the fresh driftwood, the inflowing streams, and the luxuriant foliage of the out-leaning trees on the shores make this resemblance all the more complete. The largest islands look like part of the mainland in any view to be had of them from the ship, but far the greater number are small, and appreciable as islands, scores of them being less than a mile long. These the eye easily takes in and revels in their beauty with ever fresh delight. In their relations to each other the individual members of a group have evidently been derived from the same general rock-mass, yet they never seem broken or abridged in any way as to their contour lines, however abruptly they may dip their sides. Viewed one by one, they seem detached beauties, like extracts from a poem, while, from the completeness of their lines and the way that their trees are arranged, each seems a finished stanza in itself. Contemplating the arrangement of the trees on these small islands, a distinct impression is produced of their having been sorted and harmonized as to size like a well-balanced bouquet. On some of the smaller tufted islets a group of tapering spruces is planted in the middle, and two smaller groups that evidently correspond with each other are planted on the ends at about equal distances from the central group; or the whole appears as one group with marked fringing trees that match each other spreading around the sides, like flowers leaning outward against the rim of a vase. These harmonious tree relations are so constant that they evidently are the result of design, as much so as the arrangement of the feathers of birds or the scales of fishes.

Thus perfectly beautiful are these blessed evergreen islands, and their beauty is the beauty of youth, for though the freshness of their verdure must be ascribed to the bland moisture with which they are bathed from warm ocean-currents, the very existence of the islands, their features, finish, and peculiar distribution, are all immediately referable to ice-action during the great glacial winter just now drawing to a close.

We arrived at Wrangell July 14, and after a short stop of a few hours went on to Sitka and returned on the 20th to Wrangell, the most inhospitable place at first sight I had ever seen. The little steamer that had been my home in the wonderful trip through the archipelago, after taking the mail, departed on her return to Portland, and as I watched her gliding out of sight in the dismal blurring rain, I felt strangely lonesome. The friend that had accompanied me thus far now left for his home in San Francisco, with two other interesting travelers who had made the trip for health and scenery, while my fellow passengers, the missionaries, went direct to the Presbyterian home in the old fort. There was nothing like a tavern or lodging-house in the village, nor could I find any place in the stumpy, rocky, boggy ground about it that looked dry enough to camp on until I could find a way into the wilderness to begin my studies. Every place within a mile or two of the town seemed strangely shelterless and inhospitable, for all the trees had long ago been felled for building-timber and fire-wood. At the worst, I thought, I could build a bark hut on a hill back of the village, where something like a forest loomed dimly through the draggled clouds.

I had already seen some of the high glacier-bearing mountains in distant views from the steamer, and was anxious to reach them. A few whites of the village, with whom I entered into conversation, warned me that the Indians were a bad lot, not to be trusted, that the woods were well-nigh impenetrable, and that I could go nowhere without a canoe. On the other hand, these natural difficulties made the grand wild country all the more attractive, and I determined to get into the heart of it somehow or other with a bag of hardtack, trusting to my usual good luck. My present difficulty was in finding a first base camp. My only hope was on the hill. When I was strolling past the old fort I happened to meet one of the missionaries, who kindly asked me where I was going to take up my quarters.

"I don't know," I replied. "I have not been able to find quarters of any sort. The top of that little hill over there seems the only possible place."

He then explained that every room in the mission house was full, but he thought I might obtain leave to spread my blanket in a carpenter-shop belonging to the mission. Thanking him, I ran down to the sloppy wharf for my little bundle of baggage, laid it on the shop floor, and felt glad and snug among the dry, sweet-smelling shavings.

The carpenter was at work on a new Presbyterian mission building, and when he came in I explained that Dr. Jackson* had suggested that I might be allowed to sleep on the floor, and after I assured him that I would not touch his tools or be in his way, he goodnaturedly gave me the freedom of the shop and also of his small private side room where I would find a wash-basin.

* Dr. Sheldon Jackson, 1834–1909, became Superintendent of Presbyterian Missions in Alaska in 1877, and United States General Agent of Education in 1885. [W. F. B.]

I was here only one night, however, for Mr. Vanderbilt, a merchant, who with his family occupied the best house in the fort, hearing that one of the late arrivals, whose business none seemed to know, was compelled to sleep in the carpenter-shop, paid me a good-Samaritan visit and after a few explanatory words on my glacier and forest studies, with fine hospitality offered me a room and a place at his table. Here I found a real home, with freedom to go on all sorts of excursions as opportunity offered. Annie Vanderbilt, a little doctor of divinity two years old, ruled the household with love sermons and kept it warm.

Mr. Vanderbilt introduced me to prospectors and traders and some of the most influential of the Indians. I visited the mission school and the home for Indian girls kept by Mrs. MacFarland, and made short excursions to the nearby forests and streams, and studied the rate of growth of the different species of trees and their age, counting the annual rings on stumps in the large clearings made by the military when the fort was occupied, causing wondering speculation among the Wrangell folk, as was reported by Mr. Vanderbilt.

"What can the fellow be up to?" they inquired. "He seems to spend most of his time among stumps and weeds. I saw him the other day on his knees, looking at a stump as if he expected to find gold in it. He seems to have no serious object whatever."

One night when a heavy rain-storm was blowing I unwittingly caused a lot of wondering excitement among the whites as well as the superstitious Indians. Being anxious to see how the Alaska trees behave in storms and hear the songs they sing, I stole quietly away through the gray drenching blast to the hill back of the town, without being observed. Night was falling when I set out and it was pitch dark when I reached the top. The glad, rejoicing storm in glorious voice was singing through the woods, noble compensation for mere body discomfort. But I wanted a fire, a big one, to see as well as hear how the storm and trees were behaving. After long, patient groping I found a little dry punk in a hollow trunk and carefully stored it beside my matchbox and an inch or two of candle in an inside pocket that the rain had not yet reached; then, wiping some dead twigs and whittling them into thin shavings, stored them with the punk. I then made a little conical bark hut about a foot high, and, carefully leaning over it and sheltering it as much as possible from the driving rain, I wiped and stored a lot of dead twigs, lighted the candle, and set it in the hut, carefully added pinches of punk and shavings, and at length got a little blaze, by the light of which I gradually added larger shavings, then twigs all set on end astride the inner flame, making the little hut higher and wider. Soon I had light enough to enable me to select the best dead branches and large sections of bark, which were set on end, gradually increasing the height and corresponding light of the hut fire. A considerable area was thus well lighted, from which I gathered abundance of wood, and kept adding to the fire until it had a strong, hot heart and sent up a pillar of flame thirty or forty feet high, illuminating a wide circle in spite of the rain, and casting a red glare into the flying clouds. Of all the thousands of camp-fires I have elsewhere built none was just like this one, rejoicing in triumphant strength and beauty in the heart of the rain-laden gale. It was wonderful,—the illumined rain and clouds mingled together and the trees glowing against the jet background, the colors of the mossy, lichened trunks with sparkling streams pouring down the furrows of the bark, and the gray-bearded old patriarchs bowing low and chanting in passionate worship!

My fire was in all its glory about midnight, and, having made a bark shed to shelter me from the rain and partially dry my clothing, I had nothing to do but look and listen and join the trees in their hymns and prayers.

Neither the great white heart of the fire nor the quivering enthusiastic flames shooting aloft like auroral lances could be seen from the village on account of the trees in front of it and its being back a little way over the brow of the hill; but the light in the clouds made a great show, a portentous sign in the stormy heavens unlike anything ever before seen or heard of in Wrangell. Some wakeful Indians, happening to see it about midnight, in great alarm aroused the Collector of Customs and begged him to go to the missionaries and get them to pray away the frightful omen, and inquired anxiously whether white men had ever seen anything like that sky-fire, which instead of being quenched by the rain was burning brighter and brighter. The Collector said he had heard of such strange fires, and this one he thought might perhaps be what the white man called a "volcano, or an *ignis fatuus*." When Mr. Young was called from his bed to pray, he, too, confoundedly astonished and at a loss for any sort of explanation, confessed that he had never seen anything like it in the sky or anywhere else in such cold wet weather, but that it was probably some sort of spontaneous combustion "that the white man called St. Elmo's fire, or Will-of-the-wisp." These explanations, though not convincingly clear, perhaps served to veil their own astonishment and in some measure to diminish the superstitious fears of the natives; but from what I heard, the few whites who happened to see the strange light wondered about as wildly as the Indians.

I have enjoyed thousands of camp-fires in all sorts of weather and places, warm-hearted, short-flamed, friendly little beauties glowing in the dark on open spots in high Sierra gardens, daisies and lilies circled about them, gazing like enchanted children; and large fires in silver fir forests, with spires of flame towering like the trees about them, and sending up multitudes of starry sparks to enrich the sky; and still greater fires on the mountains in winter, changing camp climate to summer, and making the frosty snow look like beds of white flowers, and oftentimes mingling their swarms of swift-flying sparks with falling snow-crystals when the clouds were in bloom. But this Wrangell camp-fire, my first in Alaska, I shall always remember for its triumphant storm-defying grandeur, and the wondrous beauty of the psalm-singing, lichen-painted trees which it brought to light.

CHAPTER III

WRANGELL ISLAND AND ALASKA SUMMERS

WRANGELL ISLAND IS about fourteen miles long, separated from the mainland by a narrow channel or fiord, and trending in the direction of the flow of the ancient ice-sheet. Like all its neighbors, it is densely forested down to the water's edge with trees that never seem to have suffered from thirst or fire or the ax of the lumberman in all their long century lives. Beneath soft, shady clouds, with abundance of rain, they flourish in wonderful strength and beauty to a good old age, while the many warm days, half cloudy, half clear, and the little groups of pure sun-days enable them to ripen their cones and send myriads of seeds flying every autumn to insure the permanence of the forests and feed the multitude of animals.

The Wrangell village was a rough place. No mining hamlet in the placer gulches of California, nor any backwoods village I ever saw, approached it in picturesque, devil-may-care *abandon*. It was a lawless draggle of wooden huts and houses, built in crooked lines, wrangling around the boggy shore of the island for a mile or so in the general form of the letter S, without the slightest subordination to the points of the compass or to building laws of any kind. Stumps and logs, like precious monuments, adorned its two streets, each stump and log, on account of the moist climate, moss-grown and tufted with grass and bushes, but muddy on the sides below the limit of the bog-line. The ground in general was an oozy, mossy bog on a foundation of jagged rocks, full of concealed pit-holes. These picturesque rock, bog, and stump obstructions, however, were not so very much in the way, for there were no wagons or carriages there. There was not a horse on the island. The domestic animals were represented by chickens, a lonely cow, a few sheep, and hogs of a breed well calculated to deepen and complicate the mud of the streets.

Most of the permanent residents of Wrangell were engaged in trade. Some little trade was carried on in fish and furs, but most of the quickening business of the place was derived from the Cassiar gold-mines, some two hundred and fifty or three hundred miles inland, by way of the Stickeen River and Dease Lake. Two sternwheel steamers plied on the river between Wrangell and Telegraph Creek at the head of navigation, a hundred and fifty miles from Wrangell, carrying freight and passengers and connecting with pack-trains for the mines. These placer mines, on tributaries of the Mackenzie River, were discovered in the year 1874. About eighteen hundred miners and prospectors were said to have passed through Wrangell that season of 1879, about half of them being Chinamen. Nearly a third of this whole number set out from here in the month of February, traveling on the Stickeen River, which usually remains safely frozen until toward the end of April. The main body of the miners, however, went up on the steamers in May and June. On account of the severe winters they were all compelled to leave the mines the end of September. Perhaps about two thirds of them passed the winter in Portland and Victoria and the towns of Puget Sound. The rest remained here in Wrangell, dozing away the long winter as best they could.

Indians, mostly of the Stickeen tribe, occupied the two ends of the town, the whites, of whom there were about forty or fifty, the middle portion; but there was no determinate line of demarcation, the dwellings of the Indians being mostly as large and solidly built of logs and planks as those of the whites. Some of them were adorned with tall totem poles.

The fort was a quadrangular stockade with a dozen block and frame buildings located upon rising ground just back of the business part of the town. It was built by our Government shortly after the purchase of Alaska, and was abandoned in 1872, reoccupied by the military in 1875, and finally abandoned and sold to private parties in 1877. In the fort and about it there were a few good, clean homes, which shone all the more brightly in their somber surroundings. The ground occupied by the fort, by being carefully leveled and drained was dry, though formerly a portion of the general swamp, showing how easily the whole town could have been improved. But in spite of disorder and squalor, shaded with clouds, washed and wiped by rain and sea winds, it was triumphantly salubrious through all the seasons. And though the houses seemed to rest uneasily among the miry rocks and stumps, squirming at all angles as if they had been tossed and twisted by earthquake shocks, and showing but little more relation to one another than may be observed among moraine boulders, Wrangell was a tranquil place. I never heard a noisy brawl in the streets, or a clap of thunder, and the waves seldom spoke much above a whisper along the beach. In summer the rain comes straight down, steamy and tepid. The clouds are usually united, filling the sky, not racing along in threatening ranks suggesting energy of an overbearing destructive kind, but forming a bland, mild, laving bath. The cloudless days are calm, pearl-gray, and brooding in tone, inclining to rest and peace; the islands seem to drowse and float on the glassy water, and in the woods scarce a leaf stirs.

The very brightest of Wrangell days are not what Californians would call bright. The tempered sunshine sifting through the moist atmosphere makes no dazzling glare, and the town, like the landscape, rests beneath a hazy, hushing, Indian-summerish spell. On the longest days the sun rises about three o'clock, but it is daybreak at midnight. The cocks crowed when they woke, without reference to the dawn, for it is never quite dark; there were only a few full-grown roosters in Wrangell, half a dozen or so, to awaken the town and give it a civilized character. After sunrise a few languid smoke-columns might be seen, telling the first stir of the people. Soon an Indian or two might be noticed here and there at the doors of their barn-like cabins, and a merchant getting ready for trade; but scarcely a sound was heard, only a dull, muffled stir gradually deepening. There were only two white babies in the town, so far as I saw, and as for Indian babies, they woke and ate and made no crying sound. Later you might hear the croaking of ravens, and the strokes of an ax on fire-wood. About eight or nine o'clock the town was awake, Indians, mostly women and children, began to gather on the front platforms of the half-dozen stores, sitting carelessly on their blankets, every other face hideously blackened, a naked circle around the eyes, and perhaps a spot on the cheek-bone and the nose where the smut has been rubbed off. Some of the little children were also blackened, and none were over-clad, their light and airy costume consisting of a calico shirt reaching only to the waist. Boys eight or ten years old sometimes had an additional garment,—a pair of castaway miner's overalls wide enough and ragged enough for extravagant ventilation. The larger girls and young women were arrayed in showy calico, and wore jaunty straw hats, gorgeously ribboned, and glowed among the blackened and blanketed old crones like scarlet tanagers in a flock of blackbirds. The women, seated on the steps and platform of the traders' shops, could hardly be called loafers, for they had berries to sell, basketfuls of huckleberries, large yellow salmon-berries, and bog raspberries that looked wondrous fresh and clean amid the surrounding squalor. After patiently waiting for purchasers until hungry, they ate what they could not sell, and went away to gather more.

Yonder you see a canoe gliding out from the shore, containing perhaps a man, a woman, and a child or two, all paddling together in natural, easy rhythm. They are going to catch a fish, no difficult matter, and when this is done their day's work is done. Another party puts out to capture bits of driftwood, for it is easier to procure fuel in this way than to drag it down from the outskirts of the woods through rocks and bushes. As the day advances, a fleet of canoes may be seen along the shore, all fashioned alike, high and long beak-like prows and sterns, with lines as fine as those of the breast of a duck. What the mustang is to the Mexican *vaquero*, the canoe is to these coast Indians. They skim along the shores to fish and hunt and trade, or merely to visit their neighbors, for they are sociable, and have family pride remarkably well developed, meeting often to inquire after each other's health, attend potlatches and dances, and gossip concerning coming marriages, births, deaths, etc. Others seem to sail for the pure pleasure of the thing, their canoes decorated with handfuls of the tall purple epilobium.

Yonder goes a whole family, grandparents and all, making a direct course for some favorite stream and campground. They are going to gather berries, as the baskets tell. Never before in all my travels, north or south, had I found so lavish an abundance of berries as here. The woods and meadows are full of them, both on the lowlands and mountains—huckleberries of many species, salmon-berries, blackberries, raspberries, with service-berries on dry open places, and cranberries in the bogs, sufficient for every bird, beast, and human being in the territory and thousands of tons to spare. The huckleberries are especially abundant. A species that grows well up on the mountains is the best and largest, a half-inch and more in diameter and delicious in flavor. These grow on bushes three or four inches to a foot high. The berries of the commonest species are smaller and grow almost everywhere on the low grounds on bushes from three to six or seven feet high. This is the species on which the Indians depend most for food, gathering them in large quantities, beating them into a paste, pressing the paste into cakes about an inch thick, and drying them over a slow fire to enrich their winter stores. Salmon-berries and service-berries are preserved in the same way.

A little excursion to one of the best huckleberry-fields adjacent to Wrangell, under the direction of the Collector of Customs, to which I was invited, I greatly enjoyed. There were nine Indians in the party, mostly women and children going to gather huckleberries. As soon as we had arrived at the chosen campground on the bank of a trout stream, all ran into the bushes and began eating berries before anything in the way of camp-making was done, laughing and chattering in natural animal enjoyment. The Collector went up the stream to examine a meadow at its head with reference to the quantity of hay it might yield for his cow, fishing by the way. All the Indians except the two eldest boys who joined the Collector, remained among the berries.

The fishermen had rather poor luck, owing, they said, to the sunny brightness of the day, a complaint seldom heard in this climate. They got good exercise, however, jumping from boulder to boulder in the brawling stream, running along slippery logs and through the bushes that fringe the bank, casting here and there into swirling pools at the foot of cascades, imitating the tempting little skips and whirls of flies so well known to fishing parsons, but perhaps still better known to Indian boys. At the lake-basin the Collector, after he had surveyed his hay-meadow, went around it to the inlet of the lake with his brown pair of attendants to try their luck, while I botanized in the delightful flora which called to mind the cool sphagnum and carex bogs of Wisconsin and Canada. Here I found many of my old favorites the heathworts—kalmia, pyrola, chiogenes, huckleberry, cranberry, etc. On the margin of the meadow darling linnæa was in its glory; purple panicled grasses in full flower reached over my head, and some of the carices and ferns were almost as tall. Here, too, on the edge of the woods I found the wild apple tree, the first I had seen in Alaska. The Indians gather the fruit, small and sour as it is, to flavor their fat salmon. I never saw a richer bog and meadow growth anywhere. The principal forest-trees are hemlock, spruce, and Nootka cypress, with a few pines (*P. contorta*) on the margin of the meadow, some of them nearly a hundred feet high, draped with gray usnea, the bark also gray with scale lichens.

We met all the berry-pickers at the lake, excepting only a small girl and the camp-keeper. In their bright colors they made a lively picture among the quivering bushes, keeping up a low pleasant chanting as if the day and the place and the berries were according to their own hearts. The children carried small baskets, holding two or three quarts; the women two large ones swung over their shoulders. In the afternoon, when the baskets were full, all started back to the campground, where the canoe was left. We parted at the lake, I choosing to follow quietly the stream through the woods. I was the first to arrive at camp. The rest of the party came in shortly afterwards, singing and humming like heavy-laden bees. It was interesting to note how kindly they held out handfuls of the best berries to the little girl, who welcomed them all in succession with smiles and merry words that I did not understand. But there was no mistaking the kindliness and serene good nature.

While I was at Wrangell the chiefs and head men of the Stickeen tribe got up a grand dinner and entertainment in honor of their distinguished visitors, three doctors of divinity and their wives, fellow passengers on the steamer with me, whose object was to organize the Presbyterian church. To both the dinner and dances I was invited, was adopted by the Stickeen tribe, and given an Indian name (Ancoutahan) said to mean adopted chief. I was inclined to regard this honor as being unlikely to have any practical value, but I was assured by Mr. Vanderbilt, Mr. Young, and others that it would be a great safeguard while I was on my travels among the different tribes of the archipelago. For travelers without an Indian name might be killed and robbed without the offender being called to account as long as the crime was kept secret from the whites; but, being adopted by the Stickeens, no one belonging to the other tribes would dare attack me, knowing that the Stickeens would hold them responsible.

The dinner-tables were tastefully decorated with flowers, and the food and general arrangements were in good taste, but there was no trace of Indian dishes. It was mostly imported canned stuff served Boston fashion. After the dinner we assembled in Chief Shakes's large block-house and were entertained with lively examples of their dances and amusements, carried on with great spirit, making a very novel barbarous durbar. The dances seemed to me wonderfully like those of the American Indians in general, a monotonous stamping accompanied by hand-clapping, head-jerking, and explosive grunts kept in time to grim drum-beats. The chief dancer and leader scattered great quantities of downy feathers like a snow-storm as blessings on everybody, while all chanted, "Hee-ee-ah-ah, hee-eeah-ah," jumping up and down until all were bathed in perspiration.

After the dancing excellent imitations were given of the gait, gestures, and behavior of several animals under different circumstances—walking, hunting, capturing, and devouring their prey, etc. While all were quietly seated, waiting to see what next was going to happen, the door of the big house was suddenly thrown open and in bounced a bear, so true to life in form and gestures we were all startled, though it was only a bear-skin nicely fitted on a man who was intimately acquainted with the animals and knew how to imitate them. The bear shuffled down into the middle of the floor and made the motion of jumping into a stream and catching a wooden salmon that was ready for him, carrying it out on to the bank, throwing his head around to listen and see if any one was coming, then tearing it to pieces, jerking his head from side to side, looking and listening in fear of hunters' rifles. Besides the bear dance, there were porpoise and deer dances with one of the party imitating the animals by stuffed specimens with an Indian inside, and the movements were so accurately imitated that they seemed the real thing.

These animal plays were followed by serious speeches, interpreted by an Indian woman: "Dear Brothers and Sisters, this is the way we used to dance. We liked it long ago when we were blind, we always danced this way, but now we are not blind. The Good Lord has taken pity upon us and sent his son, Jesus Christ, to tell us what to do. We have danced to-day only to show you how blind we were to like to dance in this foolish way. We will not dance any more."

Another speech was interpreted as follows: "'Dear Brothers and Sisters,' the chief says, 'this is the way we used to dance and play. We do not wish to do so any more. We will give away all the dance dresses you have seen us wearing, though we value them very highly.' He says he feels much honored to have so many white brothers and sisters at our dinner and plays."

Several short explanatory remarks were made all through the exercises by Chief Shakes, presiding with grave dignity. The last of his speeches concluded thus: "Dear Brothers and Sisters, we have been long, long in the dark. You have led us into strong guiding light and taught us the right way to live and the right way to die. I thank you for myself and all my people, and I give you my heart."

At the close of the amusements there was a potlatch when robes made of the skins of deer, wild sheep, marmots, and sables were distributed, and many of the fantastic head-dresses that had been worn by Shamans. One of these fell to my share.

The floor of the house was strewn with fresh hemlock boughs, bunches of showy wild flowers adorned the walls, and the hearth was filled with huckleberry branches and epilobium. Altogether it was a wonderful show.

I have found southeastern Alaska a good, healthy country to live in. The climate of the islands and shores of the mainland is remarkably bland and temperate and free from extremes of either heat or cold throughout the year. It is rainy, however,—so much so that hay-making will hardly ever be extensively engaged in here, whatever the future may show in the way of the development of mines, forests, and fisheries. This rainy weather, however, is of good quality, the best of the kind I ever experienced, mild in temperature, mostly gentle in its fall, filling the fountains of the rivers and keeping the whole land fresh and fruitful, while anything more delightful than the shining weather in the midst of the rain, the great round sun-days of July and August, may hardly be found anywhere, north or south. An Alaska summer day is a day without night. In the Far North, at Point Barrow,, the sun does not set for weeks, and even here in southeastern Alaska it is only a few degrees below the horizon at its lowest point, and the topmost colors of the sunset blend with those of the sunrise, leaving no gap of darkness between. Midnight is only a low noon, the middle point of the gloaming. The thin clouds that are almost always present are then colored yellow and red, making a striking advertisement of the sun's progress beneath the horizon. The day opens slowly. The low arc of light steals around to the northeastward with gradual increase of height and span and intensity of tone; and when at length the sun appears, it is without much of that stirring, impressive pomp, of flashing, awakening, triumphant energy, suggestive of the Bible imagery, a bridegroom coming out of his chamber and rejoicing like a strong man to run a race. The red clouds with yellow edges dissolve in hazy dimness; the islands, with grayish-white ruffs of mist about them, cast ill-defined shadows on the glistening waters, and the whole down-bending firmament becomes pearl-gray. For three or four hours after sunrise there is nothing especially impressive in the landscape. The sun, though seemingly unclouded, may almost be looked in the face, and the islands and mountains, with their wealth of woods and snow and varied beauty of architecture, seem comparatively sleepy and uncommunicative.

As the day advances toward high noon, the sun-flood streaming through the damp atmosphere lights the water levels and the sky to glowing silver. Brightly play the ripples about the bushy edges of the islands and on the plume-shaped streaks between them, ruffled by gentle passing wind-currents. The warm air throbs and makes itself felt as a life-giving, energizing ocean, embracing all the landscape, quickening the imagination, and bringing to mind the life and motion about us—the tides, the rivers, the flood of light streaming through the satiny sky; the marvelous abundance of fishes feeding in the lower ocean; the misty flocks of insects in the air; wild sheep and goats on a thousand grassy ridges; beaver and mink far back on many a rushing stream; Indians floating and basking along the shores; leaves and crystals drinking the sunbeams; and glaciers on the mountains, making valleys and basins for new rivers and lakes and fertile beds of soil.

Through the afternoon, all the way down to the sunset, the day grows in beauty. The light seems to thicken and become yet more generously fruitful without losing its soft mellow brightness. Everything seems to settle into conscious repose. The winds breathe gently or are wholly at rest. The few clouds visible are downy and luminous and combed out fine on the edges. Gulls here and there, winnowing the air on easy wing, are brought into striking relief; and every stroke of the paddles of Indian hunters in their canoes is told by a quick, glancing flash. Bird choirs in the grove are scarce heard as they sweeten the brooding stillness; and the sky, land, and water meet and blend in one inseparable scene of enchantment. Then comes the sunset with its purple and gold, not a narrow arch on the horizon, but oftentimes filling all the sky. The level cloud-bars usually present are fired on the edges, and the spaces of clear sky between them are greenish-yellow or pale amber, while the orderly flocks of small overlapping clouds, often seen higher up, are mostly touched with crimson like the out-leaning sprays of maple-groves in the beginning of an Eastern Indian Summer. Soft, mellow purple flushes the sky to the zenith and fills the air, fairly steeping and transfiguring the islands and making all the water look like wine. After the sun goes down, the glowing gold vanishes, but because it descends on a curve nearly in the same plane with the horizon, the glowing portion of the display lasts much longer than in more southern latitudes, while the upper colors with gradually lessening intensity of tone sweep around to the north, gradually increase to the eastward, and unite with those of the morning.

The most extravagantly colored of all the sunsets I have yet seen in Alaska was one I enjoyed on the voyage from Portland to Wrangell, when we were in the midst of one of the most thickly islanded parts of the Alexander Archipelago. The day had been showery, but late in the afternoon the clouds melted away from the west, all save a few that settled down in narrow level bars near the horizon. The evening was calm and the sunset colors came on gradually, increasing in extent and richness of tone by slow degrees as if requiring more time than usual to ripen. At a height of about thirty degrees there was a heavy cloud-bank, deeply reddened on its lower edge and the projecting parts of its face. Below this were three horizontal belts of purple edged with gold, while a vividly defined, spreading fan of flame streamed upward across the purple bars and faded in a feather edge of dull red. But beautiful and impressive as was this painting on the sky, the most novel and exciting effect was in the body of the atmosphere itself, which, laden with moisture, became one mass of color—a fine translucent purple haze in which the islands with softened outlines seemed to float, while a dense red ring lay around the base of each of them as a fitting border. The peaks, too, in the distance, and the snow-fields and glaciers and fleecy rolls of mist that lay in the hollows, were flushed with a deep, rosy alpenglow of ineffable loveliness. Everything near and far, even the ship, was comprehended in the glorious picture and the general color effect. The mission divines we had aboard seemed then to be truly divine as they gazed transfigured in the celestial glory. So also seemed our bluff, storm-fighting old captain, and his tarry sailors and all.

About one third of the summer days I spent in the Wrangell region were cloudy with very little or no rain, one third decidedly rainy, and one third clear. According to a record kept here of a hundred and forty-seven days beginning May 17 of that year there were sixty-five on which rain fell, forty-three cloudy with no rain, and thirty-nine clear. In June rain fell on eighteen days, in July eight days, in August fifteen days, in September twenty days. But on some of these days there was only a few minutes' rain, light showers scarce enough to count, while as a general thing the rain fell so gently and the temperature was so mild, very few of them could be called stormy or dismal; even the bleakest, most bedraggled of them all usually had a flush of late or early color to cheer them, or some white illumination about the noon hours. I never before saw so much rain fall with so little noise. None of the summer winds make roaring storms, and thunder is seldom heard. I heard none at all. This wet, misty weather seems perfectly healthful. There is no mildew in the houses, as far as I have seen, or any tendency toward moldiness in nooks hidden from the sun; and neither among the people nor the plants do we find anything flabby or dropsical.

In September clear days were rare, more than three fourths of them were either decidedly cloudy or rainy, and the rains of this month were, with one wild exception, only moderately heavy, and the clouds between showers drooped and crawled in a ragged, unsettled way without betraying hints of violence such as one often sees in the gestures of mountain storm-clouds.

July was the brightest month of the summer, with fourteen days of sunshine, six of them in uninterrupted succession, with a temperature at 7 A.M. of about 60°, at 12 m., 70°. The average 7 A.M. temperature for June was 54.3°; the average 7 A.M. temperature for July was 55.3°; at 12 m. the average temperature was 61.45°; the average 7 A.M. temperature for August was 54.12°; 12 m., 61.48°; the average 7 A.M. temperature for September was 52.14°; and 12 m., 56.12°.

The highest temperature observed here during the summer was seventy-six degrees. The most remarkable characteristic of this summer weather, even the brightest of it, is the velvet softness of the atmosphere. On the mountains of California, throughout the greater part of the year, the presence of an atmosphere is hardly recognized, and the thin, white, bodiless light of the morning comes to the peaks and glaciers as a pure spiritual essence, the most impressive of all the terrestrial manifestations of God. The clearest of

Alaskan air is always appreciably substantial, so much so that it would seem as if one might test its quality by rubbing it between the thumb and finger. I never before saw summer days so white and so full of subdued luster.

The winter storms, up to the end of December when I left Wrangell, were mostly rain at a temperature of thirty-five or forty degrees, with strong winds which sometimes roughly lash the shores and carry scud far into the woods. The long nights are then gloomy enough and the value of snug homes with crackling yellow cedar fires may be finely appreciated. Snow falls frequently, but never to any great depth or to lie long. It is said that only once since the settlement of Fort Wrangell has the ground been covered to a depth of four feet. The mercury seldom falls more than five or six degrees below the freezing-point, unless the wind blows steadily from the mainland. Back from the coast, however, beyond the mountains, the winter months are very cold. On the Stickeen River at Glenora, less than a thousand feet above the level of the sea, a temperature of from thirty to forty degrees below zero is not uncommon.

CHAPTER IV

THE STICKEEN RIVER

THE MOST INTERESTING of the short excursions we made from Fort Wrangell was the one up the Stickeen River to the head of steam navigation. From Mt. St. Elias the coast range extends in a broad, lofty chain beyond the southern boundary of the territory, gashed by stupendous cañons, each of which carries a lively river, though most of them are comparatively short, as their highest sources lie in the icy solitudes of the range within forty or fifty miles of the coast. A few, however, of these foaming, roaring streams—the Alsek, Chilcat, Chilcoot, Taku, Stickeen, and perhaps others—head beyond the range with some of the southwest branches of the Mackenzie and Yukon.

The largest side branches of the main-trunk cañons of all these mountain streams are still occupied by glaciers which descend in showy ranks, their massy, bulging snouts lying back a little distance in the shadows of the walls, or pushing forward among the cottonwoods that line the banks of the rivers, or even stretching all the way across the main cañons, compelling the rivers to find a channel beneath them.

The Stickeen was, perhaps, the best known of the rivers that cross the Coast Range, because it was the best way to the Mackenzie River Cassiar gold-mines. It is about three hundred and fifty miles long, and is navigable for small steamers a hundred and fifty miles to Glenora, and sometimes to Telegraph Creek, fifteen miles farther. It first pursues a westerly course through grassy plains darkened here and there with groves of spruce and pine; then, curving southward and receiving numerous tributaries from the north, it enters the Coast Range, and sweeps across it through a magnificent cañon three thousand to five thousand feet deep, and more than a hundred miles long. The majestic cliffs and mountains forming the cañon-walls display endless variety of form and sculpture, and are wonderfully adorned and enlivened with glaciers and waterfalls, while throughout almost its whole extent the floor is a flowery landscape garden, like Yosemite. The most striking features are the glaciers, hanging over the cliffs, descending the side cañons and pushing forward to the river, greatly enhancing the wild beauty of all the others.

Gliding along the swift-flowing river, the views change with bewildering rapidity. Wonderful, too, are the changes dependent on the seasons and the weather. In spring, when the snow is melting fast, you enjoy the countless rejoicing waterfalls; the gentle breathing of warm winds; the colors of the young leaves and flowers when the bees are busy and wafts of fragrance are drifting hither and thither from miles of wild roses, clover, and honeysuckle; the swaths of birch and willow on the lower slopes following the melting of the winter avalanche snow-banks; the bossy cumuli swelling in white and purple piles above the highest peaks; gray rain-clouds wreathing the outstanding brows and battlements of the walls; and the breaking-forth of the sun after the rain; the shining of the leaves and streams and crystal architecture of the glaciers; the rising of fresh fragrance; the song of the happy birds; and the serene color-grandeur of the morning and evening sky. In summer you find the groves and gardens in full dress; glaciers melting rapidly under sunshine and rain; waterfalls in all their glory; the river rejoicing in its strength; young birds trying their wings; bears enjoying salmon and berries; all the life of the cañon brimming full like the streams. In autumn comes rest, as if the year's work were done. The rich hazy sunshine streaming over the cliffs calls forth the last of the gentians and goldenrods; the groves and thickets and meadows bloom again as their leaves change to red and yellow petals; the rocks also, and the glaciers, seem to bloom like the plants in the mellow golden light. And so goes the song, change succeeding change in sublime harmony through all the wonderful seasons and weather.

My first trip up the river was made in the spring with the missionary party soon after our arrival at Wrangell. We left Wrangell in the afternoon and anchored for the night above the river delta, and started up the river early next morning when the heights above the "Big Stickeen" Glacier and the smooth domes and copings and arches of solid snow along the tops of the cañon walls were glowing in the early beams. We arrived before noon at the old trading-post called "Buck's" in front of the Stickeen Glacier, and remained long enough to allow the few passengers who wished a nearer view to cross the river to the terminal moraine. The sunbeams streaming through the ice pinnacles along its terminal wall produced a wonderful glory of color, and the broad, sparkling crystal prairie and the distant snowy fountains were wonderfully attractive and made me pray for opportunity to explore them.

Of the many glaciers, a hundred or more, that adorn the walls of the great Stickeen River Cañon, this is the largest. It draws its sources from snowy mountains within fifteen or twenty miles of the coast, pours through a comparatively narrow cañon about two miles in width in a magnificent cascade, and expands in a broad fan five or six miles in width, separated from the Stickeen River by its broad terminal moraine, fringed with spruces and willows. Around the beautifully drawn curve of the moraine the Stickeen River flows, having evidently been shoved by the glacier out of its direct course. On the opposite side of the cañon another somewhat smaller glacier, which now terminates four or five miles from the river, was once united front to front with the greater glacier, though at first both were tributaries of the main Stickeen Glacier which once filled the whole grand cañon. After the main trunk cañon was melted out, its side branches, drawing their sources from a height of three or four to five or six thousand feet, were cut off, and of course became separate glaciers, occupying cirques and branch cañons along the tops and sides of the walls. The Indians have a tradition that the river used to run through a tunnel under the united fronts of the two large tributary glaciers mentioned above, which entered the main cañon from either side; and that on one occasion an Indian, anxious to get rid of his wife, had her sent adrift in a canoe down through the ice tunnel, expecting that she would trouble him no more. But to his surprise she floated through under the ice in safety. All the evidence connected with the present appearance of these two glaciers indicates that they were united and formed a dam across the river after the smaller tributaries had been melted off and had receded to a greater or lesser height above the valley floor.

The big Stickeen Glacier is hardly out of sight ere you come upon another that pours a majestic crystal flood through the evergreens, while almost every hollow and tributary cañon contains a smaller one, the size, of course, varying with the extent of the area drained. Some are like mere snow-banks; others, with the blue ice apparent, depend in massive bulging curves and swells, and graduate into the river-like forms that maze through the lower forested regions and are so striking and beautiful that they are admired even by the passing miners with gold-dust in their eyes.

Thirty-five miles above the Big Stickeen Glacier is the "Dirt Glacier," the second in size. Its outlet is a fine stream, abounding in trout. On the opposite side of the river there is a group of five glaciers, one of them descending to within a hundred feet of the river.

Near Glenora, on the northeastern flank of the main Coast Range, just below a narrow gorge called "The Cañon," terraces first make their appearance, where great quantities of moraine material have been swept through the flood-choked gorge and of course outspread and deposited on the first open levels below. Here, too, occurs a marked change in climate and consequently in forests and general appearance of the face of the country. On account of destructive fires the woods are younger and are composed of smaller trees about a foot to eighteen inches in diameter and seventy-five feet high, mostly two-leaved pines which hold their seeds for several years after they are ripe. The woods here are without a trace of those deep accumulations of mosses, leaves, and decaying trunks which make so damp and unclearable a mass in the coast forests. Whole mountainsides are covered with gray moss and lichens where the forest has been utterly destroyed. The river-bank cottonwoods are also smaller, and the birch and contorta pines mingle freely with the coast hemlock and spruce. The birch is common on the lower slopes and is very effective, its round, leafy, pale-green head contrasting with the dark, narrow spires of the conifers and giving a striking character to the forest. The "tamarac pine" or black pine, as the variety of *P. contorta* is called here, is yellowish-green, in marked contrast with the dark lichen-draped spruce which grows above the pine at a height of about two thousand feet, in groves and belts where it has escaped fire and snow avalanches. There is another handsome spruce hereabouts, *Picea alba*, very slender and graceful in habit, drooping at the top like a mountain hemlock. I saw fine specimens a hundred and twenty-five feet high on deep bottom land a few miles below Glenora. The tops of some of them were almost covered with dense clusters of yellow and brown cones.

We reached the old Hudson's Bay trading-post at Glenora about one o'clock, and the captain informed me that he would stop here until the next morning, when he would make an early start for Wrangell.

At a distance of about seven or eight miles to the northeastward of the landing, there is an outstanding group of mountains crowning a spur from the main chain of the Coast Range, whose highest point rises about eight thousand feet above the level of the sea; and as Glenora is only a thousand feet above the sea, the height to be overcome in climbing this peak is about seven thousand feet. Though the time was short I determined to climb it, because of the advantageous position it occupied for general views of the peaks and glaciers of the east side of the great range.

Although it was now twenty minutes past three and the days were getting short, I thought that by rapid climbing I could reach the summit before sunset, in time to get a general view and a few pencil sketches, and make my way back to the steamer in the night. Mr. Young, one of the missionaries, asked permission to accompany me, saying that he was a good walker and climber and would not delay me or cause any trouble. I strongly advised him not to go, explaining that it involved a walk, coming and going, of fourteen or sixteen miles, and a climb through brush and boulders of seven thousand feet, a fair day's work for a seasoned mountaineer to be done in less than half a day and part of a night. But he insisted that he was a strong walker, could do a mountaineer's day's work in half a day, and would not hinder me in any way.

"Well, I have warned you," I said, "and will not assume responsibility for any trouble that may arise."

He proved to be a stout walker, and we made rapid progress across a brushy timbered flat and up the mountain slopes, open in some places, and in others thatched with dwarf firs, resting a minute here and there to refresh ourselves with huckleberries, which grew in abundance in open spots. About half an hour before sunset, when we were near a cluster of crumbling pinnacles that formed the summit, I had ceased to feel anxiety about the mountaineering strength and skill of my companion, and pushed rapidly on. In passing around the shoulder of the highest pinnacle, where the rock was rapidly disintegrating and the danger of slipping was great, I shouted in a warning voice, "Be very careful here, this is dangerous."

Mr. Young was perhaps a dozen or two yards behind me, but out of sight. I afterwards reproached myself for not stopping and lending him a steadying hand, and showing him the slight footsteps I had made by kicking out little blocks of the crumbling surface, instead of simply warning him to be careful. Only a few seconds after giving this warning, I was startled by a scream for help, and hurrying back, found the missionary face downward, his arms outstretched, clutching little crumbling knobs on the brink of a gully that plunges down a thousand feet or more to a small residual glacier. I managed to get below him, touched one of his feet, and tried to encourage him by saying, "I am below you. You are in no danger. You can't slip past me and I will soon get you out of this."

He then told me that both of his arms were dislocated. It was almost impossible to find available footholds on the treacherous rock, and I was at my wits' end to know how to get him rolled or dragged to a place where I could get about him, find out how much he was hurt, and a way back down the mountain. After narrowly scanning the cliff and making footholds, I managed to roll and lift him a few yards to a place where the slope was less steep, and there I attempted to set his arms. I found, however, that this was impossible in such a place. I therefore tied his arms to his sides with my suspenders and necktie, to prevent as much as possible inflammation from movement. I then left him, telling him to lie still, that I would be back in a few minutes, and that he was now safe from slipping. I hastily examined the ground and saw no way of getting him down except by the steep glacier gully. After scrambling to an outstanding point that commands a view of it from top to bottom, to make sure that it was not interrupted by sheer precipices, I concluded that with great care and the digging of slight footholds he could be slid down to the glacier, where I could lay him on his back and perhaps be able to set his arms. Accordingly, I cheered him up, telling him I had found a way, but that it would require lots of time and patience. Digging a footstep in the sand or crumbling rock five or six feet beneath him, I reached up, took hold of him by one of his feet, and gently slid him down on his back, placed his heels in the step, then descended another five or six feet, dug heel notches, and slid him down to them. Thus the whole distance was made by a succession of narrow steps at very short intervals, and the glacier was reached perhaps about midnight. Here I took off one of my boots, tied a handkerchief around his wrist for a good hold, placed my heel in his armpit, and succeeded in getting one of his arms into place, but my utmost strength was insufficient to reduce the dislocation of the other. I therefore bound it closely to his side, and asked him if in his exhausted and trembling condition he was still able to walk.

"Yes," he bravely replied.

So, with a steadying arm around him and many stops for rest, I marched him slowly down in the star-light on the comparatively smooth, unfissured surface of the little glacier to the terminal moraine, a distance of perhaps a mile, crossed the moraine, bathed his head at one of the outlet streams, and after many rests reached a dry place and made a brush fire. I then went ahead looking for an open way through the bushes to where larger wood could be had, made a good lasting fire of resiny silver-fir roots, and a leafy bed beside it. I now told him I would run down the mountain, hasten back with help from the boat, and carry him down in comfort. But he would not hear of my leaving him.

"No, no," he said, "I can walk down. Don't leave me."

I reminded him of the roughness of the way, his nerve-shaken condition, and assured him I would not be gone long. But he insisted on trying, saying on no account whatever must I leave him. I therefore concluded to try to get him to the ship by short walks from one fire and resting-place to another. While he was resting I went ahead, looking for the best way through the brush and rocks, then returning, got him on his feet and made him lean on my shoulder while I steadied him to prevent his falling. This slow, staggering struggle from fire to fire lasted until long after sunrise. When at last we reached the ship and stood at the foot of the narrow single plank without side rails that reached from the bank to the deck at a considerable angle, I briefly explained to Mr. Young's companions, who stood looking down at us, that he had been hurt in an accident, and requested one of them to assist me in getting him aboard. But strange to say, instead of coming down to help, they made haste to reproach him for having gone on a "wild-goose chase" with Muir.

"These foolish adventures are well enough for Mr. Muir," they said, "but you, Mr. Young, have work to do; you have a family; you have a church, and you have no right to risk your life on treacherous peaks and precipices."

The captain, Nat Lane, son of Senator Joseph Lane, had been swearing in angry impatience for being compelled to make so late a start and thus encounter a dangerous wind in a narrow gorge, and was threatening to put the missionaries ashore to seek their lost companion, while he went on down the river about his business. But when he heard my call for help, he hastened forward, and elbowed the divines away from the end of the gangplank, shouting in angry irreverence, "Oh, blank! This is no time for preaching! Don't you see the man is hurt?"

He ran down to our help, and while I steadied my trembling companion from behind, the captain kindly led him up the plank into the saloon, and made him drink a large glass of brandy. Then, with a man holding down his shoulders, we succeeded in getting the bone into its socket, notwithstanding the inflammation and contraction of the muscles and ligaments. Mr. Young was then put to bed, and he slept all the way back to Wrangell.

In his mission lectures in the East, Mr. Young oftentimes told this story. I made no record of it in my note-book and never intended to write a word about it; but after a miserable, sensational caricature of the story had appeared in a respectable magazine, I thought it but fair to my brave companion that it should be told just as it happened.

CHAPTER V

A CRUISE IN THE CASSIAR

SHORTLY AFTER OUR return to Wrangell the missionaries planned a grand mission excursion up the coast of the mainland to the Chilcat country, which I gladly joined, together with Mr. Vanderbilt, his wife, and a friend from Oregon. The river steamer Cassiar was chartered, and we had her all to ourselves, ship and officers at our command to sail and stop where and when we would, and of course everybody felt important and hopeful. The main object of the missionaries was to ascertain the spiritual wants of the war-like Chilcat tribe, with a view to the establishment of a church and school in their principal village; the merchant and his party were bent on business and scenery; while my mind was on the mountains, glaciers, and forests.

This was toward the end of July, in the very brightest and best of Alaska summer weather, when the icy mountains towering in the pearly sky were displayed in all their glory, and the islands at their feet seemed to float and drowse on the shining mirror-waters.

After we had passed through the Wrangell Narrows, the mountains of the mainland came in full view, gloriously arrayed in snow and ice, some of the largest and most river-like of the glaciers flowing through wide, high-walled valleys like Yosemite, their sources far back and concealed, others in plain sight, from their highest fountains to the level of the sea.

Cares of every kind were quickly forgotten, and though the Cassiar engines soon began to wheeze and sigh with doleful solemnity, suggesting coming trouble, we were too happy to mind them. Every face glowed with natural love of wild beauty. The islands were seen in long perspective, their forests dark green in the foreground, with varying tones of blue growing more and more tender in the distance; bays full of hazy shadows, graduating into open, silvery fields of light, and lofty headlands with fine arching insteps dipping their feet in the shining water. But every eye was turned to the mountains. Forgotten now were the Chilcats and missions while the word of God was being read in these majestic hieroglyphics blazoned along the sky. The earnest, childish wonderment with which this glorious page of Nature's Bible was contemplated was delightful to see. All evinced eager desire to learn.

"Is that a glacier," they asked, "down in that cañon? And is it all solid ice?"

"Yes."

"How deep is it?"

"Perhaps five hundred or a thousand feet."

"You say it flows. How can hard ice flow?"

"It flows like water, though invisibly slow."

"And where does it come from?"

"From snow that is heaped up every winter on the mountains."

"And how, then, is the snow changed into ice?"

"It is welded by the pressure of its own weight."

"Are these white masses we see in the hollows glaciers also?"

"Yes."

"Are those bluish draggled masses hanging down from beneath the snow-fields what you call the snouts of the glaciers?"

"Yes."

"What made the hollows they are in?"

"The glaciers themselves, just as traveling animals make their own tracks."

"How long have they been there?"

"Numberless centuries," etc. I answered as best I could, keeping up a running commentary on the subject in general, while busily engaged in sketching and noting my own observations, preaching glacial gospel in a rambling way, while the Cassiar, slowly wheezing and creeping along the shore, shifted our position so that the icy cañons were opened to view and closed again in regular succession, like the leaves of a book.

About the middle of the afternoon we were directly opposite a noble group of glaciers some ten in number, flowing from a chain of crater-like snow-fountains, guarded around their summits and well down their sides by jagged peaks and cols and curving mural ridges. From each of the larger clusters of fountains, a wide, sheer-walled cañon opens down to the sea. Three of the trunk glaciers descend to within a few feet of the sea-level. The largest of the three, probably about fifteen miles long, terminates in a magnificent valley like Yosemite, in an imposing wall of ice about two miles long, and from three to five hundred feet high, forming a barrier across the valley from wall to wall. It was to this glacier that the ships of the Alaska Ice Company resorted for the ice they carried to San Francisco and the Sandwich Islands, and, I believe, also to China and Japan. To load, they had only to sail up the fiord within a short distance of the front and drop anchor in the terminal moraine.

Another glacier, a few miles to the south of this one, receives two large tributaries about equal in size, and then flows down a forested valley to within a hundred feet or so of sea-level. The third of this low-descending group is four or five miles farther south, and, though less imposing than either of the two sketched above, is still a truly noble object, even as imperfectly seen from the channel, and would of itself be well worth a visit to Alaska to any lowlander so unfortunate as never to have seen a glacier.

The boilers of our little steamer were not made for sea water, but it was hoped that fresh water would be found at available points along our course where streams leap down the cliffs. In this particular we failed, however, and were compelled to use salt water an hour or two before reaching Cape Fanshawe, the supply of fifty tons of fresh water brought from Wrangell having then given out. To make matters worse, the captain and engineer were not in accord concerning the working of the engines. The captain repeatedly called for more steam, which the engineer refused to furnish, cautiously keeping the pressure low because the salt water foamed in the boilers and some of it passed over into the cylinders, causing heavy thumping at the end of each piston stroke, and threatening to knock out the cylinder-heads. At seven o'clock in the evening we had made only about seventy miles, which caused dissatisfaction, especially among the divines, who thereupon called a meeting in the cabin to consider what had better be done. In the discussions that followed much indignation and economy were brought to light. We had chartered the boat for sixty dollars per day, and the round trip was to have been made in four or five days. But at the present rate of speed it was found that the cost of the trip for each passenger would be five or ten dollars above the first estimate. Therefore, the majority ruled that we must return next day to Wrangell, the extra dollars outweighing the mountains and missions as if they had suddenly become dust in the balance.

Soon after the close of this economical meeting, we came to anchor in a beautiful bay, and as the long northern day had still hours of good light to offer, I gladly embraced the opportunity to go ashore to see the rocks and plants. One of the Indians, employed as a deck hand on the steamer, landed me at the mouth of a stream. The tide was low, exposing a luxuriant growth of algæ, which sent up a fine, fresh sea smell. The shingle was composed of slate, quartz, and granite, named in the order of abundance. The first land plant met was a tall grass, nine feet high, forming a meadow-like margin in front of the forest. Pushing my way well back into the forest, I found it composed almost entirely of spruce and two hemlocks (*Picea sitchensis, Tsuga heterophylla*, and *T. mertensiana*) with a few specimens of yellow cypress. The ferns were developed in remarkable beauty and size—aspidiums, one of which is about six feet high, a woodsia, lomaria, and several species of polypodium. The underbrush is chiefly alder, rubus, ledum, three species of vaccinium, and *Echinopanax horrida*, the whole about from six to eight feet high, and in some places closely intertangled and hard to penetrate. On the opener spots beneath the trees the ground is covered to a depth of two or three feet with mosses of indescribable freshness and beauty, a few dwarf cornels often planted on their rich furred bosses, together with pyrola, coptis, and Solomon's-seal. The tallest of the trees are about a hundred and fifty feet high, with a diameter of about four or five feet, their branches mingling together and making a perfect shade. As the twilight began to fall, I sat down on the mossy instep of a spruce. Not a bush or tree was moving; every leaf seemed hushed in brooding repose. One bird, a thrush, embroidered the silence with cheery notes, making the solitude familiar and sweet, while the solemn monotone of the stream sifting through the woods seemed like the very voice of God, humanized, terrestrialized, and entering one's heart as to a home prepared for it. Go where we will, all the world over, we seem to have been there before.

The stream was bridged at short intervals with picturesque, moss-embossed logs, and the trees on its banks, leaning over from side to side, made high embowering arches. The log bridge I crossed was, I think, the most beautiful of the kind I ever saw. The massive log is plushed to a depth of six inches or more with mosses of three or four species, their different tones of yellow shading finely into each other, while their delicate fronded branches and foliage lie in exquisite order, inclining outward and down the sides in rich, furred, clasping sheets overlapping and felted together until the required thickness is attained. The pedicels and spore-cases give a purplish tinge, and the whole bridge is enriched with ferns and a row of small seedling trees and currant bushes with colored leaves, every one of which seems to have been culled from the woods for this special use, so perfectly do they harmonize in size, shape, and color with the mossy cover, the width of the span, and the luxuriant, brushy abutments.

Sauntering back to the beach, I found four or five Indian deck hands getting water, with whom I returned aboard the steamer, thanking the Lord for so noble an addition to my life as was this one big mountain, forest, and glacial day.

Next morning most of the company seemed uncomfortably conscience-stricken, and ready to do anything in the way of compensation for our broken excursion that would not cost too much. It was not found difficult, therefore, to convince the captain and disappointed passengers that instead of creeping back to Wrangell direct we should make an expiatory branch-excursion to the largest of the three low-descending glaciers we had passed. The Indian pilot, well acquainted with this part of the coast, declared himself willing to guide us. The water in these fiord channels is generally deep and safe, and though at wide intervals rocks rise abruptly here and there, lacking only a few feet in height to enable them to take rank as islands, the flat-bottomed Cassiar drew but little more water than a duck, so that even the most timid raised no objection on this score. The cylinder-heads of our engines were the main source of anxiety; provided they could be kept on all might yet be well. But in this matter there was evidently some distrust, the engineer having imprudently informed some of the passengers that in consequence of using salt water in his frothing boilers the cylinder-heads might fly off at any moment. To the glacier, however, it was at length decided we should venture.

Arriving opposite the mouth of its fiord, we steered straight inland between beautiful wooded shores, and the grand glacier came in sight in its granite valley, glowing in the early sunshine and extending a noble invitation to come and see. After we passed between the two mountain rocks that guard the gate of the fiord, the view that was unfolded fixed every eye in wondering admiration. No words can convey anything like an adequate conception of its sublime grandeur—the noble simplicity and fineness of the sculpture of the walls; their magnificent proportions; their cascades, gardens, and forest adornments; the placid fiord between them; the great white and blue ice wall, and the snow-laden mountains beyond. Still more impotent are words in telling the peculiar awe one experiences in entering these mansions of the icy North, notwithstanding it is only the natural effect of appreciable manifestations of the presence of God.

Standing in the gateway of this glorious temple, and regarding it only as a picture, its outlines may be easily traced, the water foreground of a pale-green color, a smooth mirror sheet sweeping back five or six miles like one of the lower reaches of a great river, bounded at the head by a beveled barrier wall of blueish-white ice four or five hundred feet high. A few snowy mountain-tops appear beyond it, and on either hand rise a series of majestic, pale-gray granite rocks from three to four thousand feet high, some of them thinly forested and striped with bushes and flowery grass on narrow shelves, especially about halfway up, others severely sheer and bare and built together into walls like those of Yosemite, extending far beyond the ice barrier, one immense brow appearing beyond another with their bases buried in the glacier. This is a Yosemite Valley in process of formation, the modeling and sculpture of the walls nearly completed and well planted, but no groves as yet or gardens or meadows on the raw and unfinished bottom. It is as if the explorer, in entering the Merced Yosemite, should find the walls nearly in their present condition, trees and flowers in the warm nooks and along the sunny portions of the moraine-covered brows, but the bottom of the valley still covered with water and beds of gravel and mud, and the grand glacier that formed it slowly receding but still filling the upper half of the valley.

Sailing directly up to the edge of the low, outspread, water-washed terminal moraine, scarce noticeable in a general view, we seemed to be separated from the glacier only by a bed of gravel a hundred yards or so in width; but on so grand a scale are all the main features of the valley, we afterwards found the distance to be a mile or more.

The captain ordered the Indian deck hands to get out the canoe, take as many of us ashore as wished to go, and accompany us to the glacier in case we should need their help. Only three of the company, in the first place, availed themselves of this rare opportunity of meeting a glacier in the flesh,—Mr. Young, one of the doctors, and myself. Paddling to the nearest and driest-looking part of the moraine flat, we stepped ashore, but gladly wallowed back into the canoe; for the gray mineral mud, a paste made of fine-ground mountain meal kept unstable by the tides, at once began to take us in, swallowing us feet foremost with becoming glacial deliberation. Our next attempt, made nearer the middle of the valley, was successful, and we soon found ourselves on firm gravelly ground, and made haste to the huge ice-wall, which seemed to recede as we advanced. The only difficulty we met was a network of icy streams, at the largest of which we halted, not willing to get wet in fording. The Indian attendant promptly carried us over on his back. When my turn came I told him I would ford, but he bowed his shoulders in so ludicrously persuasive a manner I thought I would try the queer mount, the only one of the kind I had enjoyed since boyhood days in playing leapfrog. Away staggered my perpendicular mule over the boulders into the brawling torrent, and in spite of top-heavy predictions to the contrary, crossed without a fall. After being ferried in this way over several more of these glacial streams, we at length reached the foot of the glacier wall. The doctor simply played tag on it, touched it gently as if it were a dangerous wild beast, and hurried back to the boat, taking the portage Indian with him for safety, little knowing what he was missing. Mr. Young and I traced the glorious crystal wall, admiring its wonderful architecture, the play of light in the rifts and caverns, and the structure of the ice as displayed in the less fractured sections, finding fresh beauty everywhere and facts for study. We then tried to climb it, and by dint of patient zigzagging and doubling among the crevasses, and cutting steps here and there, we made our way up over the brow and back a mile or two to a height of about seven hundred feet. The whole front of the glacier is gashed and sculptured into a maze of shallow caves and crevasses, and a bewildering variety of novel architectural forms, clusters of glittering lance-tipped spires, gables, and obelisks, bold outstanding bastions and plain mural cliffs, adorned along the top with fretted cornice and battlement, while every gorge and crevasse, groove and hollow, was filled with light, shimmering and throbbing in pale-blue tones of ineffable tenderness and beauty. The day was warm, and back on the broad melting bosom of the glacier beyond the crevassed front, many streams were rejoicing, gurgling, ringing, singing, in frictionless channels worn down through the white disintegrated ice of the surface into the quick and living blue, in which they flowed with a grace of motion and flashing of light to be found only on the crystal hillocks and ravines of a glacier.

Along the sides of the glacier we saw the mighty flood grinding against the granite walls with tremendous pressure, rounding outswelling bosses, and deepening the retreating hollows into the forms they are destined to have when, in the fullness of appointed time, the huge ice tool shall be withdrawn by the sun. Every feature glowed with intention, reflecting the plans of God. Back a few miles from the front, the glacier is now probably but little more than a thousand feet deep; but when we examine the records on the walls, the rounded, grooved, striated, and polished features so surely glacial, we learn that in the earlier days of the ice age they were all overswept, and that this glacier has flowed at a height of from three to four thousand feet above its present level, when it was at least a mile deep.

Standing here, with facts so fresh and telling and held up so vividly before us, every seeing observer, not to say geologist, must readily apprehend the earth-sculpturing, landscape-making action of flowing ice. And here, too, one learns that the world, though made, is yet being made; that this is still the morning of creation; that mountains long conceived are now being born, channels traced for coming rivers, basins hollowed for lakes; that moraine soil is being ground and outspread for coming plants,—coarse boulders and gravel for forests, finer soil for grasses and flowers,—while the finest part of the grist, seen hastening out to sea in the draining streams, is being stored away in darkness and builded particle on particle, cementing and crystallizing, to make the mountains and valleys and plains of other predestined landscapes, to be followed by still others in endless rhythm and beauty.

Gladly would we have camped out on this grand old landscape mill to study its ways and works; but we had no bread and the captain was keeping the Cassiar whistle screaming for our return. Therefore, in mean haste, we threaded our way back through the crevasses and down the blue cliffs, snatched a few flowers from a warm spot on the edge of the ice, plashed across the moraine streams, and were paddled aboard, rejoicing in the possession of so blessed a day, and feeling that in very foundational truth we had been in one of God's own temples and had seen Him and heard Him working and preaching like a man.

Steaming solemnly out of the fiord and down the coast, the islands and mountains were again passed in review; the clouds that so often hide the mountain-tops even in good weather were now floating high above them, and the transparent shadows they cast were scarce perceptible on the white glacier fountains. So abundant and novel are the objects of interest in a pure wilderness that unless you are pursuing special studies it matters little where you go, or how often to the same place. Wherever you chance to be always seems at the moment of all places the best; and you feel that there can be no happiness in this world or in any other for those who may not be happy here. The bright hours were spent in making notes and sketches and getting more of the wonderful region into memory. In particular a second view of the mountains made me raise my first estimate of their height. Some of them must be seven or eight thousand feet at the least. Also the glaciers seemed larger and more numerous. I counted nearly a hundred, large and small, between a point ten or fifteen miles to the north of Cape Fanshawe and the mouth of the Stickeen River. We made no more landings, however, until we had passed through the Wrangell Narrows and dropped anchor for the night in a small sequestered bay. This was about sunset, and I eagerly seized the opportunity to go ashore in the canoe and see what I could learn. It is here only a step from the marine algæ to terrestrial vegetation of almost tropical luxuriance. Parting the alders and huckleberry bushes and the crooked stems of the prickly panax, I made my way into the woods, and lingered in the twilight doing nothing in particular, only measuring a few of the trees, listening to learn what birds and animals might be about, and gazing along the dusky aisles.

In the mean time another excursion was being invented, one of small size and price. We might have reached Fort Wrangell this evening instead of anchoring here; but the owners of the Cassiar would then receive only ten dollars fare from each person, while they had incurred considerable expense in fitting up the boat for this special trip, and had treated us well. No, under the circumstances, it would never do to return to Wrangell so meanly soon.

It was decided, therefore, that the Cassiar Company should have the benefit of another day's hire, in visiting the old deserted

Stickeen village fourteen miles to the south of Wrangell.

"We shall have a good time," one of the most influential of the party said to me in a semi-apologetic tone, as if dimly recognizing my disappointment in not going on to Chilcat. "We shall probably find stone axes and other curiosities. Chief Kadachan is going to guide us, and the other Indians aboard will dig for us, and there are interesting old buildings and totem poles to be seen."

It seemed strange, however, that so important a mission to the most influential of the Alaskan tribes should end in a deserted village. But divinity abounded nevertheless; the day was divine and there was plenty of natural religion in the new-born landscapes that were being baptized in sunshine, and sermons in the glacial boulders on the beach where we landed.

The site of the old village is on an outswelling strip of ground about two hundred yards long and fifty wide, sloping gently to the water with a strip of gravel and tall grass in front, dark woods back of it, and charming views over the water among the islands—a delightful place. The tide was low when we arrived, and I noticed that the exposed boulders on the beach—granite erratics that had been dropped by the melting ice toward the close of the glacial period—were piled in parallel rows at right angles to the shore-line, out of the way of the canoes that had belonged to the village.

Most of the party sauntered along the shore; for the ruins were overgrown with tall nettles, elder bushes, and prickly rubus vines through which it was difficult to force a way. In company with the most eager of the relic-seekers and two Indians, I pushed back among the dilapidated dwellings. They were deserted some sixty or seventy years before, and some of them were at least a hundred years old. So said our guide, Kadachan, and his word was corroborated by the venerable aspect of the ruins. Though the damp climate is destructive, many of the house timbers were still in a good state of preservation, particularly those hewn from the yellow cypress, or cedar as it is called here. The magnitude of the ruins and the excellence of the workmanship manifest in them was astonishing as belonging to Indians. For example, the first dwelling we visited was about forty feet square, with walls built of planks two feet wide and six inches thick. The ridgepole of yellow cypress was two feet in diameter, forty feet long, and as round and true as if it had been turned in a lathe; and, though lying in the damp weeds, it was still perfectly sound. The nibble marks of the stone adze were still visible, though crusted over with scale lichens in most places. The pillars that had supported the ridgepole were still standing in some of the ruins. They were all, as far as I observed, carved into life-size figures of men, women, and children, fishes, birds, and various other animals, such as the beaver, wolf, or bear. Each of the wall planks had evidently been hewn out of a whole log, and must have required sturdy deliberation as well as skill. Their geometrical truthfulness was admirable. With the same tools not one in a thousand of our skilled mechanics could do as good work. Compared with it the bravest work of civilized backwoodsmen is feeble and bungling. The completeness of form, finish, and proportion of these timbers suggested skill of a wild and positive kind, like that which guides the woodpecker in drilling round holes, and the bee in making its cells.

The carved totem-pole monuments are the most striking of the objects displayed here. The simplest of them consisted of a smooth, round post fifteen or twenty feet high and about eighteen inches in diameter, with the figure of some animal on top—a bear, porpoise, eagle, or raven, about life-size or larger. These were the totems of the families that occupied the houses in front of which they stood. Others supported the figure of a man or woman, life-size or larger, usually in a sitting posture, said to resemble the dead whose ashes were contained in a closed cavity in the pole. The largest were thirty or forty feet high, carved from top to bottom into human and animal totem figures, one above another, with their limbs grotesquely doubled and folded. Some of the most imposing were said to commemorate some event of an historical character. But a telling display of family pride seemed to have been the prevailing motive. All the figures were more or less rude, and some were broadly grotesque, but there was never any feebleness or obscurity in the expression. On the contrary, every feature showed grave force and decision; while the childish audacity displayed in the designs, combined with manly strength in their execution, was truly wonderful.

The colored lichens and mosses gave them a venerable air, while the larger vegetation often found on such as were most decayed produced a picturesque effect. Here, for example, is a bear five or six feet long, reposing on top of his lichen-clad pillar, with paws comfortably folded, a tuft of grass growing in each ear and rubus bushes along his back. And yonder is an old chief poised on a taller pillar, apparently gazing out over the landscape in contemplative mood, a tuft of bushes leaning back with a jaunty air from the top of his weatherbeaten hat, and downy mosses about his massive lips. But no rudeness or grotesqueness that may appear, however combined with the decorations that nature has added, may possibly provoke mirth. The whole work is serious in aspect and brave and true in execution.

Similar monuments are made by other Thlinkit tribes. The erection of a totem pole is made a grand affair, and is often talked of for a year or two beforehand. A feast, to which many are invited, is held, and the joyous occasion is spent in eating, dancing, and the distribution of gifts. Some of the larger specimens cost a thousand dollars or more. From one to two hundred blankets, worth three dollars apiece, are paid to the genius who carves them, while the presents and feast usually cost twice as much, so that only the wealthy families can afford them. I talked with an old Indian who pointed out one of the carvings he had made in the Wrangell village, for which he told me he had received forty blankets, a gun, a canoe, and other articles, all together worth about $170. Mr. Swan, who has contributed much information concerning the British Columbian and Alaskan tribes, describes a totem pole that cost $2500. They are always planted firmly in the ground and stand fast, showing the sturdy erectness of their builders.

While I was busy with my pencil, I heard chopping going on at the north end of the village, followed by a heavy thud, as if a tree had fallen. It appeared that after digging about the old hearth in the first dwelling visited without finding anything of consequence, the archaeological doctor called the steamer deck hands to one of the most interesting of the totems and directed them to cut it down, saw off the principal figure,—a woman measuring three feet three inches across the shoulders,—and convey it aboard the steamer, with a view to taking it on East to enrich some museum or other. This sacrilege came near causing trouble and would have cost us dear had the totem not chanced to belong to the Kadachan family, the representative of which is a member of the newly organized Wrangell Presbyterian Church. Kadachan looked very seriously into the face of the reverend doctor and pushed home the pertinent question: "How would you like to have an Indian go to a graveyard and break down and carry away a monument belonging to your family?"

However, the religious relations of the parties and a few trifling presents embedded in apologies served to hush and mend the matter.

Some time in the afternoon the steam whistle called us together to finish our memorable trip. There was no trace of decay in the sky; a glorious sunset gilded the water and cleared away the shadows of our meditations among the ruins. We landed at the Wrangell wharf at dusk, pushed our way through a group of inquisitive Indians, across the two crooked streets, and up to our homes in the fort. We had been away only three days, but they were so full of novel scenes and impressions the time seemed indefinitely long, and our broken Chilcat excursion, far from being a failure as it seemed to some, was one of the most memorable of my life.

CHAPTER VI

THE CASSIAR TRAIL

I MADE A second trip up the Stickeen in August and from the head of navigation pushed inland for general views over dry grassy hills and plains on the Cassiar trail.

Soon after leaving Telegraph Creek I met a merry trader who encouragingly assured me that I was going into the most wonderful region in the world, that "the scenery up the river was full of the very wildest freaks of nature, surpassing all other sceneries either natural or artificial, on paper or in nature. And give yourself no bothering care about provisions, for wild food grows in prodigious abundance everywhere. A man was lost four days up there, but he feasted on vegetables and berries and got back to camp in good condition. A mess of wild parsnips and pepper, for example, will actually do you good. And here's my advice—go slow and take the pleasures and sceneries as you go."

At the confluence of the first North Fork of the Stickeen I found a band of Toltan or Stick Indians catching their winter supply of salmon in willow traps, set where the fish are struggling in swift rapids on their way to the spawning-grounds. A large supply had already been secured, and of course the Indians were well fed and merry. They were camping in large booths made of poles set on end in the ground, with many binding cross-pieces on which tons of salmon were being dried. The heads were strung on separate poles and the roes packed in willow baskets, all being well smoked from fires in the middle of the floor. The largest of the booths near the bank of the river was about forty feet square. Beds made of spruce and pine boughs were spread all around the walls, on which some of the Indians lay asleep; some were braiding ropes, others sitting and lounging, gossiping and courting, while a little baby was swinging in a hammock. All seemed to be light-hearted and jolly, with work enough and wit enough to maintain health and comfort. In the winter they are said to dwell in substantial huts in the woods, where game, especially caribou, is abundant. They are pale copper-colored, have small feet and hands, are not at all negroish in lips or cheeks like some of the coast tribes, nor so thickset, short-necked, or heavy-featured in general.

One of the most striking of the geological features of this region are immense gravel deposits displayed in sections on the walls of the river gorges. About two miles above the North Fork confluence there is a bluff of basalt three hundred and fifty feet high, and above this a bed of gravel four hundred feet thick, while beneath the basalt there is another bed at least fifty feet thick.

From "Ward's," seventeen miles beyond Telegraph, and about fourteen hundred feet above sea-level, the trail ascends a gravel ridge to a pine-and-fir-covered plateau twenty-one hundred feet above the sea. Thence for three miles the trail leads through a forest of short, closely planted trees to the second North Fork of the Stickeen, where a still greater deposit of stratified gravel is displayed, a section at least six hundred feet thick resting on a red jaspery formation.

Nine hundred feet above the river there is a slightly dimpled plateau diversified with aspen and willow groves and mossy meadows. At "Wilson's," one and a half miles from the river, the ground is carpeted with dwarf manzanita and the blessed *Linnæa borealis*, and forested with small pines, spruces, and aspens, the tallest fifty to sixty feet high.

From Wilson's to "Caribou," fourteen miles, no water was visible, though the nearly level, mossy ground is swampy-looking. At "Caribou Camp," two miles from the river, I saw two fine dogs, a Newfoundland and a spaniel. Their owner told me that he paid only twenty dollars for the team and was offered one hundred dollars for one of them a short time afterwards. The Newfoundland, he said, caught salmon on the ripples, and could be sent back for miles to fetch horses. The fine jet-black curly spaniel helped to carry the dishes from the table to the kitchen, went for water when ordered, took the pail and set it down at the stream-side, but could not be taught to dip it full. But their principal work was hauling camp-supplies on sleds up the river in winter. These two were said to be able to haul a load of a thousand pounds when the ice was in fairly good condition. They were fed on dried fish and oatmeal boiled together.

The timber hereabouts is mostly willow or poplar on the low ground, with here and there pine, birch, and spruce about fifty feet high. None seen much exceeded a foot in diameter. Thousand-acre patches have been destroyed by fire. Some of the green trees had been burned off at the root, the raised roots, packed in dry moss, being readily attacked from beneath. A range of mountains about five thousand to six thousand feet high trending nearly north and south for sixty miles is forested to the summit. Only a few cliff-faces and one of the highest points patched with snow are treeless. No part of this range as far as I could see is deeply sculptured, though the general denudation of the country must have been enormous as the gravel-beds show.

At the top of a smooth, flowery pass about four thousand feet above the sea, beautiful Dease Lake comes suddenly in sight, shining like a broad tranquil river between densely forested hills and mountains. It is about twenty-seven miles long, one to two miles wide, and its waters, tributary to the Mackenzie, flow into the Arctic Ocean by a very long, roundabout, romantic way, the exploration of which in 1789 from Great Slave Lake to the Arctic Ocean must have been a glorious task for the heroic Scotchman, Alexander Mackenzie, whose name it bears.

Dease Creek, a fine rushing stream about forty miles long and forty or fifty feet wide, enters the lake from the west, drawing its sources from grassy mountain-ridges. Thibert Creek, about the same size, and McDames and Defot Creeks, with their many branches, head together in the same general range of mountains or on moor-like table-lands on the divide between the Mackenzie and Yukon and Stickeen. All these Mackenzie streams had proved rich in gold. The wing-dams, flumes, and sluice-boxes on the lower five or ten miles of their courses showed wonderful industry, and the quantity of glacial and perhaps pre-glacial gravel displayed was enormous. Some of the beds were not unlike those of the so-called Dead Rivers of California. Several ancient drift-filled channels on Thibert Creek, blue at bed rock, were exposed and had been worked. A considerable portion of the gold, though mostly coarse, had no doubt come from considerable distances, as boulders included in some of the deposits show. The deepest beds, though known to be rich, had not yet been worked to any great depth on account of expense. Diggings that yield less than five dollars a day to the man were considered worthless. Only three of the claims on Defot Creek, eighteen miles from the mouth of Thibert Creek, were then said to pay. One of the nuggets from this creek weighed forty pounds.

While wandering about the banks of these gold-besprinkled streams, looking at the plants and mines and miners, I was so fortunate as to meet an interesting French Canadian, an old *coureur de bois*, who after a few minutes' conversation invited me to accompany him to his gold-mine on the head of Defot Creek, near the summit of a smooth, grassy mountain-ridge which he assured me commanded extensive views of the region at the heads of Stickeen, Taku, Yukon, and Mackenzie tributaries. Though heavy-laden with flour and bacon, he strode lightly along the rough trails as if his load was only a natural balanced part of his body. Our way at first lay along Thibert Creek, now on gravel benches, now on bed rock, now close down on the bouldery edge of the stream. Above the mines the stream is clear and flows with a rapid current. Its banks are embossed with moss and grass and sedge well mixed with flowers—daisies, larkspurs, solidagos, parnassia, potentilla, strawberry, etc. Small strips of meadow occur here and there, and belts of slender arrowy fir and spruce with moss-clad roots grow close to the water's edge. The creek is about forty-five miles long, and the richest of its gold-bearing beds so far discovered were on the lower four miles of the creek; the higher four-or-five-dollars-a-day diggings were considered very poor on account of the high price of provisions and shortness of the season. After crossing many smaller streams with their strips of trees and meadows, bogs and bright wild gardens, we arrived at the Le Claire cabin about the middle of the afternoon. Before entering it he threw down his burden and made haste to show me his favorite flower, a blue forget-me-not, a specimen of which he found within a few rods of the cabin, and proudly handed it to me with the finest respect, and telling its many charms and lifelong associations, showed in every endearing look and touch and gesture that the tender little plant of the mountain wilderness was truly his best-loved darling.

After luncheon we set out for the highest point on the dividing ridge about a mile above the cabin, and sauntered and gazed until sundown, admiring the vast expanse of open rolling prairie-like highlands dotted with groves and lakes, the fountain-heads of countless cool, glad streams.

Le Claire's simple, child-like love of nature, preserved undimmed through a hard wilderness life, was delightful to see. The grand landscapes with their lakes and streams, plants and animals, all were dear to him. In particular he was fond of the birds that nested near his cabin, watched the young, and in stormy weather helped their parents to feed and shelter them. Some species were so confiding they learned to perch on his shoulders and take crumbs from his hand.

A little before sunset snow began to fly, driven by a cold wind, and by the time we reached the cabin, though we had not far to go, everything looked wintry. At half-past nine we ate supper, while a good fire crackled cheerily in the ingle and a wintry wind blew hard. The little log cabin was only ten feet long, eight wide, and just high enough under the roof peak to allow one to stand upright. The bedstead was not wide enough for two, so Le Claire spread the blankets on the floor, and we gladly lay down after our long, happy walk, our heads under the bedstead, our feet against the opposite wall, and though comfortably tired, it was long ere we fell asleep, for Le Claire, finding me a good listener, told many stories of his adventurous life with Indians, bears and wolves, snow and hunger, and of his many camps in the Canadian woods, hidden like the nests and dens of wild animals; stories that have a singular interest to everybody, for they awaken inherited memories of the lang, lang syne when we were all wild. He had nine children, he told me, the youngest eight years of age, and several of his daughters were married. His home was in Victoria.

Next morning was cloudy and windy, snowy and cold, dreary December weather in August, and I gladly ran out to see what I might learn. A gray ragged-edged cloud capped the top of the divide, its snowy fringes drawn out by the wind. The flowers, though most of them were buried or partly so, were to some extent recognizable, the bluebells bent over, shining like eyes through the snow, and the gentians, too, with their corollas twisted shut; cassiope I could recognize under any disguise; and two species of dwarf willow with their seeds already ripe, one with comparatively small leaves, were growing in mere cracks and crevices of rock-ledges where the dry snow could not lie. Snowbirds and ptarmigan were flying briskly in the cold wind, and on the edge of a grove I saw a spruce from which a bear had stripped large sections of bark for food.

About nine o'clock the clouds lifted and I enjoyed another wide view from the summit of the ridge of the vast grassy fountain region with smooth rolling features. A few patches of forest broke the monotony of color, and the many lakes, one of them about five miles long, were glowing like windows. Only the highest ridges were whitened with snow, while rifts in the clouds showed beautiful bits of yellow-green sky. The limit of tree growth is about five thousand feet.

Throughout all this region from Glenora to Cassiar the grasses grow luxuriantly in openings in the woods and on dry hillsides where the trees seem to have been destroyed by fire, and over all the broad prairies above the timber line. A kind of bunch-grass in particular is often four or five feet high, and close enough to be mowed for hay. I never anywhere saw finer or more bountiful wild pasture. Here the caribou feed and grow fat, braving the intense winter cold, often forty to sixty degrees below zero. Winter and summer seem to be the only seasons here. What may fairly be called summer lasts only two or three months, winter nine or ten, for of pure well-defined spring or autumn there is scarcely a trace. Were it not for the long severe winters, this would be a capital stock country, equaling Texas and the prairies of the old West. From my outlook on the Defot ridge I saw thousands of square miles of this prairie-like region drained by tributaries of the Stickeen, Taku, Yukon, and Mackenzie Rivers.

Le Claire told me that the caribou, or reindeer, were very abundant on this high ground. A flock of fifty or more was seen a short time before at the head of Defot Creek,—fine, hardy, able animals like their near relatives the reindeer of the Arctic tundras. The Indians hereabouts, he said, hunted them with dogs, mostly in the fall and winter. On my return trip I met several bands of these Indians on the march, going north to hunt. Some of the men and women were carrying puppies on top of their heavy loads of dried salmon, while the grown dogs had saddle-bags filled with odds and ends strapped on their backs. Small puppies, unable to carry more than five or six pounds, were thus made useful. I overtook another band going south, heavy laden with furs and skins to trade. An old woman, with short dress and leggings, was carrying a big load of furs and skins, on top of which was perched a little girl about three years old.

A brown, speckled marmot, one of Le Claire's friends, was getting ready for winter. The entrance to his burrow was a little to one side of the cabin door. A well-worn trail led to it through the grass and another to that of his companion, fifty feet away. He was a most amusing pet, always on hand at meal times for bread-crumbs and bits of bacon-rind, came when called, answering in a shrill whistle, moving like a squirrel with quick, nervous impulses, jerking his short flat tail. His fur clothing was neat and clean, fairly shining in the wintry light. The snowy weather that morning must have called winter to mind; for as soon as he got his breakfast, he ran to a tuft of dry grass, chewed it into fuzzy mouthfuls, and carried it to his nest, coming and going with admirable industry, forecast, and confidence. None watching him as we did could fail to sympathize with him; and I fancy that in practical weather wisdom no government forecaster with all his advantages surpasses this little Alaska rodent, every hair and nerve a weather instrument.

I greatly enjoyed this little inland side trip—the wide views; the miners along the branches of the great river, busy as moles and beavers; young men dreaming and hoping to strike it rich and rush home to marry their girls faithfully waiting; others hoping to clear off weary farm mortgages, and brighten the lives of the anxious home folk; but most, I suppose, just struggling blindly for gold enough to make them indefinitely rich to spend their lives in aimless affluence, honor, and ease. I enjoyed getting acquainted with the trees, especially the beautiful spruce and silver fir; the flower gardens and great grassy caribou pastures; the cheery, able marmot mountaineer; and above all the friendship and kindness of Mr. Le Claire, whom I shall never forget. Bidding good-bye, I sauntered back to the head of navigation on the Stickeen, happy and rich without a particle of obscuring gold-dust care.

CHAPTER VII

GLENORA PEAK

ON THE TRAIL to the steamboat-landing at the foot of Dease Lake, I met a Douglas squirrel, nearly as red and rusty in color as his Eastern relative the chickaree. Except in color he differs but little from the California Douglas squirrel. In voice, language, gestures, temperament, he is the same fiery, indomitable little king of the woods. Another darker and probably younger specimen met near the Caribou House, barked, chirruped, and showed off in fine style on a tree within a few feet of us.

"What does the little rascal mean?" said my companion, a man I had fallen in with on the trail. "What is he making such a fuss about? I cannot frighten him."

"Never mind," I replied; "just wait until I whistle 'Old Hundred' and you will see him fly in disgust." And so he did, just as his California brethren do. Strange that no squirrel or spermophile I yet have found ever seemed to have anything like enough of Scotch religion to enjoy this grand old tune.

The taverns along the Cassiar gold trail were the worst I had ever seen, rough shacks with dirt floors, dirt roofs, and rough meals. The meals are all alike—a potato, a slice of something like bacon, some gray stuff called bread, and a cup of muddy, semi-liquid coffee like that which the California miners call "slickens" or "slumgullion." The bread was terrible and sinful. How the Lord's good wheat could be made into stuff so mysteriously bad is past finding out. The very de'il, it would seem, in wicked anger and ingenuity, had been the baker.

On our walk from Dease Lake to Telegraph Creek we had one of these rough luncheons at three o'clock in the afternoon of the first day, then walked on five miles to Ward's, where we were solemnly assured that we could not have a single bite of either supper or breakfast, but as a great favor we might sleep on his best gray bunk. We replied that, as we had lunched at the lake, supper would not be greatly missed, and as for breakfast we would start early and walk eight miles to the next road-house. We set out at half-past four, glad to escape into the fresh air, and reached the breakfast place at eight o'clock. The landlord was still abed, and when at length he came to the door, he scowled savagely at us as if our request for breakfast was preposterous and criminal beyond anything ever heard of in all goldful Alaska. A good many in those days were returning from the mines dead broke, and he probably regarded us as belonging to that disreputable class. Anyhow, we got nothing and had to tramp on.

As we approached the next house, three miles ahead, we saw the tavern-keeper keenly surveying us, and, as we afterwards learned, taking me for a certain judge whom for some cause he wished to avoid, he hurriedly locked his door and fled. Half a mile farther on we discovered him in a thicket a little way off the trail, explained our wants, marched him back to his house, and at length obtained a little sour bread, sour milk, and old salmon, our only lonely meal between the Lake and Telegraph Creek.

We arrived at Telegraph Creek, the end of my two-hundred-mile walk, about noon. After luncheon I went on down the river to Glenora in a fine canoe owned and manned by Kitty, a stout, intelligent-looking Indian woman, who charged her passengers a dollar for the fifteen-mile trip. Her crew was four Indian paddlers. In the rapids she also plied the paddle, with stout, telling strokes, and a keen-eyed old man, probably her husband, sat high in the stern and steered. All seemed exhilarated as we shot down through the narrow gorge on the rushing, roaring, throttled river, paddling all the more vigorously the faster the speed of the stream, to hold good steering way. The canoe danced lightly amid gray surges and spray as if alive and enthusiastically enjoying the adventure. Some of the passengers were pretty thoroughly drenched. In unskillful hands the frail dugout would surely have been wrecked or upset. Most of the season goods for the Cassiar gold camps were carried from Glenora to Telegraph Creek in canoes, the steamers not being able to overcome the rapids except during high water. Even then they had usually to line two of the rapids—that is, take a line ashore, make it fast to a tree on the bank, and pull up on the capstan. The freight canoes carried about three or four tons, for which fifteen dollars per ton was charged. Slow progress was made by poling along the bank out of the swiftest part of the current. In the rapids a tow line was taken ashore, only one of the crew remaining aboard to steer. The trip took a day unless a favoring wind was blowing, which often happened.

Next morning I set out from Glenora to climb Glenora Peak for the general view of the great Coast Range that I failed to obtain on my first ascent on account of the accident that befell Mr. Young when we were within a minute or two of the top. It is hard to fail in reaching a mountain-top that one starts for, let the cause be what it may. This time I had no companion to care for, but the sky was threatening. I was assured by the local weather-prophets that the day would be rainy or snowy because the peaks in sight were muffled in clouds that seemed to be getting ready for work. I determined to go ahead, however, for storms of any kind are well worth while, and if driven back I could wait and try again.

With crackers in my pocket and a light rubber coat that a kind Hebrew passenger on the steamer Gertrude loaned me, I was ready for anything that might offer, my hopes for the grand view rising and falling as the clouds rose and fell. Anxiously I watched them as they trailed their draggled skirts across the glaciers and fountain peaks as if thoughtfully looking for the places where they could do the most good. From Glenora there is first a terrace two hundred feet above the river covered mostly with bushes, yellow apocynum on the open spaces, together with carpets of dwarf manzanita, bunch-grass, and a few of the compositæ, galiums, etc. Then comes a flat stretch a mile wide, extending to the foot-hills, covered with birch, spruce, fir, and poplar, now mostly killed by fire and the ground strewn with charred trunks. From this black forest the mountain rises in rather steep slopes covered with a luxuriant growth of bushes, grass, flowers, and a few trees, chiefly spruce and fir, the firs gradually dwarfing into a beautiful chaparral, the most beautiful, I think, I have ever seen, the flat fan-shaped plumes thickly foliaged and imbricated by snow pressure, forming a smooth, handsome thatch which bears cones and thrives as if this repressed condition were its very best. It extends up to an elevation of about fifty-five hundred feet. Only a few trees more than a foot in diameter and more than fifty feet high are found higher than four thousand feet above the sea. A few poplars and willows occur on moist places, gradually dwarfing like the conifers. Alder is the most generally distributed of the chaparral bushes, growing nearly everywhere; its crinkled stems an inch or two thick form a troublesome tangle to the mountaineer. The blue geranium, with leaves red and showy at this time of the year, is perhaps the most telling of the flowering plants. It grows up to five thousand feet or more. Larkspurs are common, with epilobium, senecio, erigeron, and a few solidagos. The harebell appears at about four thousand feet and extends to the summit, dwarfing in stature but maintaining the size of its handsome bells until they seem to be lying loose and detached on the ground as if like snow flowers they had fallen from the sky; and, though frail and delicate-looking, none of its companions is more enduring or rings out the praises of beauty-loving Nature in tones more appreciable to mortals, not forgetting even Cassiope, who also is here and her companion, Bryanthus, the loveliest and most widely distributed of the alpine shrubs. Then come crowberry, and two species of huckleberry, one of them from about six inches to a foot high with delicious berries, the other a most lavishly prolific and contented-looking dwarf, few of the bushes being more than two inches high, counting to the topmost leaf, yet each bearing from ten to twenty or more large berries. Perhaps more than half the bulk of the whole plant is fruit, the largest and finest-flavored of all the huckleberries or blueberries I ever tasted, spreading fine feasts for the grouse and ptarmigan and many others of Nature's mountain people. I noticed three species of dwarf willows, one with narrow leaves, growing at the very summit of the mountain in cracks of the rocks, as well as on patches of soil, another with large, smooth leaves now turning yellow. The third species grows between the others as to elevation; its leaves, then orange-colored, are strikingly pitted and reticulated. Another alpine shrub, a species of sericocarpus, covered with handsome heads of feathery achenia, beautiful dwarf echiverias with flocks of purple flowers pricked into their bright grass-green, cushion-like bosses of moss-like foliage, and a fine forget-me-not reach to the summit. I may also mention a large mertensia, a fine anemone, a veratrum, six feet high, a large blue daisy, growing up to three to four thousand feet, and at the summit a dwarf species, with dusky, hairy involucres, and a few ferns, aspidium, gymnogramma, and small rock cheilanthes, leaving scarce a foot of ground bare, though the mountain looks bald and brown in the distance like those of the desert ranges of the Great Basin in Utah and Nevada.

Charmed with these plant people, I had almost forgotten to watch the sky until I reached the top of the highest peak, when one of the greatest and most impressively sublime of all the mountain views I have ever enjoyed came full in sight—more than three hundred miles of closely packed peaks of the great Coast Range, sculptured in the boldest manner imaginable, their naked tops and dividing ridges dark in color, their sides and the cañons, gorges, and valleys between them loaded with glaciers and snow. From this standpoint I counted upwards of two hundred glaciers, while dark-centred luminous clouds with fringed edges hovered and crawled over them, now slowly descending, casting transparent shadows on the ice and snow, now rising high above them, lingering like loving angels guarding the crystal gifts they had bestowed. Although the range as seen from this Glenora mountain-top seems regular in its trend, as if the main axis were simple and continuous, it is, on the contrary, far from simple. In front of the highest ranks of peaks are others of the same form with their own glaciers, and lower peaks before these, and yet lower ones with their ridges and cañons, valleys and foot-hills. Alps rise beyond alps as far as the eye can reach, and clusters of higher peaks here and there closely crowded together; clusters, too, of needles and pinnacles innumerable like trees in groves. Everywhere the peaks seem comparatively slender and closely packed, as if Nature had here been trying to see how many noble well-dressed mountains could be crowded into one grand range.

The black rocks, too steep for snow to lie upon, were brought into sharp relief by white clouds and snow and glaciers, and these again were outlined and made tellingly plain by the rocks. The glaciers so grandly displayed are of every form, some crawling through gorge and valley like monster glittering serpents; others like broad cataracts pouring over cliffs into shadowy gulfs; others, with their main trunks winding through narrow cañons, display long, white finger-like tributaries descending from the summits of pinnacled ridges. Others lie back in fountain cirques walled in all around save at the lower edge, over which they pour in blue cascades. Snow, too, lay in folds and patches of every form on blunt, rounded ridges in curves, arrowy lines, dashes, and narrow ornamental flutings among the summit peaks and in broad radiating wings on smooth slopes. And on many a bulging headland and lower ridge there lay heavy, over-curling copings and smooth, white domes where wind-driven snow was pressed and wreathed and packed into every form and in every possible place and condition. I never before had seen so richly sculptured a range or so many awe-inspiring inaccessible mountains crowded together. If a line were drawn east and west from the peak on which I stood, and extended both ways to the horizon, cutting the whole round landscape in two equal parts, then all of the south half would be bounded by these icy peaks, which would seem to curve around half the horizon and about twenty degrees more, though extending in a general straight, or but moderately curved, line. The deepest and thickest and highest of all this wilderness of peaks lie to the southwest. They are probably from about nine to twelve thousand feet high, springing to this elevation from near the sea-level. The peak on which these observations were made is somewhere about seven thousand feet high, and from here I estimated the height of the range. The highest peak of all, or that seemed so to me, lies to the westward at an estimated distance of about one hundred and fifty or two hundred miles. Only its solid white summit was visible. Possibly it may be the topmost peak of St. Elias. Now look northward around the other half of the horizon, and instead of countless peaks crowding into the sky, you see a low brown region, heaving and swelling in gentle curves, apparently scarcely more waved than a rolling prairie. The so-called cañons of several forks of the upper Stickeen are visible, but even where best seen in the foreground and middle ground of the picture, they are like mere sunken gorges, making scarce perceptible marks on the landscape, while the tops of the highest mountain-swells show only small patches of snow and no glaciers.

Glenora Peak, on which I stood, is the highest point of a spur that puts out from the main range in a northerly direction. It seems to have been a rounded, broad-backed ridge which has been sculptured into its present irregular form by short residual glaciers, some of which, a mile or two long, are still at work.

As I lingered, gazing on the vast show, luminous shadowy clouds seemed to increase in glory of color and motion, now fondling the highest peaks with infinite tenderness of touch, now hovering above them like eagles over their nests.

When night was drawing near, I ran down the flowery slopes exhilarated, thanking God for the gift of this great day. The setting sun fired the clouds. All the world seemed new-born. Every thing, even the commonest, was seen in new light and was looked at with new interest as if never seen before. The plant people seemed glad, as if rejoicing with me, the little ones as well as the trees, while every feature of the peak and its traveled boulders seemed to know what I had been about and the depth of my joy, as if they could read faces.

CHAPTER VIII

EXPLORATION OF THE STICKEEN GLACIERS

NEXT DAY I planned an excursion to the so-called Dirt Glacier, the most interesting to Indians and steamer men of all the Stickeen glaciers from its mysterious floods. I left the steamer Gertrude for the glacier delta an hour or two before sunset. The captain kindly loaned me his canoe and two of his Indian deck hands, who seemed much puzzled to know what the rare service required of them might mean, and on leaving bade a merry adieu to their companions. We camped on the west side of the river opposite the front of the glacier, in a spacious valley surrounded by snowy mountains. Thirteen small glaciers were in sight and four waterfalls. It was a fine, serene evening, and the highest peaks were wearing turbans of flossy, gossamer cloud-stuff. I had my supper before leaving the steamer, so I had only to make a camp-fire, spread my blanket, and lie down. The Indians had their own bedding and lay beside their own fire.

The Dirt Glacier is noted among the river men as being subject to violent flood outbursts once or twice a year, usually in the late summer. The delta of this glacier stream is three or four miles wide where it fronts the river, and the many rough channels with which it is guttered and the uprooted trees and huge boulders that roughen its surface manifest the power of the floods that swept them to their places; but under ordinary conditions the glacier discharges its drainage water into the river through only four or five of the delta-channels.

Our camp was made on the south or lower side of the delta, below all the draining streams, so that I would not have to ford any of them on my way to the glacier. The Indians chose a sand-pit to sleep in; I chose a level spot back of a drift log. I had but little to say to my companions as they could speak no English, nor I much Thlinkit or Chinook. In a few minutes after landing they retired to their pit and were soon asleep and asnore. I lingered by the fire until after ten o'clock, for the night sky was clear, and the great white mountains in the starlight seemed nearer than by day and to be looking down like guardians of the valley, while the waterfalls, and the torrents escaping from beneath the big glacier, roared in a broad, low monotone, sounding as if close at hand, though, as it proved next day, the nearest was three miles away. After wrapping myself in my blankets, I still gazed into the marvelous sky and made out to sleep only about two hours. Then, without waking the noisy sleepers, I arose, ate a piece of bread, and set out in my shirtsleeves, determined to make the most of the time at my disposal. The captain was to pick us up about noon at a woodpile about a mile from here; but if in the mean time the steamer should run aground and he should need his canoe, a three-whistle signal would be given.

Following a dry channel for about a mile, I came suddenly upon the main outlet of the glacier, which in the imperfect light seemed as large as the river, about one hundred and fifty feet wide, and perhaps three or four feet deep. A little farther up it was only about fifty feet wide and rushing on with impetuous roaring force in its rocky channel, sweeping forward sand, gravel, cobblestones, and boulders, the bump and rumble sounds of the largest of these rolling stones being readily heard in the midst of the roaring. It was too swift and rough to ford, and no bridge tree could be found, for the great floods had cleared everything out of their way. I was therefore compelled to keep on up the right bank, however difficult the way. Where a strip of bare boulders lined the margin, the walking was easy, but where the current swept close along the ragged edge of the forest, progress was difficult and slow on account of snow-crinkled and interlaced thickets of alder and willow, reinforced with fallen trees and thorny devil's-club (*Echinopanax horridum*), making a jungle all but impenetrable. The mile of this extravagantly difficult growth through which I struggled, inch by inch, will not soon be forgotten. At length arriving within a few hundred yards of the glacier, full of panax barbs, I found that both the glacier and its unfordable stream were pressing hard against a shelving cliff, dangerously steep, leaving no margin, and compelling me to scramble along its face before I could get on to the glacier. But by sunrise all these cliff, jungle, and torrent troubles were overcome and I gladly found myself free on the magnificent ice-river.

The curving, out-bulging front of the glacier is about two miles wide, two hundred feet high, and its surface for a mile or so above the front is strewn with moraine detritus, giving it a strangely dirty, dusky look, hence its name, the "Dirt Glacier," this detritus-laden portion being all that is seen in passing up the river. A mile or two beyond the moraine-covered part I was surprised to find alpine plants growing on the ice, fresh and green, some of them in full flower. These curious glacier gardens, the first I had seen, were evidently planted by snow avalanches from the high walls. They were well watered, of course, by the melting surface of the ice and fairly well nourished by humus still attached to the roots, and in some places formed beds of considerable thickness. Seedling trees and bushes also were growing among the flowers. Admiring these novel floating gardens, I struck out for the middle of the pure white glacier, where the ice seemed smoother, and then held straight on for about eight miles, where I reluctantly turned back to meet the steamer, greatly regretting that I had not brought a week's supply of hardtack to allow me to explore the glacier to its head, and then trust to some passing canoe to take me down to Buck Station, from which I could explore the Big Stickeen Glacier.

Altogether, I saw about fifteen or sixteen miles of the main trunk. The grade is almost regular, and the walls on either hand are about from two to three thousand feet high, sculptured like those of Yosemite Valley. I found no difficulty of an extraordinary kind. Many a crevasse had to be crossed, but most of them were narrow and easily jumped, while the few wide ones that lay in my way were crossed on sliver bridges or avoided by passing around them. The structure of the glacier was strikingly revealed on its melting surface. It is made up of thin vertical or inclined sheets or slabs set on edge and welded together. They represent, I think, the successive snow-falls from heavy storms on the tributaries. One of the tributaries on the right side, about three miles above the front, has been entirely melted off from the trunk and has receded two or three miles, forming an independent glacier. Across the mouth of this abandoned part of its channel the main glacier flows, forming a dam which gives rise to a lake. On the head of the detached tributary there are some five or six small residual glaciers, the drainage of which, with that of the snowy mountain slopes above them, discharges into the lake, whose outlet is through a channel or channels beneath the damming glacier. Now these sub-channels are occasionally blocked and the water rises until it flows alongside of the glacier, but as the dam is a moving one, a grand outburst is sometimes made, which, draining the large lake, produces a flood of amazing power, sweeping down immense quantities of moraine material and raising the river all the way down to its mouth, so that several trips may occasionally be made by the steamers after the season of low water has laid them up for the year. The occurrence of these floods are, of course, well known to the Indians and steamboat men, though they know nothing of their cause. They simply remark, "The Dirt Glacier has broken out again."

I greatly enjoyed my walk up this majestic ice-river, charmed by the pale-blue, ineffably fine light in the crevasses, moulins, and wells, and the innumerable azure pools in basins of azure ice, and the network of surface streams, large and small, gliding, swirling with wonderful grace of motion in their frictionless channels, calling forth devout admiration at almost every step and filling the mind with a sense of Nature's endless beauty and power. Looking ahead from the middle of the glacier, you see the broad white flood, though apparently rigid as iron, sweeping in graceful curves between its high mountain-like walls, small glaciers hanging in the hollows on either side, and snow in every form above them, and the great down-plunging granite buttresses and headlands of the walls marvelous in bold massive sculpture; forests in side cañons to within fifty feet of the glacier; avalanche pathways overgrown with alder and willow; innumerable cascades keeping up a solemn harmony of water sounds blending with those of the glacier moulins and rills; and as far as the eye can reach, tributary glaciers at short intervals silently descending from their high, white fountains to swell the grand central ice-river.

In the angle formed by the main glacier and the lake that gives rise to the river floods, there is a massive granite dome sparsely feathered with trees, and just beyond this yosemitic rock is a mountain, perhaps ten thousand feet high, laden with ice and snow which seemed pure pearly white in the morning light. Last evening as seen from camp it was adorned with a cloud streamer, and both the streamer and the peak were flushed in the alpenglow. A mile or two above this mountain, on the opposite side of the glacier, there is a rock like the Yosemite Sentinel; and in general all the wall rocks as far as I saw them are more or less yosemitic in form and color and streaked with cascades.

But wonderful as this noble ice-river is in size and depth and in power displayed, far more wonderful was the vastly greater glacier three or four thousand feet, or perhaps a mile, in depth, whose size and general history is inscribed on the sides of the walls and over the tops of the rocks in characters which have not yet been greatly dimmed by the weather. Comparing its present size with that when it was in its prime, is like comparing a small rivulet to the same stream when it is a roaring torrent.

The return trip to the camp past the shelving cliff and through the weary devil's-club jungle was made in a few hours. The Indians had gone off picking berries, but were on the watch for me and hailed me as I approached. The captain had called for me, and, after waiting three hours, departed for Wrangell without leaving any food, to make sure, I suppose, of a quick return of his Indians and canoe. This was no serious matter, however, for the swift current swept us down to Buck Station, some thirty-five miles distant, by eight o'clock. Here I remained to study the "Big Stickeen Glacier," but the Indians set out for Wrangell soon after supper, though I invited them to stay till morning.

The weather that morning, August 27, was dark and rainy, and I tried to persuade myself that I ought to rest a day before setting out on new ice work. But just across the river the "Big Glacier" was staring me in the face, pouring its majestic flood through a broad mountain gateway and expanding in the spacious river valley to a width of four or five miles, while dim in the gray distance loomed its high mountain fountains. So grand an invitation displayed in characters so telling was of course irresistible, and body-care and weather-care vanished.

Mr. Choquette, the keeper of the station, ferried me across the river, and I spent the day in getting general views and planning the work that had been long in mind. I first traced the broad, complicated terminal moraine to its southern extremity, climbed up the west side along the lateral moraine three or four miles, making my way now on the glacier, now on the moraine-covered bank, and now compelled to climb up through the timber and brush in order to pass some rocky headland, until I reached a point commanding a good general view of the lower end of the glacier. Heavy, blotting rain then began to fall, and I retraced my steps, oftentimes stopping to admire the blue ice-caves into which glad, rejoicing streams from the mountainside were hurrying as if going home, while the glacier seemed to open wide its crystal gateways to welcome them.

The following morning blotting rain was still falling, but time and work was too precious to mind it. Kind Mr. Choquette put me across the river in a canoe, with a lot of biscuits his Indian wife had baked for me and some dried salmon, a little sugar and tea, a blanket, and a piece of light sheeting for shelter from rain during the night, all rolled into one bundle.

"When shall I expect you back?" inquired Choquette, when I bade him good-bye.

"Oh, any time," I replied. "I shall see as much as possible of the glacier, and I know not how long it will hold me."

"Well, but when will I come to look for you, if anything happens? Where are you going to try to go? Years ago Russian officers from Sitka went up the glacier from here and none ever returned. It's a mighty dangerous glacier, all full of damn deep holes and cracks. You've no idea what ticklish deceiving traps are scattered over it."

"Yes, I have," I said. "I have seen glaciers before, though none so big as this one. Do not look for me until I make my appearance on the river-bank. Never mind me. I am used to caring for myself." And so, shouldering my bundle, I trudged off through the moraine boulders and thickets.

My general plan was to trace the terminal moraine to its extreme north end, pitch my little tent, leave the blanket and most of the hardtack, and from this main camp go and come as hunger required or allowed.

After examining a cross-section of the broad moraine, roughened by concentric masses, marking interruptions in the recession of the glacier of perhaps several centuries, in which the successive moraines were formed and shoved together in closer or wider order, I traced the moraine to its northeastern extremity and ascended the glacier for several miles along the left margin, then crossed it at the grand cataract and down the right side to the river, and along the moraine to the point of beginning.

On the older portions of this moraine I discovered several kettles in process of formation and was pleased to find that they conformed in the most striking way with the theory I had already been led to make from observations on the old kettles which form so curious a feature of the drift covering Wisconsin and Minnesota and some of the larger moraines of the residual glaciers in the California Sierra. I found a pit eight or ten feet deep with raw shifting sides countersunk abruptly in the rough moraine material, and at the bottom, on sliding down by the aid of a lithe spruce tree that was being undermined, I discovered, after digging down a foot or two, that the bottom was resting on a block of solid blue ice which had been buried in the moraine perhaps a century or more, judging by the age of the tree that had grown above it. Probably more than another century will be required to complete the formation of this kettle by the slow melting of the buried ice-block. The moraine material of course was falling in as the ice melted, and the sides maintained an angle as steep as the material would lie. All sorts of theories have been advanced for the formation of these kettles, so abundant in the drift over a great part of the United States, and I was glad to be able to set the question at rest, at least as far as I was concerned.

The glacier and the mountains about it are on so grand a scale and so generally inaccessible in the ordinary sense, it seemed to matter but little what course I pursued. Everything was full of interest, even the weather, though about as unfavorable as possible for wide views, and scrambling through the moraine jungle brush kept one as wet as if all the way was beneath a cascade.

I pushed on, with many a rest and halt to admire the bold and marvelously sculptured ice-front, looking all the grander and more striking in the gray mist with all the rest of the glacier shut out, until I came to a lake about two hundred yards wide and two miles long with scores of small bergs floating in it, some aground, close inshore against the moraine, the light playing on their angles and shimmering in their blue caves in ravishing tones. This proved to be the largest of the series of narrow lakelets that lie in shallow troughs between the moraine and the glacier, a miniature Arctic Ocean, its ice-cliffs played upon by whispering, rippling wavelets and its small berg floes drifting in its currents or with the wind, or stranded here and there along its rocky moraine shore.

Hundreds of small rills and good-sized streams were falling into the lake from the glacier, singing in low tones, some of them pouring in sheer falls over blue cliffs from narrow ice-valleys, some spouting from pipe-like channels in the solid front of the glacier, others gurgling out of arched openings at the base. All these water-streams were riding on the parent ice-stream, their voices joined in one grand anthem telling the wonders of their near and far-off fountains. The lake itself is resting in a basin of ice, and the forested moraine, though seemingly cut off from the glacier and probably more than a century old, is in great part resting on buried ice left behind as the glacier receded, and melting slowly on account of the protection afforded by the moraine detritus, which keeps shifting and falling on the inner face long after it is overgrown with lichens, mosses, grasses, bushes, and even good-sized trees; these changes going on with marvelous deliberation until in fullness of time the whole moraine settles down upon its bedrock foundation.

The outlet of the lake is a large stream, almost a river in size, one of the main draining streams of the glacier. I attempted to ford it where it begins to break in rapids in passing over the moraine, but found it too deep and rough on the bottom. I then tried to ford at its head, where it is wider and glides smoothly out of the lake, bracing myself against the current with a pole, but found it too deep, and when the icy water reached my shoulders I cautiously struggled back to the moraine. I next followed it down through the rocky jungle to a place where in breaking across the moraine dam it was only about thirty-five feet wide. Here I found a spruce tree, which I felled for a bridge; it reached across, about ten feet of the top holding in the bank brush. But the force of the torrent, acting on the submerged branches and the slender end of the trunk, bent it like a bow and made it very unsteady, and after testing it by going out about a third of the way over, it seemed likely to be carried away when bent deeper into the current by my weight. Fortunately, I discovered another larger tree well situated a little farther down, which I felled, and though a few feet in the middle was submerged, it seemed perfectly safe.

As it was now getting late, I started back to the lakeside where I had left my bundle, and in trying to hold a direct course found the interlaced jungle still more difficult than it was along the bank of the torrent. For over an hour I had to creep and struggle close to the rocky ground like a fly in a spider-web without being able to obtain a single glimpse of any guiding feature of the landscape. Finding a little willow taller than the surrounding alders, I climbed it, caught sight of the glacier-front, took a compass bearing, and sunk again into the dripping, blinding maze of brush, and at length emerged on the lakeshore seven hours after leaving it, all this time as wet as though I had been swimming, thus completing a trying day's work. But everything was deliciously fresh, and I found new and old plant friends, and lessons on Nature's Alaska moraine landscape-gardening that made everything bright and light.

It was now near dark, and I made haste to make up my flimsy little tent. The ground was desperately rocky. I made out, however, to level down a strip large enough to lie on, and by means of slim alder stems bent over it and tied together soon had a home. While thus busily engaged I was startled by a thundering roar across the lake. Running to the top of the moraine, I discovered that the tremendous noise was only the outcry of a new-born berg about fifty or sixty feet in diameter, rocking and wallowing in the waves it had raised as if enjoying its freedom after its long grinding work as part of the glacier. After this fine last lesson I managed to make a small fire out of wet twigs, got a cup of tea, stripped off my dripping clothing, wrapped myself in a blanket and lay brooding on the gains of the day and plans for the morrow, glad, rich, and almost comfortable.

It was raining hard when I awoke, but I made up my mind to disregard the weather, put on my dripping clothing, glad to know it was fresh and clean; ate biscuits and a piece of dried salmon without attempting to make a tea fire; filled a bag with hardtack, slung it over my shoulder, and with my indispensable ice-ax plunged once more into the dripping jungle. I found my bridge holding bravely in place against the swollen torrent, crossed it and beat my way around pools and logs and through two hours of tangle back to the moraine on the north side of the outlet,—a wet, weary battle but not without enjoyment. The smell of the washed ground and vegetation made every breath a pleasure, and I found *Calypso borealis*, the first I had seen on this side of the continent, one of my darlings, worth any amount of hardship; and I saw one of my Douglas squirrels on the margin of a grassy pool. The drip of the rain on the various leaves was pleasant to hear. More especially marked were the flat low-toned bumps and splashes of large drops from the trees on the broad horizontal leaves of *Echinopanax horridum*, like the drumming of thunder-shower drops on veratrum and palm leaves, while the mosses were indescribably beautiful, so fresh, so bright, so cheerily green, and all so low and calm and silent, however heavy and wild the wind and the rain blowing and pouring above them. Surely never a particle of dust has touched leaf or crown of all these blessed mosses; and how bright were the red rims of the cladonia cups beside them, and the fruit of the dwarf cornel! And the wet berries, Nature's precious jewelry, how beautiful they were!—huckleberries with pale bloom and a crystal drop on each; red and yellow salmon-berries, with clusters of smaller drops; and the glittering, berry-like rain-drops adorning the interlacing arches of bent grasses and sedges around the edges of the pools, every drop a mirror with all the landscape in it. A' that and a' that and twice as muckle's a' that in this glorious Alaska day, recalling, however different, George Herbert's "Sweet day, so cool, so calm, so bright."

In the gardens and forests of this wonderful moraine one might spend a whole joyful life.

When I at last reached the end of the great moraine and the front of the mountain that forms the north side of the glacier basin, I tried to make my way along its side, but, finding the climbing tedious and difficult, took to the glacier and fared well, though a good deal of step-cutting was required on its ragged, crevassed margin. When night was drawing nigh, I scanned the steep mountainside in search of an accessible bench, however narrow, where a bed and a fire might be gathered for a camp. About dark great was my delight to find a little shelf with a few small mountain hemlocks growing in cleavage joints. Projecting knobs below it enabled me to build a platform for a fireplace and a bed, and by industrious creeping from one fissure to another, cutting bushes and small trees and sliding them down to within reach of my rock-shelf, I made out to collect wood enough to last through the night. In an hour or two I had a cheery fire, and spent the night in turning from side to side, steaming and drying after being wet two days and a night. Fortunately this night it did not rain, but it was very cold.

Pushing on next day, I climbed to the top of the glacier by ice-steps and along its side to the grand cataract two miles wide where the whole majestic flood of the glacier pours like a mighty surging river down a steep declivity in its channel. After gazing a long time on the glorious show, I discovered a place beneath the edge of the cataract where it flows over a hard, resisting granite rib, into which I crawled and enjoyed the novel and instructive view of a glacier pouring over my head, showing not only its grinding, polishing action, but how it breaks off large angular boulder-masses—a most telling lesson in earth-sculpture, confirming many I had already learned in the glacier basins of the High Sierra of California. I then crossed to the south side, noting the forms of the huge blocks into which the glacier was broken in passing over the brow of the cataract, and how they were welded.

The weather was now clear, opening views according to my own heart far into the high snowy fountains. I saw what seemed the farthest mountains, perhaps thirty miles from the front, everywhere winter-bound, but thick forested, however steep, for a distance of at least fifteen miles from the front, the trees, hemlock and spruce, clinging to the rock by root-holds among cleavage joints. The greatest discovery was in methods of denudation displayed beneath the glacier.

After a few more days of exhilarating study I returned to the river-bank opposite Choquette's landing. Promptly at sight of the signal I made, the kind Frenchman came across for me in his canoe. At his house I enjoyed a rest while writing out notes; then examined the smaller glacier fronting the one I had been exploring, until a passing canoe bound for Fort Wrangell took me aboard.

CHAPTER IX

A CANOE VOYAGE TO NORTHWARD

I ARRIVED AT Wrangell in a canoe with a party of Cassiar miners in October while the icy regions to the northward still burned in my mind. I had met several prospectors who had been as far as Chilcat at the head of Lynn Canal, who told wonderful stories about the great glaciers they had seen there. All the high mountains up there, they said, seemed to be made of ice, and if glaciers "are what you are after, that's the place for you," and to get there "all you have to do is to hire a good canoe and Indians who know the way."

But it now seemed too late to set out on so long a voyage. The days were growing short and winter was drawing nigh when all the land would be buried in snow. On the other hand, though this wilderness was new to me, I was familiar with storms and enjoyed them. The main channels extending along the coast remain open all winter, and, their shores being well forested, I knew that it would be easy to keep warm in camp, while abundance of food could be carried. I determined, therefore, to go ahead as far north as possible, to see and learn what I could, especially with reference to future work. When I made known my plans to Mr. Young, he offered to go with me, and, being acquainted with the Indians, procured a good canoe and crew, and with a large stock of provisions and blankets, we left Wrangell October 14, eager to welcome weather of every sort, as long as food lasted.

I was anxious to make an early start, but it was half-past two in the afternoon before I could get my Indians together—Toyatte, a grand old Stickeen nobleman, who was made captain, not only because he owned the canoe, but for his skill in woodcraft and seamanship; Kadachan, the son of a Chilcat chief; John, a Stickeen, who acted as interpreter; and Sitka Charley. Mr. Young, my companion, was an adventurous evangelist, and it was the opportunities the trip might afford to meet the Indians of the different tribes on our route with reference to future missionary work, that induced him to join us.

When at last all were aboard and we were about to cast loose from the wharf, Kadachan's mother, a woman of great natural dignity and force of character, came down the steps alongside the canoe oppressed with anxious fears for the safety of her son. Standing silent for a few moments, she held the missionary with her dark, bodeful eyes, and with great solemnity of speech and gesture accused him of using undue influence in gaining her son's consent to go on a dangerous voyage among unfriendly tribes; and like an ancient sibyl foretold a long train of bad luck from storms and enemies, and finished by saying, "If my son comes not back, on you will be his blood, and you shall pay. I say it."

Mr. Young tried in vain to calm her fears, promising Heaven's care as well as his own for her precious son, assuring her that he would faithfully share every danger that he encountered, and if need be die in his defense.

"We shall see whether or not you die," she said, and turned away.

Toyatte also encountered domestic difficulties. When he stepped into the canoe I noticed a cloud of anxiety on his grand old face, as if his doom now drawing near was already beginning to overshadow him. When he took leave of his wife, she refused to shake hands with him, wept bitterly, and said that his enemies, the Chilcat chiefs, would be sure to kill him in case he reached their village. But it was not on this trip that the old hero was to meet his fate, and when we were fairly free in the wilderness and a gentle breeze pressed us joyfully over the shining waters these gloomy forebodings vanished.

We first pursued a westerly course, through Sumner Strait, between Kupreanof and Prince of Wales Islands, then, turning northward, sailed up the Kiku Strait through the midst of innumerable picturesque islets, across Prince Frederick's Sound, up Chatham Strait, thence northwestward through Icy Strait and around the then uncharted Glacier Bay. Thence returning through Icy Strait, we sailed up the beautiful Lynn Canal to the Davidson Glacier and the lower village of the Chilcat tribe and returned to Wrangell along the coast of the mainland, visiting the icy Sum Dum Bay and the Wrangell Glacier on our route. Thus we made a journey more than eight hundred miles long, and though hardships and perhaps dangers were encountered, the great wonderland made compensation beyond our most extravagant hopes. Neither rain nor snow stopped us, but when the wind was too wild, Kadachan and the old captain stayed on guard in the camp and John and Charley went into the woods deer-hunting, while I examined the adjacent rocks and woods. Most of our campgrounds were in sheltered nooks where good fire-wood was abundant, and where the precious canoe could be safely drawn up beyond reach of the waves. After supper we sat long around the fire, listening to the Indians' stories about the wild animals, their hunting-adventures, wars, traditions, religion, and customs. Every Indian party we met we interviewed, and visited every village we came to.

Our first camp was made at a place called the Island of the Standing Stone, on the shore of a shallow bay. The weather was fine. The mountains of the mainland were unclouded, excepting one, which had a horizontal ruff of dull slate color, but its icy summit covered with fresh snow towered above the cloud, flushed like its neighbors in the alpenglow. All the large islands in sight were densely forested, while many small rock islets in front of our camp were treeless or nearly so. Some of them were distinctly glaciated even below the tide-line, the effects of wave washing and general weathering being scarce appreciable as yet. Some of the larger islets had a few trees, others only grass. One looked in the distance like a two-masted ship flying before the wind under press of sail.

Next morning the mountains were arrayed in fresh snow that had fallen during the night down to within a hundred feet of the sea-level. We made a grand fire, and after an early breakfast pushed merrily on all day along beautiful forested shores embroidered with autumn-colored bushes. I noticed some pitchy trees that had been deeply hacked for kindling-wood and torches, precious conveniences to belated voyagers on stormy nights. Before sundown we camped in a beautiful nook of Deer Bay, shut in from every wind by gray-bearded trees and fringed with rose bushes, rubus, potentilla, asters, etc. Some of the lichen tresses depending from the branches were six feet in length.

A dozen rods or so from our camp we discovered a family of Kake Indians snugly sheltered in a portable bark hut, a stout middle-aged man with his wife, son, and daughter, and his son's wife. After our tent was set and fire made, the head of the family paid us a visit and presented us with a fine salmon, a pair of mallard ducks, and a mess of potatoes. We paid a return visit with gifts of rice and tobacco, etc. Mr. Young spoke briefly on mission affairs and inquired whether their tribe would be likely to welcome a teacher or missionary. But they seemed unwilling to offer an opinion on so important a subject. The following words from the head of the family was the only reply:—

"We have not much to say to you fellows. We always do to Boston men as we have done to you, give a little of whatever we have, treat everybody well and never quarrel. This is all we have to say."

Our Kake neighbors set out for Fort Wrangell next morning, and we pushed gladly on toward Chilcat. We passed an island that had lost all its trees in a storm, but a hopeful crop of young ones was springing up to take their places. I found no trace of fire in these woods. The ground was covered with leaves, branches, and fallen trunks perhaps a dozen generations deep, slowly decaying, forming a grand mossy mass of ruins, kept fresh and beautiful. All that is repulsive about death was here hidden beneath abounding life. Some rocks along the shore were completely covered with crimson-leafed huckleberry bushes; one species still in fruit might well be called the winter huckleberry. In a short walk I found vetches eight feet high leaning on raspberry bushes, and tall ferns and *Smilacina unifolia* with leaves six inches wide growing on yellow-green moss, producing a beautiful effect.

Our Indians seemed to be enjoying a quick and merry reaction from the doleful domestic dumps in which the voyage was begun. Old and young behaved this afternoon like a lot of truant boys on a lark. When we came to a pond fenced off from the main channel by a moraine dam, John went ashore to seek a shot at ducks. Creeping up behind the dam, he killed a mallard fifty or sixty feet from the shore and attempted to wave it within reach by throwing stones back of it. Charley and Kadachan went to his help, enjoying the sport, especially enjoying their own blunders in throwing in front of it and thus driving the duck farther out. To expedite the business John then tried to throw a rope across it, but failed after repeated trials, and so did each in turn, all laughing merrily at their awkward bungling. Next they tied a stone to the end of the rope to carry it further and with better aim, but the result was no better. Then majestic old Toyatte tried his hand at the game. He tied the rope to one of the canoe-poles, and taking aim threw it, harpoon fashion, beyond the duck, and the general merriment was redoubled when the pole got loose and floated out to the middle of the pond. At length John stripped, swam to the duck, threw it ashore, and brought in the pole in his teeth, his companions meanwhile making merry at his expense by splashing the water in front of him and making the dead duck go through the motions of fighting and biting him in the face as he landed.

The morning after this delightful day was dark and threatening. A high wind was rushing down the strait dead against us, and just as we were about ready to start, determined to fight our way by creeping close inshore, pelting rain began to fly. We concluded therefore to wait for better weather. The hunters went out for deer and I to see the forests. The rain brought out the fragrance of the drenched trees, and the wind made wild melody in their tops, while every brown bole was embroidered by a network of rain rills. Perhaps the most delightful part of my ramble was along a stream that flowed through a leafy arch beneath overleaning trees which met at the top. The water was almost black in the deep pools and fine clear amber in the shallows. It was the pure, rich wine of the woods with a pleasant taste, bringing spicy spruce groves and widespread bog and beaver meadows to mind. On this amber stream I discovered an interesting fall. It is only a few feet high, but remarkably fine in the curve of its brow and blending shades of color, while the mossy, bushy pool into which it plunges is inky black, but wonderfully brightened by foam bells larger than common that drift in clusters on the smooth water around the rim, each of them carrying a picture of the overlooking trees leaning together at the tips like the teeth of moss capsules before they rise.

I found most of the trees here fairly loaded with mosses. Some broadly palmated branches had beds of yellow moss so wide and deep that when wet they must weigh a hundred pounds or even more. Upon these moss-beds ferns and grasses and even good-sized seedling trees grow, making beautiful hanging gardens in which the curious spectacle is presented of old trees holding hundreds of their own children in their arms, nourished by rain and dew and the decaying leaves showered down to them by their parents. The branches upon which these beds of mossy soil rest become flat and irregular like weathered roots or the antlers of deer, and at length die; and when the whole tree has thus been killed it seems to be standing on its head with roots in the air. A striking example of this sort stood near the camp and I called the missionary's attention to it.

"Come, Mr. Young," I shouted. "Here's something wonderful, the most wonderful tree you ever saw; it is standing on its head."

"How in the world," said he in astonishment, "could that tree have been plucked up by the roots, carried high in the air, and dropped down head foremost into the ground. It must have been the work of a tornado."

Toward evening the hunters brought in a deer. They had seen four others, and at the camp-fire talk said that deer abounded on all the islands of considerable size and along the shores of the mainland. But few were to be found in the interior on account of wolves that ran them down where they could not readily take refuge in the water. The Indians, they said, hunted them on the islands with trained dogs which went into the woods and drove them out, while the hunters lay in wait in canoes at the points where they were likely to take to the water. Beaver and black bear also abounded on this large island. I saw but few birds there, only ravens, jays, and wrens. Ducks, gulls, bald eagles, and jays are the commonest birds hereabouts. A flock of swans flew past, sounding their startling human-like cry which seemed yet more striking in this lonely wilderness. The Indians said that geese, swans, cranes, etc., making their long journeys in regular order thus called aloud to encourage each other and enable them to keep stroke and time like men in rowing or marching (a sort of "Row, brothers, row," or "Hip, hip" of marching soldiers).

October 18 was about half sunshine, half rain and wet snow, but we paddled on through the midst of the innumerable islands in more than half comfort, enjoying the changing effects of the weather on the dripping wilderness. Strolling a little way back into the woods when we went ashore for luncheon, I found fine specimens of cedar, and here and there a birch, and small thickets of wild apple. A hemlock, felled by Indians for bread-bark, was only twenty inches thick at the butt, a hundred and twenty feet long, and about five hundred and forty years old at the time it was felled. The first hundred of its rings measured only four inches, showing that for a century it had grown in the shade of taller trees and at the age of one hundred years was yet only a sapling in size. On the mossy trunk of an old prostrate spruce about a hundred feet in length thousands of seedlings were growing. I counted seven hundred on a length of eight feet, so favorable is this climate for the development of tree seeds and so fully do these trees obey the command to multiply and replenish the earth.

No wonder these islands are densely clothed with trees. They grow on solid rocks and logs as well as on fertile soil. The surface is first covered with a plush of mosses in which the seeds germinate; then the interlacing roots form a sod, fallen leaves soon cover their feet, and the young trees, closely crowded together, support each other, and the soil becomes deeper and richer from year to year.

I greatly enjoyed the Indians' camp-fire talk this evening on their ancient customs, how they were taught by their parents ere the whites came among them, their religion, ideas connected with the next world, the stars, plants, the behavior and language of animals under different circumstances, manner of getting a living, etc. When our talk was interrupted by the howling of a wolf on the opposite side of the strait, Kadachan puzzled the minister with the question, "Have wolves souls?" The Indians believe that they have, giving as foundation for their belief that they are wise creatures who know how to catch seals and salmon by swimming slyly upon them with their heads hidden in a mouthful of grass, hunt deer in company, and always bring forth their young at the same and most favorable time of the year. I inquired how it was that with enemies so wise and powerful the deer were not all killed. Kadachan replied that wolves knew better than to kill them all and thus cut off their most important food-supply. He said they were numerous on all the large islands, more so than on the mainland, that Indian hunters were afraid of them and never ventured far into the woods alone, for these large gray and black wolves attacked man whether they were hungry or not. When attacked, the Indian hunter, he said, climbed a tree or stood with his back against a tree or rock as a wolf never attacks face to face. Wolves, and not bears, Indians regard as masters of the woods, for they sometimes attack and kill bears, but the wolverine they never attack, "for," said John, "wolves and wolverines are companions in sin and equally wicked and cunning."

On one of the small islands we found a stockade, sixty by thirty-five feet, built, our Indians said, by the Kake tribe during one of their many war-like quarrels. Toyatte and Kadachan said these forts were common throughout the canoe waters, showing that in this foodful, kindly wilderness, as in all the world beside, man may be man's worst enemy.

We discovered small bits of cultivation here and there, patches of potatoes and turnips, planted mostly on the cleared sites of deserted villages. In spring the most industrious families sailed to their little farms of perhaps a quarter of an acre or less, and ten or fifteen miles from their villages. After preparing the ground, and planting it, they visited it again in summer to pull the weeds and speculate on the size of the crop they were likely to have to eat with their fat salmon. The Kakes were then busy digging their potatoes, which they complained were this year injured by early frosts.

We arrived at Klugh-Quan, one of the Kupreanof Kake villages, just as a funeral party was breaking up. The body had been burned and gifts were being distributed—bits of calico, handkerchiefs, blankets, etc., according to the rank and wealth of the deceased. The death ceremonies of chiefs and head men, Mr. Young told me, are very weird and imposing, with wild feasting, dancing, and singing. At this little place there are some eight totem poles of bold and intricate design, well executed, but smaller than those of the Stickeens. As elsewhere throughout the archipelago, the bear, raven, eagle, salmon, and porpoise are the chief figures. Some of the poles have square cavities, mortised into the back, which are said to contain the ashes of members of the family. These recesses are closed by a plug. I noticed one that was caulked with a rag where the joint was imperfect.

Strolling about the village, looking at the tangled vegetation, sketching the totems, etc., I found a lot of human bones scattered on the surface of the ground or partly covered. In answer to my inquiries, one of our crew said they probably belonged to Sitka Indians slain in war. These Kakes are shrewd, industrious, and rather good-looking people. It was at their largest village that an American schooner was seized and all the crew except one man murdered. A gunboat sent to punish them burned the village. I saw the anchor of the ill-fated vessel lying near the shore.

Though all the Thlinkit tribes believe in witchcraft, they are less superstitious in some respects than many of the lower classes of whites. Chief Yana Taowk seemed to take pleasure in kicking the Sitka bones that lay in his way, and neither old nor young showed the slightest trace of superstitious fear of the dead at any time.

It was at the northmost of the Kupreanof Kake villages that Mr. Young held his first missionary meeting, singing hymns, praying, and preaching, and trying to learn the number of the inhabitants and their readiness to receive instruction. Neither here nor in any of the other villages of the different tribes that we visited was there anything like a distinct refusal to receive school-teachers or ministers. On the contrary, with but one or two exceptions, all with apparent good faith declared their willingness to receive them, and many seemed heartily delighted at the prospect of gaining light on subjects so important and so dark to them. All had heard ere this of the wonderful work of the Reverend Mr. Duncan at Metlakatla, and even those chiefs who were not at all inclined to anything like piety were yet anxious to procure schools and churches that their people should not miss the temporal advantages of knowledge, which with their natural shrewdness they were not slow to recognize. "We are all children," they said, "groping in the dark. Give us this light and we will do as you bid us."

The chief of the first Kupreanof Kake village we came to was a venerable-looking man, perhaps seventy years old, with massive head and strongly marked features, a bold Roman nose, deep, tranquil eyes, shaggy eyebrows, a strong face set in a halo of long gray hair. He seemed delighted at the prospect of receiving a teacher for his people. "This is just what I want," he said. "I am ready to bid him welcome."

"This," said Yana Taowk, chief of the larger north village, "is a good word you bring us. We will be glad to come out of our darkness into your light. You Boston men must be favorites of the Great Father. You know all about God, and ships and guns and the growing of things to eat. We will sit quiet and listen to the words of any teacher you send us."

While Mr. Young was preaching, some of the congregation smoked, talked to each other, and answered the shouts of their companions outside, greatly to the disgust of Toyatte and Kadachan, who regarded the Kakes as mannerless barbarians. A little girl, frightened at the strange exercises, began to cry and was turned out of doors. She cried in a strange, low, wild tone, quite unlike the screech crying of the children of civilization.

The following morning we crossed Prince Frederick's Sound to the west coast of Admiralty Island. Our frail shell of a canoe was tossed like a bubble on the swells coming in from the ocean. Still, I suppose, the danger was not so great as it seemed. In a good canoe, skillfully handled, you may safely sail from Victoria to Chilcat, a thousand-mile voyage frequently made by Indians in their trading operations before the coming of the whites. Our Indians, however, dreaded this crossing so late in the season. They spoke of it repeatedly before we reached it as the one great danger of our voyage.

John said to me just as we left the shore, "You and Mr. Young will be scared to death on this broad water."

"Never mind us, John," we merrily replied, "perhaps some of you brave Indian sailors may be the first to show fear."

Toyatte said he had not slept well a single night thinking of it, and after we rounded Cape Gardner and entered the comparatively smooth Chatham Strait, they all rejoiced, laughing and chatting like frolicsome children.

We arrived at the first of the Hootsenoo villages on Admiralty Island shortly after noon and were welcomed by everybody. Men, women, and children made haste to the beach to meet us, the children staring as if they had never before seen a Boston man. The chief, a remarkably good-looking and intelligent fellow, stepped forward, shook hands with us Boston fashion, and invited us to his house. Some of the curious children crowded in after us and stood around the fire staring like half-frightened wild animals. Two old women drove them out of the house, making hideous gestures, but taking good care not to hurt them. The merry throng poured through the round door, laughing and enjoying the harsh gestures and threats of the women as all a joke, indicating mild parental government in general. Indeed, in all my travels I never saw a child, old or young, receive a blow or even a harsh word. When our cook began to prepare luncheon our host said through his interpreter that he was sorry we could not eat Indian food, as he was anxious to entertain us. We thanked him, of course, and expressed our sense of his kindness. His brother, in the mean time, brought a dozen turnips, which he peeled and sliced and served in a clean dish. These we ate raw as dessert, reminding me of turnip-field feasts when I was a boy in Scotland. Then a box was brought from some corner and opened. It seemed to be full of tallow or butter. A sharp stick was thrust into it, and a lump of something five or six inches long, three or four wide, and an inch thick was dug up, which proved to be a section of the back fat of a deer, preserved in fish oil and seasoned with boiled spruce and other spicy roots. After stripping off the lard-like oil, it was cut into small pieces and passed round. It seemed white and wholesome, but I was unable to taste it even for manner's sake. This disgust, however, was not noticed, as the rest of the company did full justice to the precious tallow and smacked their lips over it as a great delicacy. A lot of potatoes about the size of walnuts, boiled and peeled and added to a potful of salmon, made a savory stew that all seemed to relish. An old, cross-looking, wrinkled crone presided at the steaming chowder-pot, and as she peeled the potatoes with her fingers she, at short intervals, quickly thrust one of the best into the mouth of a little wild-eyed girl that crouched beside her, a spark of natural love which charmed her withered face and made all the big gloomy house shine. In honor of our visit, our host put on a genuine white shirt. His wife also dressed in her best and put a pair of dainty trousers on her two-year-old boy, who seemed to be the pet and favorite of the large family and indeed of the whole village. Toward evening messengers were sent through the village to call everybody to a meeting. Mr. Young delivered the usual missionary sermon and I also was called on to say something. Then the chief arose and made an eloquent reply, thanking us for our good words and for the hopes we had inspired of obtaining a teacher for their children. In particular, he said, he wanted to hear all we could tell him about God.

This village was an offshoot of a larger one, ten miles to the north, called Killisnoo. Under the prevailing patriarchal form of government each tribe is divided into comparatively few families; and because of quarrels, the chief of this branch moved his people to this little bay, where the beach offered a good landing for canoes. A stream which enters it yields abundance of salmon, while in the adjacent woods and mountains berries, deer, and wild goats abound.

"Here," he said, "we enjoy peace and plenty; all we lack is a church and a school, particularly a school for the children." His dwelling so much with benevolent aspect on the children of the tribe showed, I think, that he truly loved them and had a right intelligent insight concerning their welfare. We spent the night under his roof, the first we had ever spent with Indians, and I never felt more at home. The loving kindness bestowed on the little ones made the house glow.

Next morning, with the hearty good wishes of our Hootsenoo friends, and encouraged by the gentle weather, we sailed gladly up the coast, hoping soon to see the Chilcat glaciers in their glory. The rock hereabouts is mostly a beautiful blue marble, waveworn into a multitude of small coves and ledges. Fine sections were thus revealed along the shore, which with their colors, brightened with showers and late-blooming leaves and flowers, beguiled the weariness of the way. The shingle in front of these marble cliffs is also mostly marble, well polished and rounded and mixed with a small percentage of glacier-borne slate and granite erratics.

We arrived at the upper village about half-past one o'clock. Here we saw Hootsenoo Indians in a very different light from that which illumined the lower village. While we were yet half a mile or more away, we heard sounds I had never before heard—a storm of strange howls, yells, and screams rising from a base of gasping, bellowing grunts and groans. Had I been alone, I should have fled as from a pack of fiends, but our Indians quietly recognized this awful sound, if such stuff could be called sound, simply as the "whiskey howl" and pushed quietly on. As we approached the landing, the demoniac howling so greatly increased I tried to dissuade Mr. Young from attempting to say a single word in the village, and as for preaching one might as well try to preach in Tophet. The whole village was afire with bad whiskey. This was the first time in my life that I learned the meaning of the phrase "a howling drunk." Even our Indians hesitated to venture ashore, notwithstanding whiskey storms were far from novel to them. Mr. Young, however, hoped that in this Indian Sodom at least one man might be found so righteous as to be in his right mind and able to give trustworthy information. Therefore I was at length prevailed on to yield consent to land. Our canoe was drawn up on the beach and one of the crew left to guard it. Cautiously we strolled up the hill to the main row of houses, now a chain of alcoholic volcanoes. The largest house, just opposite the landing, was about forty feet square, built of immense planks, each hewn from a whole log, and, as usual, the only opening was a mere hole about two and a half feet in diameter, closed by a massive hinged plug like the breach of a cannon. At the dark door-hole a few black faces appeared and were suddenly withdrawn. Not a single person was to be seen on the street. At length a couple of old, crouching men, hideously blackened, ventured out and stared at us, then, calling to their companions, other black and burning heads appeared, and we began to fear that like the Alloway Kirk witches the whole legion was about to sally forth. But, instead, those outside suddenly crawled and tumbled in again. We were thus allowed to take a general view of the place and return to our canoe unmolested. But ere we could get away, three old women came swaggering and grinning down to the beach, and Toyatte was discovered by a man with whom he had once had a business misunderstanding, who, burning for revenge, was now jumping and howling and threatening as only a drunken Indian may, while our heroic old captain, in severe icy majesty, stood erect and motionless, uttering never a word. Kadachan, on the contrary, was well nigh smothered with the drunken caresses of one of his father's *tillicums* (friends), who insisted on his going back with him into the house. But reversing the words of St. Paul in his account of his shipwreck, it came to pass that we all at length got safe to sea and by hard rowing managed to reach a fine harbor before dark, fifteen sweet, serene miles from the howlers.

Our camp this evening was made at the head of a narrow bay bordered by spruce and hemlock woods. We made our beds beneath a grand old Sitka spruce five feet in diameter, whose broad, wing-like branches were outspread immediately above our heads. The night picture as I stood back to see it in the firelight was this one great tree, relieved against the gloom of the woods back of it, the light on the low branches revealing the shining needles, the brown, sturdy trunk grasping an outswelling mossy bank, and a fringe of illuminated bushes within a few feet of the tree with the firelight on the tips of the sprays.

Next morning, soon after we left our harbor, we were caught in a violent gust of wind and dragged over the seething water in a passionate hurry, though our sail was close-reefed, flying past the gray headlands in most exhilarating style, until fear of being capsized made us drop our sail and run into the first little nook we came to for shelter. Captain Toyatte remarked that in this kind of wind no Indian would dream of traveling, but since Mr. Young and I were with him he was willing to go on, because he was sure that the Lord loved us and would not allow us to perish.

We were now within a day or two of Chilcat. We had only to hold a direct course up the beautiful Lynn Canal to reach the large Davidson and other glaciers at its head in the cañons of the Chilcat and Chilcoot Rivers. But rumors of trouble among the Indians there now reached us. We found a party taking shelter from the stormy wind in a little cove, who confirmed the bad news that the Chilcats were drinking and fighting, that Kadachan's father had been shot, and that it would be far from safe to venture among them until blood-money had been paid and the quarrels settled. I decided, therefore, in the mean time, to turn westward and go in search of the wonderful "ice-mountains" that Sitka Charley had been telling us about. Charley, the youngest of my crew, noticing my interest in glaciers, said that when he was a boy he had gone with his father to hunt seals in a large bay full of ice, and that though it was long since he had been there, he thought he could find his way to it. Accordingly, we pushed eagerly on across Chatham Strait to the north end of Icy Strait, toward the new and promising ice-field.

On the south side of Icy Strait we ran into a picturesque bay to visit the main village of the Hoona tribe. Rounding a point on the north shore of the bay, the charmingly located village came in sight, with a group of the inhabitants gazing at us as we approached. They evidently recognized us as strangers or visitors from the shape and style of our canoe, and perhaps even determining that white men were aboard, for these Indians have wonderful eyes. While we were yet half a mile off, we saw a flag unfurled on a tall mast in front of the chief's house. Toyatte hoisted his United States flag in reply, and thus arrayed we made for the landing. Here we were met and received by the chief, Kashoto, who stood close to the water's edge, barefooted and bareheaded, but wearing so fine a robe and standing so grave, erect, and serene, his dignity was complete. No white man could have maintained sound dignity under circumstances so disadvantageous. After the usual formal salutations, the chief, still standing as erect and motionless as a tree, said that he was not much acquainted with our people and feared that his house was too mean for visitors so distinguished as we were. We hastened of course to assure him that we were not proud of heart, and would be glad to have the honor of his hospitality and friendship. With a smile of relief he then led us into his large fort house to the seat of honor prepared for us. After we had been allowed to rest unnoticed and unquestioned for fifteen minutes or so, in accordance with good Indian manners in case we should be weary or embarrassed, our cook began to prepare luncheon; and the chief expressed great concern at his not being able to entertain us in Boston fashion.

Luncheon over, Mr. Young as usual requested him to call his people to a meeting. Most of them were away at outlying camps gathering winter stores. Some ten or twelve men, however, about the same number of women, and a crowd of wondering boys and girls were gathered in, to whom Mr. Young preached the usual gospel sermon. Toyatte prayed in Thlinkit, and the other members of the crew joined in the hymn-singing. At the close of the mission exercises the chief arose and said that he would now like to hear what the other white chief had to say. I directed John to reply that I was not a missionary, that I came only to pay a friendly visit and see the forests and mountains of their beautiful country. To this he replied, as others had done in the same circumstances, that he would like to hear me on the subject of their country and themselves; so I had to get on my feet and make some sort of a speech, dwelling principally on the brotherhood of all races of people, assuring them that God loved them and that some of their white brethren were beginning to know them and become interested in their welfare; that I seemed this evening to be among old friends with whom I had long been acquainted, though I had never been here before; that I would always remember them and the kind reception they had given us; advised them to heed the instructions of sincere self-denying mission men who wished only to do them good and desired nothing but their friendship and welfare in return. I told them that in some far-off countries, instead of receiving the missionaries with glad and thankful hearts, the Indians killed and ate them; but I hoped, and indeed felt sure, that his people would find a better use for missionaries than putting them, like salmon, in pots for food. They seemed greatly interested, looking into each other's faces with emphatic nods and a-ahs and smiles.

The chief then slowly arose and, after standing silent a minute or two, told us how glad he was to see us; that he felt as if his heart had enjoyed a good meal; that we were the first to come humbly to his little out-of-the-way village to tell his people about God; that they were all like children groping in darkness, but eager for light; that they would gladly welcome a missionary and teacher and use them well; that he could easily believe that whites and Indians were the children of one Father just as I had told them in my speech; that they differed little and resembled each other a great deal, calling attention to the similarity of hands, eyes, legs, etc., making telling gestures in the most natural style of eloquence and dignified composure. "Oftentimes," he said, "when I was on the high mountains in the fall, hunting wild sheep for meat, and for wool to make blankets, I have been caught in snow-storms and held in camp until there was nothing to eat, but when I reached my home and got warm, and had a good meal, then my body felt good. For a long time my heart has been hungry and cold, but to-night your words have warmed my heart, and given it a good meal, and now my heart feels good."

The most striking characteristic of these people is their serene dignity in circumstances that to us would be novel and embarrassing. Even the little children behave with natural dignity, come to the white men when called, and restrain their wonder at the strange prayers, hymn-singing, etc. This evening an old woman fell asleep in the meeting and began to snore; and though both old and young were shaken with suppressed mirth, they evidently took great pains to conceal it. It seems wonderful to me that these so-called savages can make one feel at home in their families. In good breeding, intelligence, and skill in accomplishing whatever they try to do with tools they seem to me to rank above most of our uneducated white laborers. I have never yet seen a child ill-used, even to the extent of an angry word. Scolding, so common a curse in civilization, is not known here at all. On the contrary the young are fondly indulged without being spoiled. Crying is very rarely heard.

In the house of this Hoona chief a pet marmot (Parry's) was a great favorite with old and young. It was therefore delightfully confiding and playful and human. Cats were petted, and the confidence with which these cautious, thoughtful animals met strangers showed that they were kindly treated.

There were some ten or a dozen houses, all told, in the village. The count made by the chief for Mr. Young showed some seven hundred and twenty-five persons in the tribe.

CHAPTER X

THE DISCOVERY OF GLACIER BAY

FROM HERE, ON October 24, we set sail for Guide Charley's ice-mountains. The handle of our heaviest ax was cracked, and as Charley declared that there was no fire-wood to be had in the big ice-mountain bay, we would have to load the canoe with a store for cooking at an island out in the Strait a few miles from the village. We were therefore anxious to buy or trade for a good sound ax in exchange for our broken one. Good axes are rare in rocky Alaska. Soon or late an unlucky stroke on a stone concealed in moss spoils the edge. Finally one in almost perfect condition was offered by a young Hoona for our broken-handled one and a half-dollar to boot; but when the broken ax and money were given he promptly demanded an additional twenty-five cents' worth of tobacco. The tobacco was given him, then he required a half-dollar's worth more of tobacco, which was also given; but when he still demanded something more, Charley's patience gave way and we sailed in the same condition as to axes as when we arrived. This was the only contemptible commercial affair we encountered among these Alaskan Indians.

We reached the wooded island about one o'clock, made coffee, took on a store of wood, and set sail direct for the icy country, finding it very hard indeed to believe the woodless part of Charley's description of the Icy Bay, so heavily and uniformly are all the shores forested wherever we had been. In this view we were joined by John, Kadachan, and Toyatte, none of them on all their lifelong canoe travels having ever seen a woodless country.

We held a northwesterly course until long after dark, when we reached a small inlet that sets in near the mouth of Glacier Bay, on the west side. Here we made a cold camp on a desolate snow-covered beach in stormy sleet and darkness. At daybreak I looked eagerly in every direction to learn what kind of place we were in; but gloomy rain-clouds covered the mountains, and I could see nothing that would give me a clue, while Vancouver's chart, hitherto a faithful guide, here failed us altogether. Nevertheless, we made haste to be off; and fortunately, for just as we were leaving the shore, a faint smoke was seen across the inlet, toward which Charley, who now seemed lost, gladly steered. Our sudden appearance so early that gray morning had evidently alarmed our neighbors, for as soon as we were within hailing distance an Indian with his face blackened fired a shot over our heads, and in a blunt, bellowing voice roared, "Who are you?"

Our interpreter shouted, "Friends and the Fort Wrangell missionary."

Then men, women, and children swarmed out of the hut, and awaited our approach on the beach. One of the hunters having brought his gun with him, Kadachan sternly rebuked him, asking with superb indignation whether he was not ashamed to meet a missionary with a gun in his hands. Friendly relations, however, were speedily established, and as a cold rain was falling, they invited us to enter their hut. It seemed very small and was jammed full of oily boxes and bundles; nevertheless, twenty-one persons managed to find shelter in it about a smoky fire. Our hosts proved to be Hoona seal-hunters laying in their winter stores of meat and skins. The packed hut was passably well ventilated, but its heavy, meaty smells were not the same to our noses as those we were accustomed to in the sprucy nooks of the evergreen woods. The circle of black eyes peering at us through a fog of reek and smoke made a novel picture. We were glad, however, to get within reach of information, and of course asked many questions concerning the ice-mountains and the strange bay, to most of which our inquisitive Hoona friends replied with counter-questions as to our object in coming to such a place, especially so late in the year. They had heard of Mr. Young and his work at Fort Wrangell, but could not understand what a missionary could be doing in such a place as this. Was he going to preach to the seals and gulls, they asked, or to the ice-mountains? And could they take his word? Then John explained that only the friend of the missionary was seeking ice-mountains, that Mr. Young had already preached many good words in the villages we had visited, their own among the others, that our hearts were good and every Indian was our friend. Then we gave them a little rice, sugar, tea, and tobacco, after which they began to gain confidence and to speak freely. They told us that the big bay was called by them Sit-a-da-kay, or Ice Bay;

that there were many large ice-mountains in it, but no gold-mines; and that the ice-mountain they knew best was at the head of the bay, where most of the seals were found.

Notwithstanding the rain, I was anxious to push on and grope our way beneath the clouds as best we could, in case worse weather should come; but Charley was ill at ease, and wanted one of the seal-hunters to go with us, for the place was much changed. I promised to pay well for a guide, and in order to lighten the canoe proposed to leave most of our heavy stores in the hut until our return. After a long consultation one of them consented to go. His wife got ready his blanket and a piece of cedar matting for his bed, and some provisions—mostly dried salmon, and seal sausage made of strips of lean meat plaited around a core of fat. She followed us to the beach, and just as we were pushing off said with a pretty smile, "It is my husband that you are taking away. See that you bring him back."

We got under way about 10 A.M. The wind was in our favor, but a cold rain pelted us, and we could see but little of the dreary, treeless wilderness which we had now fairly entered. The bitter blast, however, gave us good speed; our bedraggled canoe rose and fell on the waves as solemnly as a big ship. Our course was northwestward, up the southwest side of the bay, near the shore of what seemed to be the mainland, smooth marble islands being on our right. About noon we discovered the first of the great glaciers, the one I afterward named for James Geikie, the noted Scotch geologist. Its lofty blue cliffs, looming through the draggled skirts of the clouds, gave a tremendous impression of savage power, while the roar of the new-born icebergs thickened and emphasized the general roar of the storm. An hour and a half beyond the Geikie Glacier we ran into a slight harbor where the shore is low, dragged the canoe beyond the reach of drifting icebergs, and, much against my desire to push ahead, encamped, the guide insisting that the big ice-mountain at the head of the bay could not be reached before dark, that the landing there was dangerous even in daylight, and that this was the only safe harbor on the way to it. While camp was being made, I strolled along the shore to examine the rocks and the fossil timber that abounds here. All the rocks are freshly glaciated, even below the sea-level, nor have the waves as yet worn off the surface polish, much less the heavy scratches and grooves and lines of glacial contour.

The next day being Sunday, the minister wished to stay in camp;

and so, on account of the weather, did the Indians. I therefore set out on an excursion, and spent the day alone on the mountain-slopes above the camp, and northward, to see what I might learn. Pushing on through rain and mud and sludgy snow, crossing many brown, boulder-choked torrents, wading, jumping, and wallowing in snow up to my shoulders was mountaineering of the most trying kind. After crouching cramped and benumbed in the canoe, poulticed in wet or damp clothing night and day, my limbs had been asleep. This day they were awakened and in the hour of trial proved that they had not lost the cunning learned on many a mountain peak of the High Sierra. I reached a height of fifteen hundred feet, on the ridge that bounds the second of the great glaciers. All the landscape was smothered in clouds and I began to fear that as far as wide views were concerned I had climbed in vain. But at length the clouds lifted a little, and beneath their gray fringes I saw the berg-filled expanse of the bay, and the feet of the mountains that stand about it, and the imposing fronts of five huge glaciers, the nearest being immediately beneath me. This was my first general view of Glacier Bay, a solitude of ice and snow and new-born rocks, dim, dreary, mysterious. I held the ground I had so dearly won for an hour or two, sheltering myself from the blast as best I could, while with benumbed fingers I sketched what I could see of the landscape, and wrote a few lines in my note-book. Then, breasting the snow again, crossing the shifting avalanche slopes and torrents, I reached camp about dark, wet and weary and glad.

While I was getting some coffee and hardtack, Mr. Young told me that the Indians were discouraged, and had been talking about turning back, fearing that I would be lost, the canoe broken, or in some other mysterious way the expedition would come to grief if I persisted in going farther. They had been asking him what possible motive I could have in climbing mountains when storms were blowing; and when he replied that I was only seeking knowledge, Toyatte said, "Muir must be a witch to seek knowledge in such a place as this and in such miserable weather."

After supper, crouching about a dull fire of fossil wood, they became still more doleful, and talked in tones that accorded well with the wind and waters and growling torrents about us, telling sad old stories of crushed canoes, drowned Indians, and hunters frozen in snow-storms. Even brave old Toyatte, dreading the treeless, forlorn appearance of the region, said that his heart was not strong, and that he feared his canoe, on the safety of which our lives depended, might be entering a skookum-house (jail) of ice, from which there might be no escape; while the Hoona guide said bluntly that if I was so fond of danger, and meant to go close up to the noses of the ice-mountains, he would not consent to go any farther; for we should all be lost, as many of his tribe had been, by the sudden rising of bergs from the bottom. They seemed to be losing heart with every howl of the wind, and, fearing that they might fail me now that I was in the midst of so grand a congregation of glaciers, I made haste to reassure them, telling them that for ten years I had wandered alone among mountains and storms, and good luck always followed me; that with me, therefore, they need fear nothing. The storm would soon cease and the sun would shine to show us the way we should go, for God cares for us and guides us as long as we are trustful and brave, therefore all childish fear must be put away. This little speech did good. Kadachan, with some show of enthusiasm, said he liked to travel with good-luck people; and dignified old Toyatte declared that now his heart was strong again, and he would venture on with me as far as I liked for my "wawa" was "delait" (my talk was very good). The old warrior even became a little sentimental, and said that even if the canoe was broken he would not greatly care, because on the way to the other world he would have good companions.

Next morning it was still raining and snowing, but the south wind swept us bravely forward and swept the bergs from our course. In about an hour we reached the second of the big glaciers, which I afterwards named for Hugh Miller. We rowed up its fiord and landed to make a slight examination of its grand frontal wall. The berg-producing portion we found to be about a mile and a half wide, and broken into an imposing array of jagged spires and pyramids, and flat-topped towers and battlements, of many shades of blue, from pale, shimmering, limpid tones in the crevasses and hollows, to the most startling, chilling, almost shrieking vitriol blue on the plain mural spaces from which bergs had just been discharged. Back from the front for a few miles the glacier rises in a series of wide steps, as if this portion of the glacier had sunk in successive sections as it reached deep water, and the sea had found its way beneath it. Beyond this it extends indefinitely in a gently rising prairie-like expanse, and branches along the slopes and cañons of the Fairweather Range.

From here a run of two hours brought us to the head of the bay, and to the mouth of the northwest fiord, at the head of which lie the Hoona sealing-grounds, and the great glacier now called the Pacific, and another called the Hoona. The fiord is about five miles long, and two miles wide at the mouth. Here our Hoona guide had a store of dry wood, which we took aboard. Then, setting sail, we were driven wildly up the fiord, as if the storm-wind were saying, "Go, then, if you will, into my icy chamber; but you shall stay in until I am ready to let you out." All this time sleety rain was falling on the bay, and snow on the mountains; but soon after we landed the sky began to open. The camp was made on a rocky bench near the front of the Pacific Glacier, and the canoe was carried beyond the reach of the bergs and berg-waves. The bergs were now crowded in a dense pack against the discharging front, as if the storm-wind had determined to make the glacier take back her crystal offspring and keep them at home.

While camp affairs were being attended to, I set out to climb a mountain for comprehensive views; and before I had reached a height of a thousand feet the rain ceased, and the clouds began to rise from the lower altitudes, slowly lifting their white skirts, and lingering in majestic, wing-shaped masses about the mountains that rise out of the broad, icy sea, the highest of all the white mountains, and the greatest of all the glaciers I had yet seen. Climbing higher for a still broader outlook, I made notes and sketched, improving the precious time while sunshine streamed through the luminous fringes of the clouds and fell on the green waters of the fiord, the glittering bergs, the crystal bluffs of the vast glacier, the intensely white, far-spreading fields of ice, and the ineffably chaste and spiritual heights of the Fairweather Range, which were now hidden, now partly revealed, the whole making a picture of icy wildness unspeakably pure and sublime.

Looking southward, a broad ice-sheet was seen extending in a gently undulating plain from the Pacific Fiord in the foreground to the horizon, dotted and ridged here and there with mountains which were as white as the snow-covered ice in which they were half, or more than half, submerged. Several of the great glaciers of the bay flow from this one grand fountain. It is an instructive example of a general glacier covering the hills and dales of a country that is not yet ready to be brought to the light of day—not only covering but creating a landscape with the features it is destined to have when, in the fullness of time, the fashioning ice-sheet shall be lifted by the sun, and the land become warm and fruitful. The view to the westward is bounded and almost filled by the glorious Fairweather Mountains, the highest among them springing aloft in sublime beauty to a height of nearly sixteen thousand feet, while from base to summit every peak and spire and dividing ridge of all the mighty host was spotless white, as if painted. It would seem that snow could never be made to lie on the steepest slopes and precipices unless plastered on when wet, and then frozen. But this snow could not have been wet. It must have been fixed by being driven and set in small particles like the storm-dust of drifts, which, when in this condition, is fixed not only on sheer cliffs, but in massive, overcurling cornices. Along the base of this majestic range sweeps the Pacific Glacier, fed by innumerable cascading tributaries, and discharging into the head of its fiord by two mouths only partly separated by the brow of an island rock about one thousand feet high, each nearly a mile wide.

Dancing down the mountain to camp, my mind glowing like the sunbeaten glaciers, I found the Indians seated around a good fire, entirely happy now that the farthest point of the journey was safely reached and the long, dark storm was cleared away. How hopefully, peacefully bright that night were the stars in the frosty sky, and how impressive was the thunder of the icebergs, rolling, swelling, reverberating through the solemn stillness! I was too happy to sleep.

About daylight next morning we crossed the fiord and landed on the south side of the rock that divides the wall of the great glacier. The whiskered faces of seals dotted the open spaces between the bergs, and I could not prevent John and Charley and Kadachan from shooting at them. Fortunately, few, if any, were hurt. Leaving the Indians in charge of the canoe, I managed to climb to the top of the wall by a good deal of step-cutting between the ice and dividing rock, and gained a good general view of the glacier. At one favorable place I descended about fifty feet below the side of the glacier, where its denuding, fashioning action was clearly shown. Pushing back from here, I found the surface crevassed and sunken in steps, like the Hugh Miller Glacier, as if it were being undermined by the action of tide-waters. For a distance of fifteen or twenty miles the river-like ice-flood is nearly level, and when it recedes, the ocean water will follow it, and thus form a long extension of the fiord, with features essentially the same as those now extending into the continent farther south, where many great glaciers once poured into the sea, though scarce a vestige of them now exists. Thus the domain of the sea has been, and is being, extended in these ice-sculptured lands, and the scenery of their shores enriched. The brow of the dividing rock is about a thousand feet high, and is hard beset by the glacier. A short time ago it was at least two thousand feet below the surface of the over-sweeping ice; and under present climatic conditions it will soon take its place as a glacier-polished island in the middle of the fiord, like a thousand others in the magnificent archipelago. Emerging from its icy sepulchre, it gives a most telling illustration of the birth of a marked feature of a landscape. In this instance it is not the mountain, but the glacier, that is in labor, and the mountain itself is being brought forth.

The Hoona Glacier enters the fiord on the south side, a short distance below the Pacific, displaying a broad and far-reaching expanse, over which many lofty peaks are seen; but the front wall, thrust into the fiord, is not nearly so interesting as that of the Pacific, and I did not observe any bergs discharged from it.

In the evening, after witnessing the unveiling of the majestic peaks and glaciers and their baptism in the down-pouring sunbeams, it seemed inconceivable that Nature could have anything finer to show us. Nevertheless, compared with what was to come the next morning, all that was as nothing. The calm dawn gave no promise of anything uncommon. Its most impressive features were the frosty clearness of the sky and a deep, brooding stillness made all the more striking by the thunder of the new-born bergs. The sunrise we did not see at all, for we were beneath the shadows of the fiord cliffs; but in the midst of our studies, while the Indians were getting ready to sail, we were startled by the sudden appearance of a red light burning with a strange unearthly splendor on the topmost peak of the Fairweather Mountains. Instead of vanishing as suddenly as it had appeared, it spread and spread until the whole range down to the level of the glaciers was filled with the celestial fire. In color it was at first a vivid crimson, with a thick, furred appearance, as fine as the alpenglow, yet indescribably rich and deep—not in the least like a garment or mere external flush or bloom through which one might expect to see the rocks or snow, but every mountain apparently was glowing from the heart like molten metal fresh from a furnace. Beneath the frosty shadows of the fiord we stood hushed and awe-stricken, gazing at the holy vision; and had we seen the heavens opened and God made manifest, our attention could not have been more tremendously strained. When the highest peak began to burn, it did not seem to be steeped in sunshine, however glorious, but rather as if it had been thrust into the body of the sun itself. Then the supernal fire slowly descended, with a sharp line of demarkation separating it from the cold, shaded region beneath; peak after peak, with their spires and ridges and cascading glaciers, caught the heavenly glow, until all the mighty host stood transfigured, hushed, and thoughtful, as if awaiting the coming of the Lord. The white, rayless light of morning, seen when I was alone amid the peaks of the California Sierra, had always seemed to me the most telling of all the terrestrial manifestations of God. But here the mountains themselves were made divine, and declared His glory in terms still more impressive. How long we gazed I never knew. The glorious vision passed away in a gradual, fading change through a thousand tones of color to pale yellow and white, and then the work of the ice-world went on again in everyday beauty. The green waters of the fiord were filled with sun-spangles; the fleet of icebergs set forth on their voyages with the upspringing breeze; and on the innumerable mirrors and prisms of these bergs, and on those of the shattered crystal walls of the glaciers, common white light and rainbow light began to burn, while the mountains shone in their frosty jewelry, and loomed again in the thin azure in serene terrestrial majesty. We turned and sailed away, joining the outgoing bergs, while "Gloria in excelsis" still seemed to be sounding over all the white landscape, and our burning hearts were ready for any fate, feeling that, whatever the future might have in store, the treasures we had gained this glorious morning would enrich our lives forever.

When we arrived at the mouth of the fiord, and rounded the massive granite headland that stands guard at the entrance on the north side, another large glacier, now named the Reid, was discovered at the head of one of the northern branches of the bay. Pushing ahead into this new fiord, we found that it was not only packed with bergs, but that the spaces between the bergs were crusted with new ice, compelling us to turn back while we were yet several miles from the discharging frontal wall. But though we were not then allowed to set foot on this magnificent glacier, we obtained a fine view of it, and I made the Indians cease rowing while I sketched its principal features. Thence, after steering northeastward a few miles, we discovered still another large glacier, now named the Carroll. But the fiord into which this glacier flows was, like the last, utterly inaccessible on account of ice, and we had to be content with a general view and sketch of it, gained as we rowed slowly past at a distance of three or four miles. The mountains back of it and on each side of its inlet are sculptured in a singularly rich and striking style of architecture, in which subordinate peaks and gables appear in wonderful profusion, and an imposing conical mountain with a wide, smooth base stands out in the main current of the glacier, a mile or two back from the discharging ice-wall.

We now turned southward down the eastern shore of the bay, and in an hour or two discovered a glacier of the second class, at the head of a comparatively short fiord that winter had not yet closed. Here we landed, and climbed across a mile or so of rough boulder-beds, and back upon the wildly broken, receding front of the glacier, which, though it descends to the level of the sea, no longer sends off bergs. Many large masses, detached from the wasting front by irregular melting, were partly buried beneath mud, sand, gravel, and boulders of the terminal moraine. Thus protected, these fossil icebergs remain unmelted for many years, some of them for a century or more, as shown by the age of trees growing above them, though there are no trees here as yet. At length melting, a pit with sloping sides is formed by the falling in of the overlying moraine material into the space at first occupied by the buried ice. In this way are formed the curious depressions in drift-covered regions called kettles or sinks. On these decaying glaciers we may also find many interesting lessons on the formation of boulders and boulder-beds, which in all glaciated countries exert a marked influence on scenery, health, and fruitfulness.

Three or four miles farther down the bay, we came to another fiord, up which we sailed in quest of more glaciers, discovering one in each of the two branches into which the fiord divides. Neither of these glaciers quite reaches tide-water. Notwithstanding the apparent fruitfulness of their fountains, they are in the first stage of decadence, the waste from melting and evaporation being greater now than the supply of new ice from their snowy fountains. We reached the one in the north branch, climbed over its wrinkled brow, and gained a good view of the trunk and some of the tributaries, and also of the sublime gray cliffs of its channel.

Then we sailed up the south branch of the inlet, but failed to reach the glacier there, on account of a thin sheet of new ice. With the tent-poles we broke a lane for the canoe for a little distance; but it was slow work, and we soon saw that we could not reach the glacier before dark. Nevertheless, we gained a fair view of it as it came sweeping down through its gigantic gateway of massive Yosemite rocks three or four thousand feet high. Here we lingered until sundown, gazing and sketching; then turned back, and encamped on a bed of cobblestones between the forks of the fiord.

We gathered a lot of fossil wood and after supper made a big fire, and as we sat around it the brightness of the sky brought on a long talk with the Indians about the stars; and their eager, child-like attention was refreshing to see as compared with the death-like apathy of weary town-dwellers, in whom natural curiosity has been quenched in toil and care and poor shallow comfort.

After sleeping a few hours, I stole quietly out of the camp, and climbed the mountain that stands between the two glaciers. The ground was frozen, making the climbing difficult in the steepest places; but the views over the icy bay, sparkling beneath the stars, were enchanting. It seemed then a sad thing that any part of so precious a night had been lost in sleep. The star-light was so full that I distinctly saw not only the berg-filled bay, but most of the lower portions of the glaciers, lying pale and spirit-like amid the mountains. The nearest glacier in particular was so distinct that it seemed to be glowing with light that came from within itself. Not even in dark nights have I ever found any difficulty in seeing large glaciers; but on this mountain-top, amid so much ice, in the heart of so clear and frosty a night, everything was more or less luminous, and I seemed to be poised in a vast hollow between two skies of almost equal brightness. This exhilarating scramble made me glad and strong and I rejoiced that my studies called me before the glorious night succeeding so glorious a morning had been spent!

I got back to camp in time for an early breakfast, and by daylight we had everything packed and were again under way. The fiord was frozen nearly to its mouth, and though the ice was so thin it gave us but little trouble in breaking a way for the canoe, yet it showed us that the season for exploration in these waters was well-nigh over. We were in danger of being imprisoned in a jam of icebergs, for the water-spaces between them freeze rapidly, binding the floes into one mass. Across such floes it would be almost impossible to drag a canoe, however industriously we might ply the ax, as our Hoona guide took great pains to warn us. I would have kept straight down the bay from here, but the guide had to be taken home, and the provisions we left at the bark hut had to be got on board. We therefore crossed over to our Sunday storm-camp, cautiously boring a way through the bergs. We found the shore lavishly adorned with a fresh arrival of assorted bergs that had been left stranded at high tide. They were arranged in a curving row, looking intensely clear and pure on the gray sand, and, with the sunbeams pouring through them, suggested the jewel-paved streets of the New Jerusalem.

On our way down the coast, after examining the front of the beautiful Geikie Glacier, we obtained our first broad view of the great glacier afterwards named the Muir, the last of all the grand company to be seen, the stormy weather having hidden it when we first entered the bay. It was now perfectly clear, and the spacious, prairie-like glacier, with its many tributaries extending far back into the snowy recesses of its fountains, made a magnificent display of its wealth, and I was strongly tempted to go and explore it at all hazards. But winter had come, and the freezing of its fiords was an insurmountable obstacle. I had, therefore, to be content for the present with sketching and studying its main features at a distance.

When we arrived at the Hoona hunting-camp, men, women, and children came swarming out to welcome us. In the neighborhood of this camp I carefully noted the lines of demarkation between the forested and deforested regions. Several mountains here are only in part deforested, and the lines separating the bare and the forested portions are well defined. The soil, as well as the trees, had slid off the steep slopes, leaving the edge of the woods raw-looking and rugged.

At the mouth of the bay a series of moraine islands show that the trunk glacier that occupied the bay halted here for some time and deposited this island material as a terminal moraine; that more of the bay was not filled in shows that, after lingering here, it receded comparatively fast. All the level portions of trunks of glaciers occupying ocean fiords, instead of melting back gradually in times of general shrinking and recession, as inland glaciers with sloping channels do, melt almost uniformly over all the surface until they become thin enough to float. Then, of course, with each rise and fall of the tide, the sea water, with a temperature usually considerably above the freezing-point, rushes in and out beneath them, causing rapid waste of the nether surface, while the upper is being wasted by the weather, until at length the fiord portions of these great glaciers become comparatively thin and weak and are broken up and vanish almost simultaneously.

Glacier Bay is undoubtedly young as yet. Vancouver's chart, made only a century ago, shows no trace of it, though found admirably faithful in general. It seems probable, therefore, that even then the entire bay was occupied by a glacier of which all those described above, great though they are, were only tributaries. Nearly as great a change has taken place in Sum Dum Bay since Vancouver's visit, the main trunk glacier there having receded from eighteen to twenty-five miles from the line marked on his chart. Charley, who was here when a boy, said that the place had so changed that he hardly recognized it, so many new islands had been born in the mean time and so much ice had vanished. As we have seen, this Icy Bay is being still farther extended by the recession of the glaciers. That this whole system of fiords and channels was added to the domain of the sea by glacial action is to my mind certain.

We reached the island from which we had obtained our store of fuel about half-past six and camped here for the night, having spent only five days in Sitadaka, sailing round it, visiting and sketching all the six glaciers excepting the largest, though I landed only on three of them,—the Geikie, Hugh Miller, and Grand Pacific,—the freezing of the fiords in front of the others rendering them inaccessible at this late season.

CHAPTER XI

THE COUNTRY OF THE CHILCATS

ON OCTOBER 30 we visited a camp of Hoonas at the mouth of a salmon-chuck. We had seen some of them before, and they received us kindly. Here we learned that peace reigned in Chilcat. The reports that we had previously heard were, as usual in such cases, wildly exaggerated. The little camp hut of these Indians was crowded with the food-supplies they had gathered—chiefly salmon, dried and tied in bunches of convenient size for handling and transporting to their villages, bags of salmon-roe, boxes of fish-oil, a lot of mountain-goat mutton, and a few porcupines. They presented us with some dried salmon and potatoes, for which we gave them tobacco and rice. About 3 P.M. we reached their village, and in the best house, that of a chief, we found the family busily engaged in making whiskey. The still and mash were speedily removed and hidden away with apparent shame as soon as we came in sight. When we entered and passed the regular greetings, the usual apologies as to being unable to furnish Boston food for us and inquiries whether we could eat Indian food were gravely made. Toward six or seven o'clock Mr. Young explained the object of his visit and held a short service. The chief replied with grave deliberation, saying that he would be heartily glad to have a teacher sent to his poor ignorant people, upon whom he now hoped the light of a better day was beginning to break. Hereafter he would gladly do whatever the white teachers told him to do and would have no will of his own. This under the whiskey circumstances seemed too good to be quite true. He thanked us over and over again for coming so far to see him, and complained that Port Simpson Indians, sent out on a missionary tour by Mr. Crosby, after making a good-luck board for him and nailing it over his door, now wanted to take it away. Mr. Young promised to make him a new one, should this threat be executed, and remarked that since he had offered to do his bidding he hoped he would make no more whiskey. To this the chief replied with fresh complaints concerning the threatened loss of his precious board, saying that he thought the Port

Simpson Indians were very mean in seeking to take it away, but that now he would tell them to take it as soon as they liked for he was going to get a better one at Wrangell. But no effort of the missionary could bring him to notice or discuss the whiskey business. The luck board nailed over the door was about two feet long and had the following inscription: "The Lord will bless those who do his will. When you rise in the morning, and when you retire at night, give him thanks. Heccla Hockla Popla."

This chief promised to pray like a white man every morning, and to bury the dead as the whites do. "I often wondered," he said, "where the dead went to. Now I am glad to know"; and at last acknowledged the whiskey, saying he was sorry to have been caught making the bad stuff. The behavior of all, even the little ones circled around the fire, was very good. There was no laughter when the strange singing commenced. They only gazed like curious, intelligent animals. A little daughter of the chief with the glow of the fire-light on her eyes made an interesting picture, head held aslant. Another in the group, with upturned eyes, seeming to half understand the strange words about God, might have passed for one of Raphael's angels.

The chief's house was about forty feet square, of the ordinary fort kind, but better built and cleaner than usual. The side-room doors were neatly paneled, though all the lumber had been nibbled into shape with a small narrow Indian adze. We had our tent pitched on a grassy spot near the beach, being afraid of wee beasties; which greatly offended Kadachan and old Toyatte, who said, "If this is the way you are to do up at Chilcat, we will be ashamed of you." We promised them to eat Indian food and in every way behave like good Chilcats.

We set out direct for Chilcat in the morning against a brisk head wind. By keeping close inshore and working hard, we made about ten miles by two or three o'clock, when, the tide having turned against us, we could make scarce any headway, and therefore landed in a sheltered cove a few miles up the west side of Lynn Canal. Here I discovered a fine growth of yellow cedar, but none of the trees were very large, the tallest only seventy-five to one hundred feet high. The flat, drooping, plume-like branchlets hang edgewise, giving the trees a thin, open, airy look. Nearly every tree that I saw in a long walk was more or less marked by the knives and axes of the Indians, who use the bark for matting, for covering house-roofs, and making temporary portable huts. For this last purpose sections five or six feet long and two or three wide are pressed flat and secured from warping or splitting by binding them with thin strips of wood at the end. These they carry about with them in their canoes, and in a few minutes they can be put together against slim poles and made into a rainproof hut. Every paddle that I have seen along the coast is made of the light, tough, handsome yellow wood of this tree. It is a tree of moderately rapid growth and usually chooses ground that is rather boggy and mossy. Whether its network of roots makes the bog or not, I am unable as yet to say.

Three glaciers on the opposite side of the canal were in sight, descending nearly to sea-level, and many smaller ones that melt a little below timber line. While I was sketching these, a canoe hove in sight, coming on at a flying rate of speed before the wind. The owners, eager for news, paid us a visit. They proved to be Hoonas, a man, his wife, and four children, on their way home from Chilcat. The man was sitting in the stern steering and holding a sleeping child in his arms. Another lay asleep at his feet. He told us that Sitka Jack had gone up to the main Chilcat village the day before he left, intending to hold a grand feast and potlatch, and that whiskey up there was flowing like water. The news was rather depressing to Mr. Young and myself, for we feared the effect of the poison on Toyatte's old enemies. At 8.30 P.M. we set out again on the turn of the tide, though the crew did not relish this night work. Naturally enough, they liked to stay in camp when wind and tide were against us, but didn't care to make up lost time after dark however wooingly wind and tide might flow and blow. Kadachan, John, and Charley rowed, and Toyatte steered and paddled, assisted now and then by me. The wind moderated and almost died away, so that we made about fifteen miles in six hours, when the tide turned and snow began to fall. We ran into a bay nearly opposite Berner's Bay, where three or four families of Chilcats were camped, who shouted when they heard us landing and demanded our names. Our men ran to the huts for news before making camp. The Indians proved to be hunters, who said there were plenty of wild sheep on the mountains back a few miles from the head of the bay. This interview was held at three o'clock in the morning, a rather early hour. But Indians never resent any such disturbance provided there is anything worth while to be said or done. By four o'clock we had our tents set, a fire made and some coffee, while the snow was falling fast. Toyatte was out of humor with this night business. He wanted to land an hour or two before we did, and then, when the snow began to fall and we all wanted to find a camping-ground as soon as possible, he steered out into the middle of the canal, saying grimly that the tide was good. He turned, however, at our orders, but read us a lecture at the first opportunity, telling us to start early if we were in a hurry, but not to travel in the night like thieves.

After a few hours' sleep, we set off again, with the wind still against us and the sea rough. We were all tired after making only about twelve miles, and camped in a rocky nook where we found a family of Hoonas in their bark hut beside their canoe. They presented us with potatoes and salmon and a big bucketful of berries, salmon-roe, and grease of some sort, probably fish-oil, which the crew consumed with wonderful relish.

A fine breeze was blowing next morning from the south, which would take us to Chilcat in a few hours, but unluckily the day was Sunday and the good wind was refused. Sunday, it seemed to me, could be kept as well by sitting in the canoe and letting the Lord's wind waft us quietly on our way. The day was rainy and the clouds hung low. The trees here are remarkably well developed, tall and straight. I observed three or four hemlocks which had been struck by lightning,—the first I noticed in Alaska. Some of the species on windy outjutting rocks become very picturesque, almost as much so as old oaks, the foliage becoming dense and the branchlets tufted in heavy plume-shaped horizontal masses.

Monday was a fine clear day, but the wind was dead ahead, making hard, dull work with paddles and oars. We passed a long stretch of beautiful marble cliffs enlivened with small merry waterfalls, and toward noon came in sight of the front of the famous Chilcat or Davidson Glacier, a broad white flood reaching out two or three miles into the canal with wonderful effect. I wanted to camp beside it but the head wind tired us out before we got within six or eight miles of it. We camped on the west side of a small rocky island in a narrow cove. When I was looking among the rocks and bushes for a smooth spot for a bed, I found a human skeleton. My Indians seemed not in the least shocked or surprised, explaining that it was only the remains of a Chilcat slave. Indians never bury or burn the bodies of slaves, but just cast them away anywhere. Kind Nature was covering the poor bones with moss and leaves, and I helped in the pitiful work.

The wind was fair and joyful in the morning, and away we glided to the famous glacier. In an hour or so we were directly in front of it and beheld it in all its crystal glory descending from its white mountain fountains and spreading out in an immense fan three or four miles wide against its tree-fringed terminal moraine. But, large as it is, it long ago ceased to discharge bergs.

The Chilcats are the most influential of all the Thlinkit tribes. Whenever on our journey I spoke of the interesting characteristics of other tribes we had visited, my crew would invariably say, "Oh, yes, these are pretty good Indians, but wait till you have seen the Chilcats." We were now only five or six miles distant from their lower village, and my crew requested time to prepare themselves to meet their great rivals. Going ashore on the moraine with their boxes that had not been opened since we left Fort Wrangell, they sat on boulders and cut each other's hair, carefully washed and perfumed themselves and made a complete change in their clothing, even to white shirts, new boots, new hats, and bright neckties. Meanwhile, I scrambled across the broad, brushy, forested moraine, and on my return scarcely recognized my crew in their dress suits. Mr. Young also made some changes in his clothing, while I, having nothing dressy in my bag, adorned my cap with an eagle's feather I found on the moraine, and thus arrayed we set forth to meet the noble Thlinkits.

We were discovered while we were several miles from the village, and as we entered the mouth of the river we were hailed by a messenger from the chief, sent to find out who we were and the objects of our extraordinary visit.

"Who are you?" he shouted in a heavy, far-reaching voice. "What are your names? What do you want? What have you come for?"

On receiving replies, he shouted the information to another messenger, who was posted on the river-bank at a distance of a quarter of a mile or so, and he to another and another in succession, and by this living telephone the news was delivered to the chief as he sat by his fireside. A salute was then fired to welcome us, and a swarm of musket-bullets, flying scarce high enough for comfort, pinged over our heads. As soon as we reached the landing at the village, a dignified young man stepped forward and thus addressed us:—

"My chief sent me to meet you, and to ask if you would do him the honor to lodge in his house during your stay in our village?"

We replied, of course, that we would consider it a great honor to be entertained by so distinguished a chief.

The messenger then ordered a number of slaves, who stood behind him, to draw our canoe out of the water, carry our provisions and bedding into the chief's house, and then carry the canoe back from the river where it would be beyond the reach of floating ice. While we waited, a lot of boys and girls were playing on a meadow near the landing—running races, shooting arrows, and wading in the icy river without showing any knowledge of our presence beyond quick stolen glances. After all was made secure, he conducted us to the house, where we found seats of honor prepared for us.

The old chief sat barefooted by the fireside, clad in a calico shirt and blanket, looking down, and though we shook hands as we passed him he did not look up. After we were seated, he still gazed into the fire without taking the slightest notice of us for about ten or fifteen minutes. The various members of the chief's family, also,—men, women, and children,—went about their usual employment and play as if entirely unconscious that strangers were in the house, it being considered impolite to look at visitors or speak to them before time had been allowed them to collect their thoughts and prepare any message they might have to deliver.

At length, after the politeness period had passed, the chief slowly raised his head and glanced at his visitors, looked down again, and at last said, through our interpreter:—

"I am troubled. It is customary when strangers visit us to offer them food in case they might be hungry, and I was about to do so, when I remembered that the food of you honorable white chiefs is so much better than mine that I am ashamed to offer it."

We, of course, replied that we would consider it a great honor to enjoy the hospitality of so distinguished a chief as he was.

Hearing this, he looked up, saying, "I feel relieved"; or, in John the interpreter's words, "He feels good now, he says he feels good."

He then ordered one of his family to see that the visitors were fed. The young man who was to act as steward took up his position in a corner of the house commanding a view of all that was going on, and ordered the slaves to make haste to prepare a good meal; one to bring a lot of the best potatoes from the cellar and wash them well; another to go out and pick a basketful of fresh berries; another to broil a salmon; while others made a suitable fire, pouring oil on the wet wood to make it blaze. Speedily the feast was prepared and passed around. The first course was potatoes, the second fish-oil and salmon, next berries and rose-hips; then the steward shouted the important news, in a loud voice like a herald addressing an army, "That's all!" and left his post.

Then followed all sorts of questions from the old chief. He wanted to know what Professor Davidson had been trying to do a year or two ago on a mountain-top back of the village, with many strange things looking at the sun when it grew dark in the daytime; and we had to try to explain eclipses. He asked us if we could tell him what made the water rise and fall twice a day, and we tried to explain that the sun and moon attracted the sea by showing how a magnet attracted iron.

Mr. Young, as usual, explained the object of his visit and requested that the people might be called together in the evening to hear his message. Accordingly all were told to wash, put on their best clothing, and come at a certain hour. There was an audience of about two hundred and fifty, to whom Mr. Young preached. Toyatte led in prayer, while Kadachan and John joined in the singing of several hymns. At the conclusion of the religious exercises the chief made a short address of thanks, and finished with a request for the message of the other chief. I again tried in vain to avoid a speech by telling the interpreter to explain that I was only traveling to see the country, the glaciers, and mountains and forests, etc., but these subjects, strange to say, seemed to be about as interesting as the gospel, and I had to deliver a sort of lecture on the fine foodful country God had given them and the brotherhood of man, along the same general lines I had followed at other villages. Some five similar meetings were held here, two of them in the daytime, and we began to feel quite at home in the big block-house with our hospitable and warlike friends.

At the last meeting an old white-haired shaman of grave and venerable aspect, with a high wrinkled forehead, big, strong Roman nose and light-colored skin, slowly and with great dignity arose and spoke for the first time.

"I am an old man," he said, "but I am glad to listen to those strange things you tell, and they may well be true, for what is more wonderful than the flight of birds in the air? I remember the first white man I ever saw. Since that long, long-ago time I have seen many, but never until now have I ever truly known and felt a white man's heart. All the white men I have heretofore met wanted to get something from us. They wanted furs and they wished to pay for them as small a price as possible. They all seemed to be seeking their own good—not our good. I might say that through all my long life I have never until now heard a white man speak. It has always seemed to me while trying to speak to traders and those seeking gold-mines that it was like speaking to a person across a broad stream that was running fast over stones and making so loud a noise that scarce a single word could be heard. But now, for the first time, the Indian and the white man are on the same side of the river, eye to eye, heart to heart. I have always loved my people. I have taught them and ministered to them as well as I could. Hereafter, I will keep silent and listen to the good words of the missionaries, who know God and the places we go to when we die so much better than I do."

At the close of the exercises, after the last sermon had been preached and the last speech of the Indian chief and head men had been made, a number of the sub-chiefs were talking informally together. Mr. Young, anxious to know what impression he had made on the tribe with reference to mission work, requested John to listen and tell him what was being said.

"They are talking about Mr. Muir's speech," he reported. "They say he knows how to talk and beats the preacher far." Toyatte also, with a teasing smile, said: "Mr. Young, mika tillicum hi yu tola wawa" (your friend leads you far in speaking).

Later, when the sending of a missionary and teacher was being considered, the chief said they wanted me, and, as an inducement, promised that if I would come to them they would always do as I directed, follow my councils, give me as many wives as I liked, build a church and school, and pick all the stones out of the paths and make them smooth for my feet.

They were about to set out on an expedition to the Hootsenoos to collect blankets as indemnity or blood-money for the death of a Chilcat woman from drinking whiskey furnished by one of the Hootsenoo tribe. In case of their refusal to pay, there would be fighting, and one of the chiefs begged that we would pray them good luck, so that no one would be killed. This he asked as a favor, after begging that we would grant permission to go on this expedition, promising that they would avoid bloodshed if possible. He spoke in a very natural and easy tone and manner, always serene and so much of a polished diplomat that all polish was hidden. The younger chief stood while speaking, the elder sat on the floor. None of the congregation had a word to say, though they gave approving nods and shrugs.

The house was packed at every meeting, two a day. Some climbed on the roof to listen around the smoke opening. I tried in vain to avoid speechmaking, but, as usual, I had to say something at every meeting. I made five speeches here, all of which seemed to be gladly heard, particularly what I said on the different kinds of white men and their motives, and their own kindness and good manners in making strangers feel at home in their houses.

The chief had a slave, a young and good-looking girl, who waited on him, cooked his food, lighted his pipe for him, etc. Her servitude seemed by no means galling. In the morning, just before we left on the return trip, interpreter John overheard him telling her that after the teacher came from Wrangell, he was going to dress her well and send her to school and use her in every way as if she were his own daughter. Slaves are still owned by the richest of the Thlinkits. Formerly, many of them were sacrificed on great occasions, such as the opening of a new house or the erection of a totem pole. Kadachan ordered John to take a pair of white blankets out of his trunk and wrap them about the chief's shoulders, as he sat by the fire. This gift was presented without ceremony or saying a single word. The chief scarcely noticed the blankets, only taking a corner in his hand, as if testing the quality of the wool. Toyatte had been an inveterate enemy and fighter of the Chilcats, but now, having joined the church, he wished to forget the past and bury all the hard feuds and be universally friendly and peaceful. It was evident, however, that he mistrusted the proud and warlike Chilcats and doubted the acceptance of his friendly advances, and as we approached their village became more and more thoughtful.

"My wife said that my old enemies would be sure to kill me. Well, never mind. I am an old man and may as well die as not." He was troubled with palpitation, and oftentimes, while he suffered, he put his hand over his heart and said, "I hope the Chilcats will shoot me here."

Before venturing up the river to the principal village, located some ten miles up the river, we sent Sitka Charley and one of the young Chilcats as messengers to announce our arrival and inquire whether we would be welcome to visit them, informing the chief that both Kadachan and Toyatte were Mr. Young's friends and mine, that we were "all one meat" and any harm done them would also be done to us.

While our messengers were away, I climbed a pure-white, dome-crowned mountain about fifty-five hundred feet high and gained noble telling views to the northward of the main Chilcat glaciers and the multitude of mighty peaks from which they draw their sources. At a height of three thousand feet I found a mountain hemlock, considerably dwarfed, in company with Sitka spruce and the common hemlock, the tallest about twenty feet high, sixteen inches in diameter. A few stragglers grew considerably higher, say at about four thousand feet. Birch and two-leaf pine were common.

The messengers returned next day, bringing back word that we would all be heartily welcomed excepting Toyatte; that the guns were loaded and ready to be fired to welcome us, but that Toyatte, having insulted a Chilcat chief not long ago in Wrangell, must not come. They also informed us in their message that they were very busy merrymaking with other visitors, Sitka Jack and his friends, but that if we could get up to the village through the running ice on the river, they would all be glad to see us; they had been drinking and Kadachan's father, one of the principal chiefs, said plainly that he had just waked up out of a ten days' sleep. We were anxious to make this visit, but, taking the difficulties and untoward circumstances into account, the danger of being frozen in at so late a time, while Kadachan would not be able to walk back on account of a shot in his foot, the danger also from whiskey, the awakening of old feuds on account of Toyatte's presence, etc., we reluctantly concluded to start back on the home journey at once. This was on Friday and a fair wind was blowing, but our crew, who loved dearly to rest and eat in these big hospitable houses, all said that Monday would be *hyas klosh* for the starting-day. I insisted, however, on starting Saturday morning, and succeeded in getting away from our friends at ten o'clock. Just as we were leaving, the chief who had entertained us so handsomely requested a written document to show that he had not killed us, so in case we were lost on the way home he could not be held accountable in any way for our death.

CHAPTER XII

THE RETURN TO FORT WRANGELL

THE DAY OF our start for Wrangell was bright and the Hoon, the north wind, strong. We passed around the east side of the larger island which lies near the south extremity of the point of land between the Chilcat and the Chilcoot channels and thence held a direct course down the east shore of the canal. At sunset we encamped in a small bay at the head of a beautiful harbor three or four miles south of Berner's Bay, and the next day, being Sunday, we remained in camp as usual, though the wind was fair and it is not a sin to go home. The Indians spent most of the day in washing, mending, eating, and singing hymns with Mr. Young, who also gave them a Bible lesson, while I wrote notes and sketched. Charley made a sweat-house and all the crew got good baths. This is one of the most delightful little bays we have thus far enjoyed, girdled with tall trees whose branches almost meet, and with views of pure-white mountains across the broad, river-like canal.

Seeing smoke back in the dense woods, we went ashore to seek it and discovered a Hootsenoo whiskey-factory in full blast. The Indians said that an old man, a friend of theirs, was about to die and they were making whiskey for his funeral.

Our Indians were already out of oily flesh, which they regard as a necessity and consume in enormous quantities. The bacon was nearly gone and they eagerly inquired for flesh at every camp we passed. Here we found skinned carcasses of porcupines and a heap of wild mutton lying on the confused hut floor. Our cook boiled the porcupines in a big pot with a lot of potatoes we obtained at the same hut, and although the potatoes were protected by their skins, the awfully wild penetrating porcupine flavor found a way through the skins and flavored them to the very heart. Bread and beans and dried fruit we had in abundance, and none of these rank aboriginal dainties ever came nigh any meal of mine. The Indians eat the hips of wild roses entire like berries, and I was laughed at for eating only the outside of this fruit and rejecting the seeds.

When we were approaching the village of the Auk tribe, venerable Toyatte seemed to be unusually pensive, as if weighed down by some melancholy thought. This was so unusual that I waited attentively to find out the cause of his trouble.

When at last he broke silence it was to say, "Mr. Young, Mr. Young,"—he usually repeated the name,—"I hope you will not stop at the Auk village."

"Why, Toyatte?" asked Mr. Young.

"Because they are a bad lot, and preaching to them can do no good."

"Toyatte," said Mr. Young, "have you forgotten what Christ said to his disciples when he charged them to go forth and preach the gospel to everybody; and that we should love our enemies and do good to those who use us badly?"

"Well," replied Toyatte, "if you preach to them, you must not call on me to pray, because I cannot pray for Auks."

"But the Bible says we should pray for all men, however bad they may be."

"Oh, yes, I know that, Mr. Young; I know it very well. But Auks are not men, good or bad,—they are dogs."

It was now nearly dark and quite so ere we found a harbor, not far from the fine Auk Glacier which descends into the narrow channel that separates Douglas Island from the mainland. Two of the Auks followed us to our camp after eight o'clock and inquired into our object in visiting them, that they might carry the news to their chief. One of the chief's houses is opposite our camp a mile or two distant, and we concluded to call on him next morning.

I wanted to examine the Auk Glacier in the morning, but tried to be satisfied with a general view and sketch as we sailed around its wide fan-shaped front. It is one of the most beautiful of all the coast glaciers that are in the first stage of decadence. We called on the Auk chief at daylight, when he was yet in bed, but he arose good-naturedly, put on a calico shirt, drew a blanket around his legs, and comfortably seated himself beside a small fire that gave light enough to show his features and those of his children and the three women that one by one came out of the shadows. All listened attentively to Mr. Young's message of good-will. The chief was a serious, sharp-featured, dark-complexioned man, sensible-looking and with good manners. He was very sorry, he said, that his people had been drinking in his absence and had used us so ill; he would like to hear us talk and would call his people together if we would return to the village. This offer we had to decline. We gave him good words and tobacco and bade him good-bye.

The scenery all through the channel is magnificent, something like Yosemite Valley in its lofty avalanche-swept wall cliffs, especially on the mainland side, which are so steep few trees can find footing. The lower island side walls are mostly forested. The trees are heavily draped with lichens, giving the woods a remarkably gray, ancient look. I noticed a good many two-leafed pines in boggy spots. The water was smooth, and the reflections of the lofty walls striped with cascades were charmingly distinct.

It was not easy to keep my crew full of wild flesh. We called at an Indian summer camp on the mainland about noon, where there were three very squalid huts crowded and jammed full of flesh of many colors and smells, among which we discovered a lot of bright fresh trout, lovely creatures about fifteen inches long, their sides adorned with vivid red spots. We purchased five of them and a couple of salmon for a box of gun-caps and a little tobacco. About the middle of the afternoon we passed through a fleet of icebergs, their number increasing as we neared the mouth of the Taku Fiord, where we camped, hoping to explore the fiord and see the glaciers where the bergs, the first we had seen since leaving Icy Bay, are derived.

We left camp at six o'clock, nearly an hour before daybreak. My Indians were glad to find the fiord barred by a violent wind, against which we failed to make any headway; and as it was too late in the season to wait for better weather, I reluctantly gave up this promising work for another year, and directed the crew to go straight ahead down the coast. We sailed across the mouth of the happy inlet at fine speed, keeping a man at the bow to look out for the smallest of the bergs, not easily seen in the dim light, and another bailing the canoe as the tops of some of the white caps broke over us. About two o'clock we passed a large bay or fiord, out of which a violent wind was blowing, though the main Stephens Passage was calm. About dusk, when we were all tired and anxious to get into camp, we reached the mouth of Sum Dum Bay, but nothing like a safe landing could we find. Our experienced captain was indignant, as well he might be, because we did not see fit to stop early in the afternoon at a good campground he had chosen. He seemed determined to give us enough of night sailing as a punishment to last us for the rest of the voyage. Accordingly, though the night was dark and rainy and the bay full of icebergs, he pushed grimly on, saying that we must try to reach an Indian village on the other side of the bay or an old Indian fort on an island in the middle of it. We made slow, weary, anxious progress while Toyatte, who was well acquainted with every feature of this part of the coast and could find his way in the dark, only laughed at our misery. After a mile or two of this dismal night work we struck across toward the island, now invisible, and came near being wrecked on a rock which showed a smooth round back over which the waves were breaking. In the hurried Indian shouts that followed and while we were close against the rock, Mr. Young shouted, as he leaned over against me, "It's a whale, a whale!" evidently fearing its tail, several specimens of these animals, which were probably still on his mind, having been seen in the forenoon. While we were passing along the east shore of the island we saw a light on the opposite shore, a joyful sight, which Toyatte took for a fire in the Indian village, and steered for it. John stood in the bow, as guide through the bergs. Suddenly, we ran aground on a sand bar. Clearing this, and running back half a mile or so, we again stood for the light, which now shone brightly. I thought it strange that Indians should have so large a fire. A broad white mass dimly visible back of the fire Mr. Young took for the glow of the fire on the clouds. This proved to be the front of a glacier. After we had effected a landing and stumbled up toward the fire over a ledge of slippery, algæ-covered rocks, and through the ordinary tangle of shore grass, we were astonished to find white men instead of Indians, the first we had seen for a month. They proved to be a party of seven gold-seekers from Fort Wrangell. It was now about eight o'clock and they were in bed, but a jolly Irishman got up to make coffee for us and find out who we were, where we had come from, where going, and the objects of our travels. We unrolled our chart and asked for information as to the extent and features of the bay. But our benevolent friend took great pains to pull wool over our eyes, and made haste to say that if "ice and sceneries" were what we were looking for, this was a very poor, dull place. There were "big rocks, gulches, and sceneries" of a far better quality down the coast on the way to Wrangell. He and his party were prospecting, he said, but thus far they had found only a few colors and they proposed going over to Admiralty Island in the morning to try their luck.

In the morning, however, when the prospectors were to have gone over to the island, we noticed a smoke half a mile back on a large stream, the outlet of the glacier we had seen the night before, and an Indian told us that the white men were building a big log house up there. It appeared that they had found a promising placer mine in the moraine and feared we might find it and spread the news. Daylight revealed a magnificent fiord that brought Glacier Bay to mind. Miles of bergs lay stranded on the shores, and the waters of the branch fiords, not on Vancouver's chart, were crowded with them as far as the eye could reach. After breakfast we set out to explore an arm of the bay that trends southeastward, and managed to force a way through the bergs about ten miles. Farther we could not go. The pack was so close no open water was in sight, and, convinced at last that this part of my work would have to be left for another year, we struggled across to the west side of the fiord and camped.

I climbed a mountain next morning, hoping to gain a view of the great fruitful glaciers at the head of the fiord or, at least, of their snowy fountains. But in this also I failed; for at a distance of about sixteen miles from the mouth of the fiord a change to the northward in its general trend cut off all its upper course from sight.

Returning to camp baffled and weary, I ordered all hands to pack up and get out of the ice as soon as possible. And how gladly was that order obeyed! Toyatte's grand countenance glowed like a sun-filled glacier, as he joyfully and teasingly remarked that "the big Sum Dum ice-mountain had hidden his face from me and refused to let me pay him a visit." All the crew worked hard boring a way down the west side of the fiord, and early in the afternoon we reached comparatively open water near the mouth of the bay. Resting a few minutes among the drifting bergs, taking last lingering looks at the wonderful place I might never see again, and feeling sad over my weary failure to explore it, I was cheered by a friend I little expected to meet here. Suddenly, I heard the familiar whir of an ouzel's wings, and, looking up, saw my little comforter coming straight from the shore. In a second or two he was with me, and flew three times around my head with a happy salute, as if saying, "Cheer up, old friend, you see I am here and all's well." He then flew back to the shore, alighted on the topmost jag of a stranded iceberg, and began to nod and bow as though he were on one of his favorite rocks in the middle of a sunny California mountain cataract.

Mr. Young regretted not meeting the Indians here, but mission work also had to be left until next season. Our happy crew hoisted sail to a fair wind, shouted "Good-bye, Sum Dum!" and soon after dark reached a harbor a few miles north of Hobart Point.

We made an early start the next day, a fine, calm morning, glided smoothly down the coast, admiring the magnificent mountains arrayed in their winter robes, and early in the afternoon reached a lovely harbor on an island five or six miles north of Cape Fanshawe. Toyatte predicted a heavy winter storm, though only a mild rain was falling as yet. Everybody was tired and hungry, and as the voyage was nearing the end, I consented to stop here. While the shelter tents were being set up and our blankets stowed under cover, John went out to hunt and killed a deer within two hundred yards of the camp. When we were at the camp-fire in Sum Dum Bay, one of the prospectors, replying to Mr. Young's complaint that they were oftentimes out of meat, asked Toyatte why he and his men did not shoot plenty of ducks for the minister. "Because the duck's friend would not let us," said Toyatte; "when we want to shoot, Mr. Muir always shakes the canoe."

Just as we were passing the south headland of Port Houghton Bay, we heard a shout, and a few minutes later saw four Indians in a canoe paddling rapidly after us. In about an hour they overtook us. They were an Indian, his son, and two women with a load of fish-oil and dried salmon to sell and trade at Fort Wrangell. They camped within a dozen yards of us; with their sheets of cedar bark and poles they speedily made a hut, spread spruce boughs in it for a carpet, unloaded the canoe, and stored their goods under cover. Toward evening the old man came smiling with a gift for Toyatte,—a large fresh salmon, which was promptly boiled and eaten by our captain and crew as if it were only a light refreshment like a biscuit between meals. A few minutes after the big salmon had vanished, our generous neighbor came to Toyatte with a second gift of dried salmon, which after being toasted a few minutes tranquilly followed the fresh one as though it were a mere mouthful. Then, from the same generous hands, came a third gift,—a large milk-panful of huckleberries and grease boiled together,—and, strange to say, this wonderful mess went smoothly down to rest on the broad and deep salmon foundation. Thus refreshed, and appetite sharpened, my sturdy crew made haste to begin on the buck, beans, bread, etc., and, boiling and roasting, managed to get comfortably full on but little more than half of it by sundown, making a good deal of sport of my pity for the deer and refusing to eat any of it and nicknaming me the ice *ancou* and the deer and duck's *tillicum.*

Sunday was a wild, driving, windy day with but little rain but big promise of more. I took a walk back in the woods. The timber here is very fine, about as large as any I have seen in Alaska, much better than farther north. The Sitka spruce and the common hemlock, one hundred and fifty and two hundred feet high, are slender and handsome. The Sitka spruce makes good fire-wood even when green, the hemlock very poor. Back a little way from the sea, there was a good deal of yellow cedar, the best I had yet seen. The largest specimen that I saw and measured on the trip was five feet three inches in diameter and about one hundred and forty feet high. In the evening Mr. Young gave the Indians a lesson, calling in our Indian neighbors. He told them the story of Christ coming to save the world. The Indians wanted to know why the Jews had killed him. The lesson was listened to with very marked attention. Toyatte's generous friend caught a devil-fish about three feet in diameter to add to his stores of food. It would be very good, he said, when boiled in berry and colicon-oil soup. Each arm of this savage animal with its double row of button-like suction discs closed upon any object brought within reach with a grip nothing could escape. The Indians tell me that devil-fish live mostly on crabs, mussels, and clams, the shells of which they easily crunch with their strong, parrot-like beaks. That was a wild, stormy, rainy night. How the rain soaked us in our tents!

"Just feel that," said the minister in the night, as he took my hand and plunged it into a pool about three inches deep in which he was lying.

"Never mind," I said, "it is only water. Everything is wet now. It will soon be morning and we will dry at the fire."

Our Indian neighbors were, if possible, still wetter. Their hut had been blown down several times during the night. Our tent leaked badly, and we were lying in a mossy bog, but around the big campfire we were soon warm and half dry. We had expected to reach Wrangell by this time. Toyatte said the storm might last several days longer. We were out of tea and coffee, much to Mr. Young's distress. On my return from a walk I brought in a good big bunch of glandular ledum and boiled it in the teapot. The result of this experiment was a bright, clear amber-colored, rank-smelling liquor which I did not taste, but my suffering companion drank the whole potful and praised it. The rain was so heavy we decided not to attempt to leave camp until the storm somewhat abated, as we were assured by Toyatte that we would not be able to round Cape Fanshawe, a sheer, outjutting headland, the nose as he called it, past which the wind sweeps with great violence in these southeastern storms. With what grateful enthusiasm the trees welcomed the life-giving rain! Strong, towering spruces, hemlocks, and cedars tossed their arms, bowing, waving, in every leap, quivering and rejoicing together in the gray, roaring storm. John and Charley put on their gun-coats and went hunting for another deer, but returned later in the afternoon with clean hands, having fortunately failed to shed any more blood. The wind still held in the south, and Toyatte, grimly trying to comfort us, told us that we might be held here a week or more, which we should not have minded much, for we had abundance of provisions. Mr. Young and I shifted our tent and tried to dry blankets. The wind moderated considerably, and at 7 A.M. we started but met a rough sea and so stiff a wind we barely succeeded in rounding the cape by all hands pulling their best. Thence we struggled down the coast, creeping close to the shore and taking advantage of the shelter of protecting rocks, making slow, hard-won progress until about the middle of the afternoon, when the sky opened and the blessed sun shone out over the beautiful waters and forests with rich amber light; and the high, glacier-laden mountains, adorned with fresh snow, slowly came to view in all their grandeur, the bluish-gray clouds crawling and lingering and dissolving until every vestige of them vanished. The sunlight made the upper snow-fields pale creamy yellow, like that seen on the Chilcat mountains the first day of our return trip. Shortly after the sky cleared, the wind abated and changed around to the north, so that we ventured to hoist our sail, and then the weary Indians had rest. It was interesting to note how speedily the heavy swell that had been rolling for the last two or three days was subdued by the comparatively light breeze from the opposite direction. In a few minutes the sound was smooth and no trace of the storm was left, save the fresh snow and the discoloration of the water. All the water of the sound as far as I noticed was pale coffee-color like that of the streams in boggy woods. How much of this color was due to the inflow of the flooded streams many times increased in size and number by the rain, and how much to the beating of the waves along the shore stirring up vegetable matter in shallow bays, I cannot determine. The effect, however, was very marked.

About four o'clock we saw smoke on the shore and ran in for news. We found a company of Taku Indians, who were on their way to Fort Wrangell, some six men and about the same number of women. The men were sitting in a bark hut, handsomely reinforced and embowered with fresh spruce boughs. The women were out at the side of a stream, washing their many bits of calico. A little girl, six or seven years old, was sitting on the gravelly beach, building a playhouse of white quartz pebbles, scarcely caring to stop her work to gaze at us. Toyatte found a friend among the men, and wished to encamp beside them for the night, assuring us that this was the only safe harbor to be found within a good many miles. But we resolved to push on a little farther and make use of the smooth weather after being stormbound so long, much to Toyatte and his companion's disgust. We rowed about a couple of miles and ran into a cozy cove where wood and water were close at hand. How beautiful and home-like it was! plushy moss for mattresses decked with red cornel berries, noble spruce standing guard about us and spreading kindly protecting arms. A few ferns, aspidiums, polypodiums, with dewberry vines, coptis, pyrola, leafless huckleberry bushes, and ledum grow beneath the trees. We retired at eight o'clock, and just then Toyatte, who had been attentively studying the sky, presaged rain and another southeaster for the morrow.

The sky was a little cloudy next morning, but the air was still and the water smooth. We all hoped that Toyatte, the old weather prophet, had misread the sky signs. But before reaching Point Vanderpeut the rain began to fall and the dreaded southeast wind to blow, which soon increased to a stiff breeze, next thing to a gale, that lashed the sound into ragged white caps. Cape Vanderpeut is part of the terminal of an ancient glacier that once extended six or eight miles out from the base of the mountains. Three large glaciers that once were tributaries still descend nearly to the sea-level, though their fronts are back in narrow fiords, eight or ten miles from the sound. A similar point juts out into the sound five or six miles to the south, while the missing portion is submerged and forms a shoal.

All the cape is forested save a narrow strip about a mile long, composed of large boulders against which the waves beat with loud roaring. A bar of foam a mile or so farther out showed where the waves were breaking on a submerged part of the moraine, and I supposed that we would be compelled to pass around it in deep water, but Toyatte, usually so cautious, determined to cross it, and after giving particular directions, with an encouraging shout every oar and paddle was strained to shoot through a narrow gap. Just at the most critical point a big wave heaved us aloft and dropped us between two huge rounded boulders, where, had the canoe been a foot or two closer to either of them, it must have been smashed. Though I had offered no objection to our experienced pilot's plan, it looked dangerous, and I took the precaution to untie my shoes so they could be quickly shaken off for swimming. But after crossing the bar we were not yet out of danger, for we had to struggle hard to keep from being driven ashore while the waves were beating us broadside on. At length we discovered a little inlet, into which we gladly escaped. A pure-white iceberg, weathered to the form of a cross, stood amid drifts of kelp and the black rocks of the wave-beaten shore in sign of safety and welcome. A good fire soon warmed and dried us into common comfort. Our narrow escape was the burden of conversation as we sat around the fire. Captain Toyatte told us of two similar adventures while he was a strong young man. In both of them his canoe was smashed and he swam ashore out of the surge with a gun in his teeth. He says that if we had struck the rocks he and Mr. Young would have been drowned, all the rest of us probably would have been saved. Then, turning to me, he asked me if I could have made a fire in such a case without matches, and found a way to Wrangell without canoe or food.

We started about daybreak from our blessed white cross harbor, and, after rounding a bluff cape opposite the mouth of Wrangell Narrows, a fleet of icebergs came in sight, and of course I was eager to trace them to their source. Toyatte naturally enough was greatly excited about the safety of his canoe and begged that we should not venture to force a way through the bergs, risking the loss of the canoe and our lives now that we were so near the end of our long voyage.

"Oh, never fear, Toyatte," I replied. "You know we are always lucky—the weather is good. I only want to see the Thunder Glacier for a few minutes, and should the bergs be packed dangerously close, I promise to turn back and wait until next summer."

Thus assured, he pushed rapidly on until we entered the fiord, where we had to go cautiously slow. The bergs were close packed almost throughout the whole extent of the fiord, but we managed to reach a point about two miles from the head—commanding a good view of the down-plunging lower end of the glacier and blue, jagged ice-wall. This was one of the most imposing of the first-class glaciers I had as yet seen, and with its magnificent fiord formed a fine triumphant close for our season's ice work. I made a few notes and sketches and turned back in time to escape from the thickest packs of bergs before dark. Then Kadachan was stationed in the bow to guide through the open portion of the mouth of the fiord and across Soutchoi Strait. It was not until several hours after dark that we were finally free from ice. We occasionally encountered stranded packs on the delta, which in the starlight seemed to extend indefinitely in every direction. Our danger lay in breaking the canoe on small bergs hard to see and in getting too near the larger ones that might split or roll over.

"Oh, when will we escape from this ice?" moaned much-enduring old Toyatte.

We ran aground in several places in crossing the Stickeen delta, but finally succeeded in groping our way over muddy shallows before the tide fell, and encamped on the boggy shore of a small island, where we discovered a spot dry enough to sleep on, after tumbling about in a tangle of bushes and mossy logs.

We left our last camp November 21 at daybreak. The weather was calm and bright. Wrangell Island came into view beneath a lovely rosy sky, all the forest down to the water's edge silvery gray with a dusting of snow. John and Charley seemed to be seriously distressed to find themselves at the end of their journey while a portion of the stock of provisions remained uneaten. "What is to be done about it?" they asked, more than half in earnest. The fine, strong, and specious deliberation of Indians was well illustrated on this eventful trip. It was fresh every morning. They all behaved well, however, exerted themselves under tedious hardships without flinching for days or weeks at a time; never seemed in the least nonplussed; were prompt to act in every exigency; good as servants, fellow travelers, and even friends.

We landed on an island in sight of Wrangell and built a big smoky signal fire for friends in town, then set sail, unfurled our flag, and about noon completed our long journey of seven or eight hundred miles. As we approached the town, a large canoeful of friendly Indians came flying out to meet us, cheering and handshaking in lusty Boston fashion. The friends of Mr. Young had intended to come out in a body to welcome him back, but had not had time to complete their arrangements before we landed. Mr. Young was eager for news. I told him there could be no news of importance about a town. We only had real news, drawn from the wilderness. The mail steamer had left Wrangell eight days before, and Mr. Vanderbilt and family had sailed on her to Portland. I had to wait a month for the next steamer, and though I would have liked to go again to Nature, the mountains were locked for the winter and canoe excursions no longer safe.

So I shut myself up in a good garret alone to wait and work. I was invited to live with Mr. Young but concluded to prepare my own food and enjoy quiet work. How grandly long the nights were and short the days! At noon the sun seemed to be about an hour high, the clouds colored like sunset. The weather was rather stormy. North winds prevailed for a week at a time, sending down the temperature to near zero and chilling the vapor of the bay into white reek, presenting a curious appearance as it streamed forward on the wind, like combed wool. At Sitka the minimum was eight degrees plus; at Wrangell, near the storm-throat of the Stickeen, zero. This is said to be the coldest weather ever experienced in southeastern Alaska.

CHAPTER XIII

ALASKA INDIANS

LOOKING BACK ON my Alaska travels, I have always been glad that good luck gave me Mr. Young as a companion, for he brought me into confiding contact with the Thlinkit tribes, so that I learned their customs, what manner of men they were, how they lived and loved, fought and played, their morals, religion, hopes and fears, and superstitions, how they resembled and differed in their characteristics from our own and other races. It was easy to see that they differed greatly from the typical American Indian of the interior of this continent. They were doubtless derived from the Mongol stock. Their down-slanting oval eyes, wide cheek-bones, and rather thick, outstanding upper lips at once suggest their connection with the Chinese or Japanese. I have not seen a single specimen that looks in the least like the best of the Sioux, or indeed of any of the tribes to the east of the Rocky Mountains. They also differ from other North American Indians in being willing to work, when free from the contamination of bad whites. They manage to feed themselves well, build good substantial houses, bravely fight their enemies, love their wives and children and friends, and cherish a quick sense of honor. The best of them prefer death to dishonor, and sympathize with their neighbors in their misfortunes and sorrows. Thus when a family loses a child by death, neighbors visit them to cheer and console. They gather around the fire and smoke, talk kindly and naturally, telling the sorrowing parents not to grieve too much, reminding them of the better lot of their child in another world and of the troubles and trials the little ones escape by dying young, all this in a perfectly natural, straightforward way, wholly unlike the vacant, silent, hesitating behavior of most civilized friends, who oftentimes in such cases seem nonplussed, awkward, and afraid to speak, however sympathetic.

The Thlinkits are fond and indulgent parents. In all my travels I never heard a cross, fault-finding word, or anything like scolding inflicted on an Indian child, or ever witnessed a single case of spanking, so common in civilized communities. They consider the want of a son to bear their name and keep it alive the saddest and most deplorable ill-fortune imaginable.

The Thlinkit tribes give a hearty welcome to Christian missionaries. In particular they are quick to accept the doctrine of the atonement, because they themselves practice it, although to many of the civilized whites it is a stumbling-block and rock of offense. As an example of their own doctrine of atonement they told Mr. Young and me one evening that twenty or thirty years ago there was a bitter war between their own and the Sitka tribe, great fighters, and pretty evenly matched. After fighting all summer in a desultory, squabbling way, fighting now under cover, now in the open, watching for every chance for a shot, none of the women dared venture to the salmon-streams or berry-fields to procure their winter stock of food. At this crisis one of the Stickeen chiefs came out of his block-house fort into an open space midway between their fortified camps, and shouted that he wished to speak to the leader of the Sitkas.

When the Sitka chief appeared he said:—

"My people are hungry. They dare not go to the salmon-streams or berry-fields for winter supplies, and if this war goes on much longer most of my people will die of hunger. We have fought long enough; let us make peace. You brave Sitka warriors go home, and we will go home, and we will all set out to dry salmon and berries before it is too late."

The Sitka chief replied:—

"You may well say let us stop fighting, when you have had the best of it. You have killed ten more of my tribe than we have killed of yours. Give us ten Stickeen men to balance our blood-account; then, and not till then, will we make peace and go home."

"Very well," replied the Stickeen chief, "you know my rank. You know that I am worth ten common men and more. Take me and make peace."

This noble offer was promptly accepted; the Stickeen chief stepped forward and was shot down in sight of the fighting bands. Peace was thus established, and all made haste to their homes and ordinary work. That chief literally gave himself a sacrifice for his people. He died that they might live. Therefore, when missionaries preached the doctrine of atonement, explaining that when all mankind had gone astray, had broken God's laws and deserved to die, God's son came forward, and, like the Stickeen chief, offered himself as a sacrifice to heal the cause of God's wrath and set all the people of the world free, the doctrine was readily accepted.

"Yes, your words are good," they said. "The Son of God, the Chief of chiefs, the Maker of all the world, must be worth more than all mankind put together; therefore, when His blood was shed, the salvation of the world was made sure."

A telling illustration of the ready acceptance of this doctrine was displayed by Shakes, head chief of the Stickeens at Fort Wrangell. A few years before my first visit to the Territory, when the first missionary arrived, he requested Shakes to call his people together to hear the good word he had brought them. Shakes accordingly sent out messengers throughout the village, telling his people to wash their faces, put on their best clothing, and come to his block-house to hear what their visitor had to say. When all were assembled, the missionary preached a Christian sermon on the fall of man and the atonement whereby Christ, the Son of God, the Chief of chiefs, had redeemed all mankind, provided that this redemption was voluntarily accepted with repentance of their sins and the keeping of his commandments.

When the missionary had finished his sermon, Chief Shakes slowly arose, and, after thanking the missionary for coming so far to bring them good tidings and taking so much unselfish interest in the welfare of his tribe, he advised his people to accept the new religion, for he felt satisfied that because the white man knew so much more than the Indian, the white man's religion was likely to be better than theirs.

"The white man," said he, "makes great ships. We, like children, can only make canoes. He makes his big ships go with the wind, and he also makes them go with fire. We chop down trees with stone axes; the Boston man with iron axes, which are far better. In everything the ways of the white man seem to be better than ours. Compared with the white man we are only blind children, knowing not how best to live either here or in the country we go to after we die. So I wish you to learn this new religion and teach it to your children, that you may all go when you die into that good heaven country of the white man and be happy. But I am too old to learn a new religion, and besides, many of my people who have died were bad and foolish people, and if this word the missionary has brought us is true, and I think it is, many of my people must be in that bad country the missionary calls 'Hell,' and I must go there also, for a Stickeen chief never deserts his people in time of trouble. To that bad country, therefore, I will go, and try to cheer my people and help them as best I can to endure their misery."

Toyatte was a famous orator. I was present at the meeting at Fort Wrangell at which he was examined and admitted as a member of the Presbyterian Church. When called upon to answer the questions as to his ideas of God, and the principal doctrines of Christianity, he slowly arose in the crowded audience, while the missionary said, "Toyatte, you do not need to rise. You can answer the questions seated."

To this he paid no attention, but stood several minutes without speaking a word, never for a moment thinking of sitting down like a tired woman while making the most important of all the speeches of his life. He then explained in detail what his mother had taught him as to the character of God, the great Maker of the world; also what the shamans had taught him; the thoughts that often came to his mind when he was alone on hunting expeditions, and what he first thought of the religion which the missionaries had brought them. In all his gestures, and in the language in which he expressed himself, there was a noble simplicity and earnestness and majestic bearing which made the sermons and behavior of the three distinguished divinity doctors present seem commonplace in comparison.

Soon after our return to Fort Wrangell this grand old man was killed in a quarrel in which he had taken no other part than that of peacemaker. A number of the Taku tribe came to Fort Wrangell, camped near the Stickeen village, and made merry, manufacturing and drinking *hootchenoo*, a vile liquor distilled from a mash made of flour, dried apples, sugar, and molasses, and drunk hot from the still. The manufacture of *hootchenoo* being illegal, and several of Toyatte's tribe having been appointed deputy constables to prevent it, they went to the Taku camp and destroyed as much of the liquor as they could find. The Takus resisted, and during the quarrel one of the Stickeens struck a Taku in the face—an unpardonable offense. The next day messengers from the Taku camp gave notice to the Stickeens that they must make atonement for that blow, or fight with guns. Mr. Young, of course, was eager to stop the quarrel and so was Toyatte. They advised the Stickeen who had struck the Taku to return to their camp and submit to an equal blow in the face from the Taku. He did so; went to the camp, said he was ready to make atonement, and invited the person whom he had struck to strike him. This the Taku did with so much force that the balance of justice was again disturbed. The attention of the Takus was called to the fact that this atoning blow was far harder than the one to be atoned for, and immediately a sort of general free fist-fight began, and the quarrel was thus increased in bitterness rather than diminished.

Next day the Takus sent word to the Stickeens to get their guns ready, for to-morrow they would come up and fight them, thus boldly declaring war. The Stickeens in great excitement assembled and loaded their guns for the coming strife. Mr. Young ran hither and thither amongst the men of his congregation, forbidding them to fight, reminding them that Christ told them when they were struck to offer the other cheek instead of giving a blow in return, doing everything in his power to still the storm, but all in vain. Toyatte stood outside one of the big block-houses with his men about him, awaiting the onset of the Takus. Mr. Young tried hard to get him away to a place of safety, reminding him that he belonged to his church and no longer had any right to fight. Toyatte calmly replied:—

"Mr. Young, Mr. Young, I am not going to fight. You see I have no gun in my hand; but I cannot go inside of the fort to a place of safety like women and children while my young men are exposed to the bullets of their enemies. I must stay with them and share their dangers, but I will not fight. But you, Mr. Young, *you* must go away; you are a minister and you are an important man. It would not do for you to be exposed to bullets. Go to your home in the fort; pretty soon 'hi yu poogh'" (much shooting).

At the first fire Toyatte fell, shot through the breast. Thus died for his people the noblest old Roman of them all.

On this first Alaska excursion I saw Toyatte under all circumstances,—in rain and snow, landing at night in dark storms, making fires, building shelters, exposed to all kinds of discomfort, but never under any circumstances did I ever see him do anything, or make a single gesture, that was not dignified, or hear him say a word that might not be uttered anywhere. He often deplored the fact that he had no son to take his name at his death, and expressed himself as very grateful when I told him that his name would not be forgotten,—that I had named one of the Stickeen glaciers for him.

PART TWO

THE TRIP OF 1880

CHAPTER XIV

SUM DUM BAY

I ARRIVED EARLY on the morning of the eighth of August on the steamer California to continue my explorations of the fiords to the northward which were closed by winter the previous November. The noise of our cannon and whistle was barely sufficient to awaken the sleepy town. The morning shout of one good rooster was the only evidence of life and health in all the place. Everything seemed kindly and familiar—the glassy water; evergreen islands; the Indians with their canoes and baskets and blankets and berries; the jet ravens, prying and flying about the streets and spruce trees; and the bland, hushed atmosphere brooding tenderly over all.

How delightful it is, and how it makes one's pulses bound to get back into this reviving northland wilderness! How truly wild it is, and how joyously one's heart responds to the welcome it gives, its waters and mountains shining and glowing like enthusiastic human faces! Gliding along the shores of its network of channels, we may travel thousands of miles without seeing any mark of man, save at long intervals some little Indian village or the faint smoke of a campfire. Even these are confined to the shore. Back a few yards from the beach the forests are as trackless as the sky, while the mountains, wrapped in their snow and ice and clouds, seem never before to have been even looked at.

For those who really care to get into hearty contact with the coast region, travel by canoe is by far the better way. The larger canoes carry from one to three tons, rise lightly over any waves likely to be met on the inland channels, go well under sail, and are easily paddled alongshore in calm weather or against moderate winds, while snug harbors where they may ride at anchor or be pulled up on a smooth beach are to be found almost everywhere. With plenty of provisions packed in boxes, and blankets and warm clothing in rubber or canvas bags, you may be truly independent, and enter into partnership with Nature; to be carried with the winds and currents, accept the noble invitations offered all along your way to enter the mountain fiords, the homes of the waterfalls and glaciers, and encamp almost every night beneath hospitable trees.

I left Fort Wrangell the 16th of August, accompanied by Mr. Young, in a canoe about twenty-five feet long and five wide, carrying two small square sails and manned by two Stickeen Indians—Captain Tyeen and Hunter Joe—and a half-breed named Smart Billy. The day was calm, and bright, fleecy clouds hung about the lowest of the mountain-brows, while far above the clouds the peaks were seen stretching grandly away to the northward with their ice and snow shining in as calm a light as that which was falling on the glassy waters. Our Indians welcomed the work that lay before them, dipping their oars in exact time with hearty good-will as we glided past island after island across the delta of the Stickeen into Soutchoi Channel.

By noon we came in sight of a fleet of icebergs from Hutli Bay. The Indian name of this icy fiord is Hutli, or Thunder Bay, from the sound made by the bergs in falling and rising from the front of the inflowing glacier.

As we floated happily on over the shining waters, the beautiful islands, in ever-changing pictures, were an unfailing source of enjoyment; but chiefly our attention was turned upon the mountains. Bold granite headlands with their feet in the channel, or some broad-shouldered peak of surpassing grandeur, would fix the eye, or some one of the larger glaciers, with far-reaching tributaries clasping entire groups of peaks and its great crystal river pouring down through the forest between gray ridges and domes. In these grand picture lessons the day was spent, and we spread our blankets beneath a Menzies spruce on moss two feet deep.

Next morning we sailed around an outcurving bank of boulders and sand ten miles long, the terminal moraine of a grand old glacier on which last November we met a perilous adventure. It is located just opposite three large converging glaciers which formerly united to form the vanished trunk of the glacier to which the submerged moraine belonged. A few centuries ago it must have been the grandest feature of this part of the coast, and, so well preserved are the monuments of its greatness, the noble old ice-river may be seen again in imagination about as vividly as if present in the flesh, with snow-clouds crawling about its fountains, sunshine sparkling on its broad flood, and its ten-mile ice-wall planted in the deep waters of the channel and sending off its bergs with loud resounding thunder.

About noon we rounded Cape Fanshawe, scudding swiftly before a fine breeze, to the delight of our Indians, who had now only to steer and chat. Here we overtook two Hoona Indians and their families on their way home from Fort Wrangell. They had exchanged five sea-otter furs, worth about a hundred dollars apiece, and a considerable number of fur-seal, land-otter, marten, beaver, and other furs and skins, some $800 worth, for a new canoe valued at eighty dollars, some flour, tobacco, blankets, and a few barrels of molasses for the manufacture of whiskey. The blankets were not to wear, but to keep as money, for the almighty dollar of these tribes is a Hudson's Bay blanket. The wind died away soon after we met, and as the two canoes glided slowly side by side, the Hoonas made minute inquiries as to who we were and what we were doing so far north. Mr. Young's object in meeting the Indians as a missionary they could in part understand, but mine in searching for rocks and glaciers seemed past comprehension, and they asked our Indians whether gold-mines might not be the main object. They remembered, however, that I had visited their Glacier Bay ice-mountains a year ago, and seemed to think there might be, after all, some mysterious interest about them of which they were ignorant. Toward the middle of the afternoon they engaged our crew in a race. We pushed a little way ahead for a time, but, though possessing a considerable advantage, as it would seem, in our long oars, they at length overtook us and kept up until after dark, when we camped together in the rain on the bank of a salmon-stream among dripping grass and bushes some twenty-five miles beyond Cape Fanshawe.

These cold northern waters are at times about as brilliantly phosphorescent as those of the warm South, and so they were this evening in the rain and darkness, with the temperature of the water at forty-nine degrees, the air fifty-one. Every stroke of the oar made a vivid surge of white light, and the canoes left shining tracks.

As we neared the mouth of the well-known salmon-stream where we intended making our camp, we noticed jets and flashes of silvery light caused by the startled movement of the salmon that were on their way to their spawning-grounds. These became more and more numerous and exciting, and our Indians shouted joyfully, "Hi yu salmon! Hi yu muck-a-muck!" while the water about the canoe and beneath the canoe was churned by thousands of fins into silver fire. After landing two of our men to commence camp-work, Mr. Young and I went up the stream with Tyeen to the foot of a rapid, to see him catch a few salmon for supper. The stream was so filled with them there seemed to be more fish than water in it, and we appeared to be sailing in boiling, seething silver light marvelously relieved in the jet darkness. In the midst of the general auroral glow and the specially vivid flashes made by the frightened fish darting ahead and to right and left of the canoe, our attention was suddenly fixed by a long, steady, comet-like blaze that seemed to be made by some frightful monster that was pursuing us. But when the portentous object reached the canoe, it proved to be only our little dog, Stickeen.

After getting the canoe into a side eddy at the foot of the rapids, Tyeen caught half a dozen salmon in a few minutes by means of a large hook fastened to the end of a pole. They were so abundant that he simply groped for them in a random way, or aimed at them by the light they themselves furnished. That food to last a month or two may thus be procured in less than an hour is a striking illustration of the fruitfulness of these Alaskan waters.

Our Hoona neighbors were asleep in the morning at sunrise, lying in a row, wet and limp like dead salmon. A little boy about six years old, with no other covering than a remnant of a shirt, was lying peacefully on his back, like Tam o' Shanter, despising wind and rain and fire. He is up now looking happy and fresh, with no clothes to dry and no need of washing while this weather lasts. The two babies are firmly strapped on boards, leaving only their heads and hands free. Their mothers are nursing them, holding the boards on end, while they sit on the ground with their breasts level with the little prisoners' mouths.

This morning we found out how beautiful a nook we had got into. Besides the charming picturesqueness of its lines, the colors about it, brightened by the rain, made a fine study. Viewed from the shore, there was first a margin of dark-brown algæ, then a bar of yellowish-brown, next a dark bar on the rugged rocks marking the highest tides, then a bar of granite boulders with grasses in the seams, and above this a thick, bossy, overleaning fringe of bushes colored red and yellow and green. A wall of spruces and hemlocks draped and tufted with gray and yellow lichens and mosses embowered the campground and overarched the little river, while the camp-fire smoke, like a stranded cloud, lay motionless in their branches. Down on the beach ducks and sandpipers in flocks of hundreds were getting their breakfasts, bald eagles were seen perched on dead spars along the edge of the woods, heavy-looking and overfed, gazing stupidly like gorged vultures, and porpoises were blowing and plunging outside.

As for the salmon, as seen this morning urging their way up the swift current,—tens of thousands of them, side by side, with their backs out of the water in shallow places now that the tide was low,—nothing that I could write might possibly give anything like a fair conception of the extravagance of their numbers. There was more salmon apparently, bulk for bulk, than water in the stream. The struggling multitudes, crowding one against another, could not get out of our way when we waded into the midst of them. One of our men amused himself by seizing them above the tail and swinging them over his head. Thousands could thus be taken by hand at low tide, while they were making their way over the shallows among the stones.

Whatever may be said of other resources of the Territory, it is hardly possible to exaggerate the importance of the fisheries. Not to mention cod, herring, halibut, etc., there are probably not less than a thousand salmon-streams in southeastern Alaska as large or larger than this one (about forty feet wide) crowded with salmon several times a year. The first run commenced that year in July, while the king salmon, one of the five species recognized by the Indians, was in the Chilcat River about the middle of the November before.

From this wonderful salmon-camp we sailed joyfully up the coast to explore icy Sum Dum Bay, beginning my studies where I left off the previous November. We started about six o'clock, and pulled merrily on through fog and rain, the beautiful wooded shore on our right, passing bergs here and there, the largest of which, though not over two hundred feet long, seemed many times larger as they loomed gray and indistinct through the fog. For the first five hours the sailing was open and easy, nor was there anything very exciting to be seen or heard, save now and then the thunder of a falling berg rolling and echoing from cliff to cliff, and the sustained roar of cataracts.

About eleven o'clock we reached a point where the fiord was packed with ice all the way across, and we ran ashore to fit a block of wood on the cutwater of our canoe to prevent its being battered or broken. While Captain Tyeen, who had had considerable experience among berg ice, was at work on the canoe, Hunter Joe and Smart Billy prepared a warm lunch.

The sheltered hollow where we landed seems to be a favorite camping-ground for the Sum Dum seal-hunters. The pole-frames of tents, tied with cedar bark, stood on level spots strewn with seal bones, bits of salmon, and spruce bark.

We found the work of pushing through the ice rather tiresome. An opening of twenty or thirty yards would be found here and there, then a close pack that had to be opened by pushing the smaller bergs aside with poles. I enjoyed the labor, however, for the fine lessons I got, and in an hour or two we found zigzag lanes of water, through which we paddled with but little interruption, and had leisure to study the wonderful variety of forms the bergs presented as we glided past them. The largest we saw did not greatly exceed two hundred feet in length, or twenty-five or thirty feet in height above the water. Such bergs would draw from one hundred and fifty to two hundred feet of water. All those that have floated long undisturbed have a projecting base at the water-line, caused by the more rapid melting of the immersed portion. When a portion of the berg breaks off, another base line is formed, and the old one, sharply cut, may be seen rising at all angles, giving it a marked character. Many of the oldest bergs are beautifully ridged by the melting out of narrow furrows strictly parallel throughout the mass, revealing the bedded structure of the ice, acquired perhaps centuries ago, on the mountain snow fountains. A berg suddenly going to pieces is a grand sight, especially when the water is calm and no motion is visible save perchance the slow drift of the tide-current. The prolonged roar of its fall comes with startling effect, and heavy swells are raised that haste away in every direction to tell what has taken place, and tens of thousands of its neighbors rock and swash in sympathy, repeating the news over and over again. We were too near several large ones that fell apart as we passed them, and our canoe had narrow escapes. The seal-hunters, Tyeen says, are frequently lost in these sudden berg accidents.

In the afternoon, while we were admiring the scenery, which, as we approached the head of the fiord, became more and more sublime, one of our Indians called attention to a flock of wild goats on a mountain overhead, and soon afterwards we saw two other flocks, at a height of about fifteen hundred feet, relieved against the mountains as white spots. They are abundant here and throughout the Alaskan Alps in general, feeding on the grassy slopes above the timber line. Their long, yellowish hair is shed at this time of year and they were snowy white. None of Nature's cattle are better fed or better protected from the cold. Tyeen told us that before the introduction of guns they used to hunt them with spears, chasing them with their wolf-dogs, and thus bringing them to bay among the rocks, where they were easily approached and killed.

The upper half of the fiord is about from a mile to a mile and a half wide, and shut in by sublime Yosemite cliffs, nobly sculptured, and adorned with waterfalls and fringes of trees, bushes, and patches of flowers; but amid so crowded a display of novel beauty it was not easy to concentrate the attention long enough on any portion of it without giving more days and years than our lives could afford. I was determined to see at least the grand fountain of all this ice. As we passed headland after headland, hoping as each was rounded we should obtain a view of it, it still remained hidden.

"Ice-mountain hi yu kumtux hide,"—glaciers know how to hide extremely well,—said Tyeen, as he rested for a moment after rounding a huge granite shoulder of the wall whence we expected to gain a view of the extreme head of the fiord. The bergs, however, were less closely packed and we made good progress, and at half-past eight o'clock, fourteen and a half hours after setting out, the great glacier came in sight at the head of a branch of the fiord that comes in from the northeast.

The discharging front of this fertile, fast-flowing glacier is about three quarters of a mile wide, and probably eight or nine hundred feet deep, about one hundred and fifty feet of its depth rising above the water as a grand blue barrier wall. It is much wider a few miles farther back, the front being jammed between sheer granite walls from thirty-five hundred to four thousand feet high. It shows grandly from where it broke on our sight, sweeping boldly forward and downward in its majestic channel, swaying from side to side in graceful fluent lines around stern unflinching rocks. While I stood in the canoe making a sketch of it, several bergs came off with tremendous dashing and thunder, raising a cloud of ice-dust and spray to a height of a hundred feet or more.

"The ice-mountain is well disposed toward you," said Tyeen. "He is firing his big guns to welcome you."

After completing my sketch and entering a few notes, I directed the crew to pull around a lofty burnished rock on the west side of the channel, where, as I knew from the trend of the cañon, a large glacier once came in; and what was my delight to discover that the glacier was still there and still pouring its ice into a branch of the fiord. Even the Indians shared my joy and shouted with me. I expected only one first-class glacier here, and found two. They are only about two miles apart. How glorious a mansion that precious pair dwell in! After sunset we made haste to seek a campground. I would fain have shared these upper chambers with the two glaciers, but there was no landing-place in sight, and we had to make our way back a few miles in the twilight to the mouth of a side cañon where we had seen timber on the way up. There seemed to be a good landing as we approached the shore, but, coming nearer, we found that the granite fell directly into deep water without leading any level margin, though the slope a short distance back was not very steep.

After narrowly scanning the various seams and steps that roughened the granite, we concluded to attempt a landing rather than grope our way farther down the fiord through the ice. And what a time we had climbing on hands and knees up the slippery glacier-polished rocks to a shelf some two hundred feet above the water and dragging provisions and blankets after us! But it proved to be a glorious place, the very best campground of all the trip,—a perfect garden, ripe berries nodding from a fringe of bushes around its edges charmingly displayed in the light of our big fire. Close alongside there was a lofty mountain capped with ice, and from the blue edge of that ice-cap there were sixteen silvery cascades in a row, falling about four thousand feet, each one of the sixteen large enough to be heard at least two miles.

How beautiful was the firelight on the nearest larkspurs and geraniums and daisies of our garden! How hearty the wave greeting on the rocks below brought to us from the two glaciers! And how glorious a song the sixteen cascades sang!

The cascade songs made us sleep all the sounder, and we were so happy as to find in the morning that the berg waves had spared our canoe. We set off in high spirits down the fiord and across to the right side to explore a remarkably deep and narrow branch of the main fiord that I had noted on the way up, and that, from the magnitude of the glacial characters on the two colossal rocks that guard the entrance, promised a rich reward for our pains.

After we had sailed about three miles up this side fiord, we came to what seemed to be its head, for trees and rocks swept in a curve around from one side to the other without showing any opening, although the walls of the cañon were seen extending back indefinitely, one majestic brow beyond the other.

When we were tracing this curve, however, in a leisurely way, in search of a good landing, we were startled by Captain Tyeen shouting, "Skookum chuck! Skookum chuck!" (strong water, strong water), and found our canoe was being swept sideways by a powerful current, the roar of which we had mistaken for a waterfall. We barely escaped being carried over a rocky bar on the boiling flood, which, as we afterwards learned, would have been only a happy shove on our way. After we had made a landing a little distance back from the brow of the bar, we climbed the highest rock near the shore to seek a view of the channel beyond the inflowing tide rapids, to find out whether or not we could safely venture in. Up over rolling, mossy, bushy, burnished rock waves we scrambled for an hour or two, which resulted in a fair view of the deep-blue waters of the fiord stretching on and on along the feet of the most majestic Yosemite rocks we had yet seen. This determined our plan of shooting the rapids and exploring it to its farthest recesses. This novel interruption of the channel is a bar of exceedingly hard resisting granite, over which the great glacier that once occupied it swept, without degrading it to the general level, and over which tide-waters now rush in and out with the violence of a mountain torrent.

Returning to the canoe, we pushed off, and in a few moments were racing over the bar with lightning speed through hurrahing waves and eddies and sheets of foam, our little shell of a boat tossing lightly as a bubble. Then, rowing across a belt of back-flowing water, we found ourselves on a smooth mirror reach between granite walls of the very wildest and most exciting description, surpassing in some ways those of the far-famed Yosemite Valley.

As we drifted silent and awe-stricken beneath the shadows of the mighty cliffs, which, in their tremendous height and abruptness, seemed to overhang at the top, the Indians gazing intently, as if they, too, were impressed with the strange, awe-inspiring grandeur that shut them in, one of them at length broke the silence by saying, "This must be a good place for woodchucks; I hear them calling."

When I asked them, further on, how they thought this gorge was made, they gave up the question, but offered an opinion as to the formation of rain and soil. The rain, they said, was produced by the rapid whirling of the earth by a stout mythical being called Yek. The water of the ocean was thus thrown up, to descend again in showers, just as it is thrown off a wet grindstone. They did not, however, understand why the ocean water should be salt, while the rain from it is fresh. The soil, they said, for the plants to grow on is formed by the washing of the rain on the rocks and gradually accumulating. The grinding action of ice in this connection they had not recognized.

Gliding on and on, the scenery seemed at every turn to become more lavishly fruitful in forms as well as more sublime in dimensions—snowy falls booming in splendid dress; colossal domes and battlements and sculptured arches of a fine neutral-gray tint, their bases laved by the blue fiord water; green ferny dells; bits of flower-bloom on ledges; fringes of willow and birch; and glaciers above all. But when we approached the base of a majestic rock like the Yosemite Half Dome at the head of the fiord, where two short branches put out, and came in sight of another glacier of the first order sending off bergs, our joy was complete. I had a most glorious view of it, sweeping in grand majesty from high mountain fountains, swaying around one mighty bastion after another, until it fell into the fiord in shattered over-leaning fragments. When we had feasted awhile on this unhoped-for treasure, I directed the Indians to pull to the head of the left fork of the fiord, where we found a large cascade with a volume of water great enough to be called a river, doubtless the outlet of a receding glacier not in sight from the fiord.

This is in form and origin a typical Yosemite valley, though as yet its floor is covered with ice and water,—ice above and beneath, a noble mansion in which to spend a winter and a summer! It is about ten miles long, and from three quarters of a mile to one mile wide. It contains ten large falls and cascades, the finest one on the left side near the head. After coming in an admirable rush over a granite brow where it is first seen at a height of nine hundred or a thousand feet, it leaps a sheer precipice of about two hundred and fifty feet, then divides and reaches the tide-water in broken rapids over boulders. Another about a thousand feet high drops at once on to the margin of the glacier two miles back from the front. Several of the others are upwards of three thousand feet high, descending through narrow gorges as richly feathered with ferns as any channel that water ever flowed in, though tremendously abrupt and deep. A grander array of rocks and waterfalls I have never yet beheld in Alaska.

The amount of timber on the walls is about the same as that on the Yosemite walls, but owing to greater moisture, there is more small vegetation,—bushes, ferns, mosses, grasses, etc.; though by far the greater portion of the area of the wall-surface is bare and shining with the polish it received when occupied by the glacier that formed the fiord. The deep-green patches seen on the mountains back of the walls at the limits of vegetation are grass, where the wild goats, or chamois rather, roam and feed. The still greener and more luxuriant patches farther down in gullies and on slopes where the declivity is not excessive, are made up mostly of willows, birch, and huckleberry bushes, with a varying amount of prickly ribes and rubus and echinopanax. This growth, when approached, especially on the lower slopes near the level of the sea at the jaws of the great side cañons, is found to be the most impenetrable and tedious and toilsome combination of fighting bushes that the weary explorer ever fell into, incomparably more punishing than the buckthorn and manzanita tangles of the Sierra.

The cliff gardens of this hidden Yosemite are exceedingly rich in color. On almost every rift and bench, however small, as well as on the wider table-rocks where a little soil has lodged, we found gay multitudes of flowers, far more brilliantly colored than would be looked for in so cool and beclouded a region,—larkspurs, geraniums, painted-cups, bluebells, gentians, saxifrages, epilobiums, violets, parnassia, veratrum, spiranthes and other orchids, fritillaria, smilax, asters, daisies, bryanthus, cassiope, linnæa, and a great variety of flowering ribes and rubus and heathworts. Many of the above, though with soft stems and leaves, are yet as brightly painted as those of the warm sunlands of the south. The heathworts in particular are very abundant and beautiful, both in flower and fruit, making delicate green carpets for the rocks, flushed with pink bells, or dotted with red and blue berries. The tallest of the grasses have ribbon leaves well tempered and arched, and with no lack of bristly spikes and nodding purple panicles. The alpine grasses of the Sierra, making close carpets on the glacier meadows, I have not yet seen in Alaska.

The ferns are less numerous in species than in California, but about equal in the number of fronds. I have seen three aspidiums, two woodsias, a lomaria, polypodium, cheilanthes, and several species of pteris.

In this eastern arm of Sum Dum Bay and its Yosemite branch, I counted from my canoe, on my way up and down, thirty small glaciers back of the walls, and we saw three of the first order; also thirty-seven cascades and falls, counting only those large enough to make themselves heard several miles. The whole bay, with its rocks and woods and ice, reverberates with their roar. How many glaciers may be disclosed in the other great arm that I have not seen as yet, I cannot say, but, judging from the bergs it sends down, I guess not less than a hundred pour their turbid streams into the fiord, making about as many joyful, bouncing cataracts.

About noon we began to retrace our way back into the main fiord, and arrived at the gold-mine camp after dark, rich and weary.

On the morning of August 21 I set out with my three Indians to explore the right arm of this noble bay, Mr. Young having decided, on account of mission work, to remain at the gold-mine. So here is another fine lot of Sum Dum ice,—thirty-five or forty square miles of bergs, one great glacier of the first class descending into the fiord at the head, the fountain whence all these bergs were derived, and thirty-one smaller glaciers that do not reach tide-water; also nine cascades and falls, large size, and two rows of Yosemite rocks from three to four thousand feet high, each row about eighteen or twenty miles long, burnished and sculptured in the most telling glacier style, and well trimmed with spruce groves and flower gardens; a' that and more of a kind that cannot here be catalogued.

For the first five or six miles there is nothing excepting the icebergs that is very striking in the scenery as compared with that of the smooth unencumbered outside channels, where all is so evenly beautiful. The mountain-wall on the right as you go up is more precipitous than usual, and a series of small glaciers is seen along the top of it, extending their blue-crevassed fronts over the rims of pure-white snow fountains, and from the end of each front a hearty stream coming in a succession of falls and rapids over the terminal moraines, through patches of dwarf willows, and then through the spruce woods into the bay, singing and dancing all the way down. On the opposite side of the bay from here there is a small side bay about three miles deep, with a showy group of glacier-bearing mountains back of it. Everywhere else the view is bounded by comparatively low mountains densely forested to the very top.

After sailing about six miles from the mine, the experienced mountaineer could see some evidence of an opening from this wide lower portion, and on reaching it, it proved to be the continuation of the main west arm, contracted between stupendous walls of gray granite, and crowded with bergs from shore to shore, which seem to bar the way against everything but wings. Headland after headland, in most imposing array, was seen plunging sheer and bare from dizzy heights, and planting its feet in the ice-encumbered water without leaving a spot on which one could land from a boat, while no part of the great glacier that pours all these miles of ice into the fiord was visible. Pushing our way slowly through the packed bergs, and passing headland after headland, looking eagerly forward, the glacier and its fountain mountains were still beyond sight, cut off by other projecting headland capes, toward which I urged my way, enjoying the extraordinary grandeur of the wild unfinished Yosemite. Domes swell against the sky in fine lines as lofty and as perfect in form as those of the California valley, and rock-fronts stand forward, as sheer and as nobly sculptured. No ice-work that I have ever seen surpasses this, either in the magnitude of the features or effectiveness of composition.

On some of the narrow benches and tables of the walls rows of spruce trees and two-leaved pines were growing, and patches of considerable size were found on the spreading bases of those mountains that stand back inside the cañons, where the continuity of the walls is broken. Some of these side cañons are cut down to the level of the water and reach far back, opening views into groups of glacier fountains that give rise to many a noble stream; while all along the tops of the walls on both sides small glaciers are seen, still busily engaged in the work of completing their sculpture. I counted twenty-five from the canoe. Probably the drainage of fifty or more pours into this fiord. The average elevation at which they melt is about eighteen hundred feet above sea-level, and all of them are residual branches of the grand trunk that filled the fiord and overflowed its walls when there was only one Sum Dum glacier.

The afternoon was wearing away as we pushed on and on through the drifting bergs without our having obtained a single glimpse of the great glacier. A Sum Dum seal-hunter, whom we met groping his way deftly through the ice in a very small, unsplitable cottonwood canoe, told us that the ice-mountain was yet fifteen miles away. This was toward the middle of the afternoon, and I gave up sketching and making notes and worked hard with the Indians to reach it before dark. About seven o'clock we approached what seemed to be the extreme head of the fiord, and still no great glacier in sight—only a small one, three or four miles long, melting a thousand feet above the sea. Presently, a narrow side opening appeared between tremendous cliffs sheer to a height of four thousand feet or more, trending nearly at right angles to the general trend of the fiord, and apparently terminated by a cliff, scarcely less abrupt or high, at a distance of a mile or two. Up this bend we toiled against wind and tide, creeping closely along the wall on the right side, which, as we looked upward, seemed to be leaning over, while the waves beating against the bergs and rocks made a discouraging kind of music. At length, toward nine o'clock, just before the gray darkness of evening fell, a long, triumphant shout told that the glacier, so deeply and desperately hidden, was at last hunted back to its benmost bore. A short distance around a second bend in the cañon, I reached a point where I obtained a good view of it as it pours its deep, broad flood into the fiord in a majestic course from between the noble mountains, its tributaries, each of which would be regarded elsewhere as a grand glacier, converging from right and left from a fountain set far in the silent fastnesses of the mountains.

"There is your lost friend," said the Indians laughing; "he says, 'Sagh-a-ya'" (how do you do)? And while berg after berg was being born with thundering uproar, Tyeen said, "Your friend has klosh tumtum [good heart]. Hear! Like the other big-hearted one he is firing his guns in your honor."

I stayed only long enough to make an outline sketch, and then urged the Indians to hasten back some six miles to the mouth of a side cañon I had noted on the way up as a place where we might camp in case we should not find a better. After dark we had to move with great caution through the ice. One of the Indians was stationed in the bow with a pole to push aside the smaller fragments and look out for the most promising openings, through which he guided us, shouting, "Friday! Tucktay!" (shoreward, seaward) about ten times a minute. We reached this landing-place after ten o'clock, guided in the darkness by the roar of a glacier torrent. The ground was all boulders and it was hard to find a place among them, however small, to lie on. The Indians anchored the canoe well out from the shore and passed the night in it to guard against berg-waves and drifting waves, after assisting me to set my tent in some sort of way among the stones well back beyond the reach of the tide. I asked them as they were returning to the canoe if they were not going to eat something. They answered promptly:—

"We will sleep now, if your ice friend will let us. We will eat to-morrow, but we can find some bread for you if you want it."

"No," I said, "go to rest. I, too, will sleep now and eat tomorrow." Nothing was attempted in the way of light or fire. Camping that night was simply lying down. The boulders seemed to make a fair bed after finding the best place to take their pressure.

During the night I was awakened by the beating of the spent ends of berg-waves against the side of my tent, though I had fancied myself well beyond their reach. These special waves are not raised by wind or tide, but by the fall of large bergs from the snout of the glacier, or sometimes by the overturning or breaking of large bergs that may have long floated in perfect poise. The highest berg-waves oftentimes travel half a dozen miles or farther before they are much spent, producing a singularly impressive uproar in the far recesses of the mountains on calm dark nights when all beside is still. Far and near they tell the news that a berg is born, repeating their story again and again, compelling attention and reminding us of earthquake-waves that roll on for thousands of miles, taking their story from continent to continent.

When the Indians came ashore in the morning and saw the condition of my tent they laughed heartily and said, "Your friend [meaning the big glacier] sent you a good word last night, and his servant knocked at your tent and said, 'Sagh-a-ya, are you sleeping well?'"

I had fasted too long to be in very good order for hard work, but while the Indians were cooking, I made out to push my way up the cañon before breakfast to seek the glacier that once came into the fiord, knowing from the size and muddiness of the stream that drains it that it must be quite large and not far off. I came in sight of it after a hard scramble of two hours through thorny chaparral and across steep avalanche taluses of rocks and snow. The front reaches across the cañon from wall to wall, covered with rocky detritus, and looked dark and forbidding in the shadow cast by the cliffs, while from a low, cave-like hollow its draining stream breaks forth, a river in size, with a reverberating roar that stirs all the cañon. Beyond, in a cloudless blaze of sunshine, I saw many tributaries, pure and white as new-fallen snow, drawing their sources from clusters of peaks and sweeping down waving slopes to unite their crystal currents with the trunk glacier in the central cañon. This fine glacier reaches to within two hundred and fifty feet of the level of the sea, and would even yet reach the fiord and send off bergs but for the waste it suffers in flowing slowly through the trunk cañon, the declivity of which is very slight.

Returning, I reached camp and breakfast at ten o'clock; then had everything packed into the canoe, and set off leisurely across the fiord to the mouth of another wide and low cañon, whose lofty outer cliffs, facing the fiord, are telling glacial advertisements. Gladly I should have explored it all, traced its streams of water and streams of ice, and entered its highest chambers, the homes and fountains of the snow. But I had to wait. I only stopped an hour or two, and climbed to the top of a rock through the common underbrush, whence I had a good general view. The front of the main glacier is not far distant from the fiord, and sends off small bergs into a lake. The walls of its tributary cañons are remarkably jagged and high, cut in a red variegated rock, probably slate. On the way back to the canoe I gathered ripe salmon-berries an inch and a half in diameter, ripe huckleberries, too, in great abundance, and several interesting plants I had not before met in the Territory.

About noon, when the tide was in our favor, we set out on the return trip to the gold-mine camp. The sun shone free and warm. No wind stirred. The water spaces between the bergs were as smooth as glass, reflecting the unclouded sky, and doubling the ravishing beauty of the bergs as the sunlight streamed through their innumerable angles in rainbow colors.

Soon a light breeze sprang up, and dancing lily spangles on the water mingled their glory of light with that burning on the angles of the ice.

On days like this, true sun-days, some of the bergs show a

purplish tinge, though most are white from the disintegrating of their weathered surfaces. Now and then a new-born one is met that is pure blue crystal throughout, freshly broken from the fountain or recently exposed to the air by turning over. But in all of them, old and new, there are azure caves and rifts of ineffable beauty, in which the purest tones of light pulse and shimmer, lovely and untainted as anything on earth or in the sky.

As we were passing the Indian village I presented a little tobacco to the headmen as an expression of regard, while they gave us a few smoked salmon, after putting many questions concerning my exploration of their bay and bluntly declaring their disbelief in the ice business.

About nine o'clock we arrived at the gold camp, where we found Mr. Young ready to go on with us the next morning, and thus ended two of the brightest and best of all my Alaska days.

CHAPTER XV

FROM TAKU RIVER TO TAYLOR BAY

I NEVER SAW Alaska looking better than it did when we bade farewell to Sum Dum on August 22 and pushed on northward up the coast toward Taku. The morning was clear, calm, bright—not a cloud in all the purple sky, nor wind, however gentle, to shake the slender spires of the spruces or dew-laden grass around the shores. Over the mountains and over the broad white bosoms of the glaciers the sunbeams poured, rosy as ever fell on fields of ripening wheat, drenching the forests and kindling the glassy waters and icebergs into a perfect blaze of colored light. Every living thing seemed joyful, and Nature's work was going on in glowing enthusiasm, not less appreciable in the deep repose that brooded over every feature of the landscape, suggesting the coming fruitfulness of the icy land and showing the advance that has already been made from glacial winter to summer. The care-laden commercial lives we lead close our eyes to the operations of God as a workman, though openly carried on that all who will look may see. The scarred rocks here and the moraines make a vivid showing of the old winter-time of the glacial period, and mark the bounds of the *mer-de-glace* that once filled the bay and covered the surrounding mountains. Already that sea of ice is replaced by water, in which multitudes of fishes are fed, while the hundred glaciers lingering about the bay and the streams that pour from them are busy night and day bringing in sand and mud and stones, at the rate of tons every minute, to fill it up. Then, as the seasons grow warmer, there will be fields here for the plough.

Our Indians, exhilarated by the sunshine, were garrulous as the gulls and plovers, and pulled heartily at their oars, evidently glad to get out of the ice with a whole boat.

"Now for Taku," they said, as we glided over the shining water. "Good-bye, Ice-Mountains; good-bye, Sum Dum." Soon a light breeze came, and they unfurled the sail and laid away their oars and began, as usual in such free times, to put their goods in order, unpacking and sunning provisions, guns, ropes, clothing, etc. Joe has an old flintlock musket suggestive of Hudson's Bay times, which he wished to discharge and reload. So, stepping in front of the sail, he fired at a gull that was flying past before I could prevent him, and it fell slowly with outspread wings alongside the canoe, with blood dripping from its bill. I asked him why he had killed the bird, and followed the question by a severe reprimand for his stupid cruelty, to which he could offer no other excuse than that he had learned from the whites to be careless about taking life. Captain Tyeen denounced the deed as likely to bring bad luck.

Before the whites came most of the Thlinkits held, with Agassiz, that animals have souls, and that it was wrong and unlucky to even speak disrespectfully of the fishes or any of the animals that supplied them with food. A case illustrating their superstitious beliefs in this connection occurred at Fort Wrangell while I was there the year before. One of the sub-chiefs of the Stickeens had a little son five or six years old, to whom he was very much attached, always taking him with him in his short canoe-trips, and leading him by the hand while going about town. Last summer the boy was taken sick, and gradually grew weak and thin, whereupon his father became alarmed, and feared, as is usual in such obscure cases, that the boy had been bewitched. He first applied in his trouble to Dr. Carliss, one of the missionaries, who gave medicine, without effecting the immediate cure that the fond father demanded. He was, to some extent, a believer in the powers of missionaries, both as to material and spiritual affairs, but in so serious an exigency it was natural that he should go back to the faith of his fathers. Accordingly, he sent for one of the shamans, or medicine-men, of his tribe, and submitted the case to him, who, after going through the customary incantations, declared that he had discovered the cause of the difficulty.

"Your boy," he said, "has lost his soul, and this is the way it happened. He was playing among the stones down on the beach when he saw a crawfish in the water, and made fun of it, pointing his finger at it and saying, 'Oh, you crooked legs! Oh, you crooked legs! You can't walk straight; you go sidewise,' which made the crab so angry that he reached out his long nippers, seized the lad's soul, pulled it out of him and made off with it into deep water. And," continued the medicine-man, "unless his stolen soul is restored to him and put back in its place he will die. Your boy is really dead already; it is only his lonely, empty body that is living now, and though it may continue to live in this way for a year or two, the boy will never be of any account, not strong, nor wise, nor brave."

The father then inquired whether anything could be done about it; was the soul still in possession of the crab, and if so, could it be recovered and re-installed in his forlorn son? Yes, the doctor rather thought it might be charmed back and re-united, but the job would be a difficult one, and would probably cost about fifteen blankets.

After we were fairly out of the bay into Stephens Passage, the wind died away, and the Indians had to take to their oars again, which ended our talk. On we sped over the silvery level, close alongshore. The dark forests extending far and near, planted like a field of wheat, might seem monotonous in general views, but the appreciative observer, looking closely, will find no lack of interesting variety, however far he may go. The steep slopes on which they grow allow almost every individual tree, with its peculiarities of form and color, to be seen like an audience on seats rising above one another—the blue-green, sharply-tapered spires of the Menzies spruce, the warm yellow-green Mertens spruce with their finger-like tops all pointing in the same direction, or drooping gracefully like leaves of grass, and the airy, feathery, brownish-green Alaska cedar. The outer fringe of bushes along the shore and hanging over the brows of the cliffs, the white mountains above, the shining water beneath, the changing sky over all, form pictures of divine beauty in which no healthy eye may ever grow weary.

Toward evening at the head of a picturesque bay we came to a village belonging to the Taku tribe. We found it silent and deserted. Not a single shaman or policeman had been left to keep it. These people are so happily rich as to have but little of a perishable kind to keep, nothing worth fretting about. They were away catching salmon, our Indians said. All the Indian villages hereabout are thus abandoned at regular periods every year, just as a tent is left for a day, while they repair to fishing, berrying, and hunting stations, occupying each in succession for a week or two at a time, coming and going from the main, substantially built villages. Then, after their summer's work is done, the winter supply of salmon dried and packed, fish-oil and seal-oil stored in boxes, berries and spruce bark pressed into cakes, their trading-trips completed, and the year's stock of quarrels with the neighboring tribe patched up in some way, they devote themselves to feasting, dancing, and hootchenoo drinking.

The Takus, once a powerful and war-like tribe, were at this time, like most of the neighboring tribes, whiskied nearly out of existence. They had a larger village on the Taku River, but, according to the census taken that year by the missionaries, they numbered only 269 in all,—109 men, 79 women, and 81 children, figures that show the vanishing condition of the tribe at a glance.

Our Indians wanted to camp for the night in one of the deserted houses, but I urged them on into the clean wilderness until dark, when we landed on a rocky beach fringed with devil's-clubs, greatly to the disgust of our crew. We had to make the best of it, however, as it was too dark to seek farther. After supper was accomplished among the boulders, they retired to the canoe, which they anchored a little way out, beyond low tide, while Mr. Young and I at the expense of a good deal of scrambling and panax stinging, discovered a spot on which we managed to sleep.

The next morning, about two hours after leaving our thorny camp, we rounded a great mountain rock nearly a mile in height and entered the Taku fiord. It is about eighteen miles long and from three to five miles wide, and extends directly back into the heart of the mountains, draining hundreds of glaciers and streams. The ancient glacier that formed it was far too deep and broad and too little concentrated to erode one of those narrow cañons, usually so impressive in sculpture and architecture, but it is all the more interesting on this account when the grandeur of the ice-work accomplished is recognized. This fiord, more than any other I have examined, explains the formation of the wonderful system of channels extending along the coast from Puget Sound to about latitude 59 degrees, for it is a marked portion of the system,—a branch of Stephens Passage. Its trends and general sculpture are as distinctly glacial as those of the narrowest fiord, while the largest tributaries of the great glacier that occupied it are still in existence. I counted some forty-five altogether, big and little, in sight from the canoe in sailing up the middle of the fiord. Three of them, drawing their sources from magnificent groups of snowy mountains, came down to the level of the sea and formed a glorious spectacle. The middle one of the three belongs to the first class, pouring its majestic flood, shattered and crevassed, directly into the fiord, and crowding about twenty-five square miles of it with bergs. The next below it also sends off bergs occasionally, though a narrow strip of glacial detritus separates it from the tide-water. That forenoon a large mass fell from it, damming its draining stream, which at length broke the dam, and the resulting flood swept forward thousands of small bergs across the mud-flat into the fiord. In a short time all was quiet again; the flood-waters receded, leaving only a large blue scar on the front of the glacier and stranded bergs on the moraine flat to tell the tale.

These two glaciers are about equal in size—two miles wide—and their fronts are only about a mile and a half apart. While I sat sketching them from a point among the drifting icebergs where I could see far back into the heart of their distant fountains, two Taku seal-hunters, father and son, came gliding toward us in an extremely small canoe. Coming alongside with a goodnatured "Sagh-a-ya," they inquired who we were, our objects, etc., and gave us information about the river, their village, and two other large glaciers that descend nearly to the sea-level a few miles up the river cañon. Crouching in their little shell of a boat among the great bergs, with paddle and barbed spear, they formed a picture as arctic and remote from anything to be found in civilization as ever was sketched for us by the explorers of the Far North.

Making our way through the crowded bergs to the extreme head of the fiord, we entered the mouth of the river, but were soon compelled to turn back on account of the strength of the current. The Taku River is a large stream, nearly a mile wide at the mouth, and, like the Stickeen, Chilcat, and Chilcoot, draws its sources from far inland, crossing the mountain-chain from the interior through a majestic cañon, and draining a multitude of glaciers on its way.

The Taku Indians, like the Chilcats, with a keen appreciation of the advantages of their position for trade, hold possession of the river and compel the Indians of the interior to accept their services as middle-men, instead of allowing them to trade directly with the whites.

When we were baffled in our attempt to ascend the river, the day was nearly done, and we began to seek a campground. After sailing two or three miles along the left side of the fiord, we were so fortunate as to find a small nook described by the two Indians, where fire-wood was abundant, and where we could drag our canoe up the bank beyond reach of the berg-waves. Here we were safe, with a fine outlook across the fiord to the great glaciers and near enough to see the birth of the icebergs and the wonderful commotion they make, and hear their wild, roaring rejoicing. The sunset sky seemed to have been painted for this one mountain mansion, fitting it like a ceiling. After the fiord was in shadow the level sunbeams continued to pour through the miles of bergs with ravishing beauty, reflecting and refracting the purple light like cut crystal. Then all save the tips of the highest became dead white. These, too, were speedily quenched, the glowing points vanishing like stars sinking beneath the horizon. And after the shadows had crept higher, submerging the glaciers and the ridges between them, the divine alpenglow still lingered on their highest fountain peaks as they stood transfigured in glorious array. Now the last of the twilight purple has vanished, the stars begin to shine, and all trace of the day is gone. Looking across the fiord the water seems perfectly black, and the two great glaciers are seen stretching dim and ghostly into the shadowy mountains now darkly massed against the starry sky.

Next morning it was raining hard, everything looked dismal, and on the way down the fiord a growling head wind battered the rain in our faces, but we held doggedly on and by 10 A.M. got out of the fiord into Stephens Passage. A breeze sprung up in our favor that swept us bravely on across the passage and around the end of Admiralty Island by dark. We camped in a boggy hollow on a bluff among scraggy, usnea-bearded spruces. The rain, bitterly cold and driven by a stormy wind, thrashed us well while we floundered in the stumpy bog trying to make a fire and supper.

When daylight came we found our campground a very savage place. How we reached it and established ourselves in the thick darkness it would be difficult to tell. We crept along the shore a few miles against strong head winds, then hoisted sail and steered straight across Lynn Canal to the mainland, which we followed without great difficulty, the wind having moderated toward evening. Near the entrance to Icy Strait we met a Hoona who had seen us last year and who seemed glad to see us. He gave us two salmon, and we made him happy with tobacco and then pushed on and camped near Sitka Jack's deserted village.

Though the wind was still ahead next morning, we made about twenty miles before sundown and camped on the west end of Farewell Island. We bumped against a hidden rock and sprung a small leak that was easily stopped with resin. The salmon-berries were ripe. While climbing a bluff for a view of our course, I discovered moneses, one of my favorites, and saw many well-traveled deer-trails, though the island is cut off from the mainland and other islands by at least five or six miles of icy, berg-encumbered water.

We got under way early next day,—a gray, cloudy morning with rain and wind. Fair and head winds were about evenly balanced throughout the day. Tides run fast here, like great rivers. We rowed and paddled around Point Wimbledon against both wind and tide, creeping close to the feet of the huge, bold rocks of the north wall of Cross Sound, which here were very steep and awe-inspiring as the heavy swells from the open sea coming in past Cape Spencer dashed white against them, tossing our frail canoe up and down lightly as a feather. The point reached by vegetation shows that the surf dashes up to a height of about seventy-five or a hundred feet. We were awe-stricken and began to fear that we might be upset should the ocean waves rise still higher. But little Stickeen seemed to enjoy the storm, and gazed at the foam-wreathed cliffs like a dreamy, comfortable tourist admiring a sunset. We reached the mouth of Taylor Bay about two or three o'clock in the afternoon, when we had a view of the open ocean before we entered the bay. Many large bergs from Glacier Bay were seen drifting out to sea past Cape Spencer. We reached the head of the fiord now called Taylor Bay at five o'clock and camped near an immense glacier with a front about three miles wide stretching across from wall to wall. No icebergs are discharged from it, as it is separated from the water of the fiord at high tide by a low, smooth mass of outspread, overswept moraine material, netted with torrents and small shallow rills from the glacier-front, with here and there a lakelet, and patches of yellow mosses and garden spots bright with epilobium, saxifrage, grass-tufts, sedges, and creeping willows on the higher ground. But only the mosses were sufficiently abundant to make conspicuous masses of color to relieve the dull slaty gray of the glacial mud and gravel. The front of the glacier, like all those which do not discharge icebergs, is rounded like a brow, smooth-looking in general views, but cleft and furrowed, nevertheless, with chasms and grooves in which the light glows and shimmers in glorious beauty. The granite walls of the fiord, though very high, are not deeply sculptured. Only a few deep side cañons with trees, bushes, grassy and flowery spots interrupt their massive simplicity, leaving but few of the cliffs absolutely sheer and bare like those of Yosemite, Sum Dum, or Taku. One of the side cañons is on the left side of the fiord, the other on the right, the tributaries of the former leading over by a narrow tide-channel to the bay next to the eastward, and by a short portage over into a lake into which pours a branch glacier from the great glacier. Still another branch from the main glacier turns to the right. Counting all three of these separate fronts, the width of this great Taylor Bay Glacier must be about seven or eight miles.

While camp was being made, Hunter Joe climbed the eastern wall in search of wild mutton, but found none. He fell in with a brown bear, however, and got a shot at it, but nothing more. Mr. Young and I crossed the moraine slope, splashing through pools and streams up to the ice-wall, and made the interesting discovery that the glacier had been advancing of late years, ploughing up and shoving forward moraine soil that had been deposited long ago, and overwhelming and grinding and carrying away the forests on the sides and front of the glacier. Though not now sending off icebergs, the front is probably far below sea-level at the bottom, thrust forward beneath its wave-washed moraine.

Along the base of the mountain-wall we found an abundance of salmon-berries, the largest measuring an inch and a half in diameter. Strawberries, too, are found hereabouts. Some which visiting Indians brought us were as fine in size and color and flavor as any I ever saw anywhere. After wandering and wondering an hour or two, admiring the magnificent rock and crystal scenery about us, we returned to camp at sundown, planning a grand excursion for the morrow.

I set off early the morning of August 30 before any one else in camp had stirred, not waiting for breakfast, but only eating a piece of bread. I had intended getting a cup of coffee, but a wild storm was blowing and calling, and I could not wait. Running out against the rain-laden gale and turning to catch my breath, I saw that the minister's little dog had left his bed in the tent and was coming boring through the storm, evidently determined to follow me. I told him to go back, that such a day as this had nothing for him.

"Go back," I shouted, "and get your breakfast." But he simply stood with his head down, and when I began to urge my way again, looking around, I saw he was still following me. So I at last told him to come on if he must and gave him a piece of the bread I had in my pocket.

Instead of falling, the rain, mixed with misty shreds of clouds, was flying in level sheets, and the wind was roaring as I had never heard wind roar before. Over the icy levels and over the woods, on the mountains, over the jagged rocks and spires and chasms of the glacier it boomed and moaned and roared, filling the fiord in even, gray, structureless gloom, inspiring and awful. I first struggled up in the face of the blast to the east end of the ice-wall, where a patch of forest had been carried away by the glacier when it was advancing. I noticed a few stumps well out on the moraine flat, showing that its present bare, raw condition was not the condition of fifty or a hundred years ago. In front of this part of the glacier there is a small moraine lake about half a mile in length, around the margin of which are a considerable number of trees standing knee-deep, and of course dead. This also is a result of the recent advance of the ice.

Pushing up through the ragged edge of the woods on the left margin of the glacier, the storm seemed to increase in violence, so that it was difficult to draw breath in facing it; therefore I took shelter back of a tree to enjoy it and wait, hoping that it would at last somewhat abate. Here the glacier, descending over an abrupt rock, falls forward in grand cascades, while a stream swollen by the rain was now a torrent,—wind, rain, ice-torrent, and water-torrent in one grand symphony.

At length the storm seemed to abate somewhat, and I took off my heavy rubber boots, with which I had waded the glacial streams on the flat, and laid them with my overcoat on a log, where I might find them on my way back, knowing I would be drenched anyhow, and firmly tied my mountain shoes, tightened my belt, shouldered my ice-ax, and, thus free and ready for rough work, pushed on, regardless as possible of mere rain. Making my way up a steep granite slope, its projecting polished bosses encumbered here and there by boulders and the ground and bruised ruins of the ragged edge of the forest that had been uprooted by the glacier during its recent advance, I traced the side of the glacier for two or three miles, finding everywhere evidence of its having encroached on the woods, which here run back along its edge for fifteen or twenty miles. Under the projecting edge of this vast ice-river I could see down beneath it to a depth of fifty feet or so in some places, where logs and branches were being crushed to pulp, some of it almost fine enough for paper, though most of it stringy and coarse.

After thus tracing the margin of the glacier for three or four miles, I chopped steps and climbed to the top, and as far as the eye could reach, the nearly level glacier stretched indefinitely away in the gray cloudy sky, a prairie of ice. The wind was now almost moderate, though rain continued to fall, which I did not mind, but a tendency to mist in the drooping draggled clouds made me hesitate about attempting to cross to the opposite shore. Although the distance was only six or seven miles, no traces at this time could be seen of the mountains on the other side, and in case the sky should grow darker, as it seemed inclined to do, I feared that when I got out of sight of land and perhaps into a maze of crevasses I might find difficulty in winning a way back.

Lingering a while and sauntering about in sight of the shore, I found this eastern side of the glacier remarkably free from large crevasses. Nearly all I met were so narrow I could step across them almost anywhere, while the few wide ones were easily avoided by going up or down along their sides to where they narrowed. The dismal cloud ceiling showed rifts here and there, and, thus encouraged, I struck out for the west shore, aiming to strike it five or six miles above the front wall, cautiously taking compass bearings at short intervals to enable me to find my way back should the weather darken again with mist or rain or snow. The structure lines of the glacier itself were, however, my main guide. All went well. I came to a deeply furrowed section about two miles in width where I had to zigzag in long, tedious tacks and make narrow doublings, tracing the edges of wide longitudinal furrows and chasms until I could find a bridge connecting their sides, oftentimes making the direct distance ten times over. The walking was good of its kind, however, and by dint of patient doubling and ax-work on dangerous places, I gained the opposite shore in about three hours, the width of the glacier at this point being about seven miles. Occasionally, while making my way, the clouds lifted a little, revealing a few bald, rough mountains sunk to the throat in the broad, icy sea which encompassed them on all sides, sweeping on forever and forever as we count time, wearing them away, giving them the shape they are destined to take when in the fullness of time they shall be parts of new landscapes.

Ere I lost sight of the east-side mountains, those on the west came in sight, so that holding my course was easy, and, though making haste, I halted for a moment to gaze down into the beautiful pure blue crevasses and to drink at the lovely blue wells, the most beautiful of all Nature's water-basins, or at the rills and streams outspread over the ice-land prairie, never ceasing to admire their lovely color and music as they glided and swirled in their blue crystal channels and pot-holes, and the rumbling of the moulins, or mills, where streams poured into blue-walled pits of unknown depth, some of them as regularly circular as if bored with augers. Interesting, too, were the cascades over blue cliffs, where streams fell into crevasses or slid almost noiselessly down slopes so smooth and frictionless their motion was concealed. The round or oval wells, however, from one to ten feet wide, and from one to twenty or thirty feet deep, were perhaps the most beautiful of all, the water so pure as to be almost invisible. My widest views did not probably exceed fifteen miles, the rain and mist making distances seem greater.

On reaching the farther shore and tracing it a few miles to northward, I found a large portion of the glacier-current sweeping out westward in a bold and beautiful curve around the shoulder of a mountain as if going direct to the open sea. Leaving the main trunk, it breaks into a magnificent uproar of pinnacles and spires and upheaving, splashing wave-shaped masses, a crystal cataract incomparably greater and wilder than a score of Niagaras.

Tracing its channel three or four miles, I found that it fell into a lake, which it fills with bergs. The front of this branch of the glacier is about three miles wide. I first took the lake to be the head of an arm of the sea, but, going down to its shore and tasting it, I found it fresh, and by my aneroid perhaps less than a hundred feet above sea-level. It is probably separated from the sea only by a moraine dam. I had not time to go around its shores, as it was now near five o'clock and I was about fifteen miles from camp, and I had to make haste to recross the glacier before dark, which would come on about eight o'clock. I therefore made haste up to the main glacier, and, shaping my course by compass and the structure lines of the ice, set off from the land out on to the grand crystal prairie again. All was so silent and so concentrated, owing to the low dragging mist, the beauty close about me was all the more keenly felt, though tinged with a dim sense of danger, as if coming events were casting shadows. I was soon out of sight of land, and the evening dusk that on cloudy days precedes the real night gloom came stealing on and only ice was in sight, and the only sounds, save the low rumbling of the mills and the rattle of falling stones at long intervals, were the low, terribly earnest moanings of the wind or distant waterfalls coming through the thickening gloom. After two hours of hard work I came to a maze of crevasses of appalling depth and width which could not be passed apparently either up or down. I traced them with firm nerve developed by the danger, making wide jumps, poising cautiously on dizzy edges after cutting footholds, taking wide crevasses at a grand leap at once frightful and inspiring. Many a mile was thus traveled, mostly up and down the glacier, making but little real headway, running much of the time as the danger of having to pass the night on the ice became more and more imminent. This I could do, though with the weather and my rain-soaked condition it would be trying at best. In treading the mazes of this crevassed section I had frequently to cross bridges that were only knife-edges for twenty or thirty feet, cutting off the sharp tops and leaving them flat so that little Stickeen could follow me. These I had to straddle, cutting off the top as I progressed and hitching gradually ahead like a boy riding a rail fence. All this time the little dog followed me bravely, never hesitating on the brink of any crevasse that I had jumped, but now that it was becoming dark and the crevasses became more troublesome, he followed close at my heels instead of scampering far and wide, where the ice was at all smooth, as he had in the forenoon. No land was now in sight. The mist fell lower and darker and snow began to fly. I could not see far enough up and down the glacier to judge how best to work out of the bewildering labyrinth, and how hard I tried while there was yet hope of reaching camp that night! a hope which was fast growing dim like the sky. After dark, on such ground, to keep from freezing, I could only jump up and down until morning on a piece of flat ice between the crevasses, dance to the boding music of the winds and waters, and as I was already tired and hungry I would be in bad condition for such ice work. Many times I was put to my mettle, but with a firm-braced nerve, all the more unflinching as the dangers thickened, I worked out of that terrible ice-web, and with blood fairly up Stickeen and I ran over common danger without fatigue. Our very hardest trial was in getting across the very last of the sliver bridges. After examining the first of the two widest crevasses, I followed its edge half a mile or so up and down and discovered that its narrowest spot was about eight feet wide, which was the limit of what I was able to jump. Moreover, the side I was on—that is, the west side—was about a foot higher than the other, and I feared that in case I should be stopped by a still wider impassable crevasse ahead that I would hardly be able to take back that jump from its lower side. The ice beyond, however, as far as I could see it, looked temptingly smooth. Therefore, after carefully making a socket for my foot on the rounded brink, I jumped, but found that I had nothing to spare and more than ever dreaded having to retrace my way. Little Stickeen jumped this, however, without apparently taking a second look at it, and we ran ahead joyfully over smooth, level ice, hoping we were now leaving all danger behind us. But hardly had we gone a hundred or two yards when to our dismay we found ourselves on the very widest of all the longitudinal crevasses we had yet encountered. It was about forty feet wide. I ran anxiously up the side of it to northward, eagerly hoping that I could get around its head, but my worst fears were realized when at a distance of about a mile or less it ran into the crevasse that I had just jumped. I then ran down the edge for a mile or more below the point where I had first met it, and found that its lower end also united with the crevasse I had jumped, showing dismally that we were on an island two or three hundred yards wide and about two miles long and the only way of escape from this island was by turning back and jumping again that crevasse which I dreaded, or venturing ahead across the giant crevasse by the very worst of the sliver bridges I had ever seen. It was so badly weathered and melted down that it formed a knife-edge, and extended across from side to side in a low, drooping curve like that made by a loose rope attached at each end at the same height. But the worst difficulty was that the ends of the down-curving sliver were attached to the sides at a depth of about eight or ten feet below the surface of the glacier. Getting down to the end of the bridge, and then after crossing it getting up the other side, seemed hardly possible. However, I decided to dare the dangers of the fearful sliver rather than to attempt to retrace my steps. Accordingly I dug a low groove in the rounded edge for my knees to rest in and, leaning over, began to cut a narrow foothold on the steep, smooth side. When I was doing this, Stickeen came up behind me, pushed his head over my shoulder, looked into the crevasses and along the narrow knife-edge, then turned and looked in my face, muttering and whining as if trying to say, "Surely you are not going down there." I said, "Yes, Stickeen, this is the only way." He then began to cry and ran wildly along the rim of the crevasse, searching for a better way, then, returning baffled, of course, he came behind me and lay down and cried louder and louder.

After getting down one step I cautiously stooped and cut another and another in succession until I reached the point where the sliver was attached to the wall. There, cautiously balancing, I chipped down the upcurved end of the bridge until I had formed a small level platform about a foot wide, then, bending forward, got astride of the end of the sliver, steadied myself with my knees, then cut off the top of the sliver, hitching myself forward an inch or two at a time, leaving it about four inches wide for Stickeen. Arrived at the farther end of the sliver, which was about seventy-five feet long, I chipped another little platform on its upcurved end, cautiously rose to my feet, and with infinite pains cut narrow notch steps and finger-holds in the wall and finally got safely across. All this dreadful time poor little Stickeen was crying as if his heart was broken, and when I called to him in as reassuring a voice as I could muster, he only cried the louder, as if trying to say that he never, never could get down there—the only time that the brave little fellow appeared to know what danger was. After going away as if I was leaving him, he still howled and cried without venturing to try to follow me. Returning to the edge of the crevasse, I told him that I must go, that he could come if he only tried, and finally in despair he hushed his cries, slid his little feet slowly down into my footsteps out on the big sliver, walked slowly and cautiously along the sliver as if holding his breath, while the snow was falling and the wind was moaning and threatening to blow him off. When he arrived at the foot of the slope below me, I was kneeling on the brink ready to assist him in case he should be unable to reach the top. He looked up along the row of notched steps I had made, as if fixing them in his mind, then with a nervous spring he whizzed up and passed me out on to the level ice, and ran and cried and barked and rolled about fairly hysterical in the sudden revulsion from the depth of despair to triumphant joy. I tried to catch him and pet him and tell him how good and brave he was, but he would not be caught. He ran round and round, swirling like autumn leaves in an eddy, lay down and rolled head over heels. I told him we still had far to go and that we must now stop all nonsense and get off the ice before dark. I knew by the ice-lines that every step was now taking me nearer the shore and soon it came in sight. The headland four or five miles back from the front, covered with spruce trees, loomed faintly but surely through the mist and light fall of snow not more than two miles away. The ice now proved good all the way across, and we reached the lateral moraine just at dusk, then with trembling limbs, now that the danger was over, we staggered and stumbled down the bouldery edge of the glacier and got over the dangerous rocks by the cascades while yet a faint light lingered. We were safe, and then, too, came limp weariness such as no ordinary work ever produces, however hard it may be. Wearily we stumbled down through the woods, over logs and brush and roots, devil's-clubs pricking us at every faint blundering tumble. At last we got out on the smooth mud slope with only a mile of slow but sure dragging of weary limbs to camp. The Indians had been firing guns to guide me and had a fine supper and fire ready, though, fearing they would be compelled to seek us in the morning, a care not often applied to me. Stickeen and I were too tired to eat much, and, strange to say, too tired to sleep. Both of us, springing up in the night again and again, fancied we were still on that dreadful ice bridge in the shadow of death.

Nevertheless, we arose next morning in newness of life. Never before had rocks and ice and trees seemed so beautiful and wonderful, even the cold, biting rainstorm that was blowing seemed full of loving-kindness, wonderful compensation for all that we had endured, and we sailed down the bay through the gray, driving rain rejoicing.

CHAPTER XVI

GLACIER BAY

WHILE STICKEEN AND I were away, a Hoona, one of the head men of the tribe, paid Mr. Young a visit, and presented him with porpoise-meat and berries and much interesting information. He naturally expected a return visit, and when we called at his house, a mile or two down the fiord, he said his wives were out in the rain gathering fresh berries to complete a feast prepared for us. We remained, however, only a few minutes, for I was not aware of this arrangement or of Mr. Young's promise until after leaving the house. Anxiety to get around Cape Wimbelton was the cause of my haste, fearing the storm might increase. On account of this ignorance, no apologies were offered him, and the upshot was that the good Hoona became very angry. We succeeded, however, in the evening of the same day, in explaining our haste, and by sincere apologies and presents made peace.

After a hard struggle we got around stormy Wimbelton and into the next fiord to the northward (Klunastucksana–Dundas Bay). A cold, drenching rain was falling, darkening but not altogether hiding its extraordinary beauty, made up of lovely reaches and side fiords, feathery headlands and islands, beautiful every one and charmingly collocated. But how it rained, and how cold it was, and how weary we were pulling most of the time against the wind! The branches of this bay are so deep and so numerous that, with the rain and low clouds concealing the mountain landmarks, we could hardly make out the main trends. While groping and gazing among the islands through the misty rain and clouds, we discovered wisps of smoke at the foot of a sheltering rock in front of a mountain, where a choir of cascades were chanting their rain songs. Gladly we made for this camp, which proved to belong to a rare old Hoona sub-chief, so tall and wide and dignified in demeanor he looked grand even in the sloppy weather, and every inch a chief in spite of his bare legs and the old shirt and draggled, ragged blanket in which he was dressed. He was given to much handshaking, gripping hard, holding on and looking you gravely in the face while most emphatically speaking in Thlinkit, not a word of which we understood until interpreter John came to our help. He turned from one to the other of us, declaring, as John interpreted, that our presence did him good like food and fire, that he would welcome white men, especially teachers, and that he and all his people compared to ourselves were only children. When Mr. Young informed him that a missionary was about to be sent to his people, he said he would call them all together four times and explain that a teacher and preacher were coming and that they therefore must put away all foolishness and prepare their hearts to receive them and their words. He then introduced his three children, one a naked lad five or six years old who, as he fondly assured us, would soon be a chief, and later to his wife, an intelligent-looking woman of whom he seemed proud. When we arrived she was out at the foot of the cascade mountain gathering salmon-berries. She came in dripping and loaded. A few of the fine berries saved for the children she presented, proudly and fondly beginning with the youngest, whose only clothing was a nose-ring and a string of beads. She was lightly appareled in a cotton gown and bit of blanket, thoroughly bedraggled, but after unloading her berries she retired with a dry calico gown around the corner of a rock and soon returned fresh as a daisy and with becoming dignity took her place by the fireside. Soon two other berry-laden women came in, seemingly enjoying the rain like the bushes and trees. They put on little clothing so that they may be the more easily dried, and as for the children, a thin shirt of sheeting is the most they encumber themselves with, and get wet and half dry without seeming to notice it while we shiver with two or three dry coats. They seem to prefer being naked. The men also wear but little in wet weather. When they go out for all day they put on a single blanket, but in choring around camp, getting fire-wood, cooking, or looking after their precious canvas, they seldom wear anything, braving wind and rain in utter nakedness to avoid the bother of drying clothes. It is a rare sight to see the children bringing in big chunks of fire-wood on their shoulders, balancing in crossing boulders with firmly set bow-legs and bulging back muscles.

We gave Ka-hood-oo-shough, the old chief, some tobacco and rice and coffee, and pitched our tent near his hut among tall grass. Soon after our arrival the Taylor Bay sub-chief came in from the opposite direction from ours, telling us that he came through a cut-off passage not on our chart. As stated above, we took pains to conciliate him and soothe his hurt feelings. Our words and gifts, he said, had warmed his sore heart and made him glad and comfortable.

The view down the bay among the islands was, I thought, the finest of this kind of scenery that I had yet observed.

The weather continued cold and rainy. Nevertheless Mr. Young and I and our crew, together with one of the Hoonas, an old man who acted as guide, left camp to explore one of the upper arms of the bay, where we were told there was a large glacier. We managed to push the canoe several miles up the stream that drains the glacier to a point where the swift current was divided among rocks and the banks were overhung with alders and willows. I left the canoe and pushed up the right bank past a magnificent waterfall some twelve hundred feet high, and over the shoulder of a mountain, until I secured a good view of the lower part of the glacier. It is probably a lobe of the Taylor Bay or Brady Glacier.

On our return to camp, thoroughly drenched and cold, the old chief came to visit us, apparently as wet and cold as ourselves.

"I have been thinking of you all day," he said, "and pitying you, knowing how miserable you were, and as soon as I saw your canoe coming back I was ashamed to think that I had been sitting warm and dry at my fire while you were out in the storm; therefore I made haste to strip off my dry clothing and put on these wet rags to share your misery and show how much I love you."

I had another long talk with Ka-hood-oo-shough the next day.

"I am not able," he said, "to tell you how much good your words have done me. Your words are good, and they are strong words. Some of my people are foolish, and when they make their salmon-traps they do not take care to tie the poles firmly together, and when the big rain-floods come the traps break and are washed away because the people who made them are foolish people. But your words are strong words and when storms come to try them they will stand the storms."

There was much handshaking as we took our leave and assurances of eternal friendship. The grand old man stood on the shore watching us and waving farewell until we were out of sight.

We now steered for the Muir Glacier and arrived at the front on the east side the evening of the third, and camped on the end of the moraine, where there was a small stream. Captain Tyeen was inclined to keep at a safe distance from the tremendous threatening cliffs of the discharging wall. After a good deal of urging he ventured within half a mile of them, on the east side of the fiord, where with Mr. Young I went ashore to seek a campground on the moraine, leaving the Indians in the canoe. In a few minutes after we landed a huge berg sprung aloft with awful commotion, and the frightened Indians incontinently fled down the fiord, plying their paddles with admirable energy in the tossing waves until a safe harbor was reached around the south end of the moraine. I found a good place for a camp in a slight hollow where a few spruce stumps afforded fire-wood. But all efforts to get Tyeen out of his harbor failed. "Nobody knew," he said, "how far the angry ice-mountain could throw waves to break his canoe." Therefore I had my bedding and some provisions carried to my stump camp, where I could watch the bergs as they were discharged and get night views of the brow of the glacier and its sheer jagged face all the way across from side to side of the channel. One night the water was luminous and the surge from discharging icebergs churned the water into silver fire, a glorious sight in the darkness. I also went back up the east side of the glacier five or six miles and ascended a mountain between its first two eastern tributaries, which, though covered with grass near the top, was exceedingly steep and difficult. A bulging ridge near the top I discovered was formed of ice, a remnant of the glacier when it stood at this elevation which had been preserved by moraine material and later by a thatch of dwarf bushes and grass.

Next morning at daybreak I pushed eagerly back over the comparatively smooth eastern margin of the glacier to see as much as possible of the upper fountain region. About five miles back from the front I climbed a mountain twenty-five hundred feet high, from the flowery summit of which, the day being clear, the vast glacier and its principal branches were displayed in one magnificent view. Instead of a stream of ice winding down a mountain-walled valley like the largest of the Swiss glaciers, the Muir looks like a broad undulating prairie streaked with medial moraines and gashed with crevasses, surrounded by numberless mountains from which flow its many tributary glaciers. There are seven main tributaries from ten to twenty miles long and from two to six miles wide where they enter the trunk, each of them fed by many secondary tributaries; so that the whole number of branches, great and small, pouring from the mountain fountains perhaps number upward of two hundred, not counting the smallest. The area drained by this one grand glacier can hardly be less than seven or eight hundred miles, and probably contains as much ice as all the eleven hundred Swiss glaciers combined. Its length from the frontal wall back to the head of its farthest fountain seemed to be about forty or fifty miles, and the width just below the confluence of the main tributaries about twenty-five miles. Though apparently motionless as the mountains, it flows on forever, the speed varying in every part with the seasons, but mostly with the depth of the current, and the declivity, smoothness and directness of the different portions of the basin. The flow of the central cascading portion near the front, as determined by Professor Reid, is at the rate of from two and a half to five inches an hour, or from five to ten feet a day. A strip of the main trunk about a mile in width, extending along the eastern margin about fourteen miles to a lake filled with bergs, has so little motion and is so little interrupted by crevasses, a hundred horsemen might ride abreast over it without encountering very much difficulty.

But by far the greater portion of the vast expanse looking smooth in the distance is torn and crumpled into a bewildering network of hummocky ridges and blades, separated by yawning gulfs and crevasses, so that the explorer, crossing it from shore to shore, must always have a hard time. In hollow spots here and there in the heart of the icy wilderness are small lakelets fed by swift-glancing streams that flow without friction in blue shining channels, making delightful melody, singing and ringing in silvery tones of peculiar sweetness, radiant crystals like flowers ineffably fine growing in dazzling beauty along their banks. Few, however, will be likely to enjoy them. Fortunately to most travelers the thundering ice-wall, while comfortably accessible, is also the most strikingly interesting portion of the glacier.

The mountains about the great glacier were also seen from this standpoint in exceedingly grand and telling views, ranged and grouped in glorious array. Along the valleys of the main tributaries to the northwestward I saw far into their shadowy depths, one noble peak in its snowy robes appearing beyond another in fine perspective. One of the most remarkable of them, fashioned like a superb crown with delicately fluted sides, stands in the middle of the second main tributary, counting from left to right. To the westward the magnificent Fairweather Range is displayed in all its glory, lifting its peaks and glaciers into the blue sky. Mt. Fairweather, though not the highest, is the noblest and most majestic in port and architecture of all the sky-dwelling company. La Pérouse, at the south end of the range, is also a magnificent mountain, symmetrically peaked and sculptured, and wears its robes of snow and glaciers in noble style. Lituya, as seen from here, is an immense tower, severely plain and massive. It makes a fine and terrible and lonely impression. Crillon, though the loftiest of all (being nearly sixteen thousand feet high), presents no well-marked features. Its ponderous glaciers have ground it away into long, curling ridges until, from this point of view, it resembles a huge twisted shell. The lower summits about the Muir Glacier, like this one, the first that I climbed, are richly adorned and enlivened with flowers, though they make but a faint show in general views. Lines and dashes of bright green appear on the lower slopes as one approaches them from the glacier, and a fainter green tinge may be noticed on the subordinate summits at a height of two thousand or three thousand feet. The lower are mostly alder bushes and the top-most a lavish profusion of flowering plants, chiefly cassiope, vaccinium, pyrola, erigeron, gentiana, campanula, anemone, larkspur, and columbine, with a few grasses and ferns. Of these cassiope is at once the commonest and the most beautiful and influential. In some places its delicate stems make mattresses more than a foot thick over several acres, while the bloom is so abundant that a single handful plucked at random contains hundreds of its pale pink bells. The very thought of this Alaska garden is a joyful exhilaration. Though the storm-beaten ground it is growing on is nearly half a mile high, the glacier centuries ago flowed over it as a river flows over a boulder; but out of all the cold darkness and glacial crushing and grinding comes this warm, abounding beauty and life to teach us that what we in our faithless ignorance and fear call destruction is creation finer and finer.

When night was approaching I scrambled down out of my blessed garden to the glacier, and returned to my lonely camp, and, getting some coffee and bread, again went up the moraine to the east end of the great ice-wall. It is about three miles long, but the length of the jagged, berg-producing portion that stretches across the fiord from side to side like a huge green-and-blue barrier is only about two miles and rises above the water to a height of from two hundred and fifty to three hundred feet. Soundings made by Captain Carroll show that seven hundred and twenty feet of the wall is below the surface, and a third unmeasured portion is buried beneath the moraine detritus deposited at the foot of it. Therefore, were the water and rocky detritus cleared away, a sheer precipice of ice would be presented nearly two miles long and more than a thousand feet high. Seen from a distance, as you come up the fiord, it seems comparatively regular in form, but it is far otherwise; bold, jagged capes jut forward into the fiord, alternating with deep reëntering angles and craggy hollows with plain bastions, while the top is roughened with innumerable spires and pyramids and sharp hacked blades leaning and toppling or cutting straight into the sky.

The number of bergs given off varies somewhat with the weather and the tides, the average being about one every five or six minutes, counting only those that roar loud enough to make themselves heard at a distance of two or three miles. The very largest, however, may under favorable conditions be heard ten miles or even farther. When a large mass sinks from the upper fissured portion of the wall, there is first a keen, prolonged, thundering roar, which slowly subsides into a low muttering growl, followed by numerous smaller grating clashing sounds from the agitated bergs that dance in the waves about the newcomer as if in welcome; and these again are followed by the swash and roar of the waves that are raised and hurled up the beach against the moraines. But the largest and most beautiful of the bergs, instead of thus falling from the upper weathered portion of the wall, rise from the submerged portion with a still grander commotion, springing with tremendous voice and gestures nearly to the top of the wall, tons of water streaming like hair down their sides, plunging and rising again and again before they finally settle in perfect poise, free at last, after having formed part of the slow-crawling glacier for centuries. And as we contemplate their history, as they sail calmly away down the fiord to the sea, how wonderful it seems that ice formed from pressed snow on the far-off mountains two or three hundred years ago should still be pure and lovely in color after all its travel and toil in the rough mountain quarries, grinding and fashioning the features of predestined landscapes.

When sunshine is sifting through the midst of the multitude of icebergs that fill the fiord and through the jets of radiant spray ever rising from the tremendous dashing and splashing of the falling and upspringing bergs, the effect is indescribably glorious. Glorious, too, are the shows they make in the night when the moon and stars are shining. The berg-thunder seems far louder than by day, and the projecting buttresses seem higher as they stand forward in the pale light, relieved by gloomy hollows, while the new-born bergs are dimly seen, crowned with faint lunar rainbows in the up-dashing spray. But it is in the darkest nights when storms are blowing and the waves are phosphorescent that the most impressive displays are made. Then the long range of ice-bluffs is plainly seen stretching through the gloom in weird, unearthly splendor, luminous wave foam dashing against every bluff and drifting berg; and ever and anon amid all this wild auroral splendor some huge new-born berg dashes the living water into yet brighter foam, and the streaming torrents pouring from its sides are worn as robes of light, while they roar in awful accord with the winds and waves, deep calling unto deep, glacier to glacier, from fiord to fiord over all the wonderful bay.

After spending a few days here, we struck across to the main Hoona village on the south side of Icy Strait, thence by a long cut-off with one short portage to Chatham Strait, and thence down through Peril Strait, sailing all night, hoping to catch the mail steamer at Sitka. We arrived at the head of the strait about daybreak. The tide was falling, and rushing down with the swift current as if descending a majestic cataract was a memorable experience. We reached Sitka the same night, and there I paid and discharged my crew, making allowance for a couple of days or so for the journey back home to Fort Wrangell, while I boarded the steamer for Portland and thus ended my explorations for this season.

PART THREE

THE TRIP OF 1890

CHAPTER XVII

IN CAMP AT GLACIER BAY

I LEFT SAN FRANCISCO for Glacier Bay on the steamer City of Pueblo, June 14, 1890, at 10 A.M., this being my third trip to southeastern Alaska and fourth to Alaska, including northern and western Alaska as far as Unalaska and Pt. Barrow and the northeastern coast of Siberia. The bar at the Golden Gate was smooth, the weather cool and pleasant. The redwoods in sheltered coves approach the shore closely, their dwarfed and shorn tops appearing here and there in ravines along the coast up to Oregon. The wind-swept hills, beaten with scud, are of course bare of trees. Along the Oregon and Washington coast the trees get nearer the sea, for spruce and contorted pine endure the briny winds better than the redwoods. We took the inside passage between the shore and Race Rocks, a long range of islets on which many a good ship has been wrecked. The breakers from the deep Pacific, driven by the gale, made a glorious display of foam on the bald islet rocks, sending spray over the tops of some of them a hundred feet high or more in sublime, curving, jagged-edged and flame-shaped sheets. The gestures of these upspringing, purple-tinged waves as they dashed and broke were sublime and serene, combining displays of graceful beauty of motion and form with tremendous power—a truly glorious show. I noticed several small villages on the green slopes between the timbered mountains and the shore. Long Branch made quite a display of new houses along the beach, north of the mouth of the Columbia.

I had pleasant company on the Pueblo and sat at the chief engineer's table, who was a good and merry talker. An old San Francisco lawyer, rather stiff and dignified, knew my father-in-law, Dr. Strentzel. Three ladies, opposed to the pitching of the ship, were absent from table the greater part of the way. My best talker was an old Scandinavian sea-captain, who was having a new bark built at Port Blakely,—an interesting old salt, every sentence of his conversation flavored with sea-brine, bluff and hearty as a sea-wave, keen-eyed, courageous, self-reliant, and so stubbornly skeptical he refused to believe even in glaciers.

"After you see your bark," I said, "and find everything being done to your mind, you had better go on to Alaska and see the glaciers."

"Oh, I haf seen many glaciers already."

"But are you sure that you know what a glacier is?" I asked.

"Vell, a glacier is a big mountain all covered up with ice."

"Then a river," said I, "must be a big mountain all covered with water."

I explained what a glacier was and succeeded in exciting his interest. I told him he must reform, for a man who neither believed in God nor glaciers must be very bad, indeed the worst of all unbelievers.

At Port Townsend I met Mr. Loomis, who had agreed to go with me as far as the Muir Glacier. We sailed from here on the steamer Queen. We touched again at Victoria, and I took a short walk into the adjacent woods and gardens and found the flowery vegetation in its glory, especially the large wild rose for which the region is famous, and the spiræa and English honeysuckle of the gardens.

JUNE 18. We sailed from Victoria on the Queen at 10.30 A.M. The weather all the way to Fort Wrangell was cloudy and rainy, but the scenery is delightful even in the dullest weather. The marvelous wealth of forests, islands, and waterfalls, the cloud-wreathed heights, the many avalanche slopes and slips, the pearl-gray tones of the sky, the browns of the woods, their purple flower edges and mist fringes, the endless combinations of water and land and ever-shifting clouds—none of these greatly interest the tourists. I noticed one of the small whales that frequent these channels and mentioned the fact, then called attention to a charming group of islands, but they turned their eyes from the islands, saying, "Yes, yes, they are very fine, but where did you see the whale?"

The timber is larger and apparently better every way as you go north from Victoria, that is on the islands, perhaps on account of fires from less rain to the southward. All the islands have been overswept by the ice-sheet and are but little changed as yet, save a few of the highest summits which have been sculptured by local residual glaciers. All have approximately the form of greatest strength with reference to the overflow of an ice-sheet, excepting those mentioned above, which have been more or less eroded by local residual glaciers. Every channel also has the form of greatest strength with reference to ice-action. Islands, as we have seen, are still being born in Glacier Bay and elsewhere to the northward.

I found many pleasant people aboard, but strangely ignorant on the subject of earth-sculpture and landscape-making. Professor Niles, of the Boston Institute of Technology, is aboard; also Mr. Russell and Mr. Kerr of the Geological Survey, who are now on their way to Mt. St. Elias, hoping to reach the summit; and a granddaughter of Peter Burnett, the first governor of California.

We arrived at Wrangell in the rain at 10.30 P.M. There was a grand rush on shore to buy curiosities and see totem poles. The shops were jammed and mobbed, high prices paid for shabby stuff manufactured expressly for tourist trade. Silver bracelets hammered out of dollars and half dollars by Indian smiths are the most popular articles, then baskets, yellow cedar toy canoes, paddles, etc. Most people who travel look only at what they are directed to look at. Great is the power of the guidebook-maker, however ignorant. I inquired for my old friends Tyeen and Shakes, who were both absent.

JUNE 20. We left Wrangell early this morning and passed through the Wrangell Narrows at high tide. I noticed a few bergs near Cape Fanshawe from Wrangell Glacier. The water ten miles from Wrangell is colored with particles derived mostly from the Stickeen River glaciers and Le Conte Glacier. All the waters of the channels north of Wrangell are green or yellowish from glacier erosion. We had a good view of the glaciers all the way to Juneau, but not of their high, cloud-veiled fountains. The stranded bergs on the moraine bar at the mouth of Sum Dum Bay looked just as they did when I first saw them ten years ago.

Before reaching Juneau, the Queen proceeded up the Taku Inlet that the passengers might see the fine glacier at its head, and ventured to within half a mile of the berg-discharging front, which is about three quarters of a mile wide. Bergs fell but seldom, perhaps one in half an hour. The glacier makes a rapid descent near the front. The inlet, therefore, will not be much extended beyond its present limit by the recession of the glacier. The grand rocks on either side of its channel show ice-action in telling style. The Norris Glacier, about two miles below the Taku, is a good example of a glacier in the first stage of decadence. The Taku River enters the head of the inlet a little to the east of the glaciers, coming from beyond the main coast range. All the tourists are delighted at seeing a grand glacier in the flesh. The scenery is very fine here and in the channel at Juneau. On Douglas Island there is a large mill of 240 stamps, all run by one small water-wheel, which, however, is acted on by water at enormous pressure. The forests around the mill are being rapidly nibbled away. Wind is here said to be very violent at times, blowing away people and houses and sweeping scud far up the mountain-side. Winter snow is seldom more than a foot or two deep.

JUNE 21. We arrived at Douglas Island at five in the afternoon and went sight-seeing through the mill. Six hundred tons of low-grade quartz are crushed per day. Juneau, on the mainland opposite the Douglas Island mills, is quite a village, well supplied with stores, churches, etc. A dance-house in which Indians are supposed to show native dances of all sorts is perhaps the best-patronized of all the places of amusement. A Mr. Brooks, who prints a paper here, gave us some information on Mt. St. Elias, Mt. Wrangell, and the Cook Inlet and Prince William Sound region. He told Russell that he would never reach the summit of St. Elias, that it was inaccessible. He saw no glaciers that discharged bergs into the sea at Cook Inlet, but many in Prince William Sound.

JUNE 22. Leaving Juneau at noon, we had a good view of the Auk Glacier at the mouth of the channel between Douglas Island and the mainland, and of Eagle Glacier a few miles north of the Auk on the east side of Lynn Canal. Then the Davidson Glacier came in sight, finely curved, striped with medial moraines, and girdled in front by its magnificent tree-fringed terminal moraine; and besides these many others of every size and pattern on the mountains bounding Lynn Canal, most of them comparatively small, completing their sculpture. The mountains on either hand and at the head of the canal are strikingly beautiful at any time of the year. The sky today is mostly clear, with just clouds enough hovering about the mountains to show them to best advantage as they stretch onward in sustained grandeur like two separate and distinct ranges, each mountain with its glaciers and clouds and fine sculpture glowing bright in smooth, graded light. Only a few of them exceed five thousand feet in height; but as one naturally associates great height with ice-and-snow-laden mountains and with glacial sculpture so pronounced, they seem much higher. There are now two canneries at the head of Lynn Canal. The Indians furnish some of the salmon at ten cents each. Everybody sits up to see the midnight sky. At this time of the year there is no night here, though the sun drops a degree or two below the horizon. One may read at twelve o'clock San Francisco time.

JUNE 23. Early this morning we arrived in Glacier Bay. We passed through crowds of bergs at the mouth of the bay, though, owing to wind and tide, there were but few at the front of Muir Glacier. A fine, bright day, the last of a group of a week or two, as shown by the dryness of the sand along the shore and on the moraine—rare weather hereabouts. Most of the passengers went ashore and climbed the moraine on the east side to get a view of the glacier from a point a little higher than the top of the front wall. A few ventured on a mile or two farther. The day was delightful, and our one hundred and eighty passengers were happy, gazing at the beautiful blue of the bergs and the shattered pinnacled crystal wall, awed by the thunder and commotion of the falling and rising icebergs, which ever and anon sent spray flying several hundred feet into the air and raised swells that set all the fleet of bergs in motion and roared up the beach, telling the story of the birth of every iceberg far and near. The number discharged varies much, influenced in part no doubt by the tides and weather and seasons, sometimes one every five minutes for half a day at a time on the average, though intervals of twenty or thirty minutes may occur without any considerable fall, then three or four immense discharges will take place in as many minutes. The sound they make is like heavy thunder, with a prolonged roar after deep thudding sounds—a perpetual thunder-storm easily heard three or four miles away. The roar in our tent and the shaking of the ground one or two miles distant from points of discharge seems startlingly near.

I had to look after camp-supplies and left the ship late this morning, going with a crowd to the glacier; then, taking advantage of the fine weather, I pushed off alone into the silent icy prairie to the east, to Nunatak Island, about five hundred feet above the ice. I discovered a small lake on the larger of the two islands, and many battered and ground fragments of fossil wood, large and small. They seem to have come from trees that grew on the island perhaps centuries ago. I mean to use this island as a station in setting out stakes to measure the glacial flow. The top of Mt. Fairweather is in sight at a distance of perhaps thirty miles, the ice all smooth on the eastern border, wildly broken in the central portion. I reached the ship at 2.30 P.M. I had intended getting back at noon and sending letters and bidding friends good-bye, but could not resist this glacier saunter. The ship moved off as soon as I was seen on the moraine bluff, and Loomis and I waved our hats in farewell to the many wavings of handkerchiefs of acquaintances we had made on the trip.

Our goods—blankets, provisions, tent, etc.—lay in a rocky moraine hollow within a mile of the great terminal wall of the glacier, and the discharge of the rising and falling icebergs kept up an almost continuous thundering and echoing, while a few gulls flew about on easy wing or stood like specks of foam on the shore. These were our neighbors.

After my twelve-mile walk, I ate a cracker and planned the camp. I found that one of my boxes had been left on the steamer, but still we have more than enough of everything. We obtained two cords of dry wood at Juneau which Captain Carroll kindly had his men carry up the moraine to our campground. We piled the wood as a windbreak, then laid a floor of lumber brought from Seattle for a square tent, nine feet by nine. We set the tent, stored our provisions in it, and made our beds. This work was done by 11.30 P.M., good daylight lasting to this time. We slept well in our roomy cotton house, dreaming of California home nests in the wilderness of ice.

JUNE 25. A rainy day. For a few hours I kept count of the number of bergs discharged, then sauntered along the beach to the end of the crystal wall. A portion of the way is dangerous, the moraine bluff being capped by an overlying lobe of the glacier, which as it melts sends down boulders and fragments of ice, while the strip of sandy shore at high tide is only a few rods wide, leaving but little room to escape from the falling moraine material and the berg-waves. The view of the ice-cliffs, pinnacles, spires and ridges was very telling, a magnificent picture of nature's power and industry and love of beauty. About a hundred or a hundred and fifty feet from the shore a large stream issues from an arched, tunnel-like channel in the wall of the glacier, the blue of the ice hall being of an exquisite tone, contrasting with the strange, sooty, smoky, brown-colored stream. The front wall of the Muir Glacier is about two and a half or three miles wide. Only the central portion about two miles wide discharges icebergs. The two wings advanced over the washed and stratified moraine deposits have little or no motion, melting and receding as fast, or perhaps faster, than it advances. They have been advanced at least a mile over the old re-formed moraines, as is shown by the overlying, angular, recent moraine deposits, now being laid down, which are continuous with the medial moraines of the glacier.

In the old stratified moraine banks, trunks and branches of trees showing but little sign of decay occur at a height of about a hundred feet above tide-water. I have not yet compared this fossil wood with that of the opposite shore deposits. That the glacier was once withdrawn considerably back of its present limit seems plain. Immense torrents of water had filled in the inlet with stratified moraine material, and for centuries favorable climatic conditions allowed forests to grow upon it. At length the glacier advanced, probably three or four miles, uprooting and burying the trees which had grown undisturbed for centuries. Then came a great thaw, which produced the flood that deposited the uprooted trees. Also the trees which grew around the shores above the reach of floods were shed off, perhaps by the thawing of the soil that was resting on the buried margin of the glacier, left on its retreat and protected by a covering of moraine material from melting as fast as the exposed surface of the glacier. What appear to be remnants of the margin of the glacier when it stood at a much higher level still exist on the left side and probably all along its banks on both sides just below its present terminus.

JUNE 26. We fixed a mark on the left wing to measure the motion if any. It rained all day, but I had a grand tramp over mud, ice, and rock to the east wall of the inlet. Brown metamorphic slate, close-grained in places, dips away from the inlet, presenting edges to ice-action, which has given rise to a singularly beautiful and striking surface, polished and grooved and fluted.

All the next day it rained. The mountains were smothered in dull-colored mist and fog, the great glacier looming through the gloomy gray fog fringes with wonderful effect. The thunder of bergs booms and rumbles through the foggy atmosphere. It is bad weather for exploring but delightful nevertheless, making all the strange, mysterious region yet stranger and more mysterious.

JUNE 28. A light rain. We were visited by two parties of Indians. A man from each canoe came ashore, leaving the women in the canoe to guard against the berg-waves. I tried my Chinook and made out to say that I wanted to hire two of them in a few days to go a little way back on the glacier and around the bay. They are seal-hunters and promised to come again with "Charley," who "hi yu kumtux wawa Boston"—knew well how to speak English.

I saw three huge bergs born. Spray rose about two hundred feet. Lovely reflections showed of the pale-blue tones of the ice-wall and mountains in the calm water. Mirages are common, making the stranded bergs along the shore look like the sheer frontal wall of the glacier from which they were discharged.

I am watching the ice-wall, berg life and behavior, etc. Yesterday and today a solitary small flycatcher was feeding about camp. A sandpiper on the shore, loons, ducks, gulls, and crows, a few of each, and a bald eagle are all the birds I have noticed thus far. The glacier is thundering gloriously.

JUNE 30. Clearing clouds and sunshine. In less than a minute I saw three large bergs born. First there is usually a preliminary thundering of comparatively small masses as the large mass begins to fall, then the grand crash and boom and reverberating roaring. Oftentimes three or four heavy main throbbing thuds and booming explosions are heard as the main mass falls in several pieces, and also secondary thuds and thunderings as the mass or masses plunge and rise again and again ere they come to rest. Seldom, if ever, do the towers, battlements, and pinnacles into which the front of the glacier is broken fall forward headlong from their bases like falling trees at the water-level or above or below it. They mostly sink vertically or nearly so, as if undermined by the melting action of the water of the inlet, occasionally maintaining their upright position after sinking far below the level of the water, and rising again a hundred feet or more into the air with water streaming like hair down their sides from their crowns, then launch forward and fall flat with yet another thundering report, raising spray in magnificent, flame-like, radiating jets and sheets, occasionally to the very top of the front wall.

Illumined by the sun, the spray and angular crystal masses are indescribably beautiful. Some of the discharges pour in fragments from clefts in the wall like waterfalls, white and mealy-looking, even dusty with minute swirling ice-particles, followed by a rushing succession of thunder-tones combining into a huge, blunt, solemn roar. Most of these crumbling discharges are from the excessively shattered central part of the ice-wall; the solid deep-blue masses from the ends of the wall forming the large bergs rise from the bottom of the glacier.

Many lesser reports are heard at a distance of a mile or more from the fall of pinnacles into crevasses or from the opening of new crevasses. The berg discharges are very irregular, from three to twenty-two an hour. On one rising tide, six hours, there were sixty bergs discharged, large enough to thunder and be heard at distances of from three quarters to one and a half miles; and on one succeeding falling tide, six hours, sixty-nine were discharged.

JULY 1. We were awakened at four o'clock this morning by the whistle of the steamer George W. Elder. I went out on the moraine and waved my hand in salute and was answered by a toot from the whistle. Soon a party came ashore and asked if I was Professor Muir. The leader, Professor Harry Fielding Reid of Cleveland, Ohio, introduced himself and his companion, Mr. Cushing, also of Cleveland, and six or eight young students who had come well provided with instruments to study the glacier. They landed seven or eight tons of freight and pitched camp beside ours. I am delighted to have companions so congenial—we have now a village.

As I set out to climb the second mountain, three thousand feet high, on the east side of the glacier, I met many tourists returning from a walk on the smooth east margin of the glacier, and had to answer many questions. I had a hard climb, but wonderful views were developed and I sketched the glacier from this high point and most of its upper fountains.

Many fine alpine plants grew here, an anemone on the summit, two species of cassiope in shaggy mats, three or four dwarf willows, large blue hairy lupines eighteen inches high, parnassia, phlox, solidago, dandelion, white-flowered bryanthus, daisy, pedicularis, epilobium, etc., with grasses, sedges, mosses, and lichens, forming a delightful deep spongy sod. Woodchucks stood erect and piped dolefully for an hour "Chee-chee!" with jaws absurdly stretched to emit so thin a note—rusty-looking, seedy fellows, also a smaller striped species which stood erect and cheeped and whistled like a Douglas squirrel. I saw three or four species of birds. A finch flew from her nest at my feet; and I almost stepped on a family of young ptarmigan ere they scattered, little bunches of downy brown silk, small but able to run well. They scattered along a snow-bank, over boulders, through willows, grass, and flowers, while the mother, very lame, tumbled and sprawled at my feet. I stood still until the little ones began to peep; the mother answered "Too-too-too" and showed admirable judgment and devotion. She was in brown plumage with white on the wing primaries. She had fine grounds on which to lead and feed her young.

Not a cloud in the sky to-day; a faint film to the north vanished by noon, leaving all the sky full of soft, hazy light. The magnificent mountains around the widespread tributaries of the glacier; the great, gently undulating, prairie-like expanse of the main trunk, bluish on the east, pure white on the west and north; its trains of moraines in magnificent curving lines and many colors—black, gray, red, and brown; the stormy, cataract-like, crevassed sections; the hundred fountains; the lofty, pure white Fairweather Range; the thunder of the plunging bergs; the fleet of bergs sailing tranquilly in the inlet—formed a glowing picture of Nature's beauty and power.

JULY 2. I crossed the inlet with Mr. Reid and Mr. Adams today. The stratified drift on the west side all the way from top to base contains fossil wood. On the east side, as far as I have seen it, the wood occurs only in one stratum at a height of about a hundred and twenty feet in sand and clay. Some in a bank of the west side are rooted in clay soil. I noticed a large grove of stumps in a washed-out channel near the glacier-front but had no time to examine closely. Evidently a flood carrying great quantities of sand and gravel had overwhelmed and broken off these trees, leaving high stumps. The deposit, about a hundred feet or more above them, had been recently washed out by one of the draining streams of the glacier, exposing a part of the old forest floor certainly two or three centuries old.

I climbed along the right bank of the lowest of the tributaries and set a signal flag on a ridge fourteen hundred feet high. This tributary is about one and a fourth or one and a half miles wide and has four secondary tributaries. It reaches tide-water but gives off no bergs.

Later I climbed the large Nunatak Island, seven thousand feet high, near the west margin of the glacier. It is composed of crumbling granite draggled with washed boulders, but has some enduring bosses which on sides and top are polished and scored rigidly, showing that it had been heavily overswept by the glacier when it was thousands of feet deeper than now, like a submerged boulder in a river-channel. This island is very irregular in form, owing to the variations in the structure joints of the granite. It has several small lakelets and has been loaded with glacial drift, but by the melting of the ice about its flanks is shedding it off, together with some of its own crumbling surface. I descended a deep rock gully on the north side, the rawest, dirtiest, dustiest, most dangerous that I have seen hereabouts. There is also a large quantity of fossil wood scattered on this island, especially on the north side, that on the south side having been cleared off and carried away by the first tributary glacier, which, being lower and melting earlier, has allowed the soil of the moraine material to fall, together with its forest, and be carried off. That on the north side is now being carried off or buried. The last of the main ice foundation is melting and the moraine material re-formed over and over again, and the fallen tree-trunks, decayed or half decayed or in a fair state of preservation, are also unburied and buried again or carried off to the terminal or lateral moraine.

I found three small seedling Sitka spruces, feeble beginnings of a new forest. The circumference of the island is about seven miles. I arrived at camp about midnight, tired and cold. Sailing across the inlet in a cranky rotten boat through the midst of icebergs was dangerous, and I was glad to get ashore.

JULY 4. I climbed the east wall to the summit, about thirty-one hundred feet or so, by the northern-most ravine next to the yellow ridge, finding about a mile of snow in the upper portion of the ravine and patches on the summit. A few of the patches probably lie all the year, the ground beneath them is so plantless. On the edge of some of the snow-banks I noticed cassiope. The thin, green, moss-like patches seen from camp are composed of a rich, shaggy growth of cassiope, white-flowered bryanthus, dwarf vaccinium with bright pink flowers, saxifrages, anemones, bluebells, gentians, small erigeron, pedicularis, dwarf willow and a few species of grasses. Of these, *Cassiope tetragona* is far the most influential and beautiful. Here it forms mats a foot thick and an acre or more in area, the sections being measured by the size and drainage of the soil-patches. I saw a few plants anchored in the less crumbling parts of the steep-faced bosses and steps—parnassia, potentilla, hedysarum, lutkea, etc. The lower, rough-looking patches halfway up the mountain are mostly alder bushes ten or fifteen feet high. I had a fine view of the top of the mountain-mass which forms the boundary wall of the upper portion of the inlet on the west side, and of several glaciers, tributary to the first of the eastern tributaries of the main Muir Glacier. Five or six of these tributaries were seen, most of them now melted off from the trunk and independent. The highest peak to the eastward has an elevation of about five thousand feet or a little less. I also had glorious views of the Fairweather Range, La Pérouse, Crillon, Lituya, and Fairweather. Mt. Fairweather is the most beautiful of all the giants that stand guard about Glacier Bay. When the sun is shining on it from the east or south its magnificent glaciers and colors are brought out in most telling display. In the late afternoon its features become less distinct. The atmosphere seems pale and hazy, though around to the north and northeastward of Fairweather innumerable white peaks are displayed, the highest fountain-heads of the Muir Glacier crowded together in bewildering array, most exciting and inviting to the mountaineer. Altogether I have had a delightful day, a truly glorious celebration of the fourth.

JULY 6. I sailed three or four miles down the east coast of the inlet with the Reid party's cook, who is supposed to be an experienced camper and prospector, and landed at a stratified moraine-bank. It was here that I camped in 1880, a point at that time less than half a mile from the front of the glacier, now one and a half miles. I found my Indian's old camp made just ten years ago, and Professor Wright's of five years ago. Their alder-bough beds and fireplace were still marked and but little decayed. I found thirty-three species of plants in flower, not counting willows—a showy garden on the shore only a few feet above high tide, watered by a fine stream. Lutkea, hedysarum, parnassia, epilobium, bluebell, solidago, habenaria, strawberry with fruit half grown, arctostaphylos, mertensia, erigeron, willows, tall grasses and alder are the principal species. There are many butterflies in this garden. Gulls are breeding near here. I saw young in the water to-day.

On my way back to camp I discovered a group of monumental stumps in a washed-out valley of the moraine and went ashore to observe them. They are in the dry course of a flood-channel about eighty feet above mean tide and four or five hundred yards back from the shore, where they have been pounded and battered by boulders rolling against them and over them, making them look like gigantic shaving-brushes. The largest is about three feet in diameter and probably three hundred years old. I mean to return and examine them at leisure. A smaller stump, still firmly rooted, is standing astride of an old crumbling trunk, showing that at least two generations of trees flourished here undisturbed by the advance or retreat of the glacier or by its draining stream-floods. They are Sitka spruces and the wood is mostly in a good state of preservation. How these trees were broken off without being uprooted is dark to me at present. Perhaps most of their companions were uprooted and carried away.

JULY 7. Another fine day; scarce a cloud in the sky. The icebergs in the bay are miraged in the distance to look like the frontal wall of a great glacier. I am writing letters in anticipation of the next steamer, the Queen.

She arrived about 2.30 P.M. with two hundred and thirty tourists. What a show they made with their ribbons and kodaks! All seemed happy and enthusiastic, though it was curious to see how promptly all of them ceased gazing when the dinner-bell rang, and how many turned from the great thundering crystal world of ice to look curiously at the Indians that came alongside to sell trinkets, and how our little camp and kitchen arrangements excited so many to loiter and waste their precious time prying into our poor hut.

JULY 8. A fine clear day. I went up the glacier to observe stakes and found that a marked point near the middle of the current had flowed about a hundred feet in eight days. On the medial moraine one mile from the front there was no measureable displacement. I found a raven devouring a tom-cod that was alive on a shallow at the mouth of the creek. It had probably been wounded by a seal or eagle.

JULY 10. I have been getting acquainted with the main features of the glacier and its fountain mountains with reference to an exploration of its main tributaries and the upper part of its prairie-like trunk, a trip I have long had in mind. I have been building a sled and must now get fully ready to start without reference to the weather. Yesterday evening I saw a large blue berg just as it was detached sliding down from the front. Two of Professor Reid's party rowed out to it as it sailed past the camp, estimating it to be two hundred and forty feet in length and one hundred feet high.

CHAPTER XVIII

MY SLED-TRIP ON THE MUIR GLACIER

I STARTED OFF the morning of July 11 on my memorable sled-trip to obtain general views of the main upper part of the Muir Glacier and its seven principal tributaries, feeling sure that I would learn something and at the same time get rid of a severe bronchial cough that followed an attack of the grippe and had troubled me for three months. I intended to camp on the glacier every night, and did so, and my throat grew better every day until it was well, for no lowland microbe could stand such a trip. My sled was about three feet long and made as light as possible. A sack of hardtack, a little tea and sugar, and a sleeping-bag were firmly lashed on it so that nothing could drop off however much it might be jarred and dangled in crossing crevasses.

Two Indians carried the baggage over the rocky moraine to the clear glacier at the side of one of the eastern Nunatak Islands. Mr. Loomis accompanied me to this first camp and assisted in dragging the empty sled over the moraine. We arrived at the middle Nunatak Island about nine o'clock. Here I sent back my Indian carriers, and Mr. Loomis assisted me the first day in hauling the loaded sled to my second camp at the foot of Hemlock Mountain, returning the next morning.

JULY 13. I skirted the mountain to eastward a few miles and was delighted to discover a group of trees high up on its ragged rocky side, the first trees I had seen on the shores of Glacier Bay or on those of any of its glaciers. I left my sled on the ice and climbed the mountain to see what I might learn. I found that all the trees were mountain hemlock (*Tsuga mertensiana*), and were evidently the remnant of an old well-established forest, standing on the only ground that was stable, all the rest of the forest below it having been sloughed off with the soil from the disintegrating slate bed rock. The lowest of the trees stood at an elevation of about two thousand feet above the sea, the highest at about three thousand feet or a little higher. Nothing could be more striking than the contrast between the raw, crumbling, deforested portions of the mountain, looking like a quarry that was being worked, and the forested part with its rich, shaggy beds of cassiope and bryanthus in full bloom, and its sumptuous cushions of flower-enameled mosses. These garden-patches are full of gay colors of gentian, erigeron, anemone, larkspur, and columbine, and are enlivened with happy birds and bees and marmots. Climbing to an elevation of twenty-five hundred feet, which is about fifteen hundred feet above the level of the glacier at this point, I saw and heard a few marmots, and three ptarmigans that were as tame as barnyard fowls. The sod is sloughing off on the edges, keeping it ragged. The trees are storm-bent from the southeast. A few are standing at an elevation of nearly three thousand feet; at twenty-five hundred feet, pyrola, veratrum, vaccinium, fine grasses, sedges, willows, mountain-ash, buttercups, and acres of the most luxuriant cassiope are in bloom.

A lake encumbered with icebergs lies at the end of Divide Glacier. A spacious, level-floored valley beyond it, eight or ten miles long, with forested mountains on its west side, perhaps discharges to the southeastward into Lynn Canal. The divide of the glacier is about opposite the third of the eastern tributaries. Another berg-dotted lake into which the drainage of the Braided Glacier flows, lies a few miles to the westward and is one and a half miles long. Berg Lake is next to the remarkable Girdled Glacier to the southeastward.

When the ice-period was in its prime, much of the Muir Glacier that now flows northward into Howling Valley flowed southward into Glacier Bay as a tributary of the Muir. All the rock contours show this, and so do the medial moraines. Berg Lake is crowded with bergs because they have no outlet and melt slowly. I heard none discharged. I had a hard time crossing the Divide Glacier, on which I camped. Half a mile back from the lake I gleaned a little fossil wood and made a fire on moraine boulders for tea. I slept fairly well on the sled. I heard the roar of four cascades on a shaggy green mountain on the west side of Howling Valley and saw three wild goats fifteen hundred feet up in the steep grassy pastures.

JULY 14. I rose at four o'clock this cloudy and dismal morning and looked for my goats, but saw only one. I thought there must be wolves where there were goats, and in a few minutes heard their low, dismal, far-reaching howling. One of them sounded very near and came nearer until it seemed to be less than a quarter of a mile away on the edge of the glacier. They had evidently seen me, and one or more had come down to observe me, but I was unable to catch sight of any of them. About half an hour later, while I was eating breakfast, they began howling again, so near I began to fear they had a mind to attack me, and I made haste to the shelter of a big square boulder, where, though I had no gun, I might be able to defend myself from a front attack with my alpenstock. After waiting half an hour or so to see what these wild dogs meant to do, I ventured to proceed on my journey to the foot of Snow Dome, where I camped for the night.

There are six tributaries on the northwest side of Divide arm, counting to the Gray Glacier, next after Granite Cañon Glacier going northwest. Next is Dirt Glacier, which is dead. I saw bergs on the edge of the main glacier a mile back from here which seem to have been left by the draining of a pool in a sunken hollow. A circling rim of driftwood, back twenty rods on the glacier, marks the edge of the lakelet shore where the bergs lie scattered and stranded. It is now half past ten o'clock and getting dusk as I sit by my little fossil-wood fire writing these notes. A strange bird is calling and complaining. A stream is rushing into a glacier well on the edge of which I am camped, back a few yards from the base of the mountain for fear of falling stones. A few small ones are rattling down the steep slope. I must go to bed.

JULY 15. I climbed the dome to plan a way, scan the glacier, and take bearings, etc., in case of storms. The main divide is about fifteen hundred feet; the second divide, about fifteen hundred also, is about one and one half miles southeastward. The flow of water on the glacier noticeably diminished last night though there was no frost. It is now already increasing. Stones begin to roll into the crevasses and into new positions, sliding against each other, half turning over or falling on moraine ridges. Mud pellets with small pebbles slip and roll slowly from ice-hummocks again and again. How often and by how many ways are boulders finished and finally brought to anything like permanent form and place in beds for farms and fields, forests and gardens. Into crevasses and out again, into moraines, shifted and reinforced and re-formed by avalanches, melting from pedestals, etc. Rain, frost, and dew help in the work; they are swept in rills, caught and ground in pot-hole mills. Moraines of washed pebbles, like those on glacier margins, are formed by snow avalanches deposited in crevasses, then weathered out and projected on the ice as shallow raised moraines. There is one such at this camp.

A ptarmigan is on a rock twenty yards distant, as if on show. It has red over the eye, a white line, not conspicuous, over the red, belly white, white markings over the upper parts on ground of brown and black wings, mostly white as seen when flying, but the coverts the same as the rest of the body. Only about three inches of the folded primaries show white. The breast seems to have golden iridescent colors, white under the wings. It allowed me to approach within twenty feet. It walked down a sixty degree slope of the rock, took flight with a few whirring wing-beats, then sailed with wings perfectly motionless four hundred yards down a gentle grade, and vanished over the brow of a cliff. Ten days ago Loomis told me that he found a nest with nine eggs. On the way down to my sled I saw four more ptarmigans. They utter harsh notes when alarmed. "Crack, chuck, crack," with the *r* rolled and prolonged. I also saw fresh and old goat-tracks and some bones that suggest wolves.

There is a pass through the mountains at the head of the third glacier. Fine mountains stand at the head on each side. The one on the northeast side is the higher and finer every way. It has three glaciers, tributary to the third. The third glacier has altogether ten tributaries, five on each side. The mountain on the left side of White Glacier is about six thousand feet high. The moraines of Girdled Glacier seem scarce to run anywhere. Only a little material is carried to Berg Lake. Most of it seems to be at rest as a terminal on the main glacier-field, which here has little motion. The curves of these last as seen from this mountain-top are very beautiful.

It has been a glorious day, all pure sunshine. An hour or more before sunset the distant mountains, a vast host, seemed more softly ethereal than ever, pale blue, ineffably fine, all angles and harshness melted off in the soft evening light. Even the snow and the grinding, cascading glaciers became divinely tender and fine in this celestial amethystine light. I got back to camp at 7.15, not tired. After my hardtack supper I could have climbed the mountain again and got back before sunrise, but dragging the sled tires me. I have been out on the glacier examining a moraine-like mass about a third of a mile from camp. It is perhaps a mile long, a hundred yards wide, and is thickly strewn with wood. I think that it has been brought down the mountain by a heavy snow avalanche, loaded on the ice, then carried away from the shore in the direction of the flow of the glacier. This explains detached moraine-masses. This one seems to have been derived from a big roomy cirque or amphitheater on the northwest side of this Snow Dome Mountain.

To shorten the return journey I was tempted to glissade down what appeared to be a snow-filled ravine, which was very steep. All went well until I reached a bluish spot which proved to be ice, on which I lost control of myself and rolled into a gravel talus at the foot without a scratch. Just as I got up and was getting myself orientated, I heard a loud fierce scream, uttered in an exulting, diabolical tone of voice which startled me, as if an enemy, having seen me fall, was glorying in my death. Then suddenly two ravens came swooping from the sky and alighted on the jag of a rock within a few feet of me, evidently hoping that I had been maimed and that they were going to have a feast. But as they stared at me, studying my condition, impatiently waiting for bone-picking time, I saw what they were up to and shouted, "Not yet, not yet!"

JULY 16. At 7 A.M. I left camp to cross the main glacier. Six ravens came to the camp as soon as I left. What wonderful eyes they must have! Nothing that moves in all this icy wilderness escapes the eyes of these brave birds. This is one of the loveliest mornings I ever saw in Alaska; not a cloud or faintest hint of one in all the wide sky. There is a yellowish haze in the east, white in the west, mild and mellow as a Wisconsin Indian Summer, but finer, more ethereal, God's holy light making all divine.

In an hour or so I came to the confluence of the first of the seven grand tributaries of the main Muir Glacier and had a glorious view of it as it comes sweeping down in wild cascades from its magnificent, pure white, mountain-girt basin to join the main crystal sea, its many fountain peaks, clustered and crowded, all pouring forth their tribute to swell its grand current. I crossed its front a little below its confluence, where its shattered current, about two or three miles wide, is reunited, and many rills and good-sized brooks glide gurgling and ringing in pure blue channels, giving delightful animation to the icy solitude.

Most of the ice-surface crossed today has been very uneven, and hauling the sled and finding a way over hummocks has been fatiguing. At times I had to lift the sled bodily and to cross many narrow, nerve-trying, ice-sliver bridges, balancing astride of them, and cautiously shoving the sled ahead of me with tremendous chasms on either side. I had made perhaps not more than six or eight miles in a straight line by six o'clock this evening when I reached ice so hummocky and tedious I concluded to camp and not try to take the sled any farther. I intend to leave it here in the middle of the basin and carry my sleeping-bag and provisions the rest of the way across to the west side. I am cozy and comfortable here resting in the midst of glorious icy scenery, though very tired. I made out to get a cup of tea by means of a few shavings and splinters whittled from the bottom board of my sled, and made a fire in a little can, a small campfire, the smallest I ever made or saw, yet it answered well enough as far as tea was concerned. I crept into my sack before eight o'clock as the wind was cold and my feet wet. One of my shoes is about worn out. I may have to put on a wooden sole. This day has been cloudless throughout, with lovely sunshine, a purple evening and morning. The circumference of mountains beheld from the midst of this world of ice is marvelous, the vast plain reposing in such soft tender light, the fountain mountains so clearly cut, holding themselves aloft with their loads of ice in supreme strength and beauty of architecture. I found a skull and most of the other bones of a goat on the glacier about two miles from the nearest land. It had probably been chased out of its mountain home by wolves and devoured here. I carried its horns with me. I saw many considerable depressions in the glacial surface, also a pit-like hole, irregular, not like the ordinary wells along the slope of the many small dirt-clad hillocks, faced to the south. Now the sun is down and the sky is saffron yellow, blending and fading into purple around to the south and north. It is a curious experience to be lying in bed writing these notes, hummock waves rising in every direction, their edges marking a multitude of crevasses and pits, while all around the horizon rise peaks innumerable of most intricate style of architecture. Solemnly growling and grinding moulins contrast with the sweet low-voiced whispering and warbling of a network of rills, singing like water-ouzels, glinting, gliding with indescribable softness and sweetness of voice. They are all around, one within a few feet of my hard sled bed.

JULY 17. Another glorious cloudless day is dawning in yellow and purple and soon the sun over the eastern peak will blot out the blue peak shadows and make all the vast white ice prairie sparkle. I slept well last night in the middle of the icy sea. The wind was cold but my sleeping-bag enabled me to lie neither warm nor intolerably cold. My three-months' cough is gone. Strange that with such work and exposure one should know nothing of sore throats and of what are called colds. My heavy, thick-soled shoes, resoled just before starting on the trip six days ago, are about worn out and my feet have been wet every night. But no harm comes of it, nothing but good. I succeeded in getting a warm breakfast in bed. I reached over the edge of my sled, got hold of a small cedar stick that I had been carrying, whittled a lot of thin shavings from it, stored them on my breast, then set fire to a piece of paper in a shallow tin can, added a pinch of shavings, held the cup of water that always stood at my bedside over the tiny blaze with one hand, and fed the fire by adding little pinches of shavings until the water boiled, then pulling my bread sack within reach, made a good warm breakfast, cooked and eaten in bed. Thus refreshed, I surveyed the wilderness of crevassed, hummocky ice and concluded to try to drag my little sled a mile or two farther, then, finding encouragement, persevered, getting it across innumerable crevasses and streams and around several lakes and over and through the midst of hummocks, and at length reached the western shore between five and six o'clock this evening, extremely fatigued. This I consider a hard job well done, crossing so wildly broken a glacier, fifteen miles of it from Snow Dome Mountain, in two days with a sled weighing altogether not less than a hundred pounds. I found innumerable crevasses, some of them brimful of water. I crossed in most places just where the ice was close pressed and welded after descending cascades and was being shoved over an upward slope, thus closing the crevasses at the bottom, leaving only the upper sun-melted beveled portion open for water to collect in.

Vast must be the drainage from this great basin. The waste in sunshine must be enormous, while in dark weather rains and winds also melt the ice and add to the volume produced by the rain itself. The winds also, though in temperature they may be only a degree or two above freezing-point, dissolve the ice as fast, or perhaps faster, than clear sunshine. Much of the water caught in tight crevasses doubtless freezes during the winter and gives rise to many of the irregular veins seen in the structure of the glacier. Saturated snow also freezes at times and is incorporated with the ice, as only from the lower part of the glacier is the snow melted during the summer. I have noticed many traces of this action. One of the most beautiful things to be seen on the glacier is the myriads of minute and intensely brilliant radiant lights burning in rows on the banks of streams and pools and lakelets from the tips of crystals melting in the sun, making them look as if bordered with diamonds. These gems are rayed like stars and twinkle; no diamond radiates keener or more brilliant light. It was perfectly glorious to think of this divine light burning over all this vast crystal sea in such ineffably fine effulgence, and over how many other of icy Alaska's glaciers where nobody sees it. To produce these effects I fancy the ice must be melting rapidly, as it was being melted to-day. The ice in these pools does not melt with anything like an even surface, but in long branches and leaves, making fairy forests of points, while minute bubbles of air are constantly being set free. I am camped to-night on what I call Quarry Mountain from its raw, loose, plantless condition, seven or eight miles above the front of the glacier. I found enough fossil wood for tea. Glorious is the view to the eastward from this camp. The sun has set, a few clouds appear, and a torrent rushing down a gully and under the edge of the glacier is making a solemn roaring. No tinkling, whistling rills this night. Ever and anon I hear a falling boulder. I have had a glorious and instructive day, but am excessively weary and to bed I go.

JULY 18. I felt tired this morning and meant to rest to-day. But after breakfast at 8 A.M. I felt I must be up and doing, climbing, sketching new views up the great tributaries from the top of Quarry Mountain. Weariness vanished and I could have climbed, I think, five thousand feet. Anything seems easy after sled-dragging over hummocks and crevasses, and the constant nerve-strain in jumping crevasses so as not to slip in making the spring. Quarry Mountain is the barest I have seen, a raw quarry with infinite abundance of loose decaying granite all on the go. Its slopes are excessively steep. A few patches of epilobium make gay purple spots of color. Its seeds fly everywhere seeking homes. Quarry Mountain is cut across into a series of parallel ridges by oversweeping ice. It is still overswept in three places by glacial flows a half to three quarters of a mile wide, finely arched at the top of the divides. I have been sketching, though my eyes are much inflamed and I can scarce see. All the lines I make appear double. I fear I shall not be able to make the few more sketches I want to-morrow, but must try. The day has been gloriously sunful, the glacier pale yellow toward five o'clock. The hazy air, white with a yellow tinge, gives an Indian-summerish effect. Now the blue evening shadows are creeping out over the icy plain, some ten miles long, with sunny yellow belts between them. Boulders fall now and again with dull, blunt booming, and the gravel pebbles rattle.

JULY 19. Nearly blind. The light is intolerable and I fear I may be long unfitted for work. I have been lying on my back all day with a snow poultice bound over my eyes. Every object I try to look at seems double; even the distant mountain-ranges are doubled, the upper an exact copy of the lower, though somewhat faint. This is the first time in Alaska that I have had too much sunshine. About four o'clock this afternoon, when I was waiting for the evening shadows to enable me to get nearer the main camp, where I could be more easily found in case my eyes should become still more inflamed and I should be unable to travel, thin clouds cast a grateful shade over all the glowing landscape. I gladly took advantage of these kindly clouds to make an effort to cross the few miles of the glacier that lay between me and the shore of the inlet. I made a pair of goggles but am afraid to wear them. Fortunately the ice here is but little broken, therefore I pulled my cap well down and set off about five o'clock. I got on pretty well and camped on the glacier in sight of the main camp, which from here in a straight line is only five or six miles away. I went ashore on Granite Island and gleaned a little fossil wood with which I made tea on the ice.

JULY 20. I kept wet bandages on my eyes last night as long as I could, and feel better this morning, but all the mountains still seem to have double summits, giving a curiously unreal aspect to the landscape. I packed everything on the sled and moved three miles farther down the glacier, where I want to make measurements. Twice to-day I was visited on the ice by a humming-bird, attracted by the red lining of the bear-skin sleeping-bag.

I have gained some light on the formation of gravel-beds along the inlet. The material is mostly sifted and sorted by successive rollings and washings along the margins of the glacier-tributaries, where the supply is abundant beyond anything I ever saw elsewhere. The lowering of the surface of a glacier when its walls are not too steep leaves a part of the margin dead and buried and protected from the wasting sunshine beneath the lateral moraines. Thus a marginal valley is formed, clear ice on one side, or nearly so, buried ice on the other. As melting goes on, the marginal trough, or valley, grows deeper and wider, since both sides are being melted, the land side slower. The dead, protected ice in melting first sheds off the large boulders, as they are not able to lie on slopes where smaller ones can. Then the next larger ones are rolled off, and pebbles and sand in succession. Meanwhile this material is subjected to torrent-action, as if it were cast into a trough. When floods come it is carried forward and stratified, according to the force of the current, sand, mud, or larger material. This exposes fresh surfaces of ice and melting goes on again, until enough material has been undermined to form a veil in front; then follows another washing and carrying-away and depositing where the current is allowed to spread. In melting, protected margin terraces are oftentimes formed. Perhaps these terraces mark successive heights of the glacial surface. From terrace to terrace the grist of stone is rolled and sifted. Some, meeting only feeble streams, have only the fine particles carried away and deposited in smooth beds; others, coarser, from swifter streams, overspread the fine beds, while many of the large boulders no doubt roll back upon the glacier to go on their travels again.

It has been cloudy mostly to-day, though sunny in the afternoon, and my eyes are getting better. The steamer Queen is expected in a day or two, so I must try to get down to the inlet to-morrow and make signal to have some of the Reid party ferry me over. I must hear from home, write letters, get rest and more to eat.

Near the front of the glacier the ice was perfectly free, apparently, of anything like a crevasse, and in walking almost carelessly down it I stopped opposite the large granite Nunatak Island, thinking that I would there be partly sheltered from the wind. I had not gone a dozen steps toward the island when I suddenly dropped into a concealed water-filled crevasse, which on the surface showed not the slightest sign of its existence. This crevasse like many others was being used as the channel of a stream, and at some narrow point the small cubical masses of ice into which the glacier surface disintegrates were jammed and extended back farther and farther till they completely covered and concealed the water. Into this I suddenly plunged, after crossing thousands of really dangerous crevasses, but never before had I encountered a danger so completely concealed. Down I plunged over head and ears, but of course bobbed up again, and after a hard struggle succeeded in dragging myself out over the farther side. Then I pulled my sled over close to Nunatak cliff, made haste to strip off my clothing, threw it in a sloppy heap and crept into my sleeping-bag to shiver away the night as best I could.

JULY 21. Dressing this rainy morning was a miserable job, but might have been worse. After wringing my sloppy underclothing, getting it on was far from pleasant. My eyes are better and I feel no bad effect from my icy bath. The last trace of my three months' cough is gone. No lowland grippe microbe could survive such experiences.

I have had a fine telling day examining the ruins of the old forest of Sitka spruce that no great time ago grew in a shallow mud-filled basin near the southwest corner of the glacier. The trees were protected by a spur of the mountain that puts out here, and when the glacier advanced they were simply flooded with fine sand and overborne. Stumps by the hundred, three to fifteen feet high, rooted in a stream of fine blue mud on cobbles, still have their bark on. A stratum of decomposed bark, leaves, cones, and old trunks is still in place. Some of the stumps are on rocky ridges of gravelly soil about one hundred and twenty-five feet above the sea. The valley has been washed out by the stream now occupying it, one of the glacier's draining streams a mile long or more and an eighth of a mile wide.

I got supper early and was just going to bed, when I was startled by seeing a man coming across the moraine, Professor Reid, who had seen me from the main camp and who came with Mr. Loomis and the cook in their boat to ferry me over. I had not intended making signals for them until to-morrow but was glad to go. I had been seen also by Mr. Case and one of his companions, who were on the western mountainside above the fossil forest, shooting ptarmigans. I had a good rest and sleep and leisure to find out how rich I was in new facts and pictures and how tired and hungry I was.

CHAPTER XIX

AURORAS

A FEW DAYS later I set out with Professor Reid's party to visit some of the other large glaciers that flow into the bay, to observe what changes have taken place in them since October, 1879, when I first visited and sketched them. We found the upper half of the bay closely choked with bergs, through which it was exceedingly difficult to force a way. After slowly struggling a few miles up the east side, we dragged the whale-boat and canoe over rough rocks into a fine garden and comfortably camped for the night.

The next day was spent in cautiously picking a way across to the west side of the bay; and as the strangely scanty stock of provisions was already about done, and the ice-jam to the northward seemed impenetrable, the party decided to return to the main camp by a comparatively open, roundabout way to the southward, while with the canoe and a handful of food-scraps I pushed on northward. After a hard, anxious struggle, I reached the mouth of the Hugh Miller fiord about sundown, and tried to find a camp-spot on its steep, boulder-bound shore. But no landing-place where it seemed possible to drag the canoe above high-tide mark was discovered after examining a mile or more of this dreary, forbidding barrier, and as night was closing down, I decided to try to grope my way across the mouth of the fiord in the starlight to an open sandy spot on which I had camped in October, 1879, a distance of about three or four miles.

With the utmost caution I picked my way through the sparkling bergs, and after an hour or two of this nerve-trying work, when I was perhaps less than halfway across and dreading the loss of the frail canoe which would include the loss of myself, I came to a pack of very large bergs which loomed threateningly, offering no visible thoroughfare. Paddling and pushing to right and left, I at last discovered a sheer-walled opening about four feet wide and perhaps two hundred feet long, formed apparently by the splitting of a huge iceberg. I hesitated to enter this passage, fearing that the slightest change in the tide-current might close it, but ventured nevertheless, judging that the dangers ahead might not be greater than those I had already passed. When I had got about a third of the way in, I suddenly discovered that the smooth-walled ice-lane was growing narrower, and with desperate haste backed out. Just as the bow of the canoe cleared the sheer walls they came together with a growling crunch. Terror-stricken, I turned back, and in an anxious hour or two gladly reached the rock-bound shore that had at first repelled me, determined to stay on guard all night in the canoe or find some place where with the strength that comes in a fight for life I could drag it up the boulder wall beyond ice danger. This at last was happily done about midnight, and with no thought of sleep I went to bed rejoicing.

My bed was two boulders, and as I lay wedged and bent on their up-bulging sides, beguiling the hard, cold time in gazing into the starry sky and across the sparkling bay, magnificent upright bars of light in bright prismatic colors suddenly appeared, marching swiftly in close succession along the northern horizon from west to east as if in diligent haste, an auroral display very different from any I had ever before beheld. Once long ago in Wisconsin I saw the heavens draped in rich purple auroral clouds fringed and folded in most magnificent forms; but in this glory of light, so pure, so bright, so enthusiastic in motion, there was nothing in the least cloud-like. The short color-bars, apparently about two degrees in height, though blending, seemed to be as well defined as those of the solar spectrum.

How long these glad, eager soldiers of light held on their way I cannot tell; for sense of time was charmed out of mind and the blessed night circled away in measureless rejoicing enthusiasm.

In the early morning after so inspiring a night I launched my canoe feeling able for anything, crossed the mouth of the Hugh Miller fiord, and forced a way three or four miles along the shore of the bay, hoping to reach the Grand Pacific Glacier in front of Mt. Fairweather. But the farther I went, the ice-pack, instead of showing inviting little open streaks here and there, became so much harder jammed that on some parts of the shore the bergs, drifting south with the tide, were shoving one another out of the water beyond high-tide line. Farther progress to northward was thus rigidly stopped, and now I had to fight for a way back to my cabin, hoping that by good tide luck I might reach it before dark. But at sundown I was less than halfway home, and though very hungry was glad to land on a little rock island with a smooth beach for the canoe and a thicket of alder bushes for fire and bed and a little sleep. But shortly after sundown, while these arrangements were being made, lo and behold another aurora enriching the heavens! and though it proved to be one of the ordinary almost colorless kind, thrusting long, quivering lances toward the zenith from a dark cloud-like base, after last night's wonderful display one's expectations might well be extravagant and I lay wide awake watching.

On the third night I reached my cabin and food. Professor Reid and his party came in to talk over the results of our excursions, and just as the last one of the visitors opened the door after bidding good-night, he shouted, "Muir, come look here. Here's something fine."

I ran out in auroral excitement, and sure enough here was another aurora, as novel and wonderful as the marching rainbow-colored columns—a glowing silver bow spanning the Muir Inlet in a magnificent arch right under the zenith, or a little to the south of it, the ends resting on the top of the mountain-walls. And though colorless and steadfast, its intense, solid, white splendor, noble proportions, and fineness of finish, excited boundless admiration. In form and proportion it was like a rainbow, a bridge of one span five miles wide; and so brilliant, so fine and solid and homogeneous in every part, I fancy that if all the stars were raked together into one windrow, fused and welded and run through some celestial rolling-mill, all would be required to make this one glowing white colossal bridge.

After my last visitor went to bed, I lay down on the moraine in front of the cabin and gazed and watched. Hour after hour the wonderful arch stood perfectly motionless, sharply defined and substantial-looking as if it were a permanent addition to the furniture of the sky. At length while it yet spanned the inlet in serene unchanging splendor, a band of fluffy, pale-gray, quivering ringlets came suddenly all in a row over the eastern mountain-top, glided in nervous haste up and down the under side of the bow and over the western mountain-wall. They were about one and a half times the apparent diameter of the bow in length, maintained a vertical posture all the way across, and slipped swiftly along as if they were suspended like a curtain on rings. Had these lively auroral fairies marched across the fiord on the top of the bow instead of shuffling along the under side of it, one might have fancied they were a happy band of spirit people on a journey making use of the splendid bow for a bridge. There must have been hundreds of miles of them; for the time required for each to cross from one end of the bridge to the other seemed only a minute or less, while nearly an hour elapsed from their first appearance until the last of the rushing throng vanished behind the western mountain, leaving the bridge as bright and solid and steadfast as before they arrived. But later, half an hour or so, it began to fade. Fissures or cracks crossed it diagonally through which a few stars were seen, and gradually it became thin and nebulous until it looked like the Milky Way, and at last vanished, leaving no visible monument of any sort to mark its place.

I now returned to my cabin, replenished the fire, warmed myself, and prepared to go to bed, though too aurorally rich and happy to go to sleep. But just as I was about to retire, I thought I had better take another look at the sky, to make sure that the glorious show was over; and, contrary to all reasonable expectations, I found that the pale foundation for another bow was being laid right overhead like the first. Then losing all thought of sleep, I ran back to my cabin, carried out blankets and lay down on the moraine to keep watch until daybreak, that none of the sky wonders of the glorious night within reach of my eyes might be lost.

I had seen the first bow when it stood complete in full splendor, and its gradual fading decay. Now I was to see the building of a new one from the beginning. Perhaps in less than half an hour the silvery material was gathered, condensed, and welded into a glowing, evenly proportioned arc like the first and in the same part of the sky. Then in due time over the eastern mountain-wall came another throng of restless electric auroral fairies, the infinitely fine pale-gray garments of each lightly touching those of their neighbors as they swept swiftly along the under side of the bridge and down over the western mountain like the merry band that had gone the same way before them, all keeping quivery step and time to music too fine for mortal ears.

While the gay throng was gliding swiftly along, I watched the bridge for any change they might make upon it, but not the slightest could I detect. They left no visible track, and after all had passed the glowing arc stood firm and apparently immutable, but at last faded slowly away like its glorious predecessor.

Excepting only the vast purple aurora mentioned above, said to have been visible over nearly all the continent, these two silver bows in supreme, serene, supernal beauty surpassed everything auroral I ever beheld.

ESSAYS

TWENTY HILL HOLLOW

Overland Monthly (April, 1872)

Editor's Note: This is the hub of the region where Mr. Muir spent the greater part of the summer of 1868 and the spring of 1869.

I WISH TO say a word for the great central plain of California in general, and for Twenty Hill Hollow, in Merced County, in particular; because, in reading descriptions of California scenery, by the literary racers who annually make a trial of their speed here, one is led to fancy, that, outside the touristical see-saw of Yosemite, Geysers, and Big Trees, our State contains little else worthy of note, excepting, perhaps, certain wine-cellars and vineyards, and that our great plain is a sort of Sahara, whose narrowest and least dusty crossing they benevolently light-house. But to the few travelers who are in earnest—true lovers of the truth and beauty of wildness—we would say, Heed nothing you have heard; put no questions to "agent," or guide-book, or dearest friend; cast away your watches and almanacs, and go at once to our garden wilds—the more planless and ignorant the better. Drift away confidingly into the broad gulf-streams of Nature, helmed only by Instinct. No harsh storm, no bear, no snake, will harm you. Those who submissively allow themselves to be packed and brined down in the sweats of a stage-coach, who are hurled into Yosemite by "favorite routes," are not aware that they are crossing a grander Yosemite than that to which they are going.

The whole State of California, from Siskiyou to San Diego, is one block of beauty, one matchless valley; and our great plain, with its mountain-walls, is the true California Yosemite—exactly corresponding in its physical character and proportions to that of the Merced. Moreover, as Yosemite the less is outlined in the lesser Yosemites of Indian Cañon, Glacier Cañon, Illilouette, and Pohono, so is Yosemite the great by the Yosemites of King's River, Fresno, Merced, and Tuolumne. The only important difference between the great central Yosemite—bottomed by the plain of the Sacramento and San Joaquin, and walled by the Sierras and mountains of the coast—and the Merced Yosemite—bottomed by a glacier meadow, and walled by glacier rocks—is, that the former is double—two Yosemites in one, each proceeding from a tangle of glacier cañons, meeting opposite Suisun Bay, and sending their united waters to the sea by the Golden Gate.

Were we to cross-cut the Sierra Nevada into blocks a dozen miles or so in thickness, each section would contain a Yosemite Valley and a river, together with a bright array of lakes and meadows, rocks and forests. The grandeur and inexhaustible beauty of each block would be so vast and over-satisfying that to choose among them would be like selecting slices of bread cut from the same loaf. One bread-slice might have burnt spots, answering to craters; another would be more browned; another, more crusted or raggedly cut; but all essentially the same. In no greater degree would the Sierra slices differ in general character. Nevertheless, we all would choose the Merced slice, because, being easier of access, it has been nibbled and tasted, and pronounced very good; and because of the concentrated form of its Yosemite, caused by certain conditions of baking, yeasting, and glacier-frosting of this portion of the great Sierra loaf. In like manner, we readily perceive that the great central plain is one batch of bread—one golden cake—and we are loath to leave these magnificent loaves for crumbs, however good.

After our smoky sky has been washed in the rains of winter, the whole complex row of Sierras appear from the plain as a simple wall slightly beveled, and colored in horizontal bands laid one above another, as if entirely composed of partially straightened rainbows. So, also, the plain seen from the mountains has the same simplicity of smooth surface, colored purple and yellow, like a patchwork of irised clouds. But when we descend to this smooth-furred sheet, we discover complexity in its physical conditions equal to that of the mountains, though less strongly marked. In particular, that portion of the plain lying between the Merced and the Tuolumne, within ten miles of the slaty foothills, is most elaborately carved into valleys, hollows, and smooth undulations, and among them is laid the Merced Yosemite of the plain—Twenty Hill Hollow.

This delightful Hollow is less than a mile in length, and of just sufficient width to form a well-proportioned oval. It is situated about midway between the two rivers, and five miles from the Sierra foot-hills. Its banks are formed of twenty hemispherical hills; hence its name. They surround and enclose it on all sides, leaving only one narrow opening toward the southwest for the escape of its waters. The bottom of the Hollow is about two hundred feet below the level of the surrounding plain, and the tops of its hills are slightly below the general level. Here is no towering dome, no Tissiack, to mark its place; and one may ramble close upon its rim before he is made aware of its existence. Its twenty hills are as wonderfully regular in size and position as in form. They are like big marbles half buried in the ground, each poised and settled daintily into its place at a regular distance from its fellows, making a charming fairy-land of hills, with small, grassy valleys between, each valley having a tiny stream of its own, which leaps and sparkles out into the open hollow, uniting to form Hollow Creek.

Like all others in the immediate neighborhood, these twenty hills are composed of stratified lavas mixed with mountain drift in varying proportions. Some strata are almost wholly made up of volcanic matter—lava and cinders—thoroughly ground and mixed by the waters that deposited them; others are largely composed of slate and quartz boulders of all degrees of coarseness, forming conglomerates. A few clear, open sections occur, exposing an elaborate history of seas, and glaciers, and volcanic floods—chapters of cinders and ashes that picture dark days, when these bright snowy mountains were clouded in smoke, and rivered and laked with living fire. A fearful age, say mortals, when these bright Sierras flowed lava to the sea. What horizons of flame! What atmospheres of ashes and smoke!

The conglomerates and lavas of this region are readily denuded by water. In the time when their parent sea was removed to form this golden plain, their regular surface, in great part covered with shallow lakes, showed little variation from motionless level until torrents of rain and floods from the mountains gradually sculptured the simple page to the present diversity of bank and brae, creating, in the section between the Merced and the Tuolumne, Twenty Hill Hollow, Lily Hollow, and the lovely valley of Cascade and Castle Creeks, with many others nameless and unknown, seen only by hunters and shepherds, sunk in the wide bosom of the plain, like undiscovered gold. Twenty Hill Hollow is a fine illustration of a valley created by erosion of water. Here are no Washington columns, no angular El Capitans. The hollow cañons, cut in soft lavas, are not so deep as to require a single earthquake at the hands of science, much less a baker's dozen of those convenient tools demanded for the making of mountain Yosemites, and our moderate arithmetical standards are not outraged by a single magnitude of this simple, comprehensible hollow.

The present rate of denudation of this portion of the plain seems to be about one tenth of an inch per year. This approximation is based upon observations made upon stream-banks and perennial plants. Rains and winds remove mountains without disturbing their plant or animal inhabitants. Hovering petrels, the fishes and floating plants of ocean, sink and rise in beautiful rhythm with its waves; and, in like manner, the birds and plants of the plain sink and rise with these waves of land, the only difference being that the fluctuations are more rapid in the one case than in the other.

In March and April the bottom of the Hollow and every one of its hills are smoothly covered and plushed with yellow and purple flowers, the yellow predominating. They are mostly social *Compositæ*, with a few claytonias, gilias, eschscholtzias, white and yellow violets, blue and yellow lilies, dodecatheons, and eriogonums set in a half-floating maze of purple grasses. There is but one vine in the Hollow—the *Megarrhiza* [*Echinocystis* T. & D.] or "Big Root." The only bush within a mile of it, about four feet in height, forms so remarkable an object upon the universal smoothness that my dog barks furiously around it, at a cautious distance, as if it were a bear. Some of the hills have rock ribs that are brightly colored with red and yellow lichens, and in moist nooks there are luxuriant mosses—*Bartramia*, *Dicranum*, *Funaria*, and several *Hypnums*. In cool, sunless coves the mosses are companioned with ferns—a *Cystopteris* and the little gold-dusted rock fern, *Gymnogramma triangularis.*

The Hollow is not rich in birds. The meadowlark homes there, and the little burrowing owl, the killdeer, and a species of sparrow. Occasionally a few ducks pay a visit to its waters, and a few tall herons—the blue and the white—may at times be seen stalking along the creek; and the sparrow hawk and gray eagle come to hunt. The lark, who does nearly all the singing for the Hollow, is not identical in species with the meadowlark of the East, though closely resembling it; richer flowers and skies have inspired him with a better song than was ever known to the Atlantic lark.

I have noted three distinct lark-songs here. The words of the first, which I committed to memory at one of their special meetings, spelled as sung, are "Wee-ro spee-ro wee-o weer-ly wee-it." On the

20th of January, 1869, they sang "Queed-lix boodle," repeating it with great regularity, for hours together, to music sweet as the sky that gave it. On the 22d of the same month, they sang "Chee chool cheedildy choodildy." An inspiration is this song of the blessed lark, and universally absorbable by human souls. It seems to be the only bird-song of these hills that has been created with any direct reference to us. Music is one of the attributes of matter, into whatever forms it may be organized. Drops and sprays of air are specialized, and made to plash and churn in the bosom of a lark, as infinitesimal portions of air plash and sing about the angles and hollows of sand-grains, as perfectly composed and predestined as the rejoicing anthems of worlds; but our senses are not fine enough to catch the tones. Fancy the waving, pulsing melody of the vast flower-congregations of the Hollow flowing from myriad voices of tuned petal and pistil, and heaps of sculptured pollen. Scarce one note is for us; nevertheless, God be thanked for this blessed instrument hid beneath the feathers of a lark.

The eagle does not dwell in the Hollow; he only floats there to hunt the long-eared hare. One day I saw a fine specimen alight upon a hillside. I was at first puzzled to know what power could fetch the sky-king down into the grass with the larks. Watching him attentively, I soon discovered the cause of his earthiness. He was hungry and stood watching a long-eared hare, which stood erect at the door of his burrow, staring his winged fellow mortal full in the face. They were about ten feet apart. Should the eagle attempt to snatch the hare, he would instantly disappear in the ground. Should long-ears, tired of inaction, venture to skim the hill to some neighboring burrow, the eagle would swoop above him and strike him dead with a blow of his pinions, bear him to some favorite rock table, satisfy his hunger, wipe off all marks of grossness, and go again to the sky.

Since antelopes have been driven away, the hare is the swiftest animal of the Hollow. When chased by a dog he will not seek a burrow, as when the eagle wings in sight, but skims wavily from hill to hill across connecting curves, swift and effortless as a bird-shadow. One that I measured was twelve inches in height at the shoulders. His body was eighteen inches, from nose-tip to tail. His great ears measured six and a half inches in length and two in width. His ears—which, notwithstanding their great size, he wears gracefully and becomingly—have procured for him the homely nickname, by which he is commonly known, of "Jackass rabbit." Hares are very abundant over all the plain and up in the sunny, lightly wooded foot-hills, but their range does not extend into the close pine forests.

Coyotes, or California wolves, are occasionally seen gliding about the Hollow; but they are not numerous, vast numbers having been slain by the traps and poisons of sheep-raisers. The coyote is about the size of a small shepherd-dog, beautiful and graceful in motion, with erect ears, and a bushy tail, like a fox. Inasmuch as he is fond of mutton, he is cordially detested by "sheep-men" and nearly all cultured people.

The ground-squirrel is the most common animal of the Hollow. In several hills there is a soft stratum in which they have tunneled their homes. It is interesting to observe these rodent towns in time of alarm. Their one circular street resounds with sharp, lancing out-cries of "Seekit, seek, seek, seekit!" Near neighbors, peeping cautiously half out-of-doors, engage in low, purring chat. Others, bolt upright on the doorsill or on the rock above, shout excitedly, as if calling attention to the motions and aspects of the enemy. Like the wolf, this little animal is accursed, because of his relish for grain. What a pity that Nature should have made so many small mouths palated like our own!

All the seasons of the Hollow are warm and bright, and flowers bloom through the whole year. But the grand commencement of the annual genesis of plant and insect life is governed by the setting-in of the rains, in December or January. The air, hot and opaque, is then washed and cooled. Plant seeds, which for six months have lain on the ground dry as if garnered in a farmer's bin, at once unfold their treasured life. Flies hum their delicate tunes. Butterflies come from their coffins, like cotyledons from their husks. The network of dry water-courses, spread over valleys and hollows, suddenly gushes with bright waters, sparkling and pouring from pool to pool, like dusty mummies risen from the dead and set living and laughing with color and blood. The weather grows in beauty, like a flower. Its roots in the ground develop day-clusters a week or two in size, divided by and shaded in foliage of clouds; or round hours of ripe sunshine wave and spray in sky-shadows, like racemes of berries half hidden in leaves.

These months of so-called rainy season are not filled with rain. Nowhere else in North America, perhaps in the world, are Januarys so balmed and glowed with vital sunlight. Referring to my notes of 1868 and 1869, I find that the first heavy general rain of the season fell on the 18th of December. January yielded to the Hollow, during the day, only twenty hours of rain, which was divided among six rainy days. February had only three days on which rain fell, amounting to eighteen and one half hours in all. March had five rainy days. April had three, yielding seven hours of rain. May also had three wet days, yielding nine hours of rain, and completed the so-called "rainy season" for that year, which is probably about an average one. It must be remembered that this rain record has nothing to do with what fell in the night.

The ordinary rain-storm of this region has little of that outward pomp and sublimity of structure so characteristic of the storms of the Mississippi Valley. Nevertheless, we have experienced rain-storms out on these treeless plains, in nights of solid darkness, as impressively sublime as the noblest storms of the mountains. The wind, which in settled weather blows from the northwest, veers to the southeast; the sky curdles gradually and evenly to a grainless, seamless, homogeneous cloud; and then comes the rain, pouring steadily and often driven aslant by strong winds. In 1869, more than three fourths of the winter rains came from the southeast. One magnificent storm from the northwest occurred on the 21st of March; an immense, round-browed cloud came sailing over the flowery hills in most imposing majesty, bestowing water as from a sea. The passionate rain-gush lasted only about one minute, but was nevertheless the most magnificent cataract of the sky mountains that I ever beheld. A portion of calm sky toward the Sierras was brushed with thin, white cloud-tissue, upon which the rain-torrent showed to a great height—a cloud waterfall, which, like those of Yosemite, was neither spray, rain, nor solid water. In the same year the cloudiness of January, omitting rainy days, averaged 0.32; February, 0.13; March, 0.20; April, 0.10; May, 0.08. The greater portion of this cloudiness was gathered into a few days, leaving the others blocks of solid, universal sunshine in every chink and pore.

At the end of January, four plants were in flower: a small white cress, growing in large patches; a low-set, umbelled plant, with yellow flowers; an eriogonum, with flowers in leafless spangles; and a small boragewort. Five or six mosses had adjusted their hoods, and were in the prime of life. In February, squirrels, hares, and flowers were in springtime joy. Bright plant-constellations shone everywhere about the Hollow. Ants were getting ready for work, rubbing and sunning their limbs upon the husk-piles around their doors; fat, pollen-dusted, "burly, dozing humble-bees" were rumbling among the flowers; and spiders were busy mending up old webs, or weaving new ones. Flowers were born every day, and came gushing from the ground like gayly dressed children from a church. The bright air became daily more songful with fly-wings, and sweeter with breath of plants.

In March, plant-life is more than doubled. The little pioneer cress, by this time, goes to seed, wearing daintily embroidered silicles. Several claytonias appear; also, a large white leptosiphon [?], and two nemophilas. A small plantago becomes tall enough to wave and show silky ripples of shade. Toward the end of this month or the beginning of April, plant-life is at its greatest height. Few have any just conception of its amazing richness. Count the flowers of any portion of these twenty hills, or of the bottom of the Hollow, among the streams: you will find that there are from one to ten thousand upon every square yard, counting the heads of *Compositæ* as single flowers. Yellow *Compositæ* form by far the greater portion of this goldy-way. Well may the sun feed them with his richest light, for these shining sunlets are his very children—rays of his ray, beams of his beam! One would fancy that these California days receive more gold from the ground than they give to it. The earth has indeed become a sky; and the two cloudless skies, raying toward each other flower-beams and sunbeams, are fused and congolded into one glowing heaven. By the end of April most of the Hollow plants have ripened their seeds and died; but, undecayed, still assist the landscape with color from persistent involucres and corolla-like heads of chaffy scales.

In May, only a few deep-set lilies and eriogonums are left alive. June, July, August, and September are the season of plant rest, followed, in October, by a most extraordinary out-gush of plant-life, at the very driest time of the whole year. A small, unobtrusive plant, *Hemizonia virgata*, from six inches to three feet in height, with pale, glandular leaves, suddenly bursts into bloom, in patches miles in extent, like a resurrection of the gold of April. I have counted upward of three thousand heads upon one plant. Both leaves and pedicels are so small as to be nearly invisible among so vast a number of daisy golden-heads that seem to keep their places unsupported, like stars in the sky. The heads are about five eighths of an inch in diameter; rays and disk-flowers, yellow; stamens, purple. The rays have a rich, furred appearance, like the petals of garden pansies. The prevailing summer wind makes all the heads turn to the southeast. The waxy secretion of its leaves and involucres has suggested its grim name of "tarweed," by which it is generally known. In our estimation, it is the most delightful member of the whole Composite Family of the plain. It remains in flower until November, uniting with an eriogonum that continues the floral chain across December to the spring plants of January. Thus, although nearly all of the year's plant-life is crowded into February, March, and April, the flower circle around the Twenty Hill Hollow is never broken.

The Hollow may easily be visited by tourists *en route* for Yosemite, as it is distant only about six miles from Snelling's. It is at all seasons interesting to the naturalist; but it has little that would interest the majority of tourists earlier than January or later than April. If you wish to see how much of light, life, and joy can be got into a January, go to this blessed Hollow. If you wish to see a plant-resurrection,—myriads of bright flowers crowding from the ground, like souls to a judgment,—go to Twenty Hills in February. If you are traveling for health, play truant to doctors and friends, fill your pocket with biscuits, and hide in the hills of the Hollow, lave in its waters, tan in its golds, bask in its flower-shine, and your baptisms will make you a new creature indeed. Or, choked in the sediments of society, so tired of the world, here will your hard doubts disappear, your carnal incrustations melt off, and your soul breathe deep and free in God's shoreless atmosphere of beauty and love.

Never shall I forget my baptism in this font. It happened in January, a resurrection day for many a plant and for me. I suddenly found myself on one of its hills; the Hollow overflowed with light, as a fountain, and only small, sunless nooks were kept for mosseries and ferneries. Hollow Creek spangled and mazed like a river. The ground steamed with fragrance. Light, of unspeakable richness, was brooding the flowers. Truly, said I, is California the Golden State—in metallic gold, in sun gold, and in plant gold. The sunshine for a whole summer seemed condensed into the chambers of that one glowing day. Every trace of dimness had been washed from the sky; the mountains were dusted and wiped clean with clouds—

Pacheco Peak and Mount Diablo, and the waved blue wall between; the grand Sierra stood along the plain, colored in four horizontal bands:—the lowest, rose purple; the next higher, dark purple; the next, blue; and, above all, the white row of summits pointing to the heavens.

It may be asked, What have mountains fifty or a hundred miles away to do with Twenty Hill Hollow? To lovers of the wild, these mountains are not a hundred miles away. Their spiritual power and the goodness of the sky make them near, as a circle of friends. They rise as a portion of the hilled walls of the Hollow. You cannot feel yourself out of doors; plain, sky, and mountains ray beauty which you feel. You bathe in these spirit-beams, turning round and round, as if warming at a camp-fire. Presently you lose consciousness of your own separate existence: you blend with the landscape, and become part and parcel of Nature.

A GEOLOGIST'S WINTER WALK

Overland Monthly (April, 1873)

AFTER REACHING TURLOCK, I sped afoot over the stubble-fields, and through miles of brown *Hemizonia* and purple *Erigeron*, to Hopeton, conscious of little more than that the town was behind and beneath me, and the mountains above and before me; on through the oaks and chaparral of the foot-hills to Coulterville, and then ascended the first great mountain step upon which grows the sugar-pine. Here I slackened pace, for I drank the spicy, resiny wind, and was at home—never did pine-trees seem so dear. How sweet their breath and their song, and how grandly they winnowed the sky. I tingled my fingers among their tassels, and rustled my feet among their brown needles and burrs.

When I reached the valley, all the rocks seemed talkative, and more lovable than ever. They are dear friends, and have warm blood gushing through their granite flesh; and I love them with a love intensified by long and close companionship. After I had bathed in the bright river, sauntered over the meadows, conversed with the domes, and played with the pines, I still felt muddy, and weary, and tainted with the sticky sky of your streets; I determined, therefore, to run out to the higher temples. "The days are sunful," I said, "and though now winter, no great danger need be encountered, and a sudden storm will not block my return, if I am watchful."

The morning after this decision, I started up the Cañon of Tenaya, caring little about the quantity of bread I carried; for, I thought, a fast and a storm and a difficult cañon are just the medicine I require. When I passed Mirror Lake, I scarcely noticed it, for I was absorbed in the great Tissiack—her crown a mile away in the hushed azure; her purple drapery flowing in soft and graceful folds low as my feet, embroidered gloriously around with deep, shadowy forest. I have gazed on Tissiack a thousand times—in days of solemn storms, and when her form shone divine with jewels of winter, or was veiled in living clouds; I have heard her voice of winds, or snowy, tuneful waters; yet never did her soul reveal itself more impressively than now. I hung about her skirts, lingering timidly, till the glaciers compelled me to push up the cañon. This cañon is accessible only to determined mountaineers, and I was anxious to carry my barometer and chronometer through it, to obtain sections and altitudes. After I had passed the tall groves that stretch a mile above Mirror Lake, and scrambled around the Tenaya Fall, which is just at the head of the lake groves, and crept through the dense and spiny chaparral that plushes the roots of all the mountains here for miles, in warm, unbroken green, and was ascending a precipitous rock-front, where the foot-holds were good, I suddenly stumbled, for the first time since I touched foot to Sierra rocks. After several involuntary somersaults, I became insensible, and when consciousness returned, I found myself wedged among short, stiff bushes, not injured in the slightest. Judging by the sun, I could not have been insensible very long; probably not a minute, possibly an hour; and I could not remember what made me fall, or where I had fallen from; but I saw that if I had rolled a little further, my mountain-climbing would have been finished. "There," said I, addressing my feet, to whose separate skill I had learned to trust night and day on any mountain, "that is what you get by intercourse with stupid town stairs, and dead pavements." I felt angry and worthless. I had not reached yet the difficult portion of the cañon, but I determined to guide my humbled body over the highest practicable precipices, in the most intricate and nerve-trying places I could find; for I was now fairly awake, and felt confident that the last town-fog had been shaken from both head and feet.

I camped at the mouth of a narrow gorge, which is cut into the bottom of the main cañon, determined to take earnest exercise next day. No plush boughs did my ill-behaved bones receive that night, nor did my bumped head get any spicy cedar-plumes for pillow. I slept on a naked boulder, and when I awoke all my nervous trembling was gone.

The gorged portion of the cañon, in which I spent all the next day, is about a mile and a half in length; and I passed the time very profitably in tracing the action of the forces that determined this peculiar bottom gorge, which is an abrupt, ragged-walled, narrow-throated cañon, formed in the bottom of a wide-mouthed, smooth, and beveled cañon. I will not stop now to tell you more; some day you may see it, like a shadowy line, from Clouds' Rest. In high water, the stream occupies all the bottom of the gorge, surging and chafing glorious power from wall to wall, but the sound of the grinding was low as I entered the gorge, scarcely hoping to be able to pass through its entire length. By cool efforts, along glassy, ice-worn slopes, I reached the upper end in a little over a day, but was compelled to pass the second night in the gorge, and in the moonlight I wrote you this short pencil-letter in my note-book:

"The moon is looking down into the cañon, and how marvelously the great rocks kindle to her light—every dome, and brow, and swelling boss touched by her white rays, glows, as if lighted with snow. I am now only a mile from last night's camp; and have been climbing and sketching all day in this difficult but instructive gorge. It is formed in the bottom of the main cañon, among the roots of Clouds' Rest. It begins at the dead lake where I camped last night, and ends a few hundred yards above, in another dead lake. The walls everywhere are craggy and vertical, and in some places they overlean. It is only from twenty to sixty feet wide, and not, though black and broken enough, the thin, crooked mouth of some mysterious abyss; for in many places I saw the solid, seamless floor. I am sitting on a big stone, against which the stream divides, and goes brawling by in rapids on both sides; half my rock is white in the light, half in shadow. Looking from the opening jaws of this shadowy gorge, South Dome is immediately in front—high in the stars, her face turned from the moon, with the rest of her body gloriously muffled in waved folds of granite. On the left, cut from Clouds' Rest, by the lip of the gorge, are three magnificent rocks, sisters of the great South Dome. On the right is the massive, moonlit front of Mount Watkins, and between, low down in the furthest distance, is Sentinel Dome, girdled and darkened with forest. In the near foreground is the joyous creek, Tenaya, singing against boulders that are white with the snow. Now, look back twenty yards, and you will see a waterfall, fair as a spirit; the moonlight just touches it, bringing it in relief against the deepest, dark background. A little to the left, and a dozen steps this side of the fall, a flickering light marks my camp—and a precious camp it is. A huge, glacier-polished slab, in falling from the glassy flank of Clouds' Rest, happened to settle on edge against the wall of the gorge. I did not know that this slab was glacier-polished, until I lighted my fire. Judge of my delight. I think it was sent here by an earthquake. I wish I could take it down to the valley. It is about twelve feet square. Beneath this slab is the only place in this torrent-swept gorge where I have seen sand sufficient for a bed. I expected to sleep on the boulders, for I spent most of the afternoon on the slippery wall of the cañon, endeavoring to get around this difficult part of the gorge, and was compelled to hasten down here for water before dark. I will sleep soundly on this sand; half of it is mica. Here, wonderful to behold, are a few green stems of prickly *Rubus*, and a tiny grass. They are here to meet us. Ay, even here, in this darksome gorge, 'frightful and tormented' with raging torrents and choking avalanches of snow. Can it be? As if the *Rubus* and the grass-leaf were not enough of God's tender prattle-words of love, which we so much need in these mighty temples of power, yonder in the 'benmost bore' are two blessed *Adiantums*. Listen to them. How wholly infused with God is this one big word of love that we call the world! Good-night. Do you see the fire-glow on my ice-smoothed slab, and on my two ferns? And do you hear how sweet a sleep-song the fall and cascades are singing?"

The water-ground chips and knots that I found fastened between rocks, kept my fire alive all through the night, and I rose nerved and ready for another day of sketching and noting, and any form of climbing. I escaped from the gorge about noon, after accomplishing some of the most delicate feats of mountaineering I ever attempted; and here the cañon is all broadly open again—a dead lake, luxuriantly forested with pine, and spruce, and silver fir, and brown-trunked *Librocedrus*. The walls rise in Yosemite forms, and the stream comes down 700 feet, in a smooth brush of foam. This is a genuine Yosemite valley. It is about 2000 feet above the level of Yosemite, and about 2400 below Lake Tenaya. Lake Tenaya was frozen, and the ice was so clear and unruffled, that the mountains and the groves that looked upon it were reflected almost as perfectly as I ever beheld them in the calm evening mirrors of summer. At a little distance, it was difficult to believe the lake frozen at all; and when I walked out on it, cautiously stamping at short intervals to test the strength of the ice, I seemed to walk mysteriously, without any adequate faith, on the surface of the water. The ice was so transparent, that I could see the beautifully wave-rippled, sandy bottom, and the scales of mica glinting back the down-pouring light. When I knelt down with my face close to the ice, through which clear sunshine was pouring, I was delighted to discover myriads of Tyndall's six-sided ice-flowers, magnificently colored. A grand old mountain mansion is this Tenaya region. In the glacier period, it was a *mer de glace*, far grander than the *mer de glace* of Switzerland, which is only about half a mile broad. The Tenaya *mer de glace* was not less than two miles broad, late in the glacier epoch, when all the principal dividing crests were bare; and its depth was not less than fifteen hundred feet. Ice-streams from Mounts Lyell and Dana, and all the mountains between, and from the nearer Cathedral Peak, flowed hither, welded into one, and worked together. After accomplishing this Tenaya Lake basin, and all the splendidly-sculptured rocks and mountains that surround and adorn it, and the great Tenaya Cañon, with its wealth of all that makes mountains sublime, they were welded with the vast South Lyell and Illilouette glaciers on one side, and with those of Hoffmann on the other—thus forming a portion of a yet grander *mer de glace.*

Now your finger is raised admonishingly, and you say, "This letter-writing will not do." Therefore, I will not try to register my homeward ramblings; but since this letter is already so long, you must allow me to tell you of Clouds' Rest and Tissiack; then will I cast away my letter pen, and begin "Articles," rigid as granite and slow as glaciers.

I reached the Tenaya Cañon, on my way home, by coming in from the northeast, rambling down over the shoulders of Mount Watkins, touching bottom a mile above Mirror Lake. From thence home was but a saunter in the moonlight. After resting one day, and the weather continuing calm, I ran up over the east shoulder of the South Dome, and down in front of its grand split face, to make some measurements; completed my work, climbed to the shoulder again, and struck off along the ridge for Clouds' Rest, and reached the topmost sprays of her sunny wave in ample time for sunset. Clouds' Rest is a thousand feet higher than Tissiack. It is a wave-like crest upon a ridge, which begins at Yosemite with Tissiack, and runs continuously eastward to the thicket of peaks and crests around Tenaya. This lofty granite wall is bent this way and that by the restless and weariless action of glaciers, just as if it had been made of dough—semi-plastic, as Prof. Whitney would say. But the grand circumference of mountains and forests are coming from far and near, densing into one close assemblage; for the sun, their god and father, with love ineffable, is glowing a sunset farewell. Not one of all the assembled rocks or trees seemed remote. How impressively their faces shone with responsive love!

I ran home in the moonlight, with long, firm strides; for the sun-love made me strong. Down through the junipers—down through the firs; now in jet-shadows, now in white light; over sandy moraines and bare, clanking rock; past the huge ghost of South Dome, rising weird through the firs—past glorious Nevada—past the groves of Illilouette—through the pines of the valley; frost-crystals flashing all the sky beneath, as star-crystals on all the sky above. All of this big mountain-bread for one day! One of the rich, ripe days that enlarge one's life—so much of sun upon one side of it, so much of moon on the other.

WILD WOOL

Overland Monthly (April, 1875)

MORAL IMPROVERS HAVE calls to preach. I have a friend who has a call to plough, and woe to the daisy sod or azalea thicket that falls under the savage redemption of his keen steel shares. Not content with the so-called subjugation of every terrestrial bog, rock, and moorland, he would fain discover some method of reclamation applicable to the ocean and the sky, that in due calendar time they might be brought to bud and blossom as the rose. Our efforts are of no avail when we seek to turn his attention to wild roses, or to the fact that both ocean and sky are already about as rosy as possible—the one with stars, the other with dulse, and foam, and wild light. The practical developments of his culture are orchards and clover-fields wearing a smiling, benevolent aspect, truly excellent in their way, though a near view discloses something barbarous in them all. Wildness charms not my friend, charm it never so wisely: and whatsoever may be the character of his heaven, his earth seems only a chaos of agricultural possibilities calling for grubbing-hoes and manures.

Sometimes I venture to approach him with a plea for wildness, when he good-naturedly shakes a big mellow apple in my face, reiterating his favorite aphorism, "Culture is an orchard apple; Nature is a crab." Not all culture, however, is equally destructive and inappreciative. Azure skies and crystal waters find loving recognition, and few there be who would welcome the ax among mountain pines, or would care to apply any correction to the tones and costumes of mountain waterfalls. Nevertheless, the barbarous notion is almost universally entertained by civilized man, that there is in all the manufactures of Nature something essentially coarse which can and must be eradicated by human culture. I was, therefore, delighted in finding that the wild wool growing upon mountain sheep in the neighborhood of Mount Shasta was much finer than the average grades of cultivated wool. This *fine* discovery was made some three months ago, while hunting among the Shasta sheep between Shasta and Lower Klamath Lake. Three fleeces were obtained—one that belonged to a large ram about four years old, another to a ewe about the same age, and another to a yearling lamb. After parting their beautiful wool on the side and many places along the back, shoulders, and hips, and examining it closely with my lens, I shouted: "Well done for wildness! Wild wool is finer than tame!"

My companions stooped down and examined the fleeces for themselves, pulling out tufts and ringlets, spinning them between their fingers, and measuring the length of the staple, each in turn paying tribute to wildness. It *was* finer, and no mistake; finer than Spanish Merino. Wild wool *is* finer than tame.

"Here," said I, "is an argument for fine wildness that needs no explanation. Not that such arguments are by any means rare, for all wildness is finer than tameness, but because fine wool is appreciable by everybody alike—from the most speculative president of national wool-growers' associations all the way down to the gude-wife spinning by her ingleside."

Nature is a good mother, and sees well to the clothing of her many bairns—birds with smoothly imbricated feathers, beetles with shining jackets, and bears with shaggy furs. In the tropical south, where the sun warms like a fire, they are allowed to go thinly clad; but in the snowy northland she takes care to clothe warmly. The squirrel has socks and mittens, and a tail broad enough for a blanket; the grouse is densely feathered down to the ends of his toes; and the wild sheep, besides his undergarment of fine wool, has a thick overcoat of hair that sheds off both the snow and the rain. Other provisions and adaptations in the dresses of animals, relating less to climate than to the more mechanical circumstances of life, are made with the same consummate skill that characterizes all the love-work of Nature. Land, water, and air, jagged rocks, muddy ground, sand-beds, forests, underbrush, grassy plains, etc., are considered in all their possible combinations while the clothing of her beautiful wildlings is preparing. No matter what the circumstances of their lives may be, she never allows them to go dirty or ragged. The mole, living always in the dark and in the dirt, is yet as clean as the otter or the wave-washed seal; and our wild sheep, wading in snow, roaming through bushes, and leaping among jagged storm-beaten cliffs, wears a dress so exquisitely adapted to its mountain life that it is always found as unruffled and stainless as a bird.

On leaving the Shasta hunting-grounds I selected a few specimen tufts, and brought them away with a view to making more leisurely examinations; but, owing to the imperfectness of the instruments at my command, the results thus far obtained must be regarded only as rough approximations.

As already stated, the clothing of our wild sheep is composed of fine wool and coarse hair. The hairs are from about two to four inches long, mostly of a dull bluish-gray color, though varying somewhat with the seasons. In general characteristics they are closely related to the hairs of the deer and antelope, being light, spongy, and elastic, with a highly polished surface, and though somewhat ridged and spiraled, like wool, they do not manifest the slightest tendency to felt or become taggy. A hair two and a half inches long, which is perhaps near the average length, will stretch about one fourth of an inch before breaking. The diameter decreases rapidly both at the top and bottom, but is maintained throughout the greater portion of the length with a fair degree of regularity. The slender tapering point in which the hairs terminate is nearly black: but, owing to its fineness as compared with the main trunk, the quantity of blackness is not sufficient to affect greatly the general color. The number of hairs growing upon a square inch is about ten thousand; the number of wool fibers is about twenty-five thousand, or two and a half times that of the hairs. The wool fibers are white and glossy, and beautifully spired into ringlets. The average length of the staple is about an inch and a half. A fiber of this length, when growing undisturbed down among the hairs, measures about an inch; hence the degree of curliness may easily be inferred. I regret exceedingly that my instruments do not enable me to measure the diameter of the fibers, in order that their degrees of fineness might be definitely compared with each other and with the finest of the domestic breeds; but that the three wild fleeces under consideration are considerably finer than the average grades of Merino shipped from San Francisco is, I think, unquestionable.

When the fleece is parted and looked into with a good lens, the skin appears of a beautiful pale-yellow color, and the delicate wool fibers are seen growing up among the strong hairs, like grass among stalks of corn, every individual fiber being protected about as specially and effectively as if inclosed in a separate husk. Wild wool is too fine to stand by itself, the fibers being about as frail and invisible as the floating threads of spiders, while the hairs against which they lean stand erect like hazel wands; but, notwithstanding their great dissimilarity in size and appearance, the wool and hair are forms of the same thing, modified in just that way and to just that degree that renders them most perfectly subservient to the well-being of the sheep. Furthermore, it will be observed that these wild modifications are entirely distinct from those which are brought chancingly into existence through the accidents and caprices of culture; the former being inventions of God for the attainment of definite ends. Like the modifications of limbs—the fin for swimming, the wing for flying, the foot for walking—so the fine wool for warmth, the hair for additional warmth and to protect the wool, and both together for a fabric to wear well in mountain roughness and wash well in mountain storms.

The effects of human culture upon wild wool are analogous to those produced upon wild roses. In the one case there is an abnormal development of petals at the expense of the stamens, in the other an abnormal development of wool at the expense of the hair. Garden roses frequently exhibit stamens in which the transmutation to petals may be observed in various stages of accomplishment, and analogously the fleeces of tame sheep occasionally contain a few wild hairs that are undergoing transmutation to wool. Even wild wool presents here and there a fiber that appears to be in a state of change. In the course of my examinations of the wild fleeces mentioned above, three fibers were found that were wool at one end and hair at the other. This, however, does not necessarily imply imperfection, or any process of change similar to that caused by human culture. Water-lilies contain parts variously developed into stamens at one end, petals at the other, as the constant and normal condition. These half wool, half hair fibers may therefore subserve some fixed requirement essential to the perfection of the whole, or they may simply be the fine boundary-lines where an exact balance between the wool and the hair is attained.

I have been offering samples of mountain wool to my friends, demanding in return that the fineness of wildness be fairly recognized and confessed, but the returns are deplorably tame. The first question asked is, "Now truly, wild sheep, wild sheep, have you any wool?" while they peer curiously down among the hairs through lenses and spectacles. "Yes, wild sheep, you *have* wool; but Mary's lamb had more. In the name of use, how many wild sheep, think you, would be required to furnish wool sufficient for a pair of socks?"

I endeavor to point out the irrelevancy of the latter question, arguing that wild wool was not made for man but for sheep, and that, however deficient as clothing for other animals, it is just the thing for the brave mountain-dweller that wears it. Plain, however, as all this appears, the quantity question rises again and again in all its commonplace tameness. For in my experience it seems well-nigh impossible to obtain a hearing on behalf of Nature from any other standpoint than that of human use. Domestic flocks yield more flannel per sheep than the wild, therefore it is claimed that culture has improved upon wildness; and so it has as far as flannel is concerned, but all to the contrary as far as a sheep's dress is concerned. If every wild sheep inhabiting the Sierra were to put on tame wool, probably only a few would survive the dangers of a single season. With their fine limbs muffled and buried beneath a tangle of hairless wool, they would become short-winded, and fall an easy prey to the strong mountain wolves. In descending precipices they would be thrown out of balance and killed, by their taggy wool catching upon sharp points of rocks. Disease would also be brought on by the dirt which always finds a lodgment in tame wool, and by the draggled and water-soaked condition into which it falls during stormy weather.

No dogma taught by the present civilization seems to form so insuperable an obstacle in the way of a right understanding of the relations which culture sustains to wildness as that which regards the world as made especially for the uses of man. Every animal, plant, and crystal controverts it in the plainest terms. Yet it is taught from century to century as something ever new and precious, and in the resulting darkness the enormous conceit is allowed to go unchallenged.

I have never yet happened upon a trace of evidence that seemed to show that any one animal was ever made for another as much as it was made for itself. Not that Nature manifests any such thing as selfish isolation. In the making of every animal the presence of every other animal has been recognized. Indeed, every atom in creation may be said to be acquainted with and married to every other, but with universal union there is a division sufficient in degree for the purposes of the most intense individuality; no matter, therefore, what may be the note which any creature forms in the song of existence, it is made first for itself, then more and more remotely for all the world and worlds.

Were it not for the exercise of individualizing cares on the part of Nature, the universe would be felted together like a fleece of tame wool. But we are governed more than we know, and most when we are wildest. Plants, animals, and stars are all kept in place, bridled along appointed ways, *with* one another, and *through the midst* of one another—killing and being killed, eating and being eaten, in harmonious proportions and quantities. And it is right that we should thus reciprocally make use of one another, rob, cook, and consume, to the utmost of our healthy abilities and desires. Stars attract one another as they are able, and harmony results. Wild lambs eat as many wild flowers as they can find or desire, and men and wolves eat the lambs to just the same extent.

This consumption of one another in its various modifications is a kind of culture varying with the degree of directness with which it is carried out, but we should be careful not to ascribe to such culture any improving qualities upon those on whom it is brought to bear. The water-ouzel plucks moss from the river-bank to build its nest, but it does not improve the moss by plucking it. We pluck feathers from birds, and less directly wool from wild sheep, for the manufacture of clothing and cradle-nests, without improving the wool for the sheep, or the feathers for the bird that wore them. When a hawk pounces upon a linnet and proceeds to pull out its feathers, preparatory to making a meal, the hawk may be said to be cultivating the linnet, and he certainly does effect an improvement as far as hawk-food is concerned; but what of the songster? He ceases to be a linnet as soon as he is snatched from the woodland choir; and when, hawk-like, we snatch the wild sheep from its native rock, and, instead of eating and wearing it at once, carry it home, and breed the hair out of its wool and the bones out of its body, it ceases to be a sheep.

These breeding and plucking processes are similarly improving as regards the secondary uses aimed at; and, although the one requires but a few minutes for its accomplishment, the other many years or centuries, they are essentially alike. We eat wild oysters alive with great directness, waiting for no cultivation, and leaving scarce a second of distance between the shell and the lip; but we take wild sheep home and subject them to the many extended processes of husbandry, and finish by boiling them in a pot—a process which completes all sheep improvements as far as man is concerned. It will be seen, therefore, that wild wool and tame wool—wild sheep and tame sheep—are terms not properly comparable, nor are they in any correct sense to be considered as bearing any antagonism toward each other; they are different things, planned and accomplished for wholly different purposes.

Illustrative examples bearing upon this interesting subject may be multiplied indefinitely, for they abound everywhere in the plant and animal kingdoms wherever culture has reached. Recurring for a moment to apples. The beauty and completeness of a wild apple tree living its own life in the woods is heartily acknowledged by all those who have been so happy as to form its acquaintance. The fine wild piquancy of its fruit is unrivaled, but in the great question of quantity as human food wild apples are found wanting. Man, therefore, takes the tree from the woods, manures and prunes and grafts, plans and guesses, adds a little of this and that, selects and rejects, until apples of every conceivable size and softness are produced, like nut-galls in response to the irritating punctures of insects. Orchard apples are to me the most eloquent words that culture has ever spoken, but they reflect no imperfection upon Nature's spicy crab. Every cultivated apple is a crab, not improved, *but cooked,* variously softened and swelled out in the process, mellowed, sweetened, spiced, and rendered pulpy and foodful, but as utterly unfit for the uses of nature as a meadowlark killed and plucked and roasted. Give to Nature every cultured apple—codling, pippin, russet—and every sheep so laboriously compounded—muffled Southdowns, hairy Cotswolds, wrinkled Merinos—and she would throw the one to her caterpillars, the other to her wolves.

It is now some thirty-six hundred years since Jacob kissed his mother and set out across the plains of Padan-aram to begin his experiments upon the flocks of his uncle, Laban; and, notwithstanding the high degree of excellence he attained as a wool-grower, and the innumerable painstaking efforts subsequently made by individuals and associations in all kinds of pastures and climates, we still seem to be as far from definite and satisfactory results as we ever were. In one breed the wool is apt to wither and crinkle like hay on a sunbeaten hillside. In another, it is lodged and matted together like the lush tangled grass of a manured meadow. In one the staple is deficient in length, in another in fineness; while in all there is a constant tendency toward disease, rendering various washings and dippings indispensable to prevent its falling out. The problem of the quality and quantity of the carcass seems to be as doubtful and as far removed from a satisfactory solution as that of the wool. Desirable breeds blundered upon by long series of groping experiments are often found to be unstable and subject to disease—bots, foot-rot, blind-staggers, etc.—causing infinite trouble, both among breeders and manufacturers. Would it not be well, therefore, for some one to go back as far as possible and take a fresh start?

The source or sources whence the various breeds were derived is not positively known, but there can be hardly any doubt of their being descendants of the four or five wild species so generally distributed throughout the mountainous portions of the globe, the marked differences between the wild and domestic species being readily accounted for by the known variability of the animal, and by the long series of painstaking selection to which all its characteristics have been subjected. No other animal seems to yield so submissively to the manipulations of culture. Jacob controlled the color of his flocks merely by causing them to stare at objects of the desired hue; and possibly Merinos may have caught their wrinkles from the perplexed brows of their breeders. The California species (*Ovis montana*) is a noble animal, weighing when full-grown some three hundred and fifty pounds, and is well worthy the attention of wool-growers as a point from which to make a new departure. That it will breed with the domestic sheep I have not the slightest doubt, and I cordially recommend the experiment to the various wool-growers' associations as one of great national importance. From my knowledge of the homes and habits of our wild sheep I feel confident that several hundred could be obtained for breeding purposes from the Sierra alone, and I am ready to undertake their capture. A little pure wildness is the one great present want, both of men and sheep.

GOD'S FIRST TEMPLES

HOW SHALL WE PRESERVE OUR FORESTS?

Sacramento Daily Union (February 5, 1876)

EDS. RECORD-UNION: The forests of coniferous trees growing on our mountain-ranges are by far the most destructible of the natural resources of California. Our gold, and silver, and cinnabar are stored in the rocks, locked up in the safest of all banks, so that notwithstanding the world has been making a run upon them for the last twenty-five years, they still pay out steadily, and will probably continue to do so centuries hence, like rivers pouring from perennial mountain fountains. The riches of our magnificent soil-beds are also comparatively safe, because even the most barbarous methods of wildcat farming cannot effect complete destruction, and however great the impoverishment produced, full restoration of fertility is always possible to the enlightened farmer. But our forest belts are being burned and cut down and wasted like a field of unprotected grain, and once destroyed can never be wholly restored even by centuries of persistent and painstaking cultivation.

The practical importance of the preservation of our forests is augmented by their relations to climate, soil and streams. Strip off the woods with their underbrush from the mountain flanks, and the whole State, the lowlands as well as the highlands, would gradually change into a desert. During rainfalls, and when the winter snow was melting, every stream would become a destructive torrent, overflowing its banks, stripping off and carrying away the fertile soils, filling up the lower river channels, and overspreading the lowland fields with detritus to a vastly more destructive degree than all the washings from hydraulic mines concerning which we now hear so much. Dripping forests give rise to moist sheets and currents of air, and the sod of grasses and underbrush thus fostered, together with the roots of trees themselves, absorb and hold back rains and melting snow, yet allowing them to ooze and percolate and flow gently in useful fertilizing streams. Indeed every pine needle and rootlet, as well as fallen trunks and large clasping roots, may be regarded as dams, hoarding the bounty of storm clouds, and dispensing it as blessings all through the summer, instead of allowing it to gather and rush headlong in short-lived devastating floods. Streams taking their rise in deep woods flow unfailingly as those derived from the eternal ice and snow of the Alps. So constant indeed and apparent is the relationship between forests and never-failing springs, that effect is frequently mistaken for cause, it being often asserted that fine forests will grow only along streamsides where their roots are well watered, when in fact the forests themselves produce many of the streams flowing through them.

The main forest belt of the Sierra is restricted to the western flank, and extends unbrokenly from one end of the range to the other at an elevation of from three to eight thousand feet above sea-level. The great master-existence of these noble woods is Sequoia gigantea, or big tree. Only two species of sequoia are known to exist in the world. Both belong to California, one being found only in the Sierra, the other (Sequoia sempervirens) in the Coast Ranges, although no less than five distinct fossil species have been discovered in the tertiary and cretaceous rocks of Greenland. I would like to call attention to this noble tree, with special reference to its preservation. The species extends from the well-known Calaveras groves on the north, to the head of Deer creek on the south, near the big bend of the Kern river, a distance of about two hundred miles, at an elevation above sea-level of from about five to eight thousand feet. From the Calaveras to the South Fork of King's river it occurs only in small isolated groves, and so sparsely and irregularly distributed that two gaps occur nearly forty miles in width, the one between the Calaveras and Tuolumne groves, the other between those of the Fresno and King's rivers. From King's river the belt extends across the broad, rugged basins of the Kaweah and Tule rivers to its southern boundary on Deer creek, interrupted only by deep, rocky canyons, the width of this portion of the belt being from three to ten miles.

In the northern groves few young trees or saplings are found ready to take the places of the failing old ones, and because these ancient, childless sequoias are the only ones known to botanists, the species has been generally regarded as doomed to speedy extinction, as being nothing more than an expiring remnant of an ancient flora, and that therefore there is no use trying to save it or to prolong its few dying clays. This, however, is in the main a mistaken notion, for the Sierra as it now exists never had an ancient flora. All the species now growing on the range have been planted since the close of the glacial period, and the Big Tree has never formed a greater part of these post-glacial forests than it does today, however widely it may have been distributed throughout pre-glacial forests.

In tracing the belt southward, all the phenomena bearing upon its history goes to show that the dominion of Sequoia gigantea, as King of California trees, is not yet passing away. No tree in the woods seems more firmly established, or more safely settled in accordance with climate and soil. They fill the woods and form the principal tree, growing heartily on solid ledges, along water courses, in the deep, moist soil of meadows, and upon avalanche and glacial debris, with a multitude of thrifty seedlings and saplings crowding around the aged, ready to take their places and rule the woods.

Nevertheless Nature in her grandly deliberate way keeps up a rotation of forest crops. Species develop and die like individuals, animal as well as plant. Man himself will as surely become extinct as sequoia or mastodon, and be at length known only as a fossil. Changes of this kind are, however, exceedingly slow in their movements, and, as far as the lives of individuals are concerned, such changes have no appreciable effect. Sequoia seems scarcely further past prime as a species than its companion firs (*Picea amabilis* and *P. grandis*), and judging from its present condition and its ancient history, as far as I have been able to decipher it, our sequoia will live and flourish gloriously until A.D. 15,000 at least—probably for longer—that is, if it be allowed to remain in the hands of Nature.

But waste and pure destruction are already taking place at a terrible rate, and unless protective measures be speedily invented and enforced, in a few years this noblest tree-species in the world will present only a few hacked and scarred remnants. The great enemies of forests are fire and the ax. The destructive effects of these, as compared with those caused by the operations of nature, are instantaneous. Floods undermine and kill many a tree, storm-winds bend and break, landslips and avalanches overwhelm whole groves, lightning shatters and burns, but the combined effects of all these amount only to a wholesome beauty-producing culture. Last summer I found some five saw mills located in or near the lower edge of the Sequoia belt, all of which saw more or less of the big tree into lumber. One of these (Hyde's), situated on the north fork of the Kaweah, cut no less than 2,000,000 feet of Sequoia lumber last season. Most of the Fresno big trees are doomed to feed the mills recently erected near them, and a company has been formed by Chas. Converse to cut the noble forest on the south fork of King's river. In these milling operations waste far exceeds use. After the choice young manageable trees have been felled, the woods are cleared of limbs and refuse by burning, and in these clearing fires, made with reference to further operations, all the young seedlings and saplings are destroyed, together with many valuable fallen trees and old trees, too large to be cut, thus effectually cutting off all hopes of a renewal of the forest.

These ravages, however, of mill-fires and mill-axes are small as compared with those of the "sheep-men's" fires. Incredible numbers of sheep are driven to the mountain pastures every summer, and in order to make easy paths and to improve the pastures, running fires are set everywhere to burn off the old logs and underbrush. These fires are far more universal and destructive than would be guessed. They sweep through nearly the entire forest belt of the range from one extremity to the other, and in the dry weather, before the coming on of winter storms, are very destructive to all kinds of young trees, and especially to sequoia, whose loose, fibrous bark catches and burns at once. Excepting the Calaveras, I, last summer, examined every sequoia grove in the range, together with the main belt extending across the basins of Kaweah and Tule, and found everywhere the most deplorable waste from this cause. Indians burn off underbrush to facilitate deer-hunting. Campers of all kinds often permit fires to run, so also do mill-men, but the fires of "sheep-men" probably form more than 90 per cent. of all destructive fires that sweep the woods.

Fire, then, is the arch destroyer of our forests, and sequoia forests suffer most of all. The young trees are most easily fire killed; the old are most easily burned, and the prostrate trunks, which *never rot* and would remain valuable until our tenth centennial, are reduced to ashes.

In European countries, especially in France, Germany, Italy and Austria, the economies of forestry have been carefully studied under the auspices of Government, with the most beneficial results. Whether our loose-jointed Government is really able or willing to do anything in the matter remains to be seen. If our law makers were to discover and enforce any method tending to lessen even in a small degree the destruction going on, they would thus cover a multitude of legislative sins in the eyes of every tree lover. I am satisfied, however, that the question can be intelligently discussed only after a careful survey of our forests has been made, together with studies of the forces now acting upon them.

A law was constructed some years ago making the cutting down of sequoias over sixteen feet in diameter illegal. A more absurd and shortsighted piece of legislation could not be conceived. All the young trees might be cut and burned, and all the old ones might be burned but not cut.

THE AMERICAN FORESTS

from *Our National Parks* (1901)

THE FORESTS OF America, however slighted by man, must have been a great delight to God; for they were the best he ever planted. The whole continent was a garden, and from the beginning it seemed to be favored above all the other wild parks and gardens of the globe. To prepare the ground, it was rolled and sifted in seas with infinite loving deliberation and forethought, lifted into the light, submerged and warmed over and over again, pressed and crumpled into folds and ridges, mountains, and hills, subsoiled with heaving volcanic fires, ploughed and ground and sculptured into scenery and soil with glaciers and rivers,—every feature growing and changing from beauty to beauty, higher and higher. And in the fullness of time it was planted in groves, and belts, and broad, exuberant, mantling forests, with the largest, most varied, most fruitful, and most beautiful trees in the world. Bright seas made its border, with wave embroidery and icebergs; gray deserts were outspread in the middle of it, mossy tundras on the north, savannas on the south, and blooming prairies and plains; while lakes and rivers shone through all the vast forests and openings, and happy birds and beasts gave delightful animation. Everywhere, everywhere over all the blessed continent, there were beauty and melody and kindly, wholesome, foodful abundance.

These forests were composed of about five hundred species of trees, all of them in some way useful to man, ranging in size from twenty-five feet in height and less than one foot in diameter at the ground to four hundred feet in height and more than twenty feet in diameter,—lordly monarchs proclaiming the gospel of beauty like apostles. For many a century after the ice-ploughs were melted, Nature fed them and dressed them every day,—working like a man, a loving, devoted, painstaking gardener; fingering every leaf and flower and mossy furrowed bole; bending, trimming, modeling, balancing; painting them with the loveliest colors; bringing over them now clouds with cooling shadows and showers, now sunshine; fanning them with gentle winds and rustling their leaves; exercising them in every fiber with storms, and pruning them; loading them with flowers and fruit, loading them with snow, and ever making them more beautiful as the years rolled by. Wide-branching oak and elm in endless variety, walnut and maple, chestnut and beech, ilex and locust, touching limb to limb, spread a leafy translucent canopy along the coast of the Atlantic over the wrinkled folds and ridges of the Alleghanies,—a green billowy sea in summer, golden and purple in autumn, pearly gray like a steadfast frozen mist of interlacing branches and sprays in leafless, restful winter.

To the southward stretched dark, level-topped cypresses in knobby, tangled swamps, grassy savannas in the midst of them like lakes of light, groves of gay, sparkling spice-trees, magnolias and palms, glossy-leaved and blooming and shining continually. To the northward, over Maine and Ottawa, rose hosts of spirey, rosiny evergreens,—white pine and spruce, hemlock and cedar, shoulder to shoulder, laden with purple cones, their myriad needles sparkling and shimmering, covering hills and swamps, rocky headlands and domes, ever bravely aspiring and seeking the sky; the ground in their shade now snow-clad and frozen, now mossy and flowery; beaver meadows here and there, full of lilies and grass; lakes gleaming like eyes, and a silvery embroidery of rivers and creeks watering and brightening all the vast glad wilderness.

Thence westward were oak and elm, hickory and tupelo, gum and liriodendron, sassafras and ash, linden and laurel, spreading on ever wider in glorious exuberance over the great fertile basin of the Mississippi, over damp level bottoms, low dimpling hollows, and round dotting hills, embosoming sunny prairies and cheery park openings, half sunshine, half shade; while a dark wilderness of pines covered the region around the Great Lakes. Thence still westward swept the forests to right and left around grassy plains and deserts a thousand miles wide: irrepressible hosts of spruce and pine, aspen and willow, nut-pine and juniper, cactus and yucca, caring nothing for drought, extending undaunted from mountain to mountain, over mesa and desert, to join the darkening multitudes of pines that covered the high Rocky ranges and the glorious forests along the coast of the moist and balmy Pacific, where new species of pine, giant cedars and spruces, silver firs and Sequoias, kings of their race, growing close together like grass in a meadow, poised their brave domes and spires in the sky, three hundred feet above the ferns and the lilies that enameled the ground; towering serene through the long centuries, preaching God's forestry fresh from heaven.

Here the forests reached their highest development. Hence they went wavering northward over icy Alaska, brave spruce and fir, poplar and birch, by the coasts and the rivers, to within sight of the Arctic Ocean. American forests! the glory of the world! Surveyed thus from the east to the west, from the north to the south, they are rich beyond thought, immortal, immeasurable, enough and to spare for every feeding, sheltering beast and bird, insect and son of Adam; and nobody need have cared had there been no pines in Norway, no cedars and deodars on Lebanon and the Himalayas, no vine-clad selvas in the basin of the Amazon. With such variety, harmony, and triumphant exuberance, even Nature, it would seem, might have rested content with the forests of North America, and planted no more.

So they appeared a few centuries ago when they were rejoicing in wildness. The Indians with stone axes could do them no more harm than could gnawing beavers and browsing moose. Even the fires of the Indians and the fierce shattering lightning seemed to work together only for good in clearing spots here and there for smooth garden prairies, and openings for sunflowers seeking the light. But when the steel ax of the white man rang out on the startled air their doom was sealed. Every tree heard the bodeful sound, and pillars of smoke gave the sign in the sky.

I suppose we need not go mourning the buffaloes. In the nature of things they had to give place to better cattle, though the change might have been made without barbarous wickedness. Likewise many of Nature's five hundred kinds of wild trees had to make way for orchards and corn-fields. In the settlement and civilization of the country, bread more than timber or beauty was wanted; and in the blindness of hunger, the early settlers, claiming Heaven as their guide, regarded God's trees as only a larger kind of pernicious weeds, extremely hard to get rid of. Accordingly, with no eye to the future, these pious destroyers waged interminable forest wars; chips flew thick and fast; trees in their beauty fell crashing by millions, smashed to confusion, and the smoke of their burning has been rising to heaven more than two hundred years. After the Atlantic coast from Maine to Georgia had been mostly cleared and scorched into melancholy ruins, the overflowing multitude of bread and money seekers poured over the Alleghanies into the fertile middle West, spreading ruthless devastation ever wider and farther over the rich valley of the Mississippi and the vast shadowy pine region about the Great Lakes. Thence still westward, the invading horde of destroyers called settlers made its fiery way over the broad Rocky Mountains, felling and burning more fiercely than ever, until at last it has reached the wild side of the continent, and entered the last of the great aboriginal forests on the shores of the Pacific.

Surely, then, it should not be wondered at that lovers of their country, bewailing its baldness, are now crying aloud, "Save what is left of the forests!" Clearing has surely now gone far enough; soon timber will be scarce, and not a grove will be left to rest in or pray in. The remnant protected will yield plenty of timber, a perennial harvest for every right use, without further diminution of its area, and will continue to cover the springs of the rivers that rise in the mountains and give irrigating waters to the dry valleys at their feet, prevent wasting floods and be a blessing to everybody forever.

Every other civilized nation in the world has been compelled to care for its forests, and so must we if waste and destruction are not to go on to the bitter end, leaving America as barren as Palestine or Spain. In its calmer moments, in the midst of bewildering hunger and war and restless over-industry, Prussia has learned that the forest plays an important part in human progress, and that the advance in civilization only makes it more indispensable. It has, therefore, as shown by Mr. Pinchot, refused to deliver its forests to more or less speedy destruction by permitting them to pass into private ownership. But the state woodlands are not allowed to lie idle. On the contrary, they are made to produce as much timber as is possible without spoiling them. In the administration of its forests, the state righteously considers itself bound to treat them as a trust for the nation as a whole, and to keep in view the common good of the people for all time.

In France no government forests have been sold since 1870. On the other hand, about one half of the fifty million francs spent on forestry has been given to engineering works, to make the replanting of denuded areas possible. The disappearance of the forests in the first place, it is claimed, may be traced in most cases directly to mountain pasturage. The provisions of the Code concerning private woodlands are substantially these: no private owner may clear his woodlands without giving notice to the government at least four months in advance, and the forest service may forbid the clearing on the following grounds,—to maintain the soil on mountains, to defend the soil against erosion and flooding by rivers or torrents, to insure the existence of springs and watercourses, to protect the dunes and seashore, etc. A proprietor who has cleared his forest without permission is subject to heavy fine, and in addition may be made to replant the cleared area.

In Switzerland, after many laws like our own had been found wanting, the Swiss forest school was established in 1865, and soon after the federal forest law was enacted, which is binding over nearly two thirds of the country. Under its provisions, the cantons must appoint and pay the number of suitably educated foresters required for the fulfillment of the forest law; and in the organization of a normally stocked forest, the object of first importance must be the cutting each year of an amount of timber equal to the total annual increase, and no more.

The Russian government passed a law in 1888, declaring that clearing is forbidden in protected forests, and is allowed in others "only when its effects will not be to disturb the suitable relations which should exist between forest and agricultural lands."

Even Japan is ahead of us in the management of her forests. They cover an area of about twenty-nine million acres. The feudal lords valued the woodlands, and enacted vigorous protective laws; and when, in the latest civil war, the Mikado government destroyed the feudal system, it declared the forests that had belonged to the feudal lords to be the property of the state, promulgated a forest law binding on the whole kingdom, and founded a school of forestry in Tokio. The forest service does not rest satisfied with the present proportion of woodland, but looks to planting the best forest trees it can find in any country, if likely to be useful and to thrive in Japan.

In India systematic forest management was begun about forty years ago, under difficulties—presented by the character of the country, the prevalence of running fires, opposition from lumbermen, settlers, etc.—not unlike those which confront us now. Of the total area of government forests, perhaps seventy million acres, fifty-five million acres have been brought under the control of the forestry department,—a larger area than that of all our national parks and reservations. The chief aims of the administration are effective protection of the forests from fire, an efficient system of regeneration, and cheap transportation of the forest products; the results so far have been most beneficial and encouraging.

It seems, therefore, that almost every civilized nation can give us a lesson on the management and care of forests. So far our government has done nothing effective with its forests, though the best in the world, but is like a rich and foolish spendthrift who has inherited a magnificent estate in perfect order, and then has left his rich fields and meadows, forests and parks, to be sold and plundered and wasted at will, depending on their inexhaustible abundance. Now it is plain that the forests are not inexhaustible, and that quick measures must be taken if ruin is to be avoided. Year by year the remnant is growing smaller before the ax and fire, while the laws in existence provide neither for the protection of the timber from destruction nor for its use where it is most needed.

As is shown by Mr. E. A. Bowers, formerly Inspector of the Public Land Service, the foundation of our protective policy, which has never protected, is an act passed March 1, 1817, which authorized the Secretary of the Navy to reserve lands producing live-oak and cedar, for the sole purpose of supplying timber for the navy of the United States. An extension of this law by the passage of the act of March 2, 1831, provided that if any person should cut live-oak or red cedar trees or *other timber* from the lands of the United States for any other purpose than the construction of the navy, such person should pay a fine not less than triple the value of the timber cut, and be imprisoned for a period not exceeding twelve months. Upon this old law, as Mr. Bowers points out, having the construction of a wooden navy in view, the United States government has to-day chiefly to rely in protecting its timber throughout the arid regions of the West, where none of the naval timber which the law had in mind is to be found.

By the act of June 3, 1878, timber can be taken from public lands not subject to entry under any existing laws except for minerals, by *bona fide* residents of the Rocky Mountain states and territories and the Dakotas. Under the timber and stone act, of the same date, land in the Pacific States and Nevada, valuable mainly for timber, and unfit for cultivation if the timber is removed, can be purchased for two dollars and a half an acre, under certain restrictions. By the act of March 3, 1875, all land-grant and right-of-way railroads are authorized to take timber from the public lands adjacent to their lines for construction purposes; and they have taken it with a vengeance, destroying a hundred times more than they have used, mostly by allowing fires to run into the woods. The settlement laws, under which a settler may enter lands valuable for timber as well as for agriculture, furnish another means of obtaining title to public timber.

With the exception of the timber culture act, under which, in consideration of planting a few acres of seedlings, settlers on the treeless plains got 160 acres each, the above is the only legislation aiming to protect and promote the planting of forests. In no other way than under some one of these laws can a citizen of the United States make any use of the public forests. To show the results of the timber-planting act, it need only be stated that of the thirty-eight million acres entered under it, less than one million acres have been patented. This means that less than fifty thousand acres have been planted with stunted, woebegone, almost hopeless sprouts of trees, while at the same time the government has allowed millions of acres of the grandest forest trees to be stolen or destroyed, or sold for nothing. Under the act of June 3, 1878, settlers in Colorado and the Territories were allowed to cut timber for mining and educational purposes from mineral land, which in the practical West means both cutting and burning anywhere and everywhere, for any purpose, on any sort of public land. Thus, the prospector, the miner, and mining and railroad companies are allowed by law to take all the timber they like for their mines and roads, and the forbidden settler, if there are no mineral lands near his farm or stock-ranch or none that he knows of, can hardly be expected to forbear taking what he needs wherever he can find it. Timber is as necessary as bread, and no scheme of management failing to recognize and properly provide for this want can possibly be maintained. In any case, it will be hard to teach the pioneers that it is wrong to steal government timber. Taking from the government is with them the same as taking from Nature, and their consciences flinch no more in cutting timber from the wild forests than in drawing water from a lake or river. As for reservation and protection of forests, it seems as silly and needless to them as protection and reservation of the ocean would be, both appearing to be boundless and inexhaustible.

The special land agents employed by the General Land Office to protect the public domain from timber depredations are supposed to collect testimony to sustain prosecution and to superintend such prosecution on behalf of the government, which is represented by the district attorneys. But timber thieves of the Western class are seldom convicted, for the good reason that most of the jurors who try such cases are themselves as guilty as those on trial. The effect of the present confused, discriminating, and unjust system has been to place almost the whole population in opposition to the government; and as conclusive of its futility, as shown by Mr. Bowers, we need only state that during the seven years from 1881 to 1887 inclusive, the value of the timber reported stolen from the government lands was $36,719,935, and the amount recovered was $478,073, while the cost of the services of special agents alone was $455,000, to which must be added the expense of the trials. Thus for nearly thirty-seven million dollars' worth of timber the government got less than nothing; and the value of that consumed by running fires during the same period, without benefit even to thieves, was probably over two hundred millions of dollars. Land commissioners and Secretaries of the Interior have repeatedly called attention to this ruinous state of affairs, and asked Congress to enact the requisite legislation for reasonable reform. But, busied with tariffs, etc., Congress has given no heed to these or other appeals, and our forests, the most valuable and the most destructible of all the natural resources of the country, are being robbed and burned more rapidly than ever. The annual appropriation for so-called "protection service" is hardly sufficient to keep twenty-five timber agents in the field, and as far as any efficient protection of timber is concerned these agents themselves might as well be timber.*

That a change from robbery and ruin to a permanent rational policy is urgently needed nobody with the slightest knowledge of American forests will deny. In the East and along the northern Pacific coast, where the rainfall is abundant, comparatively few care keenly what becomes of the trees as long as fuel and lumber are not noticeably dear. But in the Rocky Mountains and California and Arizona, where the forests are inflammable, and where the fertility of the lowlands depends upon irrigation, public opinion is growing stronger every year in favor of permanent protection by the federal government of all the forests that cover the sources of the streams.

*A change for the better, compelled by public opinion, is now going on,—1901.

Even lumbermen in these regions, long accustomed to steal, are now willing and anxious to buy lumber for their mills under cover of law: some possibly from a late second growth of honesty, but most, especially the small mill-owners, simply because it no longer pays to steal where all may not only steal, but also destroy, and in particular because it costs about as much to steal timber for one mill as for ten, and, therefore, the ordinary lumberman can no longer compete with the large corporations. Many of the miners find that timber is already becoming scarce and dear on the denuded hills around their mills, and they, too, are asking for protection of forests, at least against fire. The slow-going, unthrifty farmers, also, are beginning to realize that when the timber is stripped from the mountains the irrigating streams dry up in summer, and are destructive in winter; that soil, scenery, and everything slips off with the trees: so of course they are coming into the ranks of tree-friends.

Of all the magnificent coniferous forests around the Great Lakes, once the property of the United States, scarcely any belong to it now. They have disappeared in lumber and smoke, mostly smoke, and the government got not one cent for them; only the land they were growing on was considered valuable, and two and a half dollars an acre was charged for it. Here and there in the Southern States there are still considerable areas of timbered government land, but these are comparatively unimportant. Only the forests of the West are significant in size and value, and these, although still great, are rapidly vanishing. Last summer, of the unrivaled redwood forests of the Pacific Coast Range, the United States Forestry Commission could not find a single quarter-section that remained in the hands of the government.*

Under the timber and stone act of 1878, which might well have been called the "dust and ashes act," any citizen of the United States could take up one hundred and sixty acres of timber land, and by paying two dollars and a half an acre for it obtain title. There was some virtuous effort made with a view to limit the operations of the act by requiring that the purchaser should make affidavit that he was entering the land exclusively for his own use, and by not allowing any association to enter more than one hundred and sixty acres.

*The State of California recently appropriated two hundred and fifty thousand dollars to buy a block of redwood land near Santa Cruz for a state park. A much larger national park should be made in Humboldt or Mendocino county.

Nevertheless, under this act wealthy corporations have fraudulently obtained title to from ten thousand to twenty thousand acres or more. The plan was usually as follows: A mill company, desirous of getting title to a large body of redwood or sugar-pine land, first blurred the eyes and ears of the land agents, and then hired men to enter the land they wanted, and immediately deed it to the company after a nominal compliance with the law; false swearing in the wilderness against the government being held of no account. In one case which came under the observation of Mr. Bowers, it was the practice of a lumber company to hire the entire crew of every vessel which might happen to touch at any port in the redwood belt, to enter one hundred and sixty acres each and immediately deed the land to the company, in consideration of the company's paying all expenses and giving the jolly sailors fifty dollars apiece for their trouble.

By such methods have our magnificent redwoods and much of the sugar-pine forests of the Sierra Nevada been absorbed by foreign and resident capitalists. Uncle Sam is not often called a fool in business matters, yet he has sold millions of acres of timber land at two dollars and a half an acre on which a single tree was worth more than a hundred dollars. But this priceless land has been patented, and nothing can be done now about the crazy bargain. According to the everlasting laws of righteousness, even the fraudulent buyers at less than one per cent. of its value are making little or nothing, on account of fierce competition. The trees are felled, and about half of each giant is left on the ground to be converted into smoke and ashes; the better half is sawed into choice lumber and sold to citizens of the United States or to foreigners: thus robbing the country of its glory and impoverishing it without right benefit to anybody,—a bad, black business from beginning to end.

The redwood is one of the few conifers that sprout from the stump and roots, and it declares itself willing to begin immediately to repair the damage of the lumberman and also that of the forest-burner. As soon as a redwood is cut down or burned it sends up a crowd of eager, hopeful shoots, which, if allowed to grow, would in a few decades attain a height of a hundred feet, and the strongest of them would finally become giants as great as the original tree. Gigantic second and third growth trees are found in the redwoods, forming magnificent temple-like circles around charred ruins more than a thousand years old. But not one denuded acre in a hundred is allowed to raise a new forest growth. On the contrary, all the brains, religion, and superstition of the neighborhood are brought into play to prevent a new growth. The sprouts from the roots and stumps are cut off again and again, with zealous concern as to the best time and method of making death sure. In the clearings of one of the largest mills on the coast we found thirty men at work, last summer, cutting off redwood shoots "in the dark of the moon," claiming that all the stumps and roots cleared at this auspicious time would send up no more shoots. Anyhow, these vigorous, almost immortal trees are killed at last, and black stumps are now their only monuments over most of the chopped and burned areas.

The redwood is the glory of the Coast Range. It extends along the western slope, in a nearly continuous belt about ten miles wide, from beyond the Oregon boundary to the south of Santa Cruz, a distance of nearly four hundred miles, and in massive, sustained grandeur and closeness of growth surpasses all the other timber woods of the world. Trees from ten to fifteen feet in diameter and three hundred feet high are not uncommon, and a few attain a height of three hundred and fifty feet or even four hundred, with a diameter at the base of fifteen to twenty feet or more, while the ground beneath them is a garden of fresh, exuberant ferns, lilies, gaultheria, and rhododendron. This grand tree, Sequoia sempervirens, is surpassed in size only by its near relative, Sequoia gigantea, or Big Tree, of the Sierra Nevada, if, indeed, it is surpassed. The sempervirens is certainly the taller of the two. The gigantea attains a greater girth, and is heavier, more noble in port, and more sublimely beautiful. These two Sequoias are all that are known to exist in the world, though in former geological times the genus was common and had many species. The redwood is restricted to the Coast Range, and the Big Tree to the Sierra.

As timber the redwood is too good to live. The largest sawmills ever built are busy along its seaward border, "with all the modern improvements," but so immense is the yield per acre it will be long ere the supply is exhausted. The Big Tree is also, to some extent, being made into lumber. It is far less abundant than the redwood, and is, fortunately, less accessible, extending along the western flank of the Sierra in a partially interrupted belt, about two hundred and fifty miles long, at a height of from four to eight thousand feet above the sea. The enormous logs, too heavy to handle, are blasted into manageable dimensions with gunpowder. A large portion of the best timber is thus shattered and destroyed, and, with the huge, knotty tops, is left in ruins for tremendous fires that kill every tree within their range, great and small. Still, the species is not in danger of extinction. It has been planted and is flourishing over a great part of Europe, and magnificent sections of the aboriginal forests have been reserved as national and State parks,—the Mariposa Sequoia Grove, near Yosemite, managed by the State of California, and the General Grant and Sequoia national parks on the Kings, Kaweah, and Tule rivers, efficiently guarded by a small troop of United States cavalry under the direction of the Secretary of the Interior. But there is not a single specimen of the redwood in any national park. Only by gift or purchase, so far as I know, can the government get back into its possession a single acre of this wonderful forest.

The legitimate demands on the forests that have passed into private ownership, as well as those in the hands of the government, are increasing every year with the rapid settlement and upbuilding of the country, but the methods of lumbering are as yet grossly wasteful. In most mills only the best portions of the best trees are used, while the ruins are left on the ground to feed great fires; which kill much of what is left of the less desirable timber, together with the seedlings, on which the permanence of the forest depends. Thus every mill is a center of destruction far more severe from waste and fire than from use. The same thing is true of the mines, which consume and destroy indirectly immense quantities of timber with their innumerable fires, accidental or set to make open ways, and often without regard to how far they run. The prospector deliberately sets fires to clear off the woods just where they are densest, to lay the rocks bare and make the discovery of mines easier. Sheep-owners and their shepherds also set fires everywhere through the woods in the fall to facilitate the march of their countless flocks the next summer, and perhaps in some places to improve the pasturage. The ax is not yet at the root of every tree, but the sheep is, or was before the national parks were established and guarded by the military, the only effective and reliable arm of the government free from the blight of politics. Not only do the shepherds, at the driest time of the year, set fire to everything that will burn, but the sheep consume every green leaf, not sparing even the young conifers, when they are in a starving condition from crowding, and they rake and dibble the loose soil of the mountain sides for the spring floods to wash away, and thus at last leave the ground barren.

Of all the destroyers that infest the woods, the shake-maker seems the happiest. Twenty or thirty years ago, shakes, a kind of long, boardlike shingles split with a mallet and a frow, were in great demand for covering barns and sheds, and many are used still in preference to common shingles, especially those made from the sugar-pine, which do not warp or crack in the hottest sunshine. Drifting adventurers in California, after harvest and threshing are over, oftentimes meet to discuss their plans for the winter; and their talk is interesting. Once, in a company of this kind, I heard a man say, as he peacefully smoked his pipe: "Boys, as soon as this job's done I'm goin' into the duck business. There's big money in it, and your grub costs nothing. Tule Joe made five hundred dollars last winter on mallard and teal. Shot 'em on the Joaquin, tied 'em in dozens by the neck, and shipped 'em to San Francisco. And when he was tired wading in the sloughs and touched with rheumatiz, he just knocked off on ducks, and went to the Contra Costa hills for dove and quail. It's a mighty good business; and you're your own boss, and the whole thing's fun."

Another of the company, a bushy-bearded fellow, with a trace of brag in his voice, drawled out: "Bird business is well enough for some, but bear is my game, with a deer and a California lion thrown in now and then for change. There's always market for bear grease, and sometimes you can sell the hams. They're good as hog hams any day. And you are your own boss in my business, too, if the bears ain't too big and too many for you. Old grizzlies I despise,—they want cannon to kill 'em; but the blacks and browns are beauties for grease, and when once I get 'em just right, and draw a bead on 'em, I fetch 'em every time." Another said he was going to catch up a lot of mustangs as soon as the rains set in, hitch them to a gang-plough, and go to farming on the San Joaquin plains for wheat. But most preferred the shake business, until something more profitable and as sure could be found, with equal comfort and independence.

With a cheap mustang or mule to carry a pair of blankets, a sack of flour, a few pounds of coffee, and an ax, a frow, and a cross-cut saw, the shake-maker ascends the mountains to the pine belt where it is most accessible, usually by some mine or mill road. Then he strikes off into the virgin woods, where the sugar-pine, king of all the hundred species of pines in the world in size and beauty, towers on the open sunny slopes of the Sierra in the fullness of its glory. Selecting a favorable spot for a cabin near a meadow with a stream, he unpacks his animal and stakes it out on the meadow. Then he chops into one after another of the pines, until he finds one that he feels sure will split freely, cuts this down, saws off a section four feet long, splits it, and from this first cut, perhaps seven feet in diameter, he gets shakes enough for a cabin and its furniture,—walls, roof, door, bedstead, table, and stool. Besides his labor, only a few pounds of nails are required. Sapling poles form the frame of the airy building, usually about six feet by eight in size, on which the shakes are nailed, with the edges overlapping. A few bolts from the same section that the shakes were made from are split into square sticks and built up to form a chimney, the inside and interspaces being plastered and filled in with mud. Thus, with abundance of fuel, shelter and comfort by his own fireside are secured. Then he goes to work sawing and splitting for the market, tying the shakes in bundles of fifty or a hundred. They are four feet long, four inches wide, and about one fourth of an inch thick. The first few thousands he sells or trades at the nearest mill or store, getting provisions in exchange. Then he advertises, in whatever way he can, that he has excellent sugar-pine shakes for sale, easy of access and cheap.

Only the lower, perfectly clear, free-splitting portions of the giant pines are used,—perhaps ten to twenty feet from a tree two hundred and fifty in height; all the rest is left a mass of ruins, to rot or to feed the forest fires, while thousands are hacked deeply and rejected in proving the grain. Over nearly all of the more accessible slopes of the Sierra and Cascade mountains in southern Oregon, at a height of from three to six thousand feet above the sea, and for a distance of about six hundred miles, this waste and confusion extends. Happy robbers! dwelling in the most beautiful woods, in the most salubrious climate, breathing delightful odors both day and night, drinking cool living water,—roses and lilies at their feet in the spring, shedding fragrance and ringing bells as if cheering them on in their desolating work. There is none to say them nay. They buy no land, pay no taxes, dwell in a paradise with no forbidding angel either from Washington or from heaven. Every one of the frail shake shanties is a center of destruction, and the extent of the ravages wrought in this quiet way is in the aggregate enormous.

It is not generally known that notwithstanding the immense quantities of timber cut every year for foreign and home markets and mines, from five to ten times as much is destroyed as is used, chiefly by running forest fires that only the federal government can stop. Travelers through the West in summer are not likely to forget the fire-work displayed along the various railway tracks. Thoreau, when contemplating the destruction of the forests on the east side of the continent, said that soon the country would be so bald that every man would have to grow whiskers to hide its nakedness, but he thanked God that at least the sky was safe. Had he gone West he would have found out that the sky was not safe; for all through the summer months, over most of the mountain regions, the smoke of mill and forest fires is so thick and black that no sunbeam can pierce it. The whole sky, with clouds, sun, moon, and stars, is simply blotted out. There is no real sky and no scenery. Not a mountain is left in the landscape. At least none is in sight from the lowlands, and they all might as well be on the moon, as far as scenery is concerned.

The half-dozen transcontinental railroad companies advertise the beauties of their lines in gorgeous many-colored folders, each claiming its as the "scenic route." "The route of superior desolation"—the smoke, dust, and ashes route—would be a more truthful description. Every train rolls on through dismal smoke and barbarous, melancholy ruins; and the companies might well cry in their advertisements: "Come! travel our way. Ours is the blackest. It is the only genuine Erebus route. The sky is black and the ground is black, and on either side there is a continuous border of black stumps and logs and blasted trees appealing to heaven for help as if still half alive, and their mute eloquence is most interestingly touching. The blackness is perfect. On account of the superior skill of our workmen, advantages of climate, and the kind of trees, the charring is generally deeper along our line, and the ashes are deeper, and the confusion and desolation displayed can never be rivaled. No other route on this continent so fully illustrates the abomination of desolation." Such a claim would be reasonable, as each seems the worst, whatever route you chance to take.

Of course a way had to be cleared through the woods. But the felled timber is not worked up into fire-wood for the engines and into lumber for the company's use; it is left lying in vulgar confusion, and is fired from time to time by sparks from locomotives or by the workmen camping along the line. The fires, whether accidental or set, are allowed to run into the woods as far as they may, thus assuring comprehensive destruction. The directors of a line that guarded against fires, and cleared a clean gap edged with living trees, and fringed and mantled with the grass and flowers and beautiful seedlings that are ever ready and willing to spring up, might justly boast of the beauty of their road; for nature is always ready to heal every scar. But there is no such road on the western side of the continent. Last summer, in the Rocky Mountains, I saw six fires started by sparks from a locomotive within a distance of three miles, and nobody was in sight to prevent them from spreading. They might run into the adjacent forests and burn the timber from hundreds of square miles; not a man in the State would care to spend an hour in fighting them, as long as his own fences and buildings were not threatened.

Notwithstanding all the waste and use which have been going on unchecked like a storm for more than two centuries, it is not yet too late—though it is high time—for the government to begin a rational administration of its forests. About seventy million acres it still owns,—enough for all the country, if wisely used. These residual forests are generally on mountain slopes, just where they are doing the most good, and where their removal would be followed by the greatest number of evils; the lands they cover are too rocky and high for agriculture, and can never be made as valuable for any other crop as for the present crop of trees. It has been shown over and over again that if these mountains were to be stripped of their trees and underbrush, and kept bare and sodless by hordes of sheep and the innumerable fires the shepherds set, besides those of the millmen, prospectors, shake-makers, and all sorts of adventurers, both lowlands and mountains would speedily become little better than deserts, compared with their present beneficent fertility. During heavy rainfalls and while the winter accumulations of snow were melting, the larger streams would swell into destructive torrents, cutting deep, rugged-edged gullies, carrying away the fertile humus and soil as well as sand and rocks, filling up and overflowing their lower channels, and covering the lowland fields with raw detritus. Drought and barrenness would follow.

In their natural condition, or under wise management, keeping out destructive sheep, preventing fires, selecting the trees that should be cut for lumber, and preserving the young ones and the shrubs and sod of herbaceous vegetation, these forests would be a never failing fountain of wealth and beauty. The cool shades of the forest give rise to moist beds and currents of air, and the sod of grasses and the various flowering plants and shrubs thus fostered, together with the network and sponge of tree roots, absorb and hold back the rain and the waters from melting snow, compelling them to ooze and percolate and flow gently through the soil in streams that never dry. All the pine needles and rootlets and blades of grass, and the fallen, decaying trunks of trees, are dams, storing the bounty of the clouds and dispensing it in perennial life-giving streams, instead of allowing it to gather suddenly and rush headlong in short-lived devastating floods. Everybody on the dry side of the continent is beginning to find this out, and, in view of the waste going on, is growing more and more anxious for government protection. The outcries we hear against forest reservations come mostly from thieves who are wealthy and steal timber by wholesale. They have so long been allowed to steal and destroy in peace that any impediment to forest robbery is denounced as a cruel and irreligious interference with "vested rights," likely to endanger the repose of all ungodly welfare.

Gold, gold, gold! How strong a voice that metal has!

"O wae for the siller, it is sae preva'lin'!"

Even in Congress a sizable chunk of gold, carefully concealed, will outtalk and outfight all the nation on a subject like forestry, well smothered in ignorance, and in which the money interests of only a few are conspicuously involved. Under these circumstances, the bawling, blethering oratorical stuff drowns the voice of God himself. Yet the dawn of a new day in forestry is breaking. Honest citizens see that only the rights of the government are being trampled, not those of the settlers. Only what belongs to all alike is reserved, and every acre that is left should be held together under the federal government as a basis for a general policy of administration for the public good. The people will not always be deceived by selfish opposition, whether from lumber and mining corporations or from sheepmen and prospectors, however cunningly brought forward underneath fables and gold.

Emerson says that things refuse to be mismanaged long. An exception would seem to be found in the case of our forests, which have been mismanaged rather long, and now come desperately near being like smashed eggs and spilt milk. Still, in the long run the world does not move backward. The wonderful advance made in the last few years, in creating four national parks in the West, and thirty forest reservations, embracing nearly forty million acres; and in the planting of the borders of streets and highways and spacious parks in all the great cities, to satisfy the natural taste and hunger for landscape beauty and righteousness that God has put, in some measure, into every human being and animal, shows the trend of awakening public opinion. The making of the far-famed New York Central Park was opposed by even good men, with misguided pluck, perseverance, and ingenuity; but straight right won its way, and now that park is appreciated. So we confidently believe it will be with our great national parks and forest reservations. There will be a period of indifference on the part of the rich, sleepy with wealth, and of the toiling millions, sleepy with poverty, most of whom never saw a forest; a period of screaming protest and objection from the plunderers, who are as unconscionable and enterprising as Satan. But light is surely coming, and the friends of destruction will preach and bewail in vain.

The United States government has always been proud of the welcome it has extended to good men of every nation, seeking freedom and homes and bread. Let them be welcomed still as nature welcomes them, to the woods as well as to the prairies and plains. No place is too good for good men, and still there is room. They are invited to heaven, and may well be allowed in America. Every place is made better by them. Let them be as free to pick gold and gems from the hills, to cut and hew, dig and plant, for homes and bread, as the birds are to pick berries from the wild bushes, and moss and leaves for nests. The ground will be glad to feed them, and the pines will come down from the mountains for their homes as willingly as the cedars came from Lebanon for Solomon's temple. Nor will the woods be the worse for this use, or their benign influences be diminished any more than the sun is diminished by shining. Mere destroyers, however, tree-killers, wool and mutton men, spreading death and confusion in the fairest groves and gardens ever planted,—let the government hasten to cast them out and make an end of them. For it must be told again and again, and be burningly borne in mind, that just now, while protective measures are being deliberated languidly, destruction and use are speeding on faster and farther every day. The ax and saw are insanely busy, chips are flying thick as snowflakes, and every summer thousands of acres of priceless forests, with their underbrush, soil, springs, climate, scenery, and religion, are vanishing away in clouds of smoke, while, except in the national parks, not one forest guard is employed.

All sorts of local laws and regulations have been tried and found wanting, and the costly lessons of our own experience, as well as that of every civilized nation, show conclusively that the fate of the remnant of our forests is in the hands of the federal government, and that if the remnant is to be saved at all, it must be saved quickly.

Any fool can destroy trees. They cannot run away; and if they could, they would still be destroyed,—chased and hunted down as long as fun or a dollar could be got out of their bark hides, branching horns, or magnificent bole backbones. Few that fell trees plant them; nor would planting avail much towards getting back anything like the noble primeval forests. During a man's life only saplings can be grown, in the place of the old trees—tens of centuries old—that have been destroyed. It took more than three thousand years to make some of the trees in these Western woods,—trees that are still standing in perfect strength and beauty, waving and singing in the mighty forests of the Sierra. Through all the wonderful, eventful centuries since Christ's time—and long before that—God has cared for these trees, saved them from drought, disease, avalanches, and a thousand straining, leveling tempests and floods; but he cannot save them from fools,—only Uncle Sam can do that.

THE WILD PARKS AND FOREST RESERVATIONS OF THE WEST

from *Our National Parks* (1901)

> "Keep not standing fix'd and rooted,
> Briskly venture, briskly roam;
> Head and hand, where'er thou foot it,
> And stout heart are still at home.
> In each land the sun does visit
> We are gay, whate'er betide:
> To give room for wandering is it
> That the world was made so wide."

THE TENDENCY NOWADAYS to wander in wildernesses is delightful to see. Thousands of tired, nerve-shaken, over-civilized people are beginning to find out that going to the mountains is going home; that wildness is a necessity; and that mountain parks and reservations are useful not only as fountains of timber and irrigating rivers, but as fountains of life. Awakening from the stupefying effects of the vice of over-industry and the deadly apathy of luxury, they are trying as best they can to mix and enrich their own little ongoings with those of Nature, and to get rid of rust and disease. Briskly venturing and roaming, some are washing off sins and cob-web cares of the devil's spinning in all-day storms on mountains; sauntering in rosiny pinewoods or in gentian meadows, brushing through chaparral, bending down and parting sweet, flowery sprays; tracing rivers to their sources, getting in touch with the nerves of Mother Earth; jumping from rock to rock, feeling the life of them, learning the songs of them, panting in whole-souled exercise, and rejoicing in deep, long-drawn breaths of pure wildness. This is fine and natural and full of promise. So also is the growing interest in the care and preservation of forests and wild places in general, and in the half wild parks and gardens of towns. Even the scenery habit in its most artificial forms, mixed with spectacles, silliness, and kodaks; its devotees arrayed more gorgeously than scarlet tanagers, frightening the wild game with red umbrellas,—even this is encouraging, and may well be regarded as a hopeful sign of the times.

All the Western mountains are still rich in wildness, and by means of good roads are being brought nearer civilization every year. To the sane and free it will hardly seem necessary to cross the continent in search of wild beauty, however easy the way, for they find it in abundance wherever they chance to be. Like Thoreau they see forests in orchards and patches of huckleberry brush, and oceans in ponds and drops of dew. Few in these hot, dim, strenuous times are quite sane or free; choked with care like clocks full of dust, laboriously doing so much good and making so much money,—or so little,—they are no longer good for themselves.

When, like a merchant taking a list of his goods, we take stock of our wildness, we are glad to see how much of even the most destructible kind is still unspoiled. Looking at our continent as scenery when it was all wild, lying between beautiful seas, the starry sky above it, the starry rocks beneath it, to compare its sides, the East and the West, would be like comparing the sides of a rainbow. But it is no longer equally beautiful. The rainbows of to-day are, I suppose, as bright as those that first spanned the sky; and some of our landscapes are growing more beautiful from year to year, notwithstanding the clearing, trampling work of civilization. New plants and animals are enriching woods and gardens, and many landscapes wholly new, with divine sculpture and architecture, are just now coming to the light of day as the mantling folds of creative glaciers are being withdrawn, and life in a thousand cheerful, beautiful forms is pushing into them, and new-born rivers are beginning to sing and shine in them. The old rivers, too, are growing longer, like healthy trees, gaining new branches and lakes as the residual glaciers at their highest sources on the mountains recede, while the rootlike branches in their flat deltas are at the same time spreading farther and wider into the seas and making new lands.

Under the control of the vast mysterious forces of the interior of the earth all the continents and islands are slowly rising or sinking. Most of the mountains are diminishing in size under the wearing action of the weather, though a few are increasing in height and girth, especially the volcanic ones, as fresh floods of molten rocks are piled on their summits and spread in successive layers, like the wood-rings of trees, on their sides. New mountains, also, are being created from time to time as islands in lakes and seas, or as subordinate cones on the slopes of old ones, thus in some measure balancing the waste of old beauty with new. Man, too, is making many far-reaching changes. This most influential half animal, half angel is rapidly multiplying and spreading, covering the seas and lakes with ships, the land with huts, hotels, cathedrals, and clustered city shops and homes, so that soon, it would seem, we may have to go farther than Nansen to find a good sound solitude. None of Nature's landscapes are ugly so long as they are wild; and much, we can say comfortingly, must always be in great part wild, particularly the sea and the sky, the floods of light from the stars, and the warm, unspoilable heart of the earth, infinitely beautiful, though only dimly visible to the eye of imagination. The geysers, too, spouting from the hot underworld; the steady, long-lasting glaciers on the mountains, obedient only to the sun; Yosemite domes and the tremendous grandeur of rocky cañons and mountains in general,—these must always be wild, for man can change them and mar them hardly more than can the butterflies that hover above them. But the continent's outer beauty is fast passing away, especially the plant part of it, the most destructible and most universally charming of all.

Only thirty years ago, the great Central Valley of California, five hundred miles long and fifty miles wide, was one bed of golden and purple flowers. Now it is ploughed and pastured out of existence, gone forever,—scarce a memory of it left in fence corners and along the bluffs of the streams. The gardens of the Sierra, also, and the noble forests in both the reserved and unreserved portions are sadly hacked and trampled, notwithstanding the ruggedness of the topography,—all excepting those of the parks guarded by a few soldiers. In the noblest forests of the world, the ground, once divinely beautiful, is desolate and repulsive, like a face ravaged by disease. This is true also of many other Pacific Coast and Rocky Mountain valleys and forests. The same fate, sooner or later, is awaiting them all, unless awakening public opinion comes forward to stop it. Even the great deserts in Arizona, Nevada, Utah, and New Mexico, which offer so little to attract settlers, and which a few years ago pioneers were afraid of, as places of desolation and death, are now taken as pastures at the rate of one or two square miles per cow, and of course their plant treasures are passing away,—the delicate abronias, phloxes, gilias, etc. Only a few of the bitter, thorny, unbitable shrubs are left, and the sturdy cactuses that defend themselves with bayonets and spears.

Most of the wild plant wealth of the East also has vanished, —gone into dusty history. Only vestiges of its glorious prairie and woodland wealth remain to bless humanity in boggy, rocky, unploughable places. Fortunately, some of these are purely wild, and go far to keep Nature's love visible. White water-lilies, with root-stocks deep and safe in mud, still send up every summer a Milky Way of starry, fragrant flowers around a thousand lakes, and many a tuft of wild grass waves its panicles on mossy rocks, beyond reach of trampling feet, in company with saxifrages, bluebells, and ferns. Even in the midst of farmers' fields, precious sphagnum bogs, too soft for the feet of cattle, are preserved with their charming plants unchanged,—chiogenes, Andromeda, Kalmia, Linnæa, Arethusa, etc. Calypso borealis still hides in the arbor vitæ swamps of Canada, and away to the southward there are a few unspoiled swamps, big ones, where miasma, snakes, and alligators, like guardian angels, defend their treasures and keep them as pure as paradise. And beside a' that and a' that, the East is blessed with good winters and blossoming clouds that shed white flowers over all the land, covering every scar and making the saddest landscape divine at least once a year.

The most extensive, least spoiled, and most unspoilable of the gardens of the continent are the vast tundras of Alaska. In summer they extend smooth, even, undulating, continuous beds of flowers and leaves from about lat. 62° to the shores of the Arctic Ocean; and in winter sheets of snowflowers make all the country shine, one mass of white radiance like a star. Nor are these Arctic plant people the pitiful frost-pinched unfortunates they are guessed to be by those who have never seen them. Though lowly in stature, keeping near the frozen ground as if loving it, they are bright and cheery, and speak Nature's love as plainly as their big relatives of the South. Tenderly happed and tucked in beneath downy snow to sleep through the long, white winter, they make haste to bloom in the spring without trying to grow tall, though some rise high enough to ripple and wave in the wind, and display masses of color,—yellow, purple, and blue,—so rich that they look like beds of rainbows, and are visible miles and miles away.

As early as June one may find the showy Geum glaciale in flower, and the dwarf willows putting forth myriads of fuzzy catkins, to be followed quickly, especially on the dryer ground, by mertensia, eritrichium, polemonium, oxytropis, astragalus, lathyrus, lupinus, myosotis, dodecatheon, arnica, chrysanthemum, nardosmia, saussurea, senecio, erigeron, matrecaria, caltha, valeriana, stellaria, Tofieldia, polygonum, papaver, phlox, lychnis, cheiranthus, Linnæa, and a host of drabas, saxifrages, and heathworts, with bright stars and bells in glorious profusion, particularly Cassiope, Andromeda, ledum, pyrola, and vaccinium,—Cassiope the most abundant and beautiful of them all. Many grasses also grow here, and wave fine purple spikes and panicles over the other flowers,—poa, aira, calamagrostis, alopecurus, trisetum, elymus, festuca, glyceria, etc. Even ferns are found thus far north, carefully and comfortably unrolling their precious fronds,—aspidium, cystopteris, and woodsia, all growing on a sumptuous bed of mosses and lichens; not the scaly lichens seen on rails and trees and fallen logs to the southward, but massive, round-headed, finely colored plants like corals, wonderfully beautiful, worth going round the world to see. I should like to mention all the plant friends I found in a summer's wanderings in this cool reserve, but I fear few would care to read their names, although everybody, I am sure, would love them could they see them blooming and rejoicing at home.

On my last visit to the region about Kotzebue Sound, near the middle of September, 1881, the weather was so fine and mellow that it suggested the Indian summer of the Eastern States. The winds were hushed, the tundra glowed in creamy golden sunshine, and the colors of the ripe foliage of the heathworts, willows, and birch—red, purple, and yellow, in pure bright tones—were enriched with those of berries which were scattered everywhere, as if they had been showered from the clouds like hail. When I was back a mile or two from the shore, reveling in this color-glory, and thinking how fine it would be could I cut a square of the tundra sod of conventional picture size, frame it, and hang it among the paintings on my study walls at home, saying to myself, "Such a Nature painting taken at random from any part of the thousand-mile bog would make the other pictures look dim and coarse," I heard merry shouting, and, looking round, saw a band of Eskimos—men, women, and children, loose and hairy like wild animals—running towards me. I could not guess at first what they were seeking, for they seldom leave the shore; but soon they told me, as they threw themselves down, sprawling and laughing, on the mellow bog, and began to feast on the berries. A lively picture they made, and a pleasant one, as they frightened the whirring ptarmigans, and surprised their oily stomachs with the beautiful acid berries of many kinds, and filled sealskin bags with them to carry away for festive days in winter.

Nowhere else on my travels have I seen so much warm-blooded, rejoicing life as in this grand Arctic reservation, by so many regarded as desolate. Not only are there whales in abundance along the shores, and innumerable seals, walruses, and white bears, but on the tundras great herds of fat reindeer and wild sheep, foxes, hares, mice, piping marmots, and birds. Perhaps more birds are born here than in any other region of equal extent on the continent. Not only do strong-winged hawks, eagles, and water-fowl, to whom the length of the continent is merely a pleasant excursion, come up here every summer in great numbers, but also many short-winged warblers, thrushes, and finches, repairing hither to rear their young in safety, reinforce the plant bloom with their plumage, and sweeten the wilderness with song; flying all the way, some of them, from Florida, Mexico, and Central America. In coming north they are coming home, for they were born here, and they go south only to spend the winter months, as New Englanders go to Florida. Sweet-voiced troubadours, they sing in orange groves and vine-clad magnolia woods in winter, in thickets of dwarf birch and alder in summer, and sing and chatter more or less all the way back and forth, keeping the whole country glad. Oftentimes, in New England, just as the last snow-patches are melting and the sap in the maples begins to flow, the blessed wanderers may be heard about orchards and the edges of fields where they have stopped to glean a scanty meal, not tarrying long, knowing they have far to go. Tracing the footsteps of spring, they arrive in their tundra homes in June or July, and set out on the return journey in September, or as soon as their families are able to fly well.

This is Nature's own reservation, and every lover of wildness will rejoice with me that by kindly frost it is so well defended. The discovery lately made that it is sprinkled with gold may cause some alarm; for the strangely exciting stuff makes the timid bold enough for anything, and the lazy destructively industrious. Thousands at least half insane are now pushing their way into it, some by the southern passes over the mountains, perchance the first mountains they have ever seen,—sprawling, struggling, gasping for breath, as, laden with awkward, merciless burdens of provisions and tools, they climb over rough-angled boulders and cross thin miry bogs. Some are going by the mountains and rivers to the eastward through Canada, tracing the old romantic ways of the Hudson Bay traders; others by Bering Sea and the Yukon, sailing all the way, getting glimpses perhaps of the famous fur-seals, the ice-floes, and the innumerable islands and bars of the great Alaska river. In spite of frowning hardships and the frozen ground, the Klondike gold will increase the crusading crowds for years to come, but comparatively little harm will be done. Holes will be burned and dug into the hard ground here and there, and into the quartz-ribbed mountains and hills; ragged towns like beaver and musk-rat villages will be built, and mills and locomotives will make rumbling, screeching, disenchanting noises; but the miner's pick will not be followed far by the plough, at least not until Nature is ready to unlock the frozen soil-beds with her slow-turning climate key. On the other hand, the roads of the pioneer miners will lead many a lover of wildness into the heart of the reserve, who without them would never see it.

In the meantime, the wildest health and pleasure grounds accessible and available to tourists seeking escape from care and dust and early death are the parks and reservations of the West. There are four national parks,*—the Yellowstone, Yosemite, General Grant, and Sequoia,—all within easy reach, and thirty forest reservations, a magnificent realm of woods, most of which, by railroads and trails and open ridges, is also fairly accessible, not only to the determined traveler rejoicing in difficulties, but to those (may their tribe increase) who, not tired, not sick, just naturally take wing every summer in search of wildness. The forty million acres of these reserves are in the main unspoiled as yet, though sadly wasted and threatened on their more open margins by the ax and fire of the lumberman and prospector, and by hoofed locusts, which, like the winged ones, devour every leaf within reach, while the shepherds and owners set fires with the intention of making a blade of grass grow in the place of every tree, but with the result of killing both the grass and the trees.

In the million acre Black Hills Reserve of South Dakota, the easternmost of the great forest-reserves, made for the sake of the

*There are now five parks and thirty-eight reservations.

farmers and miners, there are delightful, reviving sauntering-grounds in open parks of yellow pine, planted well apart, allowing plenty of sunshine to warm the ground. This tree is one of the most variable and most widely distributed of American pines. It grows sturdily on all kinds of soil and rocks, and, protected by a mail of thick bark, defies frost and fire and disease alike, daring every danger in firm, calm beauty and strength. It occurs here mostly on the outer hills and slopes where no other tree can grow. The ground beneath it is yellow most of the summer with showy Wythia, arnica, applopappus, solidago, and other sun-loving plants, which, though they form no heavy entangling growth, yet give abundance of color and make all the woods a garden. Beyond the yellow pine woods there lies a world of rocks of wildest architecture, broken, splintery, and spiky, not very high, but the strangest in form and style of grouping imaginable. Countless towers and spires, pinnacles and slender domed columns, are crowded together, and feathered with sharp-pointed Engelmann spruces, making curiously mixed forests,—half trees, half rocks. Level gardens here and there in the midst of them offer charming surprises, and so do the many small lakes with lilies on their meadowy borders, and bluebells, anemones, daisies, castilleias, comandras, etc., together forming landscapes delightfully novel, and made still wilder by many interesting animals,—elk, deer, beavers, wolves, squirrels, and birds. Not very long ago this was the richest of all the red man's hunting-grounds hereabout. After the season's buffalo hunts were over,—as described by Parkman, who, with a picturesque cavalcade of Sioux savages, passed through these famous hills in 1846,—every winter deficiency was here made good, and hunger was unknown until, in spite of most determined, fighting, killing opposition, the white gold-hunters entered the fat game reserve and spoiled it. The Indians are dead now, and so are most of the hardly less striking free trappers of the early romantic Rocky Mountain times. Arrows, bullets, scalping-knives, need no longer be feared; and all the wilderness is peacefully open.

The Rocky Mountain reserves are the Teton, Yellowstone, Lewis and Clark, Bitter Root, Priest River and Flathead, comprehending more than twelve million acres of mostly unclaimed, rough, forest-covered mountains in which the great rivers of the country take their rise. The commonest tree in most of them is the brave, indomitable, and altogether admirable Pinus contorta, widely distributed in all kinds of climate and soil, growing cheerily in frosty Alaska, breathing the damp salt air of the sea as well as the dry biting blasts of the Arctic interior, and making itself at home on the most dangerous flame-swept slopes and ridges of the Rocky Mountains in immeasurable abundance and variety of forms. Thousands of acres of this species are destroyed by running fires nearly every summer, but a new growth springs quickly from the ashes. It is generally small, and yields few sawlogs of commercial value, but is of incalculable importance to the farmer and miner; supplying fencing, mine timbers, and fire-wood, holding the porous soil on steep slopes, preventing landslips and avalanches, and giving kindly, nourishing shelter to animals and the widely outspread sources of the life-giving rivers. The other trees are mostly spruce, mountain pine, cedar, juniper, larch, and balsam fir; some of them, especially on the western slopes of the mountains, attaining grand size and furnishing abundance of fine timber.

Perhaps the least known of all this grand group of reserves is the Bitter Root, of more than four million acres. It is the wildest, shaggiest block of forest wildness in the Rocky Mountains, full of happy, healthy, storm-loving trees, full of streams that dance and sing in glorious array, and full of Nature's animals,—elk, deer, wild sheep, bears, cats, and innumerable smaller people.

In calm Indian summer, when the heavy winds are hushed, the vast forests covering hill and dale, rising and falling over the rough topography and vanishing in the distance, seem lifeless. No moving thing is seen as we climb the peaks, and only the low, mellow murmur of falling water is heard, which seems to thicken the silence. Nevertheless, how many hearts with warm red blood in them are beating under cover of the woods, and how many teeth and eyes are shining! A multitude of animal people, intimately related to us, but of whose lives we know almost nothing, are as busy about their own affairs as we are about ours: beavers are building and mending dams and huts for winter, and storing them with food; bears are studying winter quarters as they stand thoughtful in open spaces, while the gentle breeze ruffles the long hair on their backs; elk and deer, assembling on the heights, are considering cold pastures where they will be farthest away from the wolves; squirrels and marmots are busily laying up provisions and lining their nests against coming frost and snow foreseen; and countless thousands of birds are forming parties and gathering their young about them for flight to the southlands; while butterflies and bees, apparently with no thought of hard times to come, are hovering above the late-blooming goldenrods, and, with countless other insect folk, are dancing and humming right merrily in the sunbeams and shaking all the air into music.

Wander here a whole summer, if you can. Thousands of God's wild blessings will search you and soak you as if you were a sponge, and the big days will go by uncounted. If you are business-tangled, and so burdened with duty that only weeks can be got out of the heavy-laden year, then go to the Flathead Reserve; for it is easily and quickly reached by the Great Northern Railroad. Get off the track at Belton Station, and in a few minutes you will find yourself in the midst of what you are sure to say is the best care-killing scenery on the continent,—beautiful lakes derived straight from glaciers, lofty mountains steeped in lovely nemophila-blue skies and clad with forests and glaciers, mossy, ferny waterfalls in their hollows, nameless and numberless, and meadowy gardens abounding in the best of everything. When you are calm enough for discriminating observation, you will find the king of the larches, one of the best of the Western giants, beautiful, picturesque, and regal in port, easily the grandest of all the larches in the world. It grows to a height of one hundred and fifty to two hundred feet, with a diameter at the ground of five to eight feet, throwing out its branches into the light as no other tree does. To those who before have seen only the European larch or the Lyall species of the eastern Rocky Mountains, or the little tamarack or hackmatack of the Eastern States and Canada, this Western king must be a revelation.

Associated with this grand tree in the making of the Flathead forests is the large and beautiful mountain pine, or Western white pine (Pinus monticola), the invincible contorta or lodge-pole pine, and spruce and cedar. The forest floor is covered with the richest beds of Linnæa borealis I ever saw, thick fragrant carpets, enriched with shining mosses here and there, and with Clintonia, pyrola, moneses, and vaccinium, weaving hundred-mile beds of bloom that would have made blessed old Linnæus weep for joy.

Lake McDonald, full of brisk trout, is in the heart of this forest, and Avalanche Lake is ten miles above McDonald, at the feet of a group of glacier-laden mountains. Give a month at least to this precious reserve. The time will not be taken from the sum of your life.

Instead of shortening, it will indefinitely lengthen it and make you truly immortal. Nevermore will time seem short or long, and cares will never again fall heavily on you, but gently and kindly as gifts from heaven.

The vast Pacific Coast reserves in Washington and Oregon—the Cascade, Washington, Mount Rainier, Olympic, Bull Run, and Ashland, named in order of size—include more than 12,500,000 acres of magnificent forests of beautiful and gigantic trees. They extend over the wild, unexplored Olympic Mountains and both flanks of the Cascade Range, the wet and the dry. On the east side of the Cascades the woods are sunny and open, and contain principally yellow pine, of moderate size, but of great value as a cover for the irrigating streams that flow into the dry interior, where agriculture on a grand scale is being carried on. Along the moist, balmy, foggy, west flank of the mountains, facing the sea, the woods reach their highest development, and, excepting the California redwoods, are the heaviest on the continent. They are made up mostly of the Douglas spruce (Pseudotsuga taxifolia), with the giant arbor vitæ, or cedar; and several species of fir and hemlock in varying abundance, forming a forest kingdom unlike any other, in which limb meets limb, touching and overlapping in bright, lively, triumphant exuberance, two hundred and fifty, three hundred, and even four hundred feet above the shady, mossy ground. Over all the other species the Douglas spruce reigns supreme. It is not only a large tree, the tallest in America next to the redwood, but a very beautiful one, with bright green drooping foliage, handsome pendent cones, and a shaft exquisitely straight and round and regular. Forming extensive forests by itself in many places, it lifts its spirey tops into the sky close together with as even a growth as a well-tilled field of grain. No ground has been better tilled for wheat than these Cascade Mountains for trees: they were ploughed by mighty glaciers, and harrowed and mellowed and outspread by the broad streams that flowed from the ice-ploughs as they were withdrawn at the close of the glacial period.

In proportion to its weight when dry, Douglas spruce timber is perhaps stronger than that of any other large conifer in the country, and being tough, durable, and elastic, it is admirably suited for ship-building, piles, and heavy timbers in general; but its hardness and liability to warp when it is cut into boards render it unfit for fine work. In the lumber markets of California it is called "Oregon pine." When lumbering is going on in the best Douglas woods, especially about Puget Sound, many of the long, slender boles are saved for spars; and so superior is their quality that they are called for in almost every shipyard in the world, and it is interesting to follow their fortunes. Felled and peeled and dragged to tide-water, they are raised again as yards and masts for ships, given iron roots and canvas foliage, decorated with flags, and sent to sea, where in glad motion they go cheerily over the ocean prairie in every latitude and longitude, singing and bowing responsive to the same winds that waved them when they were in the woods. After standing in one place for centuries they thus go round the world like tourists, meeting many a friend from the old home forest; some traveling like themselves, some standing head downward in muddy harbors, holding up the platforms of wharves, and others doing all kinds of hard timber work, showy or hidden.

This wonderful tree also grows far northward in British Columbia, and southward along the coast and middle regions of Oregon and California; flourishing with the redwood wherever it can find an opening, and with the sugar-pine, yellow pine, and libocedrus in the Sierra. It extends into the San Gabriel, San Bernardino, and San Jacinto Mountains of southern California. It also grows well on the Wasatch Mountains, where it is called "red pine," and on many parts of the Rocky Mountains and short interior ranges of the Great Basin. But though thus widely distributed, only in Oregon, Washington, and some parts of British Columbia does it reach perfect development.

To one who looks from some high standpoint over its vast breadth, the forest on the west side of the Cascades seems all one dim, dark, monotonous field, broken only by the white volcanic cones along the summit of the range. Back in the untrodden wilderness a deep furred carpet of brown and yellow mosses covers the ground like a garment, pressing about the feet of the trees, and rising in rich bosses softly and kindly over every rock and moldering trunk, leaving no spot uncared for; and dotting small prairies, and fringing the meadows and the banks of streams not seen in general views, we find, besides the great conifers, a considerable number of hardwood trees,—oak, ash, maple, alder, wild apple, cherry, arbutus, Nuttall's flowering dogwood, and in some places chestnut. In a few favored spots the broad-leaved maple grows to a height of a hundred feet in forests by itself, sending out large limbs in magnificent interlacing arches covered with mosses and ferns, thus forming lofty sky-gardens, and rendering the underwoods delightfully cool. No finer forest ceiling is to be found than these maple arches, while the floor, ornamented with tall ferns and rubus vines, and cast into hillocks by the bulging, moss-covered roots of the trees, matches it well.

Passing from beneath the heavy shadows of the woods, almost anywhere one steps into lovely gardens of lilies, orchids, heathworts, and wild roses. Along the lower slopes, especially in Oregon, where the woods are less dense, there are miles of rhododendron, making glorious masses of purple in the spring, while all about the streams and the lakes and the beaver meadows there is a rich tangle of hazel, plum, cherry, crab-apple, cornel, gaultheria, and rubus, with myriads of flowers and abundance of other more delicate bloomers, such as erythronium, brodiæa, fritillaria, calochortus, Clintonia, and the lovely hider of the north, Calypso. Beside all these bloomers there are wonderful ferneries about the many misty waterfalls, some of the fronds ten feet high, others the most delicate of their tribe, the maidenhair fringing the rocks within reach of the lightest dust of the spray, while the shading trees on the cliffs above them, leaning over, look like eager listeners anxious to catch every tone of the restless waters. In the autumn berries of every color and flavor abound, enough for birds, bears, and everybody, particularly about the stream-sides and meadows where sunshine reaches the ground: huckleberries, red, blue, and black, some growing close to the ground, others on bushes ten feet high; gaultheria berries, called "sal-al" by the Indians; salmon berries, an inch in diameter, growing in dense prickly tangles, the flowers, like wild roses, still more beautiful than the fruit; raspberries, gooseberries, currants, blackberries, and strawberries. The underbrush and meadow fringes are in great part made up of these berry bushes and vines; but in the depths of the woods there is not much underbrush of any kind,—only a thin growth of rubus, huckleberry, and vine-maple.

Notwithstanding the outcry against the reservations last winter in Washington, that uncounted farms, towns, and villages were included in them, and that all business was threatened or blocked, nearly all the mountains in which the reserves lie are still covered with virgin forests. Though lumbering has long been carried on with tremendous energy along their boundaries, and home-seekers have explored the woods for openings available for farms, however small, one may wander in the heart of the reserves for weeks without meeting a human being, Indian or white man, or any conspicuous trace of one. Indians used to ascend the main streams on their way to the mountains for wild goats, whose wool furnished them clothing. But with food in abundance on the coast there was little to draw them into the woods, and the monuments they have left there are scarcely more conspicuous than those of birds and squirrels; far less so than those of the beavers, which have dammed streams and made clearings that will endure for centuries. Nor is there much in these woods to attract cattle-keepers. Some of the first settlers made farms on the small bits of prairie and in the comparatively open Cowlitz and Chehalis valleys of Washington; but before the gold period most of the immigrants from the Eastern States settled in the fertile and open Willamette Valley of Oregon. Even now, when the search for tillable land is so keen, excepting the bottom-lands of the rivers around Puget Sound, there are few cleared spots in all western Washington. On every meadow or opening of any sort some one will be found keeping cattle, raising hops, or cultivating patches of grain, but these spots are few and far between. All the larger spaces were taken long ago; therefore most of the newcomers build their cabins where the beavers built theirs. They keep a few cows, laboriously widen their little meadow openings by hacking, girdling, and burning the rim of the close-pressing forest, and scratch and plant among the huge blackened logs and stumps, girdling and killing themselves in killing the trees.

Most of the farm lands of Washington and Oregon, excepting the valleys of the Willamette and Rogue rivers, lie on the east side of the mountains. The forests on the eastern slopes of the Cascades fail altogether ere the foot of the range is reached, stayed by drought as suddenly as on the west side they are stopped by the sea; showing strikingly how dependent are these forest giants on the generous rains and fogs so often complained of in the coast climate. The lower portions of the reserves are solemnly soaked and poulticed in rain and fog during the winter months, and there is a sad dearth of sunshine, but with a little knowledge of woodcraft any one may enjoy an excursion into these woods even in the rainy season. The big, gray days are exhilarating, and the colors of leaf and branch and mossy bole are then at their best. The mighty trees getting their food are seen to be wide-awake, every needle thrilling in the welcome nourishing storms, chanting and bowing low in glorious harmony, while every raindrop and snowflake is seen as a beneficent messenger from the sky. The snow that falls on the lower woods is mostly soft, coming through the trees in downy tufts, loading their branches, and bending them down against the trunks until they look like arrows, while a strange muffled silence prevails, making everything impressively solemn. But these lowland snow-storms and their effects quickly vanish. The snow melts in a day or two, sometimes in a few hours, the bent branches spring up again, and all the forest work is left to the fog and the rain. At the same time, dry snow is falling on the upper forests and mountain-tops. Day after day, often for weeks, the big clouds give their flowers without ceasing, as if knowing how important is the work they have to do. The glinting, swirling swarms thicken the blast, and the trees and rocks are covered to a depth of ten to twenty feet. Then the mountaineer, snug in a grove with bread and fire, has nothing to do but gaze and listen and enjoy. Ever and anon the deep, low roar of the storm is broken by the booming of avalanches, as the snow slips from the overladen heights and rushes down the long white slopes to fill the fountain hollows. All the smaller streams are hushed and buried, and the young groves of spruce and fir near the edge of the timber line are gently bowed to the ground and put to sleep, not again to see the light of day or stir branch or leaf until the spring.

These grand reservations should draw thousands of admiring visitors at least in summer, yet they are neglected as if of no account, and spoilers are allowed to ruin them as fast as they like.* A few peeled spars cut here were set up in London, Philadelphia, and Chicago, where they excited wondering attention; but the countless hosts of living trees rejoicing at home on the mountains are scarce considered at all. Most travelers here are content with what they can see from car windows or the verandas of hotels, and in going from place to place cling to their precious trains and stages like wrecked

* The outlook over forest affairs is now encouraging. Popular interest, more practical than sentimental in whatever touches the welfare of the country's forests, is growing rapidly, and a hopeful beginning has been made by the Government in real protection for the reservations as well as for the parks. From July 1, 1900, there have been 9 superintendents, 39 supervisors, and from 330 to 445 rangers of reservations.

sailors to rafts. When an excursion into the woods is proposed, all sorts of dangers are imagined,—snakes, bears, Indians. Yet it is far safer to wander in God's woods than to travel on black highways or to stay at home. The snake danger is so slight it is hardly worth mentioning. Bears are a peaceable people, and mind their own business, instead of going about like the devil seeking whom they may devour. Poor fellows, they have been poisoned, trapped, and shot at until they have lost confidence in brother man, and it is not now easy to make their acquaintance. As to Indians, most of them are dead or civilized into useless innocence. No American wilderness that I know of is so dangerous as a city home "with all the modern improvements." One should go to the woods for safety, if for nothing else. Lewis and Clark, in their famous trip across the continent in 1804–1805, did not lose a single man by Indians or animals, though all the West was then wild. Captain Clark was bitten on the hand as he lay asleep. That was one bite among more than a hundred men while traveling nine thousand miles. Loggers are far more likely to be met than Indians or bears in the reserves or about their boundaries, brown weather-tanned men with faces furrowed like bark, tired-looking, moving slowly, swaying like the trees they chop. A little of everything in the woods is fastened to their clothing, rosiny and smeared with balsam, and rubbed into it, so that their scanty outer garments grow thicker with use and never wear out. Many a forest giant have these old woodmen felled, but, round-shouldered and stooping, they too are leaning over and tottering to their fall. Others, however, stand ready to take their places, stout young fellows, erect as saplings; and always the foes of trees outnumber their friends. Far up the white peaks one can hardly fail to meet the wild goat, or American chamois,—an admirable mountaineer, familiar with woods and glaciers as well as rocks,—and in leafy thickets deer will be found; while gliding about unseen there are many sleek furred animals enjoying their beautiful lives, and birds also, notwithstanding few are noticed in hasty walks. The ouzel sweetens the glens and gorges where the streams flow fastest, and every grove has its singers, however silent it seems,—thrushes, linnets, warblers; humming-birds glint about the fringing bloom of the meadows and peaks, and the lakes are stirred into lively pictures by water-fowl.

The Mount Rainier Forest Reserve should be made a national

park and guarded while yet its bloom is on;* for if in the making of the West Nature had what we call parks in mind,—places for rest, inspiration, and prayers,—this Rainier region must surely be one of them. In the center of it there is a lonely mountain capped with ice; from the ice-cap glaciers radiate in every direction, and young rivers from the glaciers; while its flanks, sweeping down in beautiful curves, are clad with forests and gardens, and filled with birds and animals. Specimens of the best of Nature's treasures have been lovingly gathered here and arranged in simple symmetrical beauty within regular bounds.

Of all the fire-mountains which, like beacons, once blazed along the Pacific Coast, Mount Rainier is the noblest in form, has the most interesting forest cover, and, with perhaps the exception of Shasta, is the highest and most flowery. Its massive white dome rises out of its forests, like a world by itself, to a height of fourteen thousand to fifteen thousand feet. The forests reach to a height of a little over six thousand feet, and above the forests there is a zone of the loveliest flowers, fifty miles in circuit and nearly two miles wide, so closely planted and luxuriant that it seems as if Nature, glad to make an open space between woods so dense and ice so deep, were economizing the precious ground, and trying to see how many of her darlings she can get together in one mountain wreath,—daisies, anemones, geraniums, columbines, erythroniums, larkspurs, etc., among which we wade knee-deep and waist-deep, the bright corollas in myriads touching petal to petal. Picturesque detached groups of the spirey Abies lasiocarpa stand like islands along the lower margin of the garden zone, while on the upper margin there are extensive beds of bryanthus, Cassiope, Kalmia, and other heathworts, and higher still saxifrages and drabas, more and more lowly, reach up to the edge of the ice. Altogether this is the richest subalpine garden I ever found, a perfect floral elysium. The icy dome needs none of man's care, but unless the reserve is guarded the flower bloom will soon be killed, and nothing of the forests will be left but black stump monuments.

* This was done shortly after the above was written. "One of the most important measures taken during the past year in connection with forest reservations was the action of Congress in withdrawing from the Mount Rainier Forest Reserve a portion of the region immediately surrounding Mount Rainier and setting it apart as a national park." (*Report of Commissioner of General Land Office*, for the year ended June, 1899.) But the park as it now stands is far too small.

The Sierra of California is the most openly beautiful and useful of all the forest reserves, and the largest, excepting the Cascade Reserve of Oregon and the Bitter Root of Montana and Idaho. It embraces over four million acres of the grandest scenery and grandest trees on the continent, and its forests are planted just where they do the most good, not only for beauty, but for farming in the great San Joaquin Valley beneath them. It extends southward from the Yosemite National Park to the end of the range, a distance of nearly two hundred miles. No other coniferous forest in the world contains so many species or so many large and beautiful trees,—Sequoia gigantea, king of conifers, "the noblest of a noble race," as Sir Joseph Hooker well says; the sugar-pine, king of all the world's pines, living or extinct; the yellow pine, next in rank, which here reaches most perfect development, forming noble towers of verdure two hundred feet high; the mountain pine, which braves the coldest blasts far up the mountains on grim, rocky slopes; and five others, flourishing each in its place, making eight species of pine in one forest, which is still further enriched by the great Douglas spruce, libocedrus, two species of silver fir, large trees and exquisitely beautiful, the Paton hemlock, the most graceful of evergreens, the curious tumion, oaks of many species, maples, alders, poplars, and flowering dogwood, all fringed with flowery underbrush, manzanita, ceanothus, wild rose, cherry, chestnut, and rhododendron. Wandering at random through these friendly, approachable woods, one comes here and there to the loveliest lily gardens, some of the lilies ten feet high, and the smoothest gentian meadows, and Yosemite valleys known only to mountaineers. Once I spent a night by a camp-fire on Mount Shasta with Asa Gray and Sir Joseph Hooker, and, knowing that they were acquainted with all the great forests of the world, I asked whether they knew any coniferous forest that rivaled that of the Sierra. They unhesitatingly said: "No. In the beauty and grandeur of individual trees, and in number and variety of species, the Sierra forests surpass all others."

This Sierra Reserve, proclaimed by the President of the United States in September, 1893, is worth the most thoughtful care of the government for its own sake, without considering its value as the fountain of the rivers on which the fertility of the great San Joaquin Valley depends. Yet it gets no care at all. In the fog of tariff, silver, and annexation politics it is left wholly unguarded, though the management of the adjacent national parks by a few soldiers shows how well and how easily it can be preserved. In the meantime, lumbermen are allowed to spoil it at their will, and sheep in uncountable ravenous hordes to trample it and devour every green leaf within reach; while the shepherds, like destroying angels, set innumerable fires, which burn not only the undergrowth of seedlings on which the permanence of the forest depends, but countless thousands of the venerable giants. If every citizen could take one walk through this reserve, there would be no more trouble about its care; for only in darkness does vandalism flourish.*

The reserves of southern California,—the San Gabriel, San Bernardino, San Jacinto, and Trabuco,—though not large, only about two million acres altogether, are perhaps the best appreciated. Their slopes are covered with a close, almost impenetrable growth of flowery bushes, beginning on the sides of the fertile coast valleys and the dry interior plains. Their higher ridges, however, and mountains are open, and fairly well forested with sugar-pine, yellow pine, Douglas spruce, libocedrus, and white fir. As timber fountains they amount to little, but as bird and bee pastures, cover for the precious streams that irrigate the lowlands, and quickly available retreats from dust and heat and care, their value is incalculable. Good roads have been graded into them, by which in a few hours lowlanders can get well up into the sky and find refuge in hospitable camps and clubhouses, where, while breathing reviving ozone, they may absorb the beauty about them, and look comfortably down on the busy towns and the most beautiful orange groves ever planted since gardening began.

The Grand Cañon Reserve of Arizona, of nearly two million acres, or the most interesting part of it, as well as the Rainier region, should be made into a national park, on account of their supreme grandeur and beauty. Setting out from Flagstaff, a station on the Atchison, Topeka, and Santa Fé Railroad, on the way to the cañon you pass through beautiful forests of yellow pine,—like those of the Black Hills, but more extensive,—and curious dwarf forests of nut pine and juniper, the spaces between the miniature trees planted with many interesting species of eriogonum, yucca, and cactus. After riding or walking seventy-five miles through these

* See note, p.779.

pleasure-grounds, the San Francisco and other mountains, abounding in flowery parklike openings and smooth shallow valleys with long vistas which in fineness of finish and arrangement suggest the work of a consummate landscape artist, watching you all the way, you come to the most tremendous cañon in the world. It is abruptly countersunk in the forest plateau, so that you see nothing of it until you are suddenly stopped on its brink, with its immeasurable wealth of divinely colored and sculptured buildings before you and beneath you. No matter how far you have wandered hitherto, or how many famous gorges and valleys you have seen, this one, the Grand Cañon of the Colorado, will seem as novel to you, as unearthly in the color and grandeur and quantity of its architecture, as if you had found it after death, on some other star; so incomparably lovely and grand and supreme is it above all the other cañons in our fire-molded, earthquake-shaken, rain-washed, wave-washed, river and glacier sculptured world. It is about six thousand feet deep where you first see it, and from rim to rim ten to fifteen miles wide. Instead of being dependent for interest upon waterfalls, depth, wall sculpture, and beauty of parklike floor, like most other great cañons, it has no waterfalls in sight, and no appreciable floor spaces. The big river has just room enough to flow and roar obscurely, here and there groping its way as best it can, like a weary, murmuring, overladen traveler trying to escape from the tremendous, bewildering labyrinthic abyss, while its roar serves only to deepen the silence. Instead of being filled with air, the vast space between the walls is crowded with Nature's grandest buildings,—a sublime city of them, painted in every color, and adorned with richly fretted cornice and battlement spire and tower in endless variety of style and architecture. Every architectural invention of man has been anticipated, and far more, in this grandest of God's terrestrial cities.

ABOUT THE INTRODUCER

TERRY TEMPEST WILLIAMS is the author of *Refuge: An Unnatural History of Family and Place*, *The Open Space of Democracy*, *Finding Beauty in a Broken World*, and most recently, *The Hour of Land: A Personal Topography of America's National Parks*. She received the Sierra Club's John Muir Award in 2014.

NON-FICTION IN EVERYMAN'S LIBRARY

JOHN JAMES AUDUBON
The Audubon Reader

AUGUSTINE
The Confessions

SIMONE DE BEAUVOIR
The Second Sex

HECTOR BERLIOZ
The Memoirs of Hector Berlioz

THE BIBLE
(King James Version)
The Old Testament
The New Testament

JAMES BOSWELL
The Life of Samuel Johnson
The Journal of a Tour to the Hebrides

JEAN ANTHELME BRILLAT-SAVARIN
The Physiology of Taste

EDMUND BURKE
Reflections on the Revolution in France and Other Writings

GIACOMO CASANOVA
History of My Life

BENVENUTO CELLINI
The Autobiography of Benvenuto Cellini

G. K. CHESTERTON
The Everyman Chesterton

CARL VON CLAUSEWITZ
On War

CONFUCIUS
The Analects

THOMAS CRANMER
The Book of Common Prayer
(UK only)

CHARLES DARWIN
The Origin of Species
The Voyage of the Beagle
(in 1 vol.)

JOAN DIDION
We Tell Ourselves Stories in Order to Live (US only)

W.E.B. DU BOIS
The Souls of Black Folk (US only)

JOHN EVELYN
The Diary of John Evelyn
(UK only)

ANNE FRANK
The Diary of a Young Girl
(US only)

BENJAMIN FRANKLIN
The Autobiography and Other Writings

EDWARD GIBBON
The Decline and Fall of the Roman Empire
Vols 1 to 3: The Western Empire
Vols 4 to 6: The Eastern Empire

KAHLIL GIBRAN
The Collected Works

HERODOTUS
The Histories

MICHAEL HERR
Dispatches (US only)

HINDU SCRIPTURES
(tr. R. C. Zaehner)

SAMUEL JOHNSON
A Journey to the Western Islands of Scotland

SØREN KIERKEGAARD
Fear and Trembling and The Book on Adler

MAXINE HONG KINGSTON
The Woman Warrior and China Men
(US only)

THE KORAN
(tr. Marmaduke Pickthall)

NICCOLÒ MACHIAVELLI
The Prince

MARCUS AURELIUS
Meditations

W. SOMERSET MAUGHAM
The Skeptical Romancer
(US only)

JOHN STUART MILL
On Liberty and Utilitarianism

MARY WORTLEY MONTAGU
Letters

MICHEL DE MONTAIGNE
The Complete Works

THOMAS MORE
Utopia

JOHN MUIR
Selected Writings

GEORGE ORWELL
Essays

THOMAS PAINE
Rights of Man
and Common Sense

PLATO
The Republic
Symposium and Phaedrus

MARCO POLO
The Travels of Marco Polo

JEAN-JACQUES ROUSSEAU
Confessions
The Social Contract and the Discourses

JOHN RUSKIN
Praeterita and Dilecta

ADAM SMITH
The Wealth of Nations

TACITUS
Annals and Histories

HENRY DAVID THOREAU
Walden

ALEXIS DE TOCQUEVILLE
Democracy in America

GIORGIO VASARI
Lives of the Painters, Sculptors and Architects (in 2 vols)

HORACE WALPOLE
Selected Letters

EVELYN WAUGH
(US only)
Waugh Abroad: Collected Travel Writing

MARY WOLLSTONECRAFT
A Vindication of the Rights of Woman

This book is set in CASLON, designed and engraved by William
Caslon of WILLIAM CASLON & SON, Letter-Founders in
London, around 1740. In England at the beginning of
the eighteenth century, Dutch type was probably
more widely used than English. The rise
of William Caslon put a stop to the
importation of Dutch types
and so changed the his-
tory of English
typecutting.

E L